Numerical Modeling of

Explosives and Propellants

Second Edition

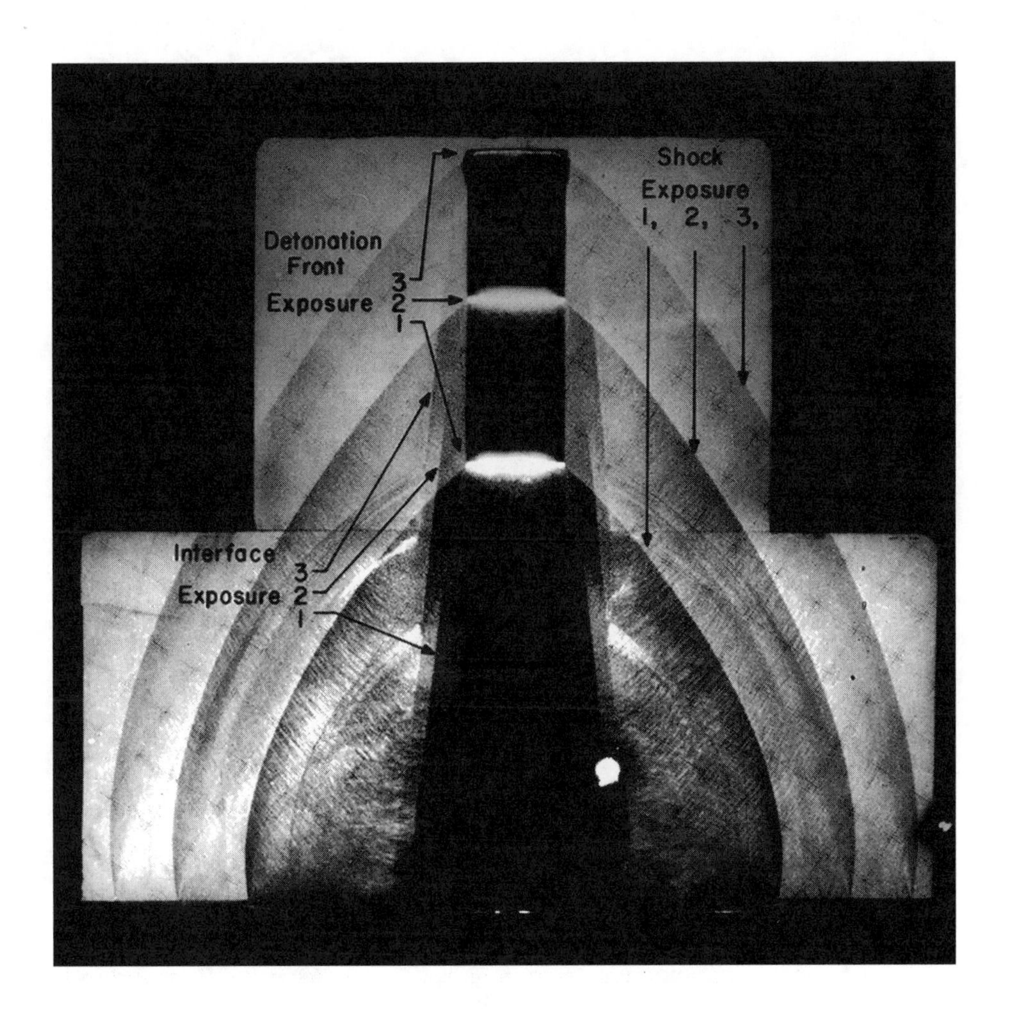

Numerical Modeling of

Explosives and Propellants

Second Edition

Charles L. Mader

CRC Press
Boca Raton London New York Washington, D.C.

Library of Congress Cataloging-in-Publication Data

Mader, Charles L.
Numerical modeling of explosives and propellants, 2nd edition / Charles L. Mader
p. cm.
Includes bibliographical references and index.
ISBN 0-8493-3149-8 (alk. paper)
1. Detonation waves. 2. Explosives. 3. Numerical analysis.
I. Title.

QC168.85.D46M34 1998
662'.2—dc21 97-21340
CIP

Direct all inquiries to CRC Press LLC, 2000 N.W. Corporate Blvd., Boca Raton, Florida 33431.

International Standard Book Number 0-8493-3149-8
Library of Congress Card Number 97-21340
Printed in the United States of America 3 4 5 6 7 8 9 0
Printed on acid-free paper

Numerical Modeling of Explosives and Propellants

Charles L. Mader

Recent advances in the development of inexpensive personal computers have resulted in computers being available to most scientists and engineers that can be used to model explosives and propellants. The author describes the numerical models he has developed for modeling the performance and vulnerability of explosives and propellants, and describes their application to scientific and engineering problems.

The mechanisms of initiation and performance of heterogeneous explosives (that is, most explosives and propellants) are dominated by the nature of the heterogeneities. The shock initiation behavior, performance, and reaction zone characteristics of explosives with voids or other density discontinuities are very different from those observed or provided by classical detonation theory for homogeneous explosives. Heterogeneous explosive performance and vulnerability have been successfully modeled numerically. The results of the numerical studies and techniques developed to describe heterogeneous explosives are described.

The computer programs for modeling explosive and propellant performance on personal computers are included on a CD-ROM along with extensive collections of the data files required for describing explosives or propellants to be modeled.

Introduction

This monograph is an updated version of the monograph *Numerical Modeling of Detonations* by the author, published in 1978 by the University of California Press. Since 1978 the numerical modeling of explosives and propellants was extended to three-dimensions, and development of powerful personal computers allowed the computer programs developed for modeling explosives and propellants to be run on personal computers. The main objective of this monograph is to describe the important physics and chemistry of explosive and propellant performance and vulnerability that has been learned using numerical and experimental methods during the last forty years. Another objective is to describe in detail the numerical methods that have been developed by the author for modeling explosives and propellants and to include on a CD-ROM the executable codes for personal computers and the required data files so that the reader may apply the numerical methods to his own particular explosive or propellant system.

A series of color video tape (VT–78–2, VT–80–2, VT–83-28) lectures featuring computer generated films of many of the applications discussed in this monograph is available from the Los Alamos National Laboratory library in one-inch format or from the author in VHS format.

Two important books that complement this monograph have been recently published. They are highly recommended for those readers interested in either more fundamental physics of explosives and shock waves or those interested in more applied explosive engineering.

Detonation of Condensed Explosives by Roger Cheret published by Springer-Verlag in 1993 is the outstanding detonation physics text of the century. Cheret describes the mechanical and thermodynamic aspects of the propagation of detonation waves, the molecular and macroscopic mechanisms of explosive decomposition, and the dynamic characterization of explosives.

Rock Blasting and Explosive Engineering by Per-Anders Persson, Roger Holmberg, and Jaimin Lee published by CRC Press in 1994 describes the science of explosive mining, petrochemical, mechanical and, civil engineering.

This monograph was carefully reviewed by Dr. Robert Cowan, Dr. John Zukas, Dr. Douglas Venable, and Mrs. Susan Henke.

The Author

Charles L. Mader is President of Mader Consulting Co. and a retired Fellow of the Los Alamos National Laboratory. He is author of *Numerical Modeling of Detonations* and *Numerical Modeling of Water Waves* . He is also the author of four Los Alamos Dynamic Material Properties volumes published by the University of California Press. His address is: Mader Consulting Company, 1049 Kamehame Drive, Honolulu, Hawaii 96825.

The Author

[illegible]

Contents

chapter one

The Detonation Wave

1.1 Steady-State Detonations

Detonations are reactive-wave phenomena whose propagation is controlled by shock waves. If the reaction rates are essentially infinite and chemical equilibrium is attained, one has a steady-state detonation whose propagation rate is governed solely by thermodynamics and hydrodynamics.

Consider a block of detonating solid explosive at a pressure, P_o, of 1 bar and a specific volume, V_o ($\rho_o = 1/V_o$ is the initial density). Advancing through this block with a velocity D is a shock front that compresses the material along the Hugoniot (single shock curve) for the solid explosive to a specific volume V_1, thereby raising the pressure to a value P_1 as shown in Figure 1.1 and initiating chemical reaction. The reaction proceeds through a zone from a few hundred angstroms to a few millimeters thick. Completion of the reaction leaves the pressure and volume at a point (P_2, V_2) on the Hugoniot for the reaction products. The reaction products then expand (the Taylor wave) along the isentrope through (P_2, V_2). The points (P_o, V_o), (P_1, V_1), and (P_2, V_2) are colinear since the slopes of the connecting lines are related to the velocities of the shock front and the rear of the reaction zone and these velocities are equal under steady-state conditions. The points define the Rayleigh line for a steady-state detonation.

The laws of conservation of mass and momentum applied across the shock front lead the relations

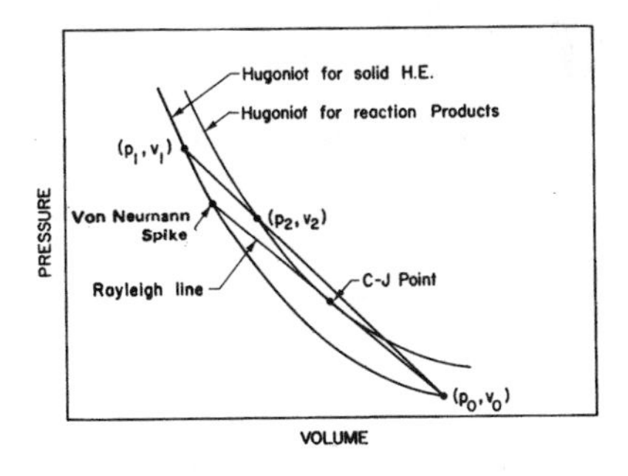

Figure 1.1 Hugoniot curves.

$$V/V_o = \rho_o/\rho = (D - U_P)/(D - U_{P_o})$$

$$P - P_o = \rho_o(D - U_{P_o})(U_P - U_{P_o})\,. \tag{1.1}$$

which may be combined to give expressions for the shock (detonation) velocity D and the particle velocity U_P at the rear of the reaction zone.

$$D - U_{P_o} = V_o\sqrt{(P - P_o)/(V_o - V)}$$

$$U_P - U_{P_o} = \sqrt{(P - P_o)(V_o - V)}\,. \tag{1.2}$$

Here P and V denote the pressure and specific volume of the reaction products - previously written (P_2, V_2) - and U_{P_o} is the particle velocity of the material ahead of the shock (and is generally zero).

The law of conservation of energy together with the equations above can be used to derive the so-called Hugoniot equation.

$$E - E_o = \frac{1}{2}(P + P_o)(V_o - V)\,, \tag{1.3}$$

where E_o is the specific internal energy of the solid explosive at (P_o, V_o) and E is the specific internal energy of the reaction products at (P, V). With the aid of an equation of state for the reaction products (commonly, but misleadingly, called an equation of state of the high explosive), E can be expressed as a function of P and V, which when substituted in Eq. (1.3) gives the equation of the Hugoniot curve for the explosion products.

For a plane detonation wave, Chapman and Jouguet's hypothesis[1*] states that the line $(P_o, V_o) \rightarrow (P, V)$ is tangent to the Hugoniot for the explosion products; in this case, the point (P, V) is called the C-J point. In the case of a concave shock front, as in an implosion, the convergence effects may raise the pressure well above the C-J point (as in the case illustrated in Figure 1.1), with a corresponding increase in the detonation velocity D. The plane-wave velocity, for which the C-J condition is satisfied, is denoted by D_{CJ}.

The C-J point has several properties of interest. First, D_{CJ} is the minimum possible value of D. Second, the energy change along the Hugoniot is given from Eq. (1.3) by

$$dE = \frac{1}{2}[(V_o - V)dP - (P + P_o)dV]\,.$$

Thus from the thermodynamic relation $TdS = dE + PdV$, it follows that for changes along the Hugoniot,

$$T\left(\frac{dS}{dV}\right)_H = \frac{(V_o - V)}{2}\left[\left(\frac{dP}{dV}\right)_H - \frac{P - P_o}{V - V_o}\right]. \tag{1.4}$$

*Note: References are keyed by number to lists at ends of chapters.

Since $(P - P_o)/(V - V_0)$ is the slope of the straight line from (P_o, V_o) to the Hugoniot point (P, V), it is evident that $(dS/dV)_H \gtreqless 0$ accordingly, as $V \gtreqless V_{CJ}$; i.e., the entropy along the Hugoniot curve has a minimum at the C-J point. This implies that the isentropic curve (adiabat) which passes through the C-J point is tangent to both the Hugoniot curve and the straight line $(P_o, V_o) \rightarrow (P_{CJ}, V_{CJ})$.

Thus if γ is defined as the negative of the logarithmic slope of the adiabat:

$$\gamma = -\left(\frac{\partial ln P}{\partial ln V}\right)_S = -\frac{V}{P}\left(\frac{\partial P}{\partial V}\right)_S \tag{1.5}$$

then at the C-J point

$$\left(\frac{\partial P}{\partial V}\right)_S = (P - P_o)/(V - V_o) ,$$

$$\gamma = -\frac{V}{P}\left(\frac{P - P_o}{V - V_o}\right) \cong \frac{V}{V_o - V} , \tag{1.6}$$

since P_o is negligible compared with P in the case of a high explosive. From this expression and Eq. (1.2), with $P_o = U_o = 0$, there follow the well-known relations

$$\frac{V_{CJ}}{V_o} = \frac{\gamma_{CJ}}{\gamma_{CJ} + 1} ,$$

$$(1.7)$$

$$P_{CJ} = \rho_o \frac{D_{CJ}^2}{\gamma_{CJ} + 1} .$$

As mentioned earlier, calculation of the Hugoniot curve from Eq. (1.3) requires a knowledge of the equation of state of the detonation products. However, high explosives produce pressures of the order of $\frac{1}{2}$ Mbar $\approx$ 500,000 atm and volumes of about 10 cc/mole (a 2- or 3-A cube per molecule), and in this range there exists neither direct experimental data nor a satisfactory self-contained theory. About the best that can be done at present is to use some empirical equation of state in which the values of the parameters are chosen so as to give the best possible agreement with the available experimental data on high explosives - in particular, the velocity of propagation and peak detonation pressure of a plane detonation wave.

As shown by Eq. (1.2), a knowledge of D tells one a little about the Hugoniot - it is tangent to a certain straight line - but gives no information at all as to where on the line the tangency occurs or the curvature of the Hugoniot at this point. However, it is possible to press solid explosives to a variety of different densities ρ_o, and D is found experimentally to increase with ρ_o. Different values of D mean different slopes of the tangents to the corresponding Hugoniots, and hence give considerably more information than just one value. In fact, if a perfect-gas equation of state is used it can be shown theoretically that D should be independent of ρ_o, so that the variation of D with ρ_o does indeed provide some measure of the gas imperfection terms in the actual equation of state. Accordingly, it has become standard practice to adjust the equation of state parameters to give a theoretical curve of D vs. ρ_o which agrees as closely as possible with the experimental one.

Experimental data on the variation of D with ρ_o are, of course, not sufficient to determine the equation of state uniquely, even in the vicinity of the C-J points, so pressure and temperature data are also needed.

1.2 Resolved Reaction Zone Detonations in One Dimension

The nature of a reaction zone depends on the chemical decomposition rates and the equation of state used to describe the undecomposed explosive, the detonation products, and the mixtures of undecomposed explosive and detonation products. As shown in section 1-D, the nature of the reaction zone also strongly depends upon whether the explosive is homogeneous or heterogeneous (whether density discontinuities are present). The Arrhenius rate law was used in the numerical modeling of the reaction zone, and the HOM equation of state described in Appendix A was used for condensed explosive, the detonation products, and any mixture of condensed explosive and detonation products, assuming pressure and temperature equilibrium.

To study the reaction zone as a function of time, one must obtain a complete solution to the Navier-Stokes equations of fluid dynamics for a reactive material. The SIN code was used and is described in Appendix A. The partial differential equations are described in Lagrangian finite-difference form appropriate to plane, one-dimensional motion. Enough viscosity was used to smear the shocks over at least three finite-difference meshes if the mesh was too large for realistic viscosity values. Use of the Navier-Stokes equations in the numerical treatment of the detonations, to avoid the difficulty of treating a shock discontinuity, gave results for ideal gases of constant heat capacity undergoing an exothermic unimolecular Arrhenius reaction that were essentially identical with results obtained using the Fickett and Wood[2] method of characteristics with sharp shock and no viscosity. The results do not depend upon viscosity.

Up to 2500 mesh points were used in the calculations. The number of mesh points in a reaction zone depends upon the initial size of the mesh, the thickness of the reaction zone, and the amount the mesh is compressed. For example, a 620-A reaction zone in nitromethane has 120 mesh points whose initial thickness was 10-A. The equation-of-state constants used are described in Table 1.1 and on the CD-ROM.

The details of the reaction zone structure are very dependent on the properties of the equation of state for the mixture of undecomposed explosive and detonation products. The amount of overdrive necessary to obtain a steady detonation and the period and magnitude of the oscillation of an unstable detonation are not as dependent upon the equation of state details, but they are very dependent on the activation energy.

TABLE 1.1 Equation of State and Rate Parameters

	Nitromethane	EXP	Liquid TNT	EXP
		Condensed		
C	+1.647	−001	+2.0	−001
S	+1.637	+000	+1.65	+000
F_s	+5.41170789261	+000	+4.23681503433	+000
G_s	−2.72959322666	+000	−8.59732701624	+000
H_s	−3.21986013188	+000	−1.60609786333	+001
I_s	−3.90757138698	+000	−1.37639638610	+001
J_s	+2.39028184133	+000	−2.05713430231	+000
C_v	+4.14	−001	+3.83	−001
α	+3.0	−004	+2.3153	−004
V_o	+8.86524823	−001	+6.91085	−001
$\gamma_s(1)$	+6.805	−001	+1.66	+000
$\gamma_s(2)$	+1.70	+ 000	+1.00	+000
T_o	+3.0	+002	+3.581	+002
		Gas		
A	−3.11585072896	+000	−3.55947543414	+000
B	−2.35968123302	+000	−2.41906238253	+000
C	+2.10663268988	−001	+2.13148095464	−001
D	+3.80357006508	−003	+5.42681375493	−002
E	−3.53454737231	−003	−1.68309952849	−002
K	−1.39936678316	+000	−1.47916342276	+000
L	+4.79350272379	−001	+5.14388037177	−001
M	+6.06707773429	−002	+9.02702839234	−002
N	+4.10672785525	−003	+8.50384728393	−003
O	+1.13326560531	−004	+3.21650937265	−004
Q	+7.79645302519	+000	+7.68605532614	+000
R	−5.33007196907	−001	−4.70441838903	−001
S	+7.09019736856	−002	+9.90648501306	−002
T	+2.06149976021	−002	+5.24087425648	−003
U	−5.66139653675	−003	−4.02187883702	−003
C'_v	+5.56	−001	+5.0	−001
Z	+1.0	−001	+1.0	−001
		Reaction		
E*	+5.36	+004	+4.11	+004
Z	+4.0	+008	+2.0	+006
		BKW CJ Parameters		
P_{CJ}	+1.30	−001	+1.57	−001
D_{CJ}	+6.463	−001	+6.406	−001
T_{CJ}	+3.120	+003	+3.126	+003

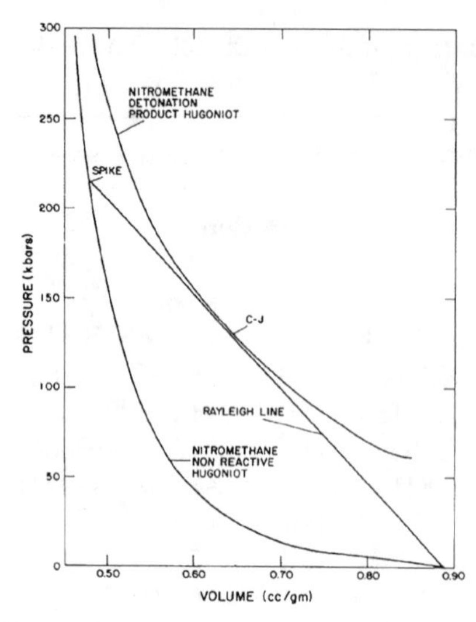

Figure 1.2 Nitromethane Hugoniots and Rayleigh line.

Nitromethane Reaction Zones

To calculate the resolved reaction zone and the stability of detonations, we wish to use a piston that is as close to the steady-state piston as possible, so that the deviations from the steady-state will be as small as permitted by the mesh used. To accomplish this for any desired detonation velocity, the piston velocity was initially set equal to the particle velocity of the undecomposed condensed explosive that corresponds to the detonation (shock) velocity. The piston compresses and heats the explosive next to it, and a certain amount of reaction occurs. The intersection of the Rayleigh line, as shown in Figure 1.2 with the corresponding Hugoniot is used to determine the proper piston velocity for the next time step. The piston velocity computed in this manner approximates the steady-state piston. A detailed description of the method used to compute the steady-state piston is presented in Appendix A.

The approximate steady-state profiles computed in the above manner for nitromethane at the BKW C-J state are shown in Figure 1.3 for a Grüneisen coefficient, γ, of 1.7 and 0.68. The partially reacted Hugoniots are very sensitive to the equation of state used for the mixture of undecomposed explosive and detonation products. A γ of 1.70 in addition to 0.68 was used to illustrate the magnitude of this effect.

For a γ of 0.68, the reaction zone thickness is 2460-A, while for a γ of 1.70 the thickness is 1620-A. The temperature profile for a γ of 0.68 has a maximum at about 900-A, while the highest temperature for a γ of 1.70 is at the end of the reaction zone.

The shock front pressure as a function of time for nitromethane with a γ of 0.68 and a mesh size of 100 A is shown in Figure 1.4 for various amounts of overdrive. For a detonation velocity of 0.6850 cm/μsec, the initial perturbation introduced by the large mesh quickly decays. For a detonation velocity of 0.6650 cm/μsec, the initial perturbation grows and reaches a steady amplitude by the fourth cycle, as shown in Figure 1.5. Using a smaller mesh size reduces the magnitude of the initial perturbation for a detonation velocity of 0.6850 cm/μsec and reduces the deviation from the steady-state solution thereafter, as is shown in Figure 1.6. The magnitude of the initial perturbation is proportional to the mesh size. Using a 25-A mesh size for a detonation velocity of 0.6650 cm/μsec reduces the magnitude

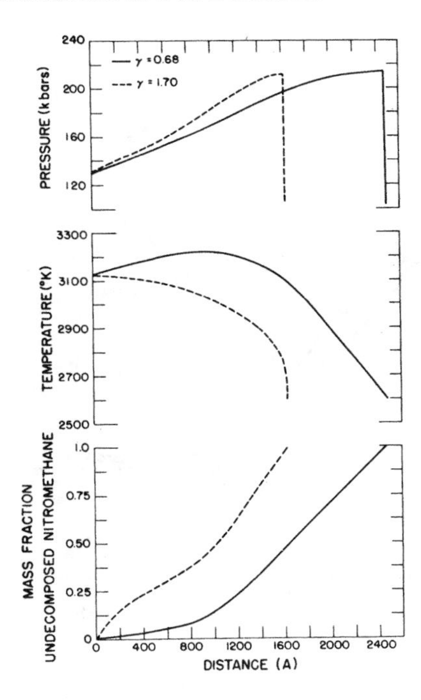

Figure 1.3 Approximate steady-state profiles for nitromethane at C-J detonation velocity of 0.6460 cm/μsec.

of the initial perturbation and, by the second cycle, it has grown to an amplitude that closely resembles the steady amplitude reached by the fourth cycle for the 100-A case shown in Figure 1.5.

The shock-front pressure as a function of time for nitromethane with a γ of 1.7 and a mesh size of 100-A is shown in Figure 1.7. As for the $\gamma = 0.68$ case, the 0.6850-cm/μsec velocity is stable. The pressure at the end of the reaction zone is not perturbed as much on the first cycle for the $\gamma = 1.70$ case, where the temperature maximum is at the end of the reaction zone, as it is for the $\gamma = 0.68$ case, where the temperature maximum is inside the reaction zone.

The period for the C-J case is about 1 x $10^{-3}\mu$sec, as estimated from the overdriven cases.

The amount of overdrive necessary to stabilize the nitromethane detonation decreases with decreasing activation energy and is independent of the frequency factor. If one assumes an activation energy of 40 kcal/mole instead of the "experimental" value of 53.6, the nitromethane detonation for a γ of 0.68 is stable at the C-J velocity of 0.6460 cm/μsec. The nitromethane detonation for a γ of 1.7 and an activation energy of 40 kcal/mole is unstable at 0.6550 cm/μsec and stable at 0.6640 cm/μsec.

The formation of an overdriven detonation in nitromethane by a 0.3-cm/μsec constant-velocity piston was computed using a 5-A mesh and realistic viscosity values. The shock thickness was 15 A, the steady-state spike pressure or shock-front pressure was 272 kbar, the pressure at the end of the reaction zone was 242 kbar, the detonation velocity was 0.7130 cm/μsec, and the reaction zone was 620-A thick (defined as the distance from mass fraction of 0.999 to 0.01). These values were determined using the steady-state piston.

The formation of the reaction zone in the pressure-volume plane is shown in Figure 1.8; that in the pressure-distance plane in Figure 1.9. The steady-state reaction zone profiles of

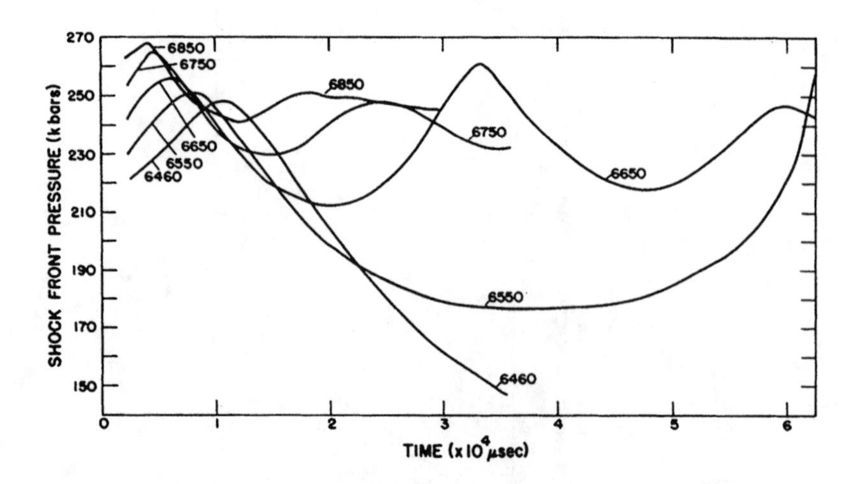

Figure 1.4 Shock-front pressure vs. time for various detonation velocities (shown in m/sec) using steady-state piston for nitromethane with $\gamma = 0.68$ and a 100-A mesh.

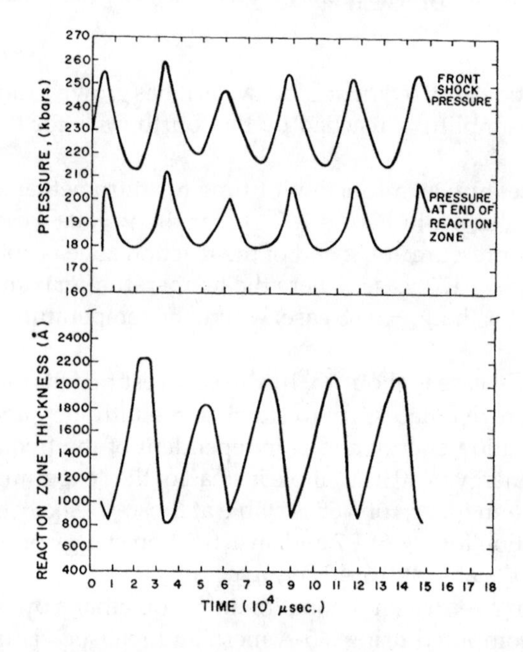

Figure 1.5 Reaction zone of nitromethane using steady-state piston with a detonation velocity of 0.6650 cm/μsec, $\gamma = 0.68$ and a 100-A mesh.

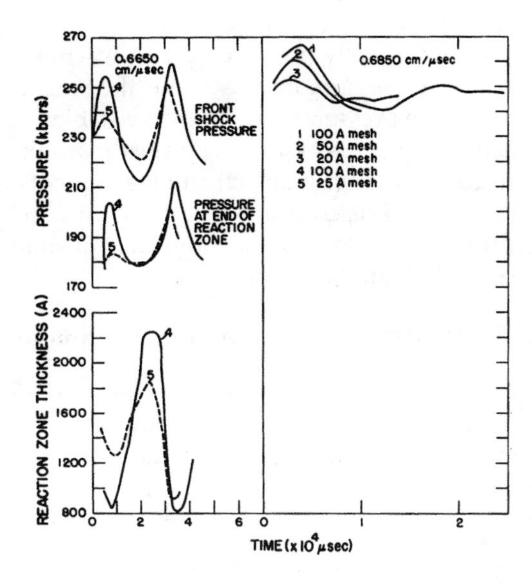

Figure 1.6 Effect of mesh size for steady-state piston for nitromethane with $\gamma = 0.68$; μ_v(viscosity) $=$ 4 x 10^{-7} for 100-A and 1 x 10^{-7} for 25-A mesh.

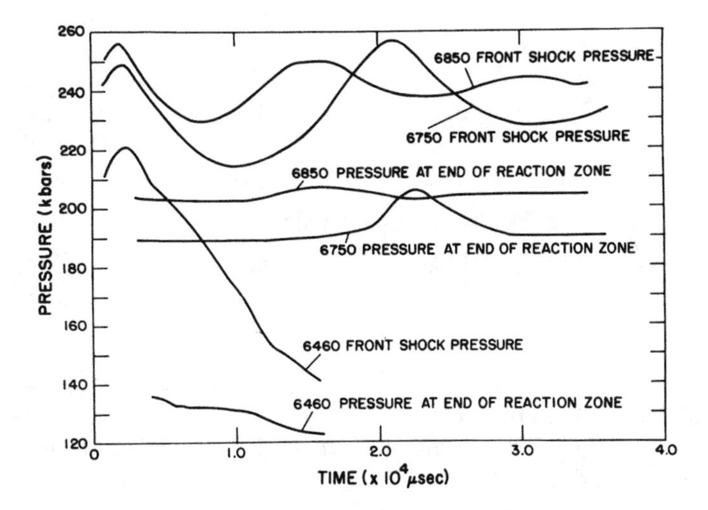

Figure 1.7 Shock-front pressure and pressure at end of reaction zone vs. time for various detonation velocities (shown in m/sec) when the steady-state piston is used for nitromethane with $\gamma = 1.7$ and a 100-A mesh.

pressure, temperature, and mass fraction are shown in Figure 1.10. The shock-front pressure and reaction zone thickness are shown as functions of time in Figure 1.11. Formation of an approximately stable reaction zone profile requires many (~ 10) reaction zone lengths.

A steady reaction zone is achieved when the rate of release of chemical energy at all points in the zone is just sufficient to sustain the wave.

The constant velocity piston initially produces a long reaction zone. As a result of the release of chemical energy, the pressure increases and the reaction zone shortens until a steadier profile is obtained. The pressure overshoots the stable pressure profile in regions of the reaction zone during the process of forming a steady profile.

As Table 1.2 shows, nitromethane with an activation energy of 40 kcal/mole and a γ of 0.68 was found to be stable at C-J detonation velocity. With a stable detonation, one can study how the reaction zone is formed by constant-velocity pistons that initially shock the unreacted nitromethane to less than the C-J pressure.

TABLE 1.2 Nitromethane and Liquid TNT Summary

Activation Energy, E^* (kcal/mole)	Frequency Factor, Z (μsec^{-1})	Grüneisen γ	Thickness of Reaction Zone	Result
		Nitromethane		
53.6	$4\text{x}10^{8a}$	0.68	2,400-A	Stable at 6750 m/sec Unstable at 6650
53.6	$4\text{x}10^{8a}$	1.70	1,600-A	Stable at 6750 m/sec Unstable at 6650
40	$1.27\text{x}10^{6b}$	0.68	77,500-A	Stable at 6460 m/sec (C-J)
40	$1.27\text{x}10^{6b}$	1.70	66,000-A	Stable at 6650 m/sec Unstable at 6550
30	$2.37\text{x}10^{4b}$	1.70	705,000-A	Stable at 6460 m/sec (C-J)
		Liquid TNT		
41.1	$2\text{x}10^{6a}$	1.66	0.0011 cm	Stable at 6600 m/sec Unstable at 6500
30	$3.14\text{x}10^{4c}$	1.66	0.0073 cm	Stable at 6406 m/sec (C-J)

[a] Experimental value (Reference 3)
[b] Z adjusted to gave an explosion time of 1.4 μsec at 1180 K (85 kbar) to agree with shock initiation experimental data (Reference 3).
[c] Z adjusted to gave an explosion time of 0.7 μsec at 1227 K (125 kbar) to agree with shock initiation experimental data (Reference 3).

The pressures at various positions as a function of time are shown in Figure 1.12 for a 0.21-cm/μsec constant-velocity piston. The calculations with a resolved reaction zone show details of the process of shock initiation of nitromethane. The basic features are identical

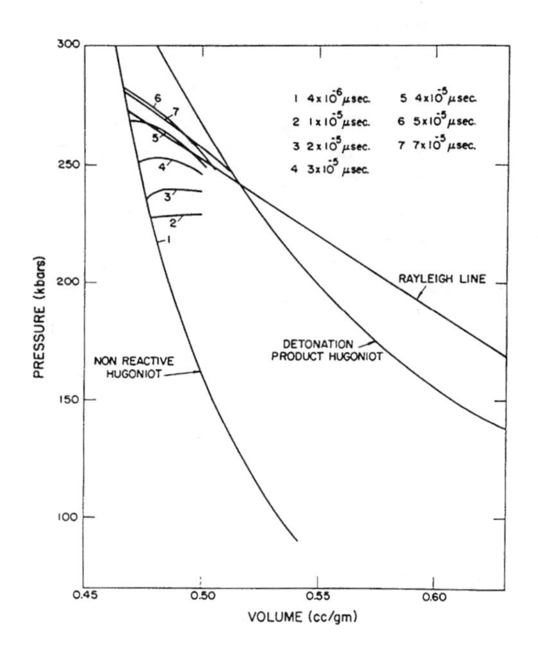

Figure 1.8 Pressure-volume profiles for formation of reaction zone of overdriven detonation in nitromethane with a 0.3-cm/μsec constant-velocity piston.

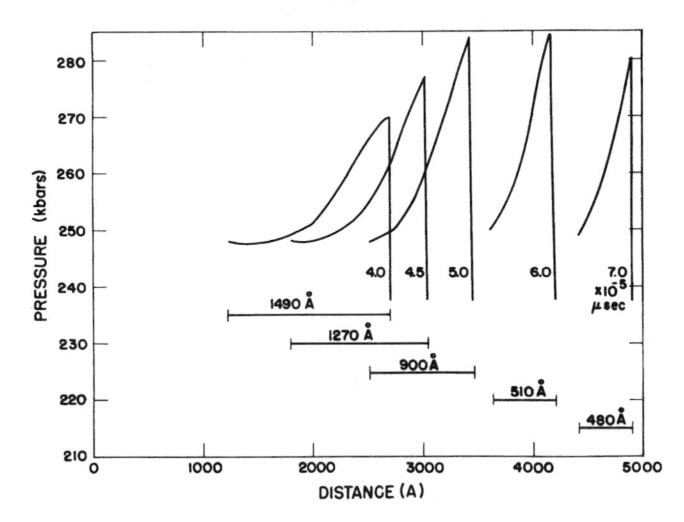

Figure 1.9 Pressure-distance profiles of reaction zone of overdriven detonation in nitromethane produced by a 0.3 cm/μsec constant-velocity piston.

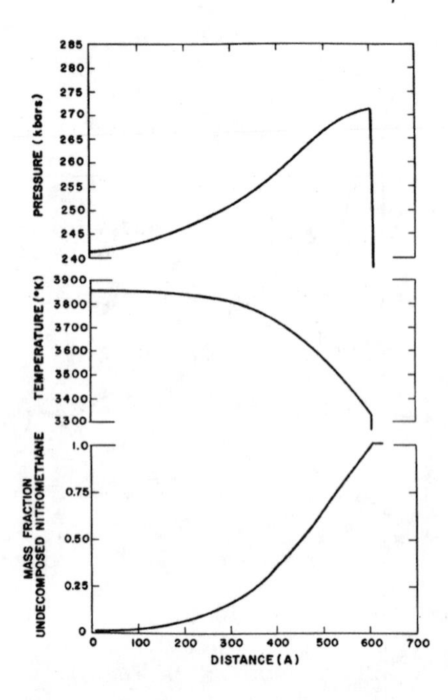

Figure 1.10 Steady state reaction zone profiles of nitromethane overdriven with 0.3-cm/μsec constant-velocity piston with a detonation velocity of 0.7130 cm/μsec.

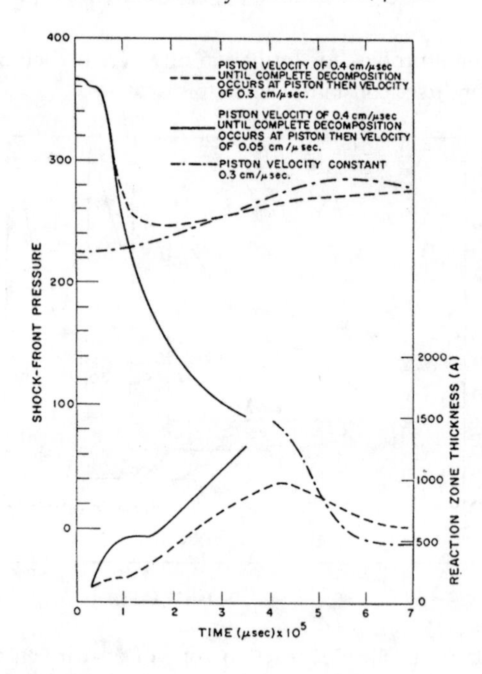

Figure 1.11 Shock-front pressure and reaction zone thickness as a function of time for nitromethane with a piston velocity of 0.4 cm/μsec (until complete decomposition occurs at the piston, then the velocity is stepped to 0.3 cm/μsec), with a 0.3 cm/μsec constant velocity piston, which form steady-state detonations, and with a 0.4/0.05-cm/μsec stepped-velocity piston which forms a failing wave. $E^* = 53.6$, $Z = 4 \times 10^8$, $\gamma = 0.68$, mesh = 5-A.

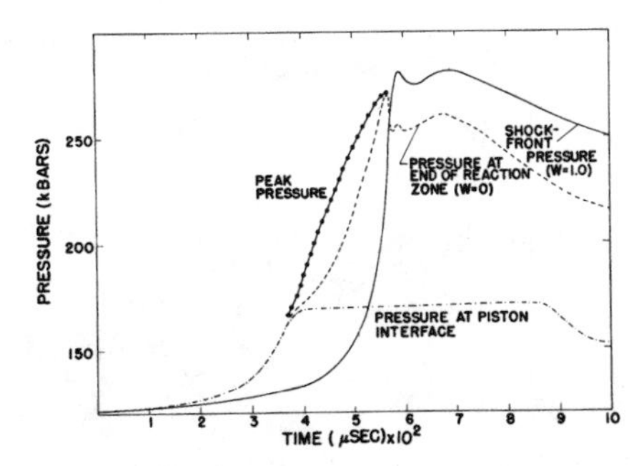

Figure 1.12 Shock initiation of nitromethane with a 0.21 cm/μsec constant-velocity piston. $E^* = 40$, $Z = 1.27 \times 10^6$, $\gamma = 0.68$, mesh = 4000-A, and reaction zone is resolved.

to those of the flow computed with an unresolved reaction zone[3] and are discussed in Chapter 3. The shocked nitromethane first completely decomposes at the piston and then achieves a detonation with a peak pressure that builds up toward the C-J pressure of the high-density shocked nitromethane. The detonation wave overtakes the shock wave, and the pressure at the end of the reaction zone decays toward the piston pressure.

The formation of an overdriven detonation by a piston, whose initial velocity of 0.4 cm/μsec is decreased to 0.3 cm/μsec when complete decomposition occurs at the piston/nitromethane interface, was computed using a 5-A mesh and realistic viscosity values. The initial reaction zone was about 100-A thick, and the shock-front pressure was 360 kbar. The steady-state reaction zone thickness for a 0.3-cm/μsec piston is 620-A; the shock-front pressure is 272 kbar as described in section 1.1. The formation of the reaction zone in the pressure-volume plane is shown in Figure 1.13. The shock-front pressure and the reaction zone thickness are shown as functions of time in Figure 1.11 for the 0.3-cm/μsec constant-velocity piston, and for the 0.4/0.3-cm/μsec stepped-velocity piston. The 0.4/0.3 stepped-velocity piston results in a reaction zone thickness of 100-A at $3 \times 10^{-6}\mu$sec which increases to a maximum of 920-A at $4.25 \times 10^{-5}\mu$sec and then decreases to 610-A by $7 \times 10^{-5}\mu$sec. The 0.4/0.3 stepped-velocity piston produces an initial shock-front pressure of 365 kbar, which decreases to 248 kbar by $1.5 \times 10^{-5}\mu$sec and remains almost constant for the next $10^{-5}\mu$sec; during this time the reaction zone thickness increases from 300 to 600-A. Once the reaction zone thickness exceeds the steady-state thickness, the shock-front pressure begins to increase toward the steady-state shock-front pressure.

Figure 1.11 shows the shock-front pressure and reaction zone thickness for 0.4/0.3- and 0.4/0.05-cm/μsec pistons. The 0.4/0.05-cm/μsec piston produces a detonation that fails to propagate in the time scale of interest.

A decreasing-step piston produces an initially short reaction zone that cannot support the overdriven wave when the second, lower, step occurs. The reaction zone lengthens, and the pressure decreases until the release of chemical energy gain exceeds that necessary to sustain the wave. If the piston has a large enough lower step, the chemical energy again exceeds that necessary to sustain the wave. If the piston has a small enough lower step, the release of chemical energy cannot exceed that necessary to sustain the wave in the time scale of interest and failure occurs. If the piston has a large enough lower step, the chemical energy is released faster than necessary to sustain the wave, the shock-wave pressure increases, and the reaction zone shortens until a steadier profile is obtained. Oscillations of decreasing magnitude about the steady-state solution occur for periods of many reaction zone lengths.

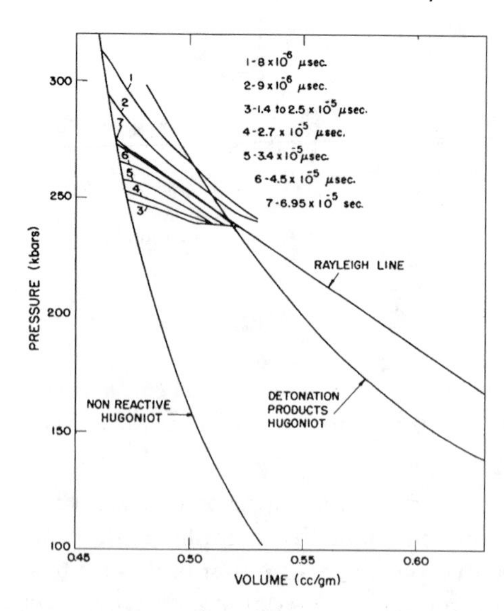

Figure 1.13 Pressure-volume profiles for formation of the reaction zone of an overdriven detonation in nitromethane with a 0.4-cm/μsec piston velocity until complete decomposition occurs at the piston, at which point the piston velocity is set equal to 0.3-cm/μsec.

The formation of an overdriven steady detonation by a "saw-tooth" piston whose velocity decreases to a minimum constant value was computed using a 10-A mesh.

An ideally programmed piston would produce a steady-state reaction zone as soon as the end of the reaction zone was reached by the piston. However, a piston that is programmed either too slow or too fast produces a stable reaction zone in about the same distance or time as the more violent step pistons, even though the derivations of the reaction zone length and pressure profile from the steady-state values may be considerably smaller. This is a result of attempting to balance the release of chemical energy at all points in the zone so that it is just sufficient to sustain the wave. The shock-front pressure and reaction zone thickness as functions of time for various saw-tooth pistons are shown in Figure 1.14. The saw-tooth-piston particle velocity decreases with increasing time at the rate of $A\partial t$, where A is a constant and ∂t is the product of the cycle number and the time increment in μsec. The time increment was set equal to 1/5 the mesh size in cm. The 2.5 x $10^3 \partial t$ piston is closest to the ideal or steady-state piston. The shock-front pressure and reaction zone thickness are within 1% of the steady-state value after 12 x 10^{-5} μsec. The velocity of the (4 x $10^3 \partial t$) piston does not decrease fast enough, so the shock pressure increases and produces a reaction zone that is initially thinner than the steady-state value.

Liquid TNT Reaction Zones

The approximate steady-state profiles for liquid TNT at the BKW C-J state are shown in Figure 1.15 for γ's of 1.66 and 1.0. The reaction zone thickness is about 1 x 10^{-3} cm for both gammas. The temperature profile for $\gamma = 1.0$ has a maximum at about 2.5 x 10^{-4} cm, while the maximum temperature for $\gamma = 1.66$ is at the end of the reaction zone.

The shock-front pressure and the pressure at the end of the reaction zone as functions of time for the steady-state piston are shown in Figure 1.16 for $\gamma = 1.0$ and in Figure 1.17 for

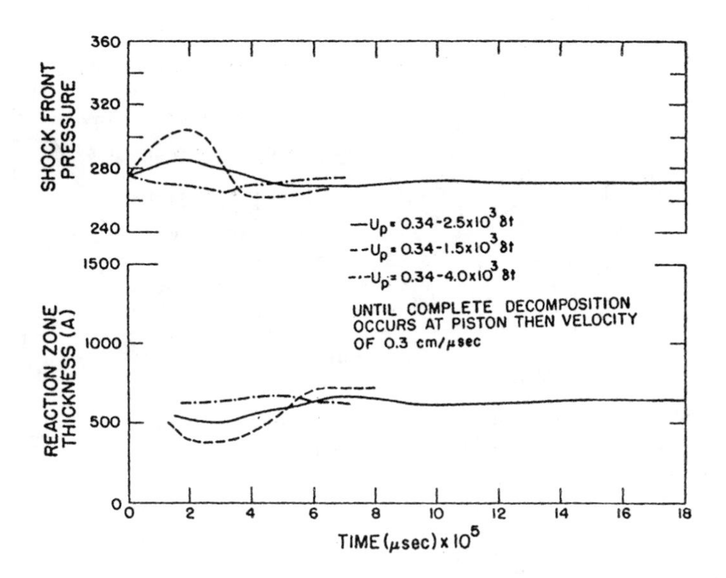

Figure 1.14 Shock-front pressure and reaction zone thickness as a function of time for nitromethane with various saw-tooth pistons that form steady-state detonations. U_p is piston particle velocity and $\delta t = n\Delta t$ where Δt is time increment and n is cycle number.

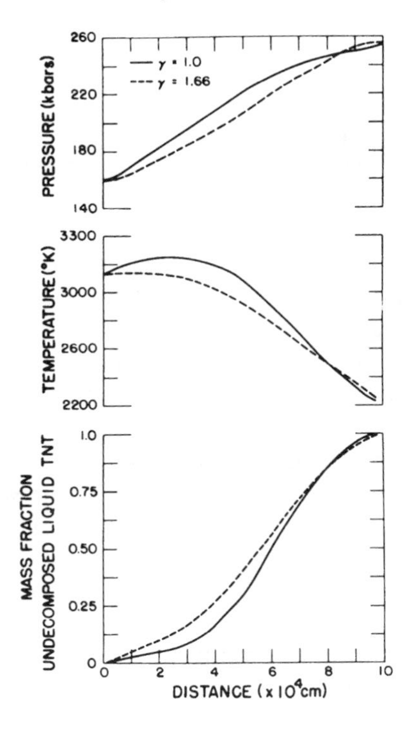

Figure 1.15 Approximate steady-state profiles for liquid TNT at C-J detonation velocity of 0.6406 cm/μsec.

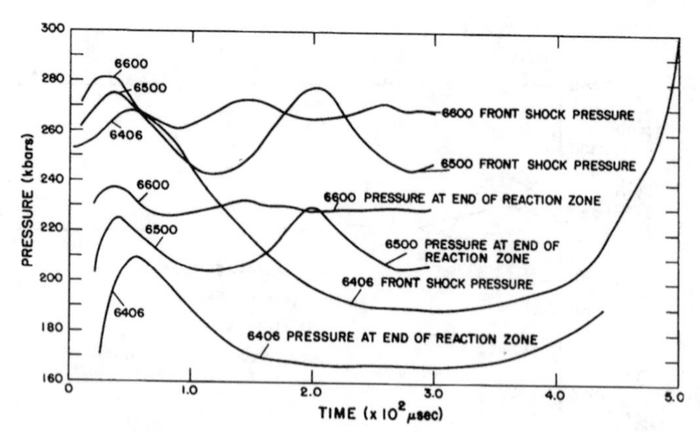

Figure 1.16 Shock-front pressure and pressure at end of reaction zone vs. time for various detonation velocities (shown in m/sec) using steady-state piston for liquid TNT with $\gamma = 1.0$ and mesh = 5000-A.

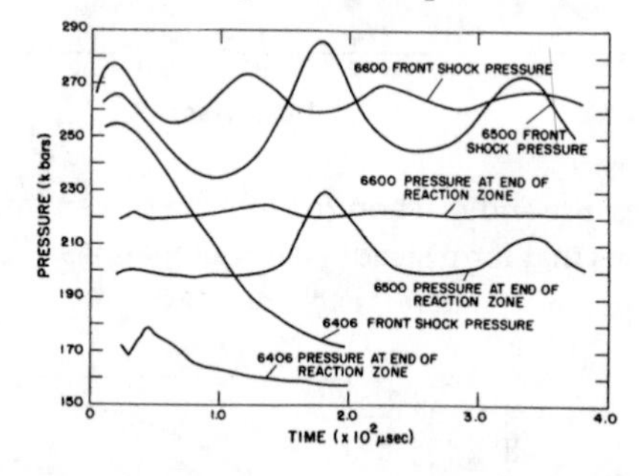

Figure 1.17 Shock-front pressure and pressure at end of reaction zone vs. time for various detonation velocities (shown in m/sec) using steady-state piston for liquid TNT with $\gamma = 1.66$ and mesh = 5000-A.

$\gamma = 1.66$. In both cases the 0.6699-cm/μsec case is stable and the 0.6500-cm/μsec case is unstable. For the C-J cases, the period from peak to peak is about 6 x $10^{-2}\mu$sec.

The amount of overdrive necessary to stabilize the liquid TNT detonation decreases with decreasing activation energy. If one assumes an activation energy of 30 kcal/mole instead of the "experimental" value of 41.4 kcal/mole, the liquid TNT detonation for a γ of 1.66 is stable at the C-J velocity of 0.6406 cm/μsec.

The calculated results for nitromethane and liquid TNT are summarized in Table 1.2.

Ideal Gas Reaction Zones

The equation of state for an ideal gas of constant heat capacity undergoing an exothermic, irreversible, unimolecular reaction was used. Given I, the internal energy; W, the mass fraction of undecomposed explosive; and V, the specific volume; the pressure P and temperature T is calculated from

$$P = \frac{[I + (1 - W)Q][\gamma - 1]}{V},$$

and

$$T = \frac{[I + (1 - W)Q]}{C_v} .$$

The values of the parameters that Erpenbeck[4] used were

$$\gamma = 1.2$$

$$Q = 50RT_o$$

$$C_v = 5 = R/(\gamma - 1)$$

$$R = 1$$

$$T_o, V_o, P_o = 1$$

$$I_o = 5 = C_v T_o$$

$$D_{CJ} = 6.809$$

$$U_o = 0$$

The Arrhenius rate law is

$$\frac{dW}{dt} = ZWe^{-E^*/RT} ,$$

where E^* is the activation energy and Z is the frequency factor. Z is adjusted so that W is one-half when the time is one.

Erpenbeck's linearized analysis of the detonation equations for ideal gases predicts that certain cases will be unstable to infinitesimal longitudinal perturbations. Fickett and Wood[2] have used the method of characteristics with sharp shocks to describe the time dependent flow and found agreement with Erpenbeck's results.

SIN calculations for the Erpenbeck and Fickett ideal gas detonation model, as shown in Table 1.3 and Figure 1.18, resulted in both stable and unstable detonations. The results agree with Erpenbeck's non-viscous linearized analysis and Fickett and Wood's sharp-shock solutions. For example, for the unstable case of $E^* = 50, \gamma = 1.2, Q = 50$, and $f = D^2/D_{CJ}^2 = 1.6$, with enough viscosity to smear the shock front over 10 cells, the same period and final amplitude was obtained as Fickett and Wood with no viscosity. Additional comparisons will be made in the next section.

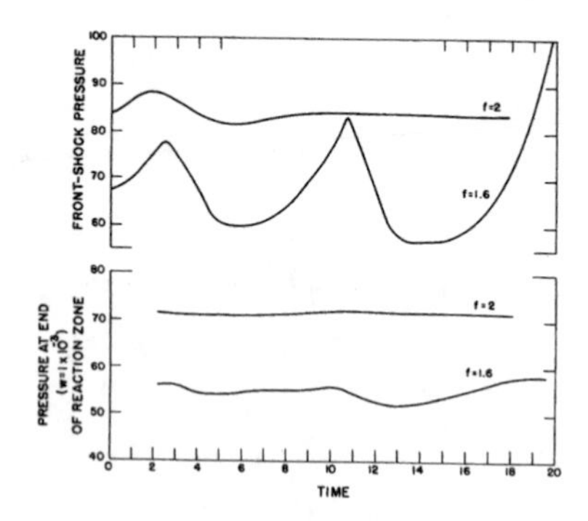

Figure 1.18 Shock-front pressure and pressure at end of reaction zone vs. time for various values of $f = D^2/D_{CJ}^2$ using steady-state piston for an ideal gas with $E^* = 50$, $\gamma = 1.2$, and Q = 50.

TABLE 1.3 Ideal Gas Reaction Zone Summary

	1	2	3	4
E^*	50	50	10	10
Z^*	206	82.5	2.07	3
$f = (D/D_{CJ})^2$	1.6	2.0	1.6	1.0
Δx^b	0.35	0.385	0.35	0.272
Δt	0.0035	0.00385	0.0035	0.00272
D	8.613	9.63	8.613	6.809
Result	Unstable	Stable	Stable	Stable
$P_{W=1}$	67.5	84.0	67.5	42.2
$P_{W=0}$	55.0	72.0	55.0	21.5

*Z is adjusted so as to have W = 0.5 when the time is 1.0.
$^b\Delta x$ is adjusted so as to have about 25 space zones for W = 1.0 - 0.5.

Fickett, Jacobson, and Schott[5] have used the characteristic method to investigate ideal gas detonations with a rate function approximating reactions dominated by a branching chain mechanism resulting in an induction period followed by the main strongly exothermic reaction. They observed pulsating detonations similar to those described above for first-order Arrhenius kinetics.

1.3 *Two-Dimensional Reaction Zones of Homogeneous Explosives*

The resolved reaction zones of nitromethane, liquid TNT, and ideal gases have also been studied using the two-dimensional Lagrangian code 2DL, described in Appendix B and the two-dimensional Eulerain code 2DE described in Appendix C.

To study the stability of detonations, one uses a piston that is as close to the steady-state piston as possible so that the one-dimensional perturbations introduced by deviations

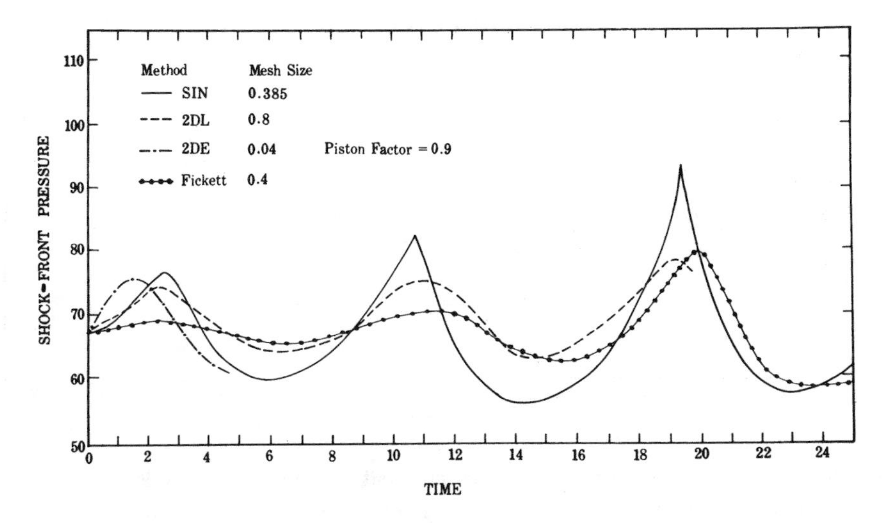

Figure 1.19 Shock-front pressure vs. time for an ideal gas detonation unstable to one-dimensional perturbations with $f = 1.6$ and Q = 50, for various numerical methods.

from the steady-state will be of tractable size. The magnitude of the one-dimensional perturbations depends upon the details of the piston profile, the numerical method, and the mesh size. For a given mesh size, one-dimensional perturbations increase with the method in the order 2DL, SIN, 2DE. While attempts were made, by adjusting the mesh size, to obtain one-dimensional perturbations of approximately the same size in all three methods, this is unimportant to the question of whether the perturbations introduced grew or decayed. To calculate the steady-state reaction zone piston in the 2DE method, it was necessary to use for the piston's W a lower mass fraction of undecomposed explosive than the W of the cell above the piston. This was accomplished by multiplying by a constant called "piston factor".

The shock-front pressure vs. time for an ideal gas detonation unstable to one-dimensional perturbations with an $f = D^2/D_{CJ}^2 = 1.6$ (where D is detonation velocity) is shown in Figure 1.19 for the various numerical methods. In Figure 1.20 the shock-front pressure vs. time is shown for ideal gas detonation stable to one-dimensional perturbations with an $f = 2$ for various numerical methods. The shock-front pressure vs. time for nitromethane detonations with velocities of 0.6850, 0.6650, and 0.6465 cm/μsec is shown in Figure 1.21. Similar results were obtained for TNT with a velocity of 0.6406 cm/μsec.

Since the magnitude of the one-dimensional perturbation introduced into the calculation varies, the shock-front pressure vs. time is not identical for the various methods. The stability results are related not to the magnitude but only to the growth or decay of the perturbations. These results show that growth or decay of the perturbation is independent of the numerical method and, incidentally, independent of the nature of the perturbation.

One can obtain identical stability results for two-dimensionally unperturbed flow using any of the four methods of numerically solving the reactive hydrodynamics. While this is not sufficient to guarantee that the stability results are independent of the numerical method, it is a necessary result. Encouraged by these results, we proceed to study two-dimensionally perturbed detonations.

To study the two-dimensionally perturbed flow, a perturbation was introduced by increasing the initial density of a center cell near the piston interface. The density was in-

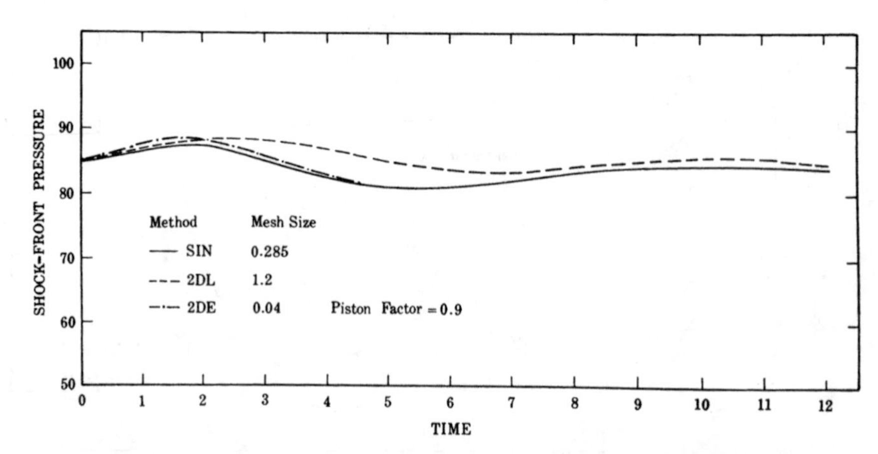

Figure 1.20 Shock-front pressure vs. time for an ideal gas detonation stable to one-dimensional perturbation, with $f = 2$ and Q = 50, for various numerical methods.

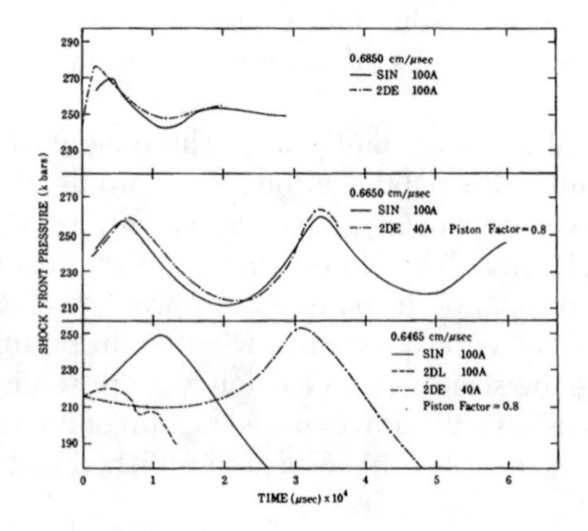

Figure 1.21 Shock-front pressure vs. time for nitromethane with various amounts of overdrive, for various numerical methods. $E^* = 53.6$, Z = 4 x 10^8, and $\gamma = 0.68$. The mesh size is given for each method, and the detonation velocity is given in cm/μsec.

creased by 1.1 and 1.3 ρ_o, without changing the stability results. Perturbations of all wavelengths are introduced by this procedure. In 2DL the viscosity constant was changed over half an order of magnitude without changing the stability results.

The 2DL calculations were performed with 50 cells in the X or R (radial) direction and 250 cells in the Z direction. The 2DE calculations were performed with 100 cells in the R and Z directions. The perturbations were initially located at the center of the piston boundary.

Erpenbeck[4] predicts that an ideal gas detonation with $f = 2$, $E^* = 50$, and $Q = 50$ will be stable to one-dimensional perturbations but unstable to two-dimensional perturbations, and that with $Q = 0.2$ it will be stable to all perturbations.

As an extra check, a perturbed constant-pressure piston was run without chemical reaction and with a pressure equal to the ideal gas spike pressure of 84 for $f = 2$. The perturbation decayed as expected. The $f = 2$, $Q = 50$ ideal gas detonation that is stable to one-dimensional perturbations is unstable to two-dimensional perturbations. The re-

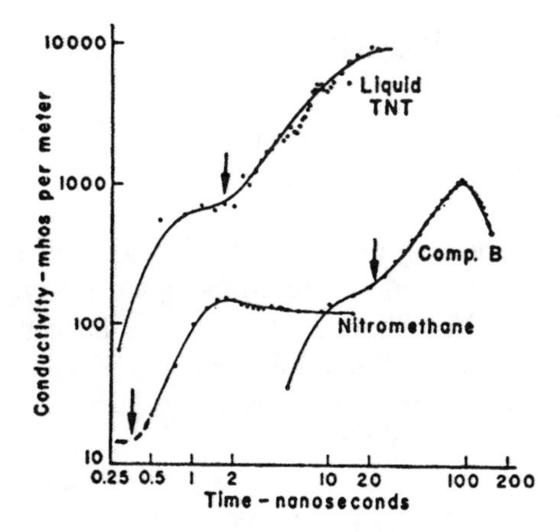

Figure 1.22 Hayes' conductivity reaction zone profiles for liquid TNT, Composition B, and Nitromethane. Arrows show probable end of reaction zone.

sults agree with Erpenbeck's predictions and give us further confidence in our numerical approach for determining detonation wave stability.

The nitromethane detonations with E^*s of 40 and 53.6 at C-J velocity and with an E^* of 53.6 at 0.7130 cm/μsec were calculated using the Langrangian and Eulerian two-dimensional codes. In all cases, the two-dimensional perturbation did not grow and the stability was identical with that observed for one-dimensional perturbations.

1.4 Discussion of Reaction Zones of Homogeneous Explosives

The reaction zones obtained by solving the one-dimensional hydrodynamic equations with Arrhenius kinetics and realistic equations of state for homogeneous explosives have been described. The partially reacted Hugoniots and, therefore, the steady-state profiles, are very sensitive to the parameters and to the details of the equations of state used for mixture of undecomposed explosive and detonation products. The amount of overdrive necessary to stabilize a detonation decreases with decreasing activation energy. The space and time scale is so small that the approximation of treating a discrete particle field as a continuum introduces large uncertainties into experimental values used in the equation of state and kinetics.

It is impossible to know how closely these idealized one-dimensional reaction zones resemble real three-dimensional reaction zones in homogeneous explosives without a detailed experimental description for comparison. Unfortunately, the time and space scale of interest is not readily accessible experimentally, and only a very limited amount of experimental evidence is available.

Hayes' reaction zone conductivity measurements[6] reproduced in Figure 1.22 are the most resolved homogeneous reaction zone data available; however, the interpretation of the results is not obvious. Since the carbon content correlates well with the second conductivity peak, it is reasonable to suppose that this peak is a result of carbon precipitation and that the first peak shown with an arrow is the end of the reaction zone. If the experimental data are interpreted thus, the reaction zone thickness of nitromethane is less than 1 x 10^{-4} cm and that of liquid TNT is 1×10^{-3} cm. Hayes has observed a 1000-megacycle ($10^{-3}\mu$sec/cycle) oscillation in liquid TNT.

Craig, from his free-surface velocity measurements with very thin metal plates, estimates that the reaction zone thickness of liquid TNT is about 1×10^{-2} cm and that of nitromethane is less than 2×10^{-3} cm.[7,8]

One may conclude that, although the experimental evidence does not conflict with the results of the modeling of homogeneous reaction zones, it is insufficient to validate them.

Since most explosives and propellants are heterogeneous (contain density discontinuities such as voids or air holes), the nature of the homogeneous reaction zone is mostly of academic interest. Experimental studies of the reaction zone of heterogeneous explosives give thicknesses and pressure profiles that vary greatly for the same explosive dependent upon the method of measurement used. While each experimentalist was certain that his values were correct and the others were wrong, in reality they all were valid, but with different averages of a complicated three-dimensional flow. As shown in the next section, the fault is in the basic assumption that heterogeneous explosives have a unique one-dimensional reaction zone with a single reaction zone thickness, Von-Neuman spike pressure and C-J pressure.

1.5 *Three-Dimensional Reaction Zones of Heterogeneous Explosives*

Detonations of the condensed homogeneous explosives nitromethane and liquid TNT were found to exhibit unstable periodic behavior. The steady-state Chapman-Jouguet theory of the detonation process will not properly describe the behavior of homogeneous explosives that exhibit such unstable behavior.

Most experimental studies of reaction zone characteristics of explosives have been performed using heterogeneous explosives rather than liquids or single crystals. Heterogeneous explosives are explosives containing density discontinuities such as voids or air holes. The shock interactions that occur when shocks interact with voids or air holes result in local high temperature and pressure regions called "hot spots." These "hot spots" decompose and add their energy to the flow and result in the process of heterogeneous shock initiation. The process has been modeled numerically and is described in detail by the hydrodynamic hot spot model described in References 9 and 10 and Chapter 3. The success of the three-dimensional numerical models in describing the interaction of shock waves with density discontinuities and of a detonation wave interacting with a matrix of tungsten particles in HMX, described in Reference 11 and Chapter 2, encouraged us to numerically examine the interaction of a resolved reactive zone in HMX with a two volume percent matrix of air holes. An objective of the study was to determine the nature of the flow being examined in experimental studies of reaction zones of heterogeneous explosives, since different methods gave greatly different values for the reaction zone parameters for the same explosive.

Experimental Observations

The heterogeneous explosive reaction zone that has been the most studied is PBX-9404 (94/3/3 HMX/Nitrocellulose/Tris-β-Chloroethyl phosphate). A summary of the estimated reaction zone thickness and Von Neumann spike pressure is given in Table 1.4 along with the calculated reaction zone parameters using the solid Arrhenius HMX constants of

34.8 kcal/mole for activation energy and 3 x 10^4 μs^{-1} for frequency factor described in Reference 9 and Chapter 4.

TABLE 1.4 PBX-9404 Reaction Zone

Experimental Technique	Reaction Zone Thickness (cm)	$P_{VN\ Spike}$ (kbar)	P_{C-J} (kbar)
Bromoform	0.02	485	
Interferometer	<0.01		
Infrared Radiometry	0.02–0.03		
Metal Free Surface	0.01	550	365
Calculated	0.07	560	365

The metal free-surface measurements of B. G. Craig used the technique described in Reference 12. The infrared radiometry measurements of W. Von Holle used the technique described in Reference 13. The interferometer measurements were performed by W. Seitz. The application of this method to reaction zone measurements is described in Reference 14. The bromoform measurements were made by R. McQueen and J. Fritz using the technique described in Reference 15. Sheffield[16] studied the reaction zone of heterogeneous explosives using a foil-water system.

The calculated reaction zone for "homogeneous" PBX-9404 is larger than any of the observed reaction zones for heterogeneous PBX-9404, although within the uncertainties associated with experimental interpretation and with the solid Arrhenius constants. The effect of heterogenities on reaction zone structure was investigated to determine if they might result in a reaction zone whose effective thickness was different than the ideal steady-state reaction zone length.

Three-Dimensional Numerical Modeling

The steady-state reaction zone for PBX-9404 was calculated using the one-dimensional reactive hydrodynamic code, SIN, the HOM equation-of-state constants and the Arrhenius constants for solid HMX described in Reference 9 and Chapter 4.

The calculated PBX-9404 reaction zone profile is shown in Figure 1.23.

The time-dependent behavior of the flow in the reaction zone of detonating PBX-9404 was investigated using one-dimensional Lagrangian and three-dimensional Eulerian numerical hydrodynamics. The steady-state solution was stable and perturbations were found to decay. This is in contrast to the time-dependent, unstable, periodic reaction zones reported for liquid TNT and nitromethane.

The stable steady-state reaction zone of PBX-9404 permits us to study the effect of heterogenities on the reaction zone profile without the complication associated with a time-dependent reaction zone.

To examine the effect of heterogenities on the reaction zone, the three-dimensional Eulerian reactive hydrodynamic code, 3DE, described in Reference 17 and Appendix D was used.

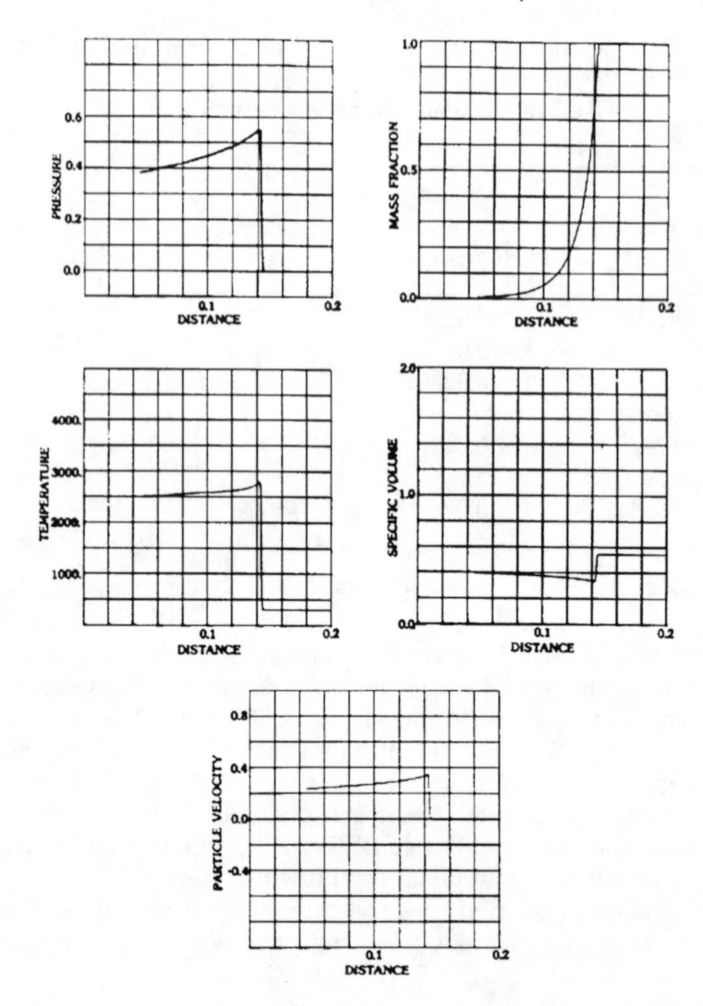

Figure 1.23 The high resolution one-dimensional reaction zone profile of PBX-9404 calculated using the SIN code.

The three-dimensional computational grid contained 30 cells in the X direction, 28 cells in the Y direction, and 57 cells in the Z direction, each 0.004 cm on a side. The time increment was $8 \times 10^{-4} \mu s$. At the bottom of the grid was a reaction zone piston, which was programmed to initialize the flow with a steady-state reaction zone. After the steady-state reaction zone had traveled one reaction zone length in solid PBX-9404, it interacted with a two percent by volume HCP (hexagonal closed packed) matrix of air holes.

Thirty-four spherical air holes, each with a diameter of 0.012 cm, occupy a region in the middle of the mesh about 25 cells high. Partial air spheres occur on the boundaries as necessary. The air hole size was chosen to be representative of the actual hole size present in pressed PBX-9404.

Numerical tests with two to six cells per air sphere diameter showed the results were independent of grid size for 3 or more cells per sphere diameter.

Figure 1.24 shows the initial configuration of spherical air holes in PBX-9404.

The low resolution necessary for the three-dimensional calculation results in a less re-

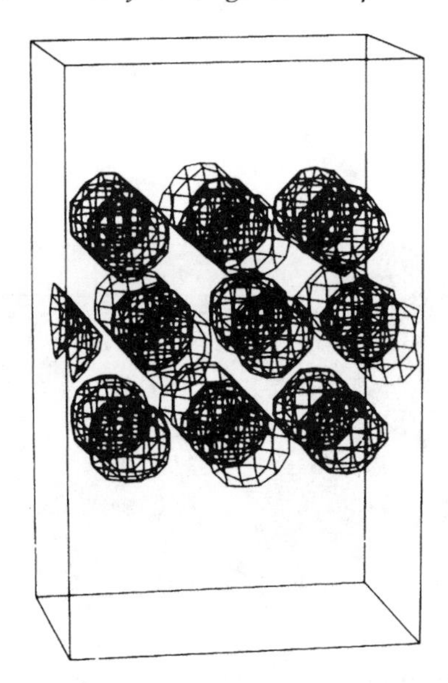

Figure 1.24 The initial configuration of air spheres in a cube of solid PBX-9404.

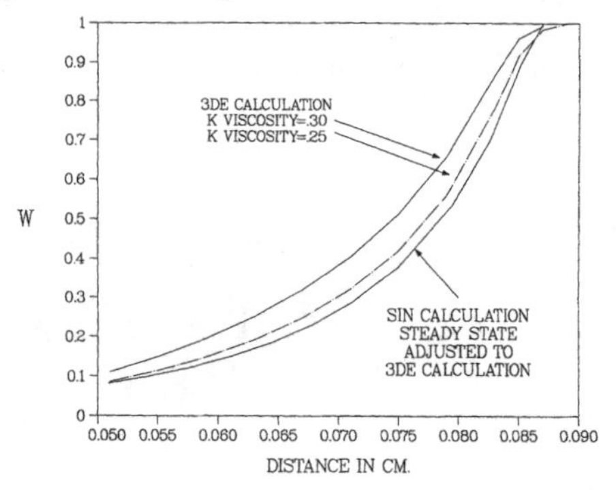

Figure 1.25 The reaction zone mass fraction as a function of distance for the high resolution SIN calculation and low resolution 3DE calculation.

solved reaction zone than described earlier using the one-dimensional SIN code. The reaction zone burn fraction as a function of distance is shown in Figure 1.25 for the SIN one-dimensional calculation and for the 3DE calculation for two viscosity coefficients. The viscosity coefficient shifts the location of the start of the burn. The profile is not significantly changed by variations in the viscosity.

The 3DE reaction zone profile for pressure, temperature, particle velocity, and mass fraction as a function of distance are shown in Figure 1.26.

The burn fraction surfaces of a PBX-9404 reaction zone after it has interacted with the region of two volume percent air spheres is shown in Figure 1.27. The heterogenities perturb the reaction zone. A complicated reaction region develops and is maintained by the reactive flow.

Cross sectional plots of pressure and burn fraction through the 15th cell in the X direction (I=15) are shown in Figure 1.28. A complicated time-dependent, multi-dimensional reaction region proceeds through the heterogeneous explosive.

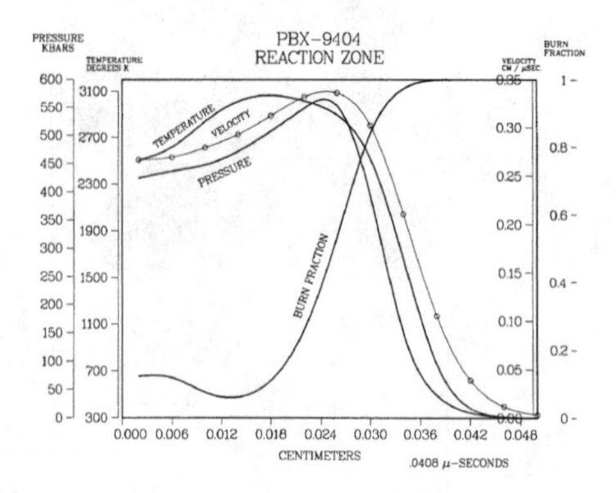

Figure 1.26 The reaction zone profiles in the 3DE calculation.

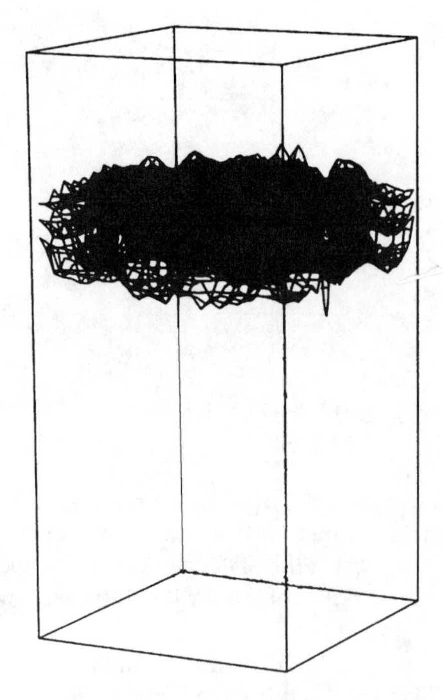

Figure 1.27 Burn fraction surface profiles after reaction zone interacts with heterogenities in PBX-9404.

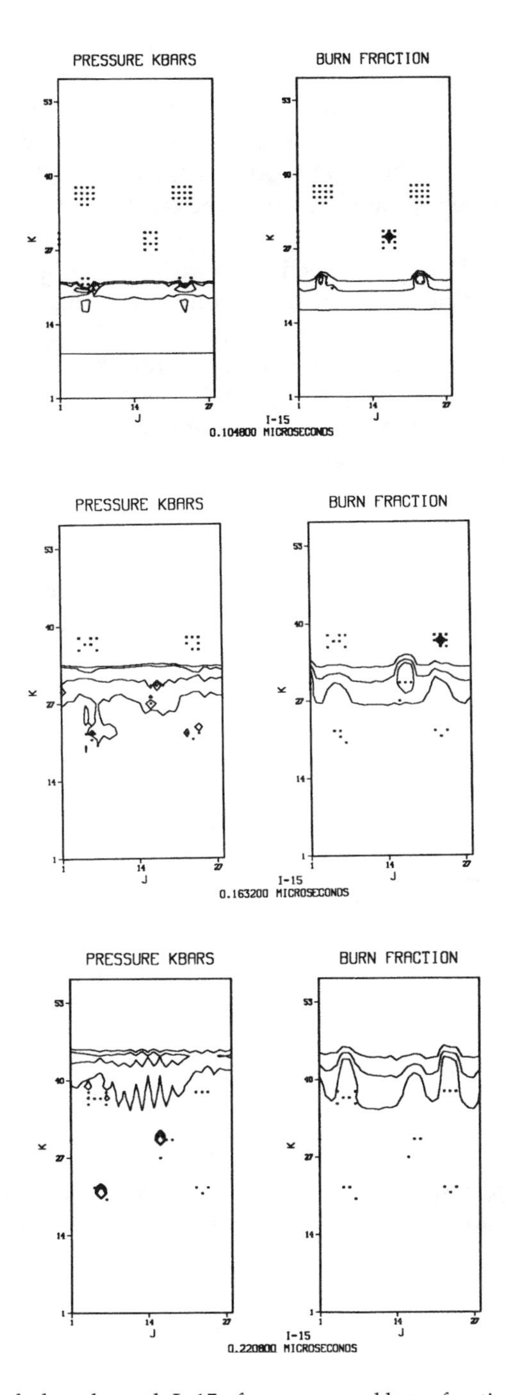

Figure 1.28 Cross sectional plots through I=15 of pressure and burn fraction showing the heterogeneous PBX-9404 reaction region.

Discussion

The calculated reaction zone of PBX-9404 using solid HMX Arrhenius kinetics is stable to perturbations, and a steady-state reaction zone profile is maintained. This is in contrast to the time-dependent, periodic reaction zone calculated for nitromethane and liquid TNT.

The effect of two volume percent spherical air holes on the reaction zone was modeled using the three-dimensional Eulerian hydrodynamic code, 3DE. The air holes perturb the reaction zone flow. A complicated reaction region develops and is maintained by the reactive fluid dynamics.

Thus, any experimental study of a reaction region in a heterogeneous explosive is actually measuring some mean value of an irregular, complicated multi-dimensional flow. It is not surprising that different experimental techniques may give quite different reaction zone "thicknesses", Von Neumann spike pressures and profiles.

As shown in Table 1.5 for PBX-9502 and Table 1.6 for Composition B, the measured reaction zone parameters for heterogeneous explosives vary considerably with the experimental technique. The reported reaction zone thickness for PBX-9502 (95/5 TATB/Kel F, $\rho = 1.894$) varies by a factor of 8 between the metal free-surface measurement of Craig[8] and the foil-water measurement reported by Sheffield.[16] As shown in Table 1.6, the reaction zone thickness of Composition B (64/36 RDX/TNT, $\rho = 1.713$) varies by a factor of 4 between the bromoform measurement and the conductivity measurement of Hayes.[6]

TABLE 1.5 PBX-9502 Reaction Zone

Experimental Technique	Reaction Zone Thickness (cm)	$P_{VN\ Spike}$ (kbar)	P_{C-J} (kbar)
Interferometer	0.08-0.16	376	
Infrared Radiometry	0.08		
Metal Free Surface	0.03		290
Foil/Water	0.21	375	
Calculated		377	290

The measured Von Neumann spike pressure can also vary with the experimental technique as shown in Table 1.6, where the reported Composition B Von Neumann spike pressure varies from 374 to 420 kilobars, and in Table 1.4, where it varies from 485 to 550 kilobars for PBX-9404.

The reactive region in heterogeneous explosives is complicated, time-dependent, and multi-dimensional (non laminar).

The reactive region has bounds which approach a steady-state condition, but the flow inside those bounds is multi-dimensional and time-dependent. The reaction zone "thickness" at various locations across the zone may vary by an order of magnitude. The Von Neuman spike pressure and the pressure at the end of the reaction zone also varies with location by 25 percent or more. The reaction zone is highly variable with time but has bounds within which it varies.

TABLE 1.6 Composition B Reaction Zone

Experimental Technique	Reaction Zone Thickness (cm)	$P_{VN\ Spike}$ (kbar)	P_{C-J} (kbar)
Bromoform	0.04	395	
Interferometer	0.02	420	
Metal Free Surface	0.014	374	285
Conductivity	0.013		
Calculated		437	285

Theoretical analytical or numerical models that assume a unique reaction zone thickness and pressure profile throughout the reaction zone may be appropriate for certain homogeneous explosives that are stable at the C-J state, but such models CANNOT describe the reaction zone of any heterogeneous explosive or propellant.

Experimental measurements of the reaction zone of heterogeneous explosives need to be made with high resolution in space and time. The space scale required must be less than the size of the smaller density discontinuities, and the time scale must be small enough to follow the variable flow associated with the density discontinuities.

References

1. D. L. Chapman, "On the Rate of Explosion in Gases," Philosophical Magazine 47, 90 (1899).

2. W. Fickett and W. W. Wood, "Flow Calculations for Pulsating One-Dimensional Detonations," The Physics of Fluids 9, 903–916 (1966).

3. Charles L. Mader, "Shock and Hot Spot Initiation of Homogeneous Explosives," Physics of Fluids 6, 375 (1963).

4. J. J. Erpenbeck, "Stability of Idealized One-Reaction Detonations," The Physics of Fluids 7, 684–696 (1964), 8, 1192 (1965), 9, 1293 (1966).

5. W. Fickett, J. D. Jacobson, and G. L. Schott, "Calculated Pulsating One-Dimensional Detonations with Induction-Zone Kinetics," A.I.A.A. Journal 10, 514–516 (1972).

6. Bernard Hayes, "Electrical Measurements in Reaction Zones of High Explosives," Tenth Symposium (International) on Combustion 869–874 (1965).

7. Bobby G. Craig, private communication.

8. B. G. Craig, "Measurements of the Detonation-Front Structure in Condensed-Phase Explosives," Tenth Symposium (International) on Combustion 863–867 (1965).

9. Charles L. Mader and James D. Kershner, "Three-Dimensional Modeling of Explosive Desensitization by Preshocking," Journal of Energetic Materials 3, 35–47 (1985).

10. Charles L. Mader and James D. Kershner, "The Three-Dimensional Hydrodynamic Hot Spot," Eighth Symposium (International) on Detonation (1985).

11. Charles L. Mader, James D. Kershner, and George H. Pimbley, "Three-Dimensional Modeling of Inert Metal-Loaded Explosives," Journal of Energetic Materials 1, 293–324 (1983).

12. B. G. Craig, "Measurements of the Detonation-Front Structure in Condensed-Phase Explosives," Tenth Symposium (International) on Combustion 863–867 (1965).

13. William G. Von Holle and Roy A. McWilliams, "Application of Fast Infrared Detectors to Detonation Science," S.P.I.E. Proceedings 386, 1–8 (1983).

14. W. L. Seitz, H. L. Stacy, and J. Wackerle, "Detonation Reaction Zone Studies on TATB Explosives," Eighth Symposium (International) on Detonation (1986).

15. R. G. McQueen and J. N. Fritz, "The Reaction Zone of High Explosives," Los Alamos National Laboratory Report M–6–530 (1983).

16. C. A. Sheffield, D. D. Bloomquist, and C. M. Tarver, "Subnanosecond Measurements of Detonation Fronts in Solid High Explosives," Journal of Chemical Physics 80, 3831–3844 (1984).

17. Charles L. Mader and James D. Kershner, "Three-Dimensional Modeling of Triple Wave Initiation of Insensitive Explosives," Los Alamos Scientific Laboratory Report LA–8206 (1980).

chapter two

Performance of Explosives and Propellants

2.1 Steady-State Detonations

The assumption of stable, one-dimensional detonations is not valid when one considers small-scale details, as Chapter 1 shows. However, although steady-state theory is invalid on a microscale, it does provide an excellent first approximation and a very useful aid in detonation performance calculations. Assumptions of chemical equilibrium in the steady-state model are incorrect. One of the interesting problems in modeling the equation of state of detonation products is finding reasonable changes in the detonation product composition which will reproduce the experimentally observed explosive and propellant performance.

A classic example of observed explosive properties which cannot be described by the usual steady-state chemical equilibrium models is the detonation velocity of TNT as a function of density reported by Urizar, James, and Smith.[1] The velocity is plotted as a function of density in Figure 2.1. It has a sharp change of slope of 3163 to 1700 m/sec/g/cc at 1.55 g/cc. This problem is discussed in more detail in the section on carbon condensation.

The Becker-Kistiakowsky-Wilson (BKW) equation of state is based upon a repulsive potential applied to the virial equation of state. The BKW equation of state has the form

$$\frac{PV}{RT} = 1 + x\, e^{\beta x}$$

$$x = \kappa \frac{\sum X_i k_i}{V T^{0.5}}$$

where x_i and k_i are the mole fractions and covolumes of the detonation products and κ and β are equation of state constants. In a parameter study of the BKW equation of state one may adjust κ to give the experimental detonation velocity for a single explosive at a single density. One may change the slope of the detonation velocity-density curve by changing β. With successive iterations on κ and β one can reproduce the experimental detonation velocities at two densities for a single explosive. The repulsive potential for each detonation product is represented by a covolume. The experimental Hugoniots for each detonation product are used to evaluate the covolume. A single set of parameters could not be found that would fit RDX and TNT data. This is attributed to the large difference in the amount

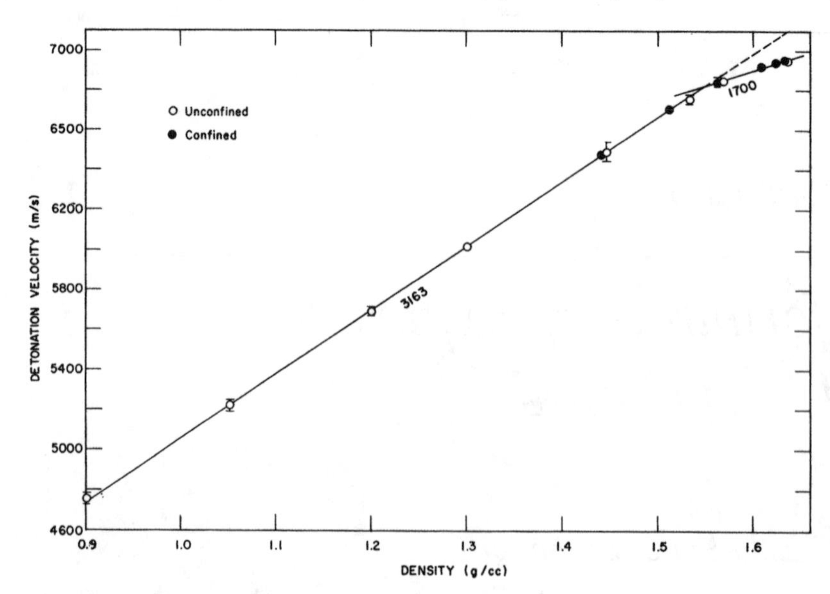

Figure 2.1 Detonation velocity of TNT as a function of density.

of solid carbon in the detonation products of RDX and TNT. Two parameter sets are used: the RDX parameter set for most explosives, and the TNT parameter set for explosives with large amounts of solid carbon in the detonation products.

The BKW code is described in detail in Appendix E and the personal computer executable code is included on the CD-ROM along with the USERBKW code, which assists in assembling the required input files for BKW. If the explosive or propellant elemental composition, density, and heat of formation are known, the BKW code can be used to compute the C-J equilibrium detonation product composition, the C-J pressure, detonation velocity, temperature, the single shock Hugoniot and the isentrope. The HOM equation of state constants are generated for use in hydrodynamic codes.

The Becker-Kistiakowsky-Wilson (BKW) equation of state described in Appendix E is the most used and best calibrated of those used to calculate detonation properties assuming steady-state and chemical equilibrium. Comparison of the calculated and experimental detonation properties permits evaluation of the errors to be expected from steady-state modeling of detonation products. Table 2.1 lists the calculated and experimental C-J properties of various explosives and mixtures. The calculated detonation product compositions of some of the explosives are given in Table 2.2.

Experimental data are required in evaluating any model of the detonation process. Data that relate to the detonation properties are detonation velocity as a function of density, C-J pressures and temperatures, and state points on the shock Hugoniot and expansion isentrope of the detonation products.

The detonation velocity can be measured easily to within 1%. It changes with the charge diameter, so one must make measurements at several diameters for a single density and extrapolate to "infinite diameter" velocity for comparison with steady-state calculations. The detonation velocity is the least sensitive detonation parameter.

TABLE 2.1 Experimental and Calculated C-J Properties

Explosive	C-J Param.	Expt'l.	Ref.	BKW RDX Param.	BKW TNT Param.
BCHNO					
1/3.75 moles Ethyl Decaborane/Tetranitromethane	D	6740	2	6945	–
$\rho = 1.40$	P	172		196	
$B_{10}H_{18}C_{5.75}N_{15}O_{30}$	T	4460		5336	
E_o =109.2	γ	2.70		2.43	
1/4.45 moles Ethyl Decaborane/Tetranitromethane	D	6820	2	6897	–
$\rho = 1.427$	P	167		196	
$B_{10}H_{18}C_{6.45}N_{17.8}O_{35.6}$	T			5409	
E_o =124.3	γ	2.97		2.46	
CHNF					
1,2-Bis(difluoroamino)propane	D	–	–	7213	6215
$\rho = 1.256$	P			209	147
$C_3H_6N_2F_4$	T			2905	2292
E_o =-27.5	γ			2.12	2.29
CHNO (Solid One-Component)					
RDX, Cyclotrimethylene trinitramine	D	8754	2	8754	8263
$\rho = 1.80$	P	347		347	324
$C_3H_6N_6O_6$	T			2587	2861
E_o =33.97	γ	2.98		2.98	2.79
RDX	D	5981	2	6128	
$\rho = 1.0$	P			108	
	T			3600	
	γ			2.48	
TNT, Trinitrotoluene	D	6950	2	7197	6950
$\rho = 1.64$	P	190		213	206
$C_7H_5N_3O_6$	T			2829	2937
E_o =-1.44	γ	3.16		2.98	2.85

TABLE 2.1 Experimental and Calculated C-J Properties (continued)

Explosive	C-J Param.	Expt'l.	Ref.	BKW RDX Param.	BKW TNT Param.
TNT	D	5254	2		5339
$\rho = 1.061$	P	110			85
	T				3175
	γ	1.66			2.54
TNT	D	4200	2		4511
$\rho = 0.732$	P	59			45
	T				3080
	γ	1.18			2.31
HMX, Cyclotetramethylene-tetranitramine	D	9100	2	9159	8556
$\rho = 1.90$	P	393		395	364
$C_4H_8N_8O_8$	T			2364	2693
E_o =43.46	γ	3.0		3.03	2.82
PETN, Pentaerythritol Tetranitrate	D	7980	2	8056	7696
$\rho = 1.67$	P	300		280	267
$C_5H_8N_4O_{12}$	T			3018	3226
E_o =-95.5	γ	2.55		2.86	2.70
PETN	D	8300	2	8421	
$\rho = 1.77$	P	335		318	
	T	3400		2833	
	γ	2.64		2.94	
PETN	D	5480	3	5947	
$\rho = 1.0$	P	87		101	
	T			3970	
	γ	2.45		2.48	
PETN	D	3600	3	4313	
$\rho = 0.5$	P	24		30.3	
	T			4493	
	γ	1.70		2.06	
PETN	D	2830	3	3414	
$\rho = 0.25$	P	7		10.5	
	T			4442	
	γ	1.86		1.78	

TABLE 2.1 Experimental and Calculated C-J Properties (continued)

Explosive	C-J Param.	Expt'l.	Ref.	BKW RDX Param.	BKW TNT Param.
TATB, 1,3,5-Triamino-2,4,6-trinitrobenzene	D	7860	2	8411	7848
$\rho = 1.895$	P	315		326	297
$C_6H_6N_6O_6$	T			1887	2128
E_o =-17.0	γ	2.72		3.11	2.92
DATB, 1,3-Diamino-2,4,6-trinitrobenzene	D	7520	2	7959	7559
$\rho = 1.788$	P	259		282	264
$C_6H_5N_5O_6$	T			2477	2667
E_o = 12.0	γ	2.90		3.02	2.86
Tetryl, N-Methyl-N, 2,4,6-tetranitroaniline	D	7560	2	7629	
$\rho = 1.70$	P			251	
$C_7H_5N_5O_8$	T			2917	
E_o = 8.0	γ			2.93	
NQ, Nitroguanidine	D	7980	2	8069	
$\rho = 1.629$	P			256	
$CH_4N_4O_2$	T			1581	
E_o =-11.1	γ			3.15	
Sorguyl, 1,3,4,6-Tetranitro-glycoluril	D			8791	
$\rho = 1.98$	P			369	
$C_4H_2N_8O_{10}$	T			2276	
E_o = 10.0	γ			3.15	
1,2,4,5-Tetranitrobenzene	D			8000	
$\rho = 1.82$	P			301	
$C_6H_2N_4O_8$	T			3473	
E_o = 34.02	γ			2.86	
PYX, 2,6-Bis(picrylamino)-3,5-dinitropyridine	D			7587	7257
$\rho = 1.77$	P			256	243
$C_{17}H_7N_{11}O_{16}$	T			2868	3040
E_o = 58.24	γ			2.97	2.83

TABLE 2.1 Experimental and Calculated C-J Properties (continued)

Explosive	C-J Param.	Expt'l.	Ref.	BKW RDX Param.	BKW TNT Param.
PATO, 3-Picrylamino-1,2,4-triazole	D			8338	7844
$\rho = 1.94$	P			333	306
$C_8H_5N_7O_6$	T			2190	2422
$E_o = 56.63$	γ			3.05	2.89
TPT, 2,4,6-Tripicryl-s-triazine	D			7143	6902
$\rho = 1.67$	P			218	210
$C_{21}H_6N_{12}O_{18}$	T			3185	3315
$E_o = 118.1$	γ			2.92	2.79
T-TACOT, 1,3,8,10-Tetranitro-benzotriazolo-[1,2-a]benzotriazole	D			7618	7309
$\rho = 1.81$	P			262	250
$C_{12}H_4N_8O_8$	T			2917	3076
$E_o = 133.3$	γ			3.00	2.86
Z-TACOT, 1,3,7, 9-Tetranitro-benzotriazolo-[2,1-a]benzotriazole	D	7250	5	7757	7424
$\rho = 1.85$	P			278	263
$C_{12}H_4N_8O_8$	T			2860	3033
$E_o = 133.7$	γ			3.00	2.87
PADP, 2,6-Bis(picrylazo)-3,5-dinitropyridine	D			7971	7616
$\rho = 1.86$	P			300	283
$C_{17}H_5N_{13}O_{16}$	T			3112	3331
$E_o = 186.59$	γ			2.93	2.80
ABH, Azobis(2,2′,4,4′,6,6′-hexanitrobiphenyl)	D	7600	5	7627	7335
$\rho = 1.78$	P			265	254
$C_{24}H_6N_{14}O_{24}$	T			3356	3538
$E_o = 165.85$	γ			2.89	2.76

TABLE 2.1 Experimental and Calculated C-J Properties (continued)

Explosive	C-J Param.	Expt'l.	Ref.	BKW RDX Param.	BKW TNT Param.
BisHNAB, Dodecanitro-3, 3-bis(phenylazo)biphenyl	D			7799	7494
$\rho = 1.81$	P			283	270
$C_{24}H_6N_{16}O_{24}$	T			3400	3599
$E_o = 242.32$	γ			2.89	2.76
HNAB, 2,2′,4,4′,6,6′-Hexanitroazobenzene	D			7697	7396
$\rho = 1.775$	P			269	257
$C_{12}H_4N_8O_{12}$	T			3265	3448
$E_o = 96.77$	γ			2.90	2.77
TPB, 1,3,5-Tripicrylbenzene	D			6896	6646
$\rho = 1.67$	P			197	189
$C_{24}H_9N_9O_{18}$	T			2772	2877
$E_o = -15.6$	γ			3.03	2.90
PENCO, 2,2′,4,4′,6-Penta-nitrobenzophenone	D			7686	7319
$\rho = 1.86$	P			272	257
$C_{13}H_5N_5O_{11}$	T			2717	2895
$E_o = -2.69$	γ			3.03	2.87
DODECA, 2,2′,2″ 2‴,4,4′,4″,4‴, 6,6′,6″,6‴-Dodecanitro-m,m′-quatraphenyl	D			7650	7337
$\rho = 1.81$	P			270	257
$C_{24}H_6N_{12}O_{24}$	T			3249	3437
$E_o = 99.87$	γ			2.92	2.78
BTX, 5,7-Dinitro-1-picryl-benzotriazole	D			7451	7168
$\rho = 1.74$	P			245	234
$C_{12}H_4N_8O_{10}$	T			3070	3223
$E_o = 97.23$	γ			2.95	2.82

TABLE 2.1 Experimental and Calculated C-J Properties (continued)

Explosive	C-J Param.	Expt'l.	Ref.	BKW RDX Param.	BKW TNT Param.
NONA, 2,2′,2″, 4,4′,4″, 6,6′,6″-Nonanitroterphenyl	D	7560	5	7568	7270
$\rho = 1.78$	P			260	249
$C_{18}H_5N_9O_{18}$	T			3252	3428
$E_o = 68.83$	γ			2.91	2.78
HNBP, 2,2′, 4,4′, 6,6′-Hexanitrobiphenyl	D			7474	7195
$\rho = 1.74$	P			249	238
$C_{12}H_4N_6O_{12}$	T			3252	3412
$E_o = 41.8$	γ			2.90	2.77
TNN, 1,4,5,8-Tetranitronapthalene	D			7528	7221
$\rho = 1.80$	P			255	243
$C_{10}H_4N_4O_8$	T			2932	3083
$E_o = 30.15$	γ			2.99	2.86
ONT, 2,2′, 4,4′,4″, 6,6′,6″-Octanitroterphenyl	D	7330	5	7562	7251
$\rho = 1.80$	P			260	247
$C_{18}H_6N_8O_{16}$	T			3060	3225
$E_o = 57.77$	γ			2.96	2.82
DIPAM, Dipicramide	D	7500	5	7738	7379
$\rho = 1.790$	P			269	255
$C_{12}H_6N_8O_{12}$	T			2781	2966
$E_o = 25.99$	γ			2.98	2.83
HNS, 2,2′, 4,4′, 6,6′-Hexanitrostilbene	D	7130	5	7410	7134
$\rho = 1.74$	P			241	231
$C_{14}H_6N_6O_{12}$	T			3059	3199
$E_o = 44.22$	γ			2.96	2.83
ATNI, Ammonium 2,4,5-Trinitroimidazole	D			8560	
$\rho = 1.835$	P			330	
$C_3H_4N_6O_6$	T			2266	
$E_o = -3.43$	γ			3.06	

TABLE 2.1 Experimental and Calculated C-J Properties (continued)

Explosive	C-J Param.	Expt'l.	Ref.	BKW RDX Param.	BKW TNT Param.
Explosive D, Ammonium Picrate	D	6850	6	6986	
$\rho = 1.55$	P			189	
$C_6H_6N_4O_7$	T			2374	
$E_o = -78.35$	γ			3.00	
1,3-Bis(picrylamino)-2,4,6-trinitrobenzene	D			7285	
$\rho = 1.79$	P			248	
$C_{18}H_7N_{11}O_{18}$	T			3031	
$E_o = 23.63$	γ			2.83	
2,4,6-Tris(picrylamino)-3,5-dinitropyridine	D			7291	
$\rho = 1.80$	P			249	
$C_{23}H_9N_{15}O_{22}$	T			2930	
$E_o = 41.5$	γ			2.84	
2,4,6-Tris(picrylamino)-5-nitropyrimidine	D			7422	
$\rho = 1.88$	P			265	
$C_{22}H_9N_{15}O_{20}$	T			2542	
$E_o = -21.54$	γ			2.90	
2,4,6-Tris(picrylamino)-s-triazine	D			7075	
$\rho = 1.75$	P			226	
$C_{21}H_9N_{15}O_{18}$	T			2700	
$E_o = 26.00$	γ			2.88	
EDNA (Haleite) Ethylene Dinitramine	D			8534	
$\rho = 1.71$	P			310	
$C_2H_6N_4O_4$	T			2162	
$E_o = -12.80$	γ			3.02	
DNPA, 2,2-Dinitropropyl Acrylate	D			6680	
$\rho = 1.47$	P			161	
$C_6H_8N_2O_6$	T			2197	
$E_o = -103.22$	γ			3.06	

TABLE 2.1 Experimental and Calculated C-J Properties (continued)

Explosive	C-J Param.	Expt'l.	Ref.	BKW RDX Param.	BKW TNT Param.
CHNO (Solid Mixture)					
Octol, 76.3/23.7 HMX/TNT	D	8476	2	8555	
$\rho = 1.809$	P	343		333	
$C_{6.835}H_{10.025}N_{9.215}O_{10.43}$	T			2578	
E_o =42.88	γ	2.79		2.98	
Cyclotol, 77/23 RDX/TNT	D	8250	2	8311	7910
$\rho = 1.743$	P	313		305	288
$C_{5.045}H_{7.461}N_{6.876}O_{7.753}$	T			2711	2928
E_o =33.55	γ	2.79		2.95	2.78
Composition B, 64/36 RDX/TNT	D	8030	2	8084	
$\rho = 1.713$	P	294		284	
$C_{6.851}H_{8.750}N_{7.650}O_{9.3}$	T			2763	
E_o =33.18	γ	2.76		2.94	
PBX-9011, 90/10 HMX/Estane	D	8500	2	8496	
$\rho = 1.767$	P	298		319	
$C_{5.696}H_{10.476}N_{8.062}O_{8.589}$	T			2393	
E_o =15.4	γ	3.28		2.99	
PBX-9205, 92/6/2	D			8125	
RDX/Polystyrene/DOP					
$\rho = 1.69$	P			281	
$C_{4.406}H_{7.5768}N_6O_{6.049}$	T			2497	
E_o =16.1	γ			2.97	
PBX-9501, 95/2.5/2.5	D	8826	2	8886	
HMX/Estane/BDNPF					
$\rho = 1.841$	P			363	
$C_{4.575}H_{8.8678}N_{8.112}O_{8.390}$	T			2455	
E_o =39.55	γ			3.00	
EDC-11, 64/4/30/1/1	D	8213	7	8384	
HMX/RDX/TNT/Wax/Trylene					
$\rho = 1.782$	P	315		315	
$C_{1.986}H_{2.7825}N_{2.233}O_{2.6293}$	T			2571	
E_o =9.43	γ	2.82		2.97	

TABLE 2.1 Experimental and Calculated C-J Properties (continued)

Explosive	C-J Param.	Expt'l.	Ref.	BKW RDX Param.	BKW TNT Param.
EDC-24, 95/5 HMX/Wax	D	8713	7	8636	
$\rho = 1.776$	P	342		334	
$C_{5.113}H_{10.252}N_8O_8$	T			2476	
E_o =40.45	γ	2.94		2.97	
LX-14, 95/5 HMX/Estane	D	8760	8	8749	
$\rho = 1.81$	P			348	
$C_{4.800}H_{9.1365}N_{8.024}O_{8.2811}$	T			2510	
E_o =41.90	γ			2.98	
Pentolite, 50/50 TNT/PETN	D	7465	6	7740	7470
$\rho = 1.65$	P			257	247
$C_{2.332}H_{2.3659}N_{1.293}O_{3.2187}$	T			3239	3390
E_o =-1.82	γ			2.84	2.72
HBX-O 48.9/46.1/5 RDX/TNT/WAX	D			7609	7333
$\rho = 1.62$	P			237	229
$C_{2.438}H_{8.0487}N_{1.930}O_{2.5386}$	T			2741	2866
E_o = 6.47	γ			2.95	2.81
PBXC-116, 86/14 RDX/Binder	D	7960	9	7930	
$\rho = 1.65$	P			258	
$C_{1.968}H_{3.7463}N_{2.356}O_{2.4744}$	T			2224	
E_o = -8.95	γ			3.01	
PBXC-119, 82/18 HMX/Binder	D	8075	9	8127	
$\rho = 1.635$	P			274	
$C_{1.817}H_{4.1073}N_{2.2149}O_{2.6880}$	T			2552	
E_o = 0.0	γ			2.94	
75/25 HMX/Hydrazine Nitrate	D	~9100	10	9240	
$\rho = 1.86$	P			388	
$C_4H_{13.192}N_{11.115}O_{11.115}$	T			2095	
E_o = -6.07	γ			3.09	

TABLE 2.1 Experimental and Calculated C-J Properties (continued)

Explosive	C-J Param.	Expt'l.	Ref.	BKW RDX Param.	BKW TNT Param.
61/21/18 Nitroguanidine /HMX/Estane	D			7703	
$\rho = 1.57$	P			229	
$C_{1.793}H_{4.2242}N_{2.940}O_{2.0641}$	T			2001	
$E_o = -5.23$	γ			3.05	
95/5 PYX/Polyethylene	D	7097	2	6897	
$\rho = 1.556$	P			188	
$C_{19.33}H_{11.663}N_{11}O_{16}$	T			2987	
$E_o = 68.76$	γ			2.93	
95/5 Nitroguanidine/ Estane	D	8300	2	8427	
$\rho = 1.705$	P	268		288	
$C_{1.281}H_{4.3993}N_4O_{2.0987}$	T			1467	
$E_o = -11.647$	γ	3.38		3.20	
ANFO 6/94, Oil/ Ammonium Nitrate	D	5500	11	5440	
$\rho = 0.88$	P			73.4	
$C_{0.365}H_{4.7129}N_2O_3$	T			2252	
$E_o = -79.51$	γ			2.55	
CHNO (Liquids)					
Nitroglycerine	D	7580	2	7700	
$\rho = 1.59$	P			246	
$C_3H_5N_3O_9$	T	3470		3216	
$E_o = -66.52$	γ			2.83	
Liquid TNT	D	6580	2	6556	6406
$\rho = 1.447$	P	172		160	157
$C_7H_5N_3O_6$	T	3030		3055	3126
$E_o = 3.41$	γ	2.64		2.87	2.77
Nitromethane	D	6290	2	6463	6390
$\rho = 1.128$	P	141		130	130
CH_3NO_2	T	3380		3120	3167
$E_o = -14.92$	γ	2.17		2.62	2.54
1/1.25 moles Acrylonitrile/ Tetranitromethane	D	6710	2	7074	
$\rho = 1.380$	P	156		192	
$C_{4.25}H_3N_6O_{10}$	T	4000		4760	
$E_o = 70.50$	γ	2.98		2.59	

TABLE 2.1 Experimental and Calculated C-J Properties (continued)

Explosive	C-J Param.	Expt'l.	Ref.	BKW RDX Param.	BKW TNT Param.
1/1.29 moles Benzene Tetranitromethane	D	6850	2	6960	
$\rho = 1.362$	P			181	
$C_{7.29}H_6N_{5.16}O_{10.32}$	T	3520		3855	
$E_o = 46.74$	γ			2.65	
6.435/2.2275/6.434 moles $HNO_3/H_2O/CH_3NO_2$	D	6540	2	6666	
$\rho = 1.29$	P	145		153	
$C_{6.434}H_{30.192}N_{12.869}O_{34.405}$	T	3400		2477	
$E_o = -474.98$	γ	2.80		2.76	
1/0.071 moles Nitromethane/ Tetranitromethane	D	6570	2	6798	
$\rho = 1.197$	P	138		153	
$C_{1.071}H_3N_{1.284}O_{2.568}$	T	3480		3354	
$E_o = -13.35$	γ	2.74		2.61	
1/0.25 moles Nitromethane/ Tetranitromethane	D	6880	2	7094	
$\rho = 1.310$	P	156		181	
$C_{1.25}H_3N_2O_4$	T	3750		3998	
$E_o = -9.51$	γ	2.98		2.63	
1/0.50 moles Nitromethane/ Tetranitromethane	D	6780	2	6908	
$\rho = 1.397$	P	168		179	
$C_{1.5}H_3N_3O_6$	T	3580		3248	
$E_o = -4.1$	γ	2.82		2.73	
2,3,4,5,6-Pentanitroaniline	D			7442	
$\rho = 1.70$	P			239	
$C_6H_2N_6O_{10}$	T			2737	
$E_o = -72.7$	γ			2.93	
14.5/85.5 Toluene/ Nitromethane	D	5840	2	5945	
$\rho = 1.088$	P	100		106	
$C_{2.503}H_{5.461}N_{1.4006}O_{2.8013}$	T			2907	
$E_o = -20.6$	γ	2.71		2.63	

TABLE 2.1 Experimental and Calculated C-J Properties (continued)

Explosive	C-J Param.	Expt'l.	Ref.	BKW RDX Param.	BKW TNT Param.
CHNO Al					
Alex 20, 44/32.2/19.8/4 RDX/TNT/Al/Wax	D	7530	2	7496	
$\rho = 1.801$	P	230		252	
$C_{1.873}H_{2.469}N_{1.613}O_{2.039}Al_{0.7335}$	T			5166	
$E_o = 5.73$	γ	3.44		3.01	
Alex 32, 37.4/27.8/30.8/4 RDX/TNT/Al/Wax	D	7300	2	7066	
$\rho = 1.88$	P	215		213	
$C_{1.647}H_{2.093}N_{1.365}O_{1.744}Al_{1.142}$	T			5928	
$E_o = 4.76$	γ	3.66		3.41	
HBX-1, 40/38/17/5 RDX/TNT/Al/Wax	D	7224	6	7270	
$\rho = 1.72$	P			229	
$C_{2.068}H_{2.83}N_{1.586}O_{2.085}Al_{0.63}$	T			4839	
$E_o = 4.92$	γ			2.98	
HBX-3, 31/29/35/5 RDX/TNT/Al/Wax	D	6917	6	6853	
$\rho = 1.81$	P			195	
$C_{1.669}H_{2.1887}N_{1.220}O_{1.603}Al_{1.2977}$	T			5690	
$E_o = 3.61$	γ			3.37	
Tritonal, 80/20 TNT/Al	D	6475	6	6583	
$\rho = 1.72$	P			191	
$C_{2.465}H_{1.76}N_{1.06}O_{2.11}Al_{0.741}$	T			5330	
$E_o = -0.51$	γ			2.90	
H-6, 45/30/20/5 RDX/TNT/Al/Wax	D	7194	6	7235	
$\rho = 1.71$	P			225	
$C_{1.888}H_{2.589}N_{1.611}O_{2.00}Al_{0.7415}$	T			5138	
$E_o = 5.7$	γ			2.96	
Torpex, 42/40/18 RDX/TNT/Al	D	7495	6	7492	
$\rho = 1.81$	P			259	
$C_{1.8}H_{2.015}N_{1.663}O_{2.191}Al_{0.6674}$	T			5261	
$E_o = 6.17$	γ			2.92	

TABLE 2.1 Experimental and Calculated C-J Properties (continued)

Explosive	C-J Param.	Expt'l.	Ref.	BKW RDX Param.	BKW TNT Param.
PBXC-117, 71/17/12	D	7700	9	7680	
RDX/Al/Binder					
$\rho = 1.75$	P			249	
$C_{1.65}H_{3.1378}N_{1.946}O_{2.048}Al_{0.6303}$	T			4237	
$E_o = -7.26$	γ			3.14	
PBXN-1, 68/20/12	D	7930	9	7693	
RDX/Al/Nylon					
$\rho = 1.77$	P	245		254	
$C_{4.892}H_{9.3473}N_{6.437}O_{6.437}Al_{2.4224}$	T			4721	
$E_o = -12.6$	γ	3.54		3.12	
Destex, 74.766/18.691/4.672/1.869	D	6650	2	6439	
TNT/Al/Wax/Graphite					
$\rho = 1.68$	P	175		174	
$C_{2.791}H_{2.3121}N_{0.987}O_{1.975}Al_{0.6930}$	T			4866	
$E_o = -1.36$	γ	3.24		3.00	
CHNO Cl					
60/40 Nitromethane/UP	D	6700	4	6897	
UP=90/10 $CO(NH_2)_2HClO_4/H_2O$					
$\rho = 1.30$	P			173	
$C_{1.207}H_{4.5135}N_{1.432}O_{3.309}Cl_{0.2341}$	T			2760	
$E_o = -43.42$	γ			2.58	
40/60 Nitromethane/UP	D	6800	4	7042	
UP=90/10 $CO(NH_2)_2HClO_4/H_2O$					
$\rho = 1.39$	P			186	
$C_{9.918}H_{4.313}N_{1.328}O_{3.325}Cl_{0.3365}$	T			2515	
$E_o = -52.91$	γ			2.71	
20/80 Nitromethane/UP	D	7220	4	7207	
UP=90/10 $CO(NH_2)_2HClO_4/H_2O$					
$\rho = 1.48$	P			200	
$C_{7.763}H_{4.112}N_{1.225}O_{3.342}Cl_{0.4487}$	T			2253	
$E_o = -62.39$	γ			2.85	
80/20 Nitromethane/UP	D	6550	4	6833	
UP=90/10 $CO(NH_2)_2HClO_4/H_2O$					
$\rho = 1.215$	P			154	
$C_{1.423}H_{5.9352}N_{1.535}O_{3.181}Cl_{1.1217}$	T			2829	
$E_o = -33.97$	γ			2.67	

TABLE 2.1 Experimental and Calculated C-J Properties (continued)

Explosive	C-J Param.	Expt'l.	Ref.	BKW RDX Param.	BKW TNT Param.
50/50 Nitromethane/ Carbon Tetrachloride	D	5200	12,13	5082	
$\rho = 1.35$	P	92		106	
$C_{1.43}H_{2.76}N_{0.92}O_{1.84}Cl_{2.05}$	T			2078	
$E_o = -28.37$	γ	2.97		2.28	
59.1/40.9 Ammonium Perchlorate/Nitromethane	D	~7000	14	6883	
$\rho = 1.52$	P			184	
$C_{6.70}H_{3.3587}N_{1.173}O_{3.352}Cl_{0.3030}$	T			1995	
$E_o = -51.96$	γ			2.91	
CHNO Cl P					
PBX-9404, 94/3/3 HMX/ Nitrocellulose/ Tris-β-Chloroethyl phosphate	D	8800	2	8879	
$\rho = 1.844$	P	365		363	
$C_{4.42}H_{8.659}N_{8.075}O_{8.470}Cl_{0.0993}P_{0.033}$	T			2466	
$E_o = 38.8$	γ	2.91		3.01	
PBX-9404	D	5905	2	5976	
$\rho = 0.969$	P	92		99	
	T			3564	
	γ	2.67		2.46	
PBX-9408, 94/3.6/2.4 HMX/ DNPA/CEF	D	8787	2	8865	
$\rho = 1.842$	P			361	
$C_{4.49}H_{8.76}N_{8.111}O_{8.440}Cl_{0.0795}P_{0.026}$	T			2427	
$E_o = 34.42$	γ			3.01	
CHNO F					
TFNA, 1,1,1-Trifluoro-3,5,5-trinitro-3-azahexane	D	7400	2	7569	
$\rho = 1.692$	P	249		242	
$C_5H_7N_4O_6F_3$	T			2204	
$E_o = -158.0$	γ	2.72		3.01	

TABLE 2.1 Experimental and Calculated C-J Properties (continued)

Explosive	C-J Param.	Expt'l.	Ref.	BKW RDX Param.	BKW TNT Param.
TFENA, 2,2,2-Trifluoro-ethylnitramine	D	6650	2	6491	
$\rho = 1.523$	P	174		162	
$C_2H_3N_2O_2F_3$	T			1827	
$E_o = -152.0$	γ	2.87		2.95	
TFTF, 1,3,5-Trifluoro-2,4,6-trinitrobenzene	D	7000	8	7049	
$\rho = 1.80$	P			224	
$C_6N_3O_6F_3$	T			2906	
$E_o = -114.0$	γ			2.98	
TFET, 2,4,6-Trinitrophenyl-2,2,2-Trifluoroethylnitramine	D	7400	8	7500	
$\rho = 1.786$	P			252	
$C_8H_4N_5O_8F_3$	T			2611	
$E_o = -112.0$	γ			2.98	
PF, 1-Fluoro-2,4,6-trinitrobenzene	D	7500	8	7612	
$\rho = 1.83$	P			268	
$C_6H_2N_3O_6F$	T			2829	
$E_o = -36.0$	γ			2.95	
DFTNB, 1,3-Difluoro-2,4,6-trinitrobenzene	D	7800	8	7450	
$\rho = 1.851$	P			257	
$C_6HN_3O_6F_2$	T			2801	
$E_o = -75.0$	γ			3.0	
LX-04, 85/15 HMX/Viton	D	8530	2	8698	
$\rho = 1.865$	P			348	
$C_{5.485}H_{9.2229}N_8O_8F_{1.747}$	T			2218	
$E_o = -35.2$	γ			3.06	
LX-07, 90/10 HMX/Viton	D	8640	2	8805	
$\rho = 1.865$	P			358	
$C_{4.935}H_{8.77}N_8O_8F_{1.1}$	T			2292	
$E_o = -5.9$	γ			3.04	
LX-9, 93.3/4.4/2.3 HMX/DNPA/FEFO	D	8840	2	8823	
$\rho = 1.838$	P			356	
$C_{1.425}H_{2.7357}N_{2.678}O_{2.635}F_{0.0144}$	T			2389	
$E_o = 10.37$	γ			3.01	

TABLE 2.1 Experimental and Calculated C-J Properties (continued)

Explosive	C-J Param.	Expt'l.	Ref.	BKW RDX Param.	BKW TNT Param.
LX-10, 95/5 HMX/Viton	D	8820	2	8890	
$\rho = 1.86$	P			364	
$C_{1.425}H_{2.6831}N_{2.566}O_{2.566}F_{0.167}$	T			2374	
$E_o = 6.4$	γ			3.03	
X0204, 83/17 HMX/Teflon	D	8440	9	8791	
$\rho = 1.909$	P			362	
$C_{1.461}H_{2.2420}N_{2.242}O_{2.242}F_{0.6799}$	T			2255	
$E_o = -124.6$	γ			3.07	
65/35 RDX/TFNA	D	8220	2	8278	
$\rho = 1.754$	P	324		302	
$C_{3.7}H_{6.35}N_{5.3}O_6F_{1.05}$	T			2446	
$E_o = -33.3$	γ	2.66		2.98	
95/5 Nitroguanidine/Viton	D			8309	
$\rho = 1.69$	P			280	
$C_{1.156}H_{4.128}N_4O_2F_{0.183}$	T			1454	
$E_o = -19.33$	γ			3.15	
CHNO F Cl					
PBX-9010, 90/10 RDX/Kel-F	D	8363	2	8371	
$\rho = 1.781$	P	319		313	
$C_{3.42}H_6N_6O_6F_{0.6354}Cl_{0.212}$	T			2490	
$E_o = 3.25$	γ	2.91		2.98	
90.54/9.46 HMX/Exon	D	8665	2	8625	
$\rho = 1.833$	P	343		340	
$C_{4.63}H_{8.469}N_8O_8F_{0.4696}Cl_{0.313}$	T			2383	
$E_o = 19.2$	γ	3.01		3.01	
PBX-9407, 94/6 RDX/Exon	D	7910	2	7886	
$\rho = 1.61$	P			262	
$C_{3.32}H_{6.238}N_6O_6F_{0.2377}Cl_{0.158}$	T			2853	
$E_o = 21.69$	γ			2.81	
PBXC-9, 75/20/5 HMX/Al/Viton	D	8503	9	8463	
$\rho = 1.975$	P			354	
$C_{4.56}H_{8.462}N_8O_8F_{0.66}Al_{2.928}$	T			5373	
$E_o = 13.76$	γ			2.99	

TABLE 2.1 Experimental and Calculated C-J Properties (continued)

Explosive	C-J Param.	Expt'l.	Ref.	BKW RDX Param.	BKW TNT Param.
X-0219, 90/10 TATB/Kel-F	D	7630	2		7638
$\rho = 1.914$	P				281
$C_{6.57}H_{6.179}N_6O_6F_{0.7731}Cl_{0.227}$	T				1975
$E_o = -58.7$	γ				2.96
PBX-9502, 95/5 TATB/Kel-F	D	7710	46		7707
$\rho = 1.894$	P	285			285
$C_{6.27}H_{6.085}N_6O_6F_{0.3662}Cl_{0.123}$	T				2063
$E_o = -36.77$	γ	2.95			2.94
45/45/10 TATB/HMX/Kel-F	D	8167	2	8553	7958
$\rho = 1.898$	P			339	310
$C_{10.76}H_{13.61}N_{12.97}O_{12.97}F_{1.59}Cl_{0.32}$	T			2018	2294
$E_o = -52.21$	γ			3.10	2.88
90.1/9.9 RDX/Exon	D	8404	2	8403	
$\rho = 1.786$	P	320		317	
$C_{3.544}H_{6.408}N_6O_6F_{0.408}Cl_{0.272}$	T			2468	
$E_o = 12.89$	γ	2.95		2.98	
CHNO S					
Bis(2,2-dinitropropyl) Sulfite	D			6650	
$\rho = 1.53$	P	185	2	174	
$C_6H_{10}N_4O_{11}S$	T			2587	
$E_o = -136.$	γ			2.88	
CHNO Si					
76/24 PETN/Silicon Rubber	D	7100	15	7037	
$\rho = 1.49$	P			173	
$C_{10.718}H_{16.169}N_4O_{12}Si_{0.8169}$	T			2814	
$E_o = -182.$	γ			3.26	
CN					
2,4,6-Triazido-s-triazine	D	5600	6	5788	
$\rho = 1.15$	P			105	
C_3N_{12}	T			3581	
$E_o = 234.6$	γ			2.66	

TABLE 2.1 Experimental and Calculated C-J Properties (continued)

Explosive	C-J Param.	Expt'l.	Ref.	BKW RDX Param.	BKW TNT Param.
CNO					
TNM Tetranitromethane	D	6360	2	6421	
$\rho = 1.64$	P	159		162	
CN_4O_8	T	2800		1341	
$E_o = 21.6$	γ	3.17		3.16	
TNTAB, 1,3,5-Triazido-2,4,6-trinitrobenzene	D	8576	2	8094	
$\rho = 1.74$	P			300	
$C_6N_{12}O_6$	T			4046	
$E_o = 290.6$	γ			2.80	
HNB, Hexanitrobenzene	D	~9300	8	8466	
$\rho = 1.973$	P			355	
$C_6N_6O_{12}$	T			3269	
$E_o = -2.54$	γ			2.98	
BTF, Benzotris(1,2,5-oxadiazole-1-oxide)	D	8485	2	8156	
$\rho = 1.859$	P			325	
$C_6N_6O_6$	T			4059	
$E_o = 157.8$	γ			2.81	
BTF	D	8262	2	7890	
$\rho = 1.76$	P			291	
	T			4214	
	γ			2.76	
1/1 moles BTF/TNM	D			7791	
$\rho = 1.673$	P			273	
$C_7N_{10}O_{14}$	T			4750	
$E_o = 179.4$	γ			2.72	
3/1 moles Cyanogen/Tetranitromethane	D			6316	
$\rho = 1.1276$	P			135	
$C_4N_7O_8$	T			5793	
$E_o = 210.6$	γ			2.33	

TABLE 2.1 Experimental and Calculated C-J Properties (continued)

Explosive	C-J Param.	Expt'l.	Ref.	BKW RDX Param.	BKW TNT Param.
HN					
Hydrazine Azide	D	~6000	16	7390	
$\rho = 1.0$	P			135	
H_5N_5	T			1400	
$E_o = 55.$	γ			3.05	
HNO					
Hydrazine Nitrate	D	8691	2	8474	
$\rho = 1.626$	P			276	
$H_5N_3O_3$	T			1347	
$E_o = -47.7$	γ			3.22	
21/79 Hydrazine/ Hydrazine Nitrate	D	8600	17	8682	
$\rho = 1.4418$	P			265	
$H_{8.154}N_{4.577}O_3$	T	2900		1500	
$E_o = -33.37$	γ			3.10	
70/30 Hydrazine/ Hydrazine Nitrate	D	8025	17	9393	
$\rho = 1.14$	P			227	
$H_{32.685}N_{16.843}O_3$	T	2180		1008	
$E_o = 78.06$	γ			3.42	
Ammonium Nitrate	D	~4500	18	5262	
$\rho = 1.05$	P			73.2	
$H_4N_2O_3$	T			1112	
$E_o = -78.1$	γ			2.97	
Ammonium Nitrate	D			7952	
$\rho = 1.725$	P			223	
$H_4N_2O_3$	T			526	
$E_o = -78.1$	γ			3.89	
HNO Al					
90/10 Ammonium Nitrate/ Aluminum	D	~5600	18	5453	
$\rho = 1.05$	P			85.3	
$H_{4.5}N_{2.25}O_{3.37}Al_{0.37}$	T			2775	
$E_o = -87.8$	γ			2.65	

TABLE 2.1 Experimental and Calculated C-J Properties (continued)

Explosive	C-J Param.	Expt'l.	Ref.	BKW RDX Param.	BKW TNT Param.
80/20 Ammonium Nitrate/ Aluminum	D	~5800	18	5370	
$\rho = 1.05$	P			85.6	
$H_4N_2O_3Al_{0.74}$	T			4754	
$E_o = -78.0$	γ			2.54	
70/30 Ammonium Nitrate/ Aluminum	D	~5400	18	5093	
$\rho = 1.05$	P			78.8	
$H_{3.5}N_{1.75}O_{2.62}Al_{1.11}$	T			5586	
$E_o = -68.3$	γ			2.46	
NO					
Nitric Oxide	D	5620	19	5607	
$\rho = 1.30$	P	103		106	
NO	T			1854	
$E_o = 19.03$	γ	2.99		2.84	
NPb					
Lead Azide	D	5000	6	4994	
$\rho = 4.0$	P			231	
PbN_3	T			2658	
$E_o = 112.11$	γ			3.30	

TABLE 2.2 Calculated C-J Composition

Explosive	ρ_o	H_2O	H_2	O_2	CO_2	CO	NH_3	CH_4	NO	N_2	$C_{(s)}$	Al_2O_3
RDX	1.80	3.00	—	—	1.49	0.022	—	—	—	3.00	1.49	—
	1.0	2.80	.111	—	0.67	1.855	.029	.021	—	2.98	0.45	—
TNT	1.64	2.50	—	—	1.66	0.188	.001	—	—	1.50	5.15	—
	0.732	1.76	.707	—	0.69	2.865	.020	—	—	1.49	3.45	—
TNT (liquid)	1.447	2.49	.005	—	1.52	0.479	.003	.001	—	1.50	5.00	—
PETN	1.77	4.00	—	—	3.89	0.223	—	—	—	2.00	0.89	—
	0.50	3.73	.167	.072	2.81	2.188	.001	—	.134	1.93	0.0	—
HMX	1.90	4.00	—	—	2.00	0.008	—	—	—	4.00	2.0	—
TATB	1.895	3.00	—	—	1.50	0.006	—	—	—	3.00	4.5	—
DATB	1.788	2.50	—	—	1.73	0.055	—	—	—	2.50	4.22	—
Nitroguanidine	1.629	2.00	—	—	—	—	—	—	—	2.0	1.0	—
1/.5 NM/TNM	1.397	1.50	—	.701	1.49	0.002	—	—	.099	1.45	0.0	—
TNM	1.64	—	—	3.0	1.0	—	—	—	.001	2.0	0.0	—
TNTAB	1.74	—	—	—	2.60	0.800	—	—	.008	6.0	2.61	—
HNB	1.973	—	—	—	5.98	0.019	—	—	.009	2.99	0.0	—
BTF	1.859	—	—	—	2.74	0.514	—	—	.005	3.00	2.75	—
Triazido-s-triazine	1.15	—	—	—	—	—	—	—	—	6.0	3.0	—
21/79 Hydrazine/ Hydrazine Nitrate	1.442	3.00	—	—	—	—	.717	—	—	1.93	—	—

TABLE 2.2 Calculated C-J Composition (continued)

Explosive	ρ_o	H_2O	H_2	O_2	CO_2	CO	NH_3	CH_4	NO	N_2	$C_{(s)}$	Al_2O_3
70/30 Hydrazine/ Hydrazine Nitrate	1.14	3.00	—	—	—	—	8.89	—	—	3.97	—	—
Ammonium Nitrate	1.05	2.0	—	.50	—	—	—	—	—	1.0	—	—
90/10 AN/Al	1.05	2.25	—	.266	—	—	—	—	.034	1.11	—	.185
80/20 AN/Al	1.05	1.87	.115	.001	—	—	.009	—	.013	0.99	—	.37
NO	1.30	—	—	.498	—	—	—	—	.003	.498	—	—
Tetryl	1.70	2.50	—	—	2.66	0.173	—	—	—	2.50	4.16	—
Octol	1.809	5.01	—	—	2.69	0.041	—	—	—	4.61	4.10	—
Cyclotol	1.743	3.73	—	—	1.98	0.061	—	—	—	3.44	3.00	—
Comp B	1.713	4.37	—	—	2.41	0.103	—	—	—	3.82	4.34	—
ANFO	0.88	2.32	.034	—	0.31	0.051	—	—	—	1.0	0.0	—
Nitroglycerine	1.59	2.50	—	.227	3.00	0.004	—	—	.049	1.48	0.0	—
Nitromethane	1.128	1.48	.012	—	0.17	0.190	.004	.002	—	0.50	0.64	—
1/1.25 Acrylo-nitrile/TNM	1.38	1.50	.002	.129	3.79	0.456	—	—	.198	2.90	—	—
1/1.29 C_6H_6/TNM	1.362	2.97	.018	—	2.63	2.086	.007	.003	.003	2.57	2.57	—
1/0.071 NM/TNM	1.197	1.48	.008	—	0.38	0.322	.003	.001	—	0.64	0.37	—
1/0.25 NM/TNM	1.310	1.50	—	.020	1.18	0.068	—	—	.028	0.99	0.0	—

The C-J state value with the least information is temperature. The temperatures are measured experimentally from the brightness of the detonation front as it proceeds toward a detector. The temperatures reported are those of a blackbody of equivalent photographic brightness with an absolute accuracy of about 200 K. The relationship between these numbers and the actual detonation temperature is unknown. It is not known whether the detonation products radiate like a blackbody. They most certainly do not if the products contain lots of free oxygen, like tetranitromethane whose observed temperature is higher than one would obtain for an ideal gas equation of state. It is not known how much radiation from the detonation products is absorbed by the shocked and partially decomposed explosive between the detector and the end of the reaction zone. In a periodic detonation wave, the observed temperature is probably an upper limit value, as contrasted with the measured velocity and pressure which are mean values. If the detonation temperature measurement is to be useful, it must be for a void-free, density discontinuity free system such as a liquid or a single crystal. Any voids or density discontinuities will lead to measurements of the brightness of the shocked air or shocked detonation products rather than the brightness of the C-J detonation products.

The C-J pressure and state points on the shock Hugoniot and expansion isentrope of the detonation products are usually obtained by measuring the shock or free-surface velocity of inerts of varying thickness and density in contact with the detonating explosive. The values usually accepted for the C-J pressure have been obtained using the geometry, first described by Deal,[20] of a 14.2-cm-long, 14.2-cm-diameter charge boosted by a Baratol lens. This geometry is two-dimensional, and for most high-performance explosives, the two-dimensional effects lower the extrapolated pressure into the 4- to 10-cm thickness range where most plane-wave experiments are performed. As is shown later in this chapter, the pressure of interest in comparison with detonation calculations is the effective C-J pressure for the geometry and booster system being considered, not the infinite-medium C-J pressure. The equation-of-state calibration performed using Deal's effective C-J pressures determines the range of experimental geometries for which one can expect the equation of state to be realistic. It is reasonable to expect the calculated and experimental C-J pressures to differ by 10 to 20% while the detonation velocities agree to within a few percent because of the nonsteady-state nature of the detonation wave.

The conservation conditions require that the pressure and particle velocity be identical across the interface between an explosive and an inert. Since the Hugoniots for many materials have been measured as shown in Figure 2.2,[21] and an experimentally determined state point for an explosive-inert interface determines a point on the detonation product shock Hugoniot if the match is above the C-J point, and a point on the detonation product expansion isentrope if it is below the C-J point. The C-J point usually is determined by the intersection of the Rayleigh line (line of constant detonation velocity) with the experimentally measured isentrope or Hugoniot.

In Figure 2.2 the dashed lines labeled "Sys. 350," etc., are for experimental systems of the explosive and a Dural plate used to produce the system number (350) pressure value in kbar in the Dural plate. These experimental systems then produce the pressures shown for the various other inerts when placed in contact with the Dural plates. This is how the experimental Hugoniots were determined. The shock Hugoniots were collected and published in a data volume entitled *LASL Hugoniot Data*[21] which is now out of print; however, the collection of data is included on the CD-ROM. The BKW Hugoniot and isentrope through the C-J point and the Rayleigh line for Composition B explosive are also shown in Figure 2.2.

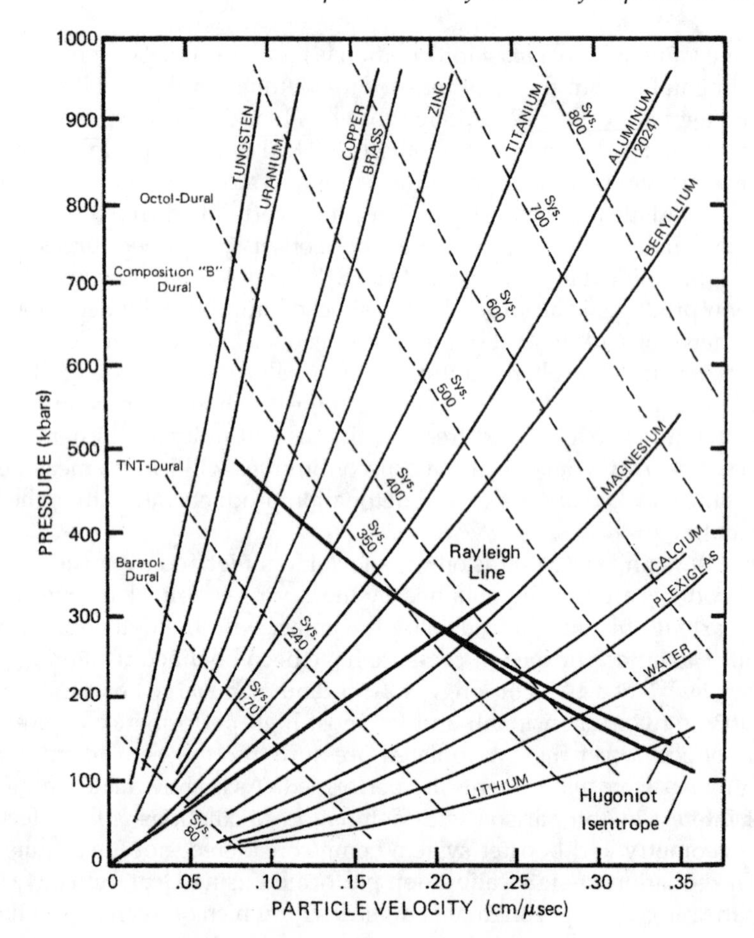

Figure 2.2 Hugoniots for several materials and BKW-calculated Composition B detonation product Hugoniot, Rayleigh line, and expansion isentrope through C-J point.

It is generally assumed that the free-surface velocity is twice the particle velocity of the shock wave just before it interacts with a free surface. This assumption permits one to make a good first approximation to state points between the shock state in an inert and states resulting from interaction with other inerts by reflecting the Hugoniot or release isentrope of the material. In Figure 2.2 the various systems are reflections of the 2024 Dural aluminum Hugoniot through the intersection of the detonation product Hugoniot with the aluminum Hugoniot.

For example, the pressure expected when a slab of Composition B explosive is detonated in slab geometry and the detonation wave then interacts with a Dural plate is 350 kbar. The particle velocity is 0.166 cm/μsec. If the Dural plate then interacts with water, one expects a shock pressure in the water of 146 kbar and a particle velocity of 0.253 cm/μsec. This result is obtained by reflecting the Dural Hugoniot through the 350-kbar point and finding the interaction of that Hugoniot with the water Hugoniot.

Using these techniques, one can experimentally map out the shock Hugoniot and isentrope of the detonation products through the C-J point for a particular explosive in a particular geometrical configuration. Such "shock matching" is useful for plane, one-dimensional systems with flat-topped detonation waves. The CD-ROM has a shock matching code called MATCH. Numerical studies are necessary to determine the effect of multidimensional geometries and of explosives with steep Taylor waves.

For practical purposes, one cannot obtain useful information about the detonation product isentrope from the interaction of the explosive products with low-pressure gaseous inerts, such as one-tenth-atmosphere air.

It has been noted that the expansion points for argon and air[22] have different pressures, given the same particle velocity. Davis[23] has concluded that the velocity depends on the sectional density of the gases rather than on the pressure at low densities. Several investigators have found velocities of several cm/μsec for low-density air shocked by heterogeneous explosives. Rapid expansion of a small amount of material is followed by slower expansion of the rest of the detonation products. Since the effect is not seen in nitromethane,[23] the suggestion has been made that the high-velocity material in heterogeneous explosives comes from interactions such as jetting at grain boundaries.

The experimental isentrope state of several explosives is shown in Table 2.3. Figures 2.3-2.10 show the BKW-calculated isentropes and experimental state points for Octol, Cyclotol, Composition B, TNT, nitromethane, EDC-11, ethyl decaborane/tetranitromethane (EDB/TNM), and 9404, respectively.

Considering that real detonations are not steady-state and that chemical equilibrium is not necessarily achieved, it is difficult to evaluate any equation of state. One can probably consider an equation of state adequate for engineering purposes if, over a wide range of density and composition, the computed and experimental pressures and temperatures agree to within 20% and the detonation velocities to within 10%. Such explosive systems usually exhibit small changes in detonation parameters with diameter, have small failure diameters, and behave like most high-energy explosive systems. It is also important that the expansion isentrope be accurately reproduced to within 5%.

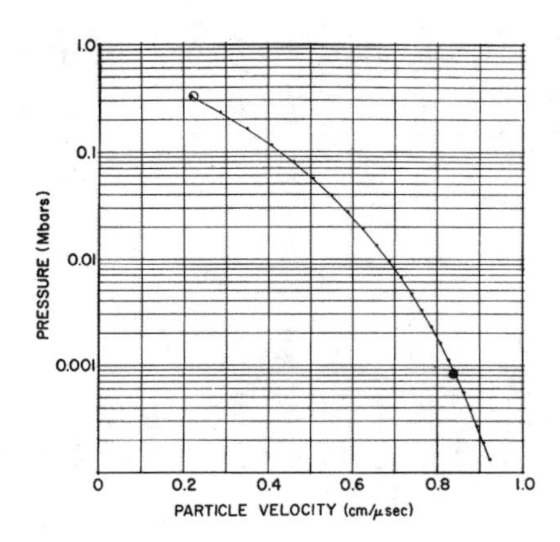

Figure 2.3 BKW isentrope for Octol through C-J point in pressure and particle velocity plane. The experimental C-J and air isentrope are shown.

TABLE 2.3 Experimental Isentrope State Values

Explosive	Inert	Pressure (kbar)	Particle Velocity cm/μsec	Shock Velocity cm/μsec	Reference
Octol	P = 0.8 bar air	0.75	0.84	0.91	24
Cyclotol	P = 0.8 bar air	0.71	0.81	0.89	24
Composition B	P = 0.8 bar air	0.70	0.80	0.87	22
Composition B	P = 10 bar air	5.8	0.67	0.75	2
Composition B	0.13 g/cc Stayfoam	44	0.516	0.65	22
Composition B	0.48 g/cc Stayfoam	113	0.396	0.58	22
Composition B	0.617 g/cc Stayfoam	144	0.345	0.68	22
Composition B	1.0 g/cc water	188	0.295	0.64	22
TNT	P = 0.8 bar air	0.52	0.69	0.75	24
Nitromethane	P = 0.8 bar air	0.57	0.74	0.81	2
EDC-11	P = 1 bar air	0.80	0.78	0.85	25
EDC-11	P = 9 bar air	5.1	0.65	0.73	25
EDC-11	P = 6.5 bar argon	5.75	0.68	0.76	25
EDC-11	P = 31 bar argon	21.8	0.59	0.69	25
EDC-11	1.18 g/cc Perspex	210	0.268	0.67	25
1/3.75 EDB/TNM	P = 0.8 bar air	0.71	0.81	0.89	2
1/3.75 EDB/TNM	0.91 g/cc polyethylene	119.1	0.216	0.61	2
9404	P = 0.8 bar air	0.76	0.85	0.92	2
Alex 20	P = 0.8 bar air	0.60	0.75	0.82	2
Alex 32	P = 0.8 bar air	0.50	0.68	0.74	2
ANFO (10-cm-dia)	P = 0.8 bar air	0.305	0.505	0.56	2

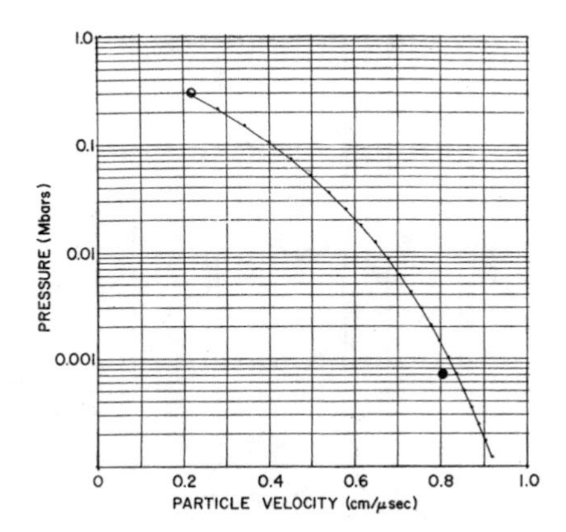

Figure 2.4 BKW isentrope for Cyclotol through C-J point in pressure and particle velocity plane. The experimental C-J and air isentrope are shown.

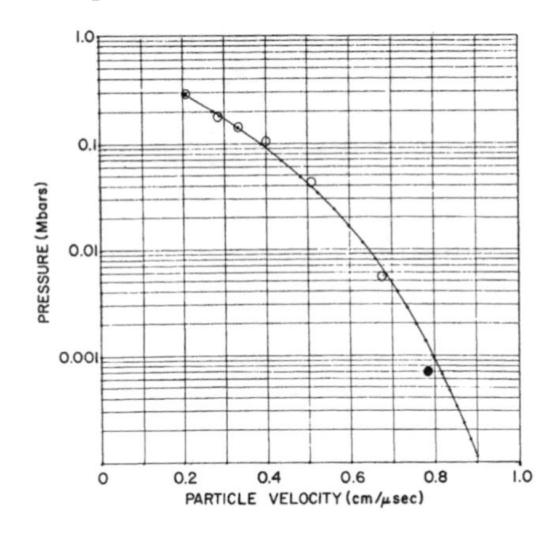

Figure 2.5 BKW isentrope for Composition B through C-J point in pressure and particle velocity plane. The experimental C-J and isentrope states obtained using water, pentene, Stayfoam, and air are shown.

As Appendix E shows, the Hugoniots of water, carbon monoxide, carbon dioxide, ammonia, nitrogen, oxygen, and hydrogen are reproduced by the BKW equation of state used in the detonation calculations. The BKW calculations adequately reproduce the detonation properties of nitrogen oxide (NO), cyanuric triazide (C_3N_{12}), hydrazine nitrate ($N_2H_4NO_3$), and tetranitromethane (CN_4O_8), in addition to various CHNO, BCHNO, CHNOAl, CHNOCl, CHNOClP, CHNOF, and CHNOFCl explosives. So BKW can be used to calculate for engineering purposes the detonation properties of explosives of almost any elemental composition over a wide range of densities, heats of formation, and composition changes.

One can find examples for almost any elemental composition where the experimental results agree poorly with the BKW or any other equation of state calculations. The experimentally observed detonation velocities of high density TNTAB, BTF, and HNB are considerably higher than the calculated values. This problem will be discussed later in this chapter. The calculated BKW temperatures for hydrazine/hydrazine nitrate mixtures and TNM are half as high as those observed experimentally.

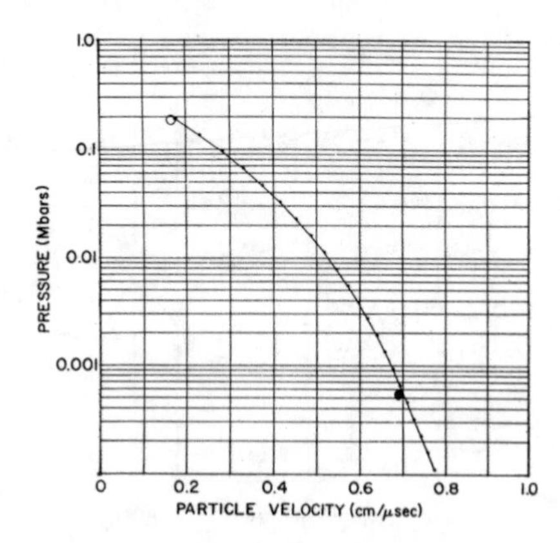

Figure 2.6 BKW isentrope for TNT through C-J point in pressure and particle velocity plane. The experimental C-J and air isentrope are shown.

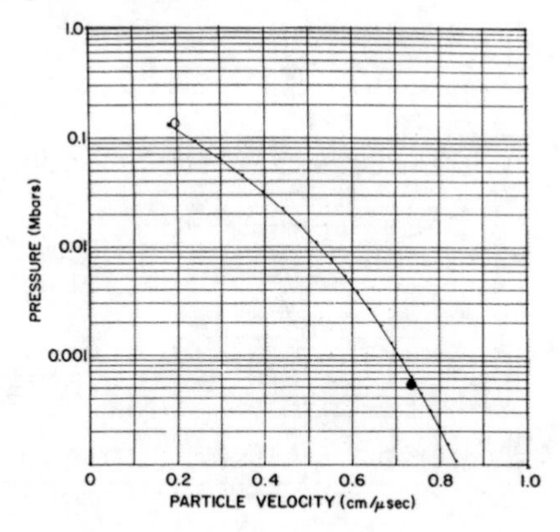

Figure 2.7 BKW isentrope for Nitromethane through C-J point in pressure and particle velocity plane. The experimental C-J and air isentrope are shown.

Since the TNM experimental temperature is higher than the ideal gas detonation temperature, it is believed that bright temperatures for oxygen-rich systems are not blackbody. Such an argument would not explain the high temperature observed for the 70/30 hydrazine/hydrazine nitrate mixture (Table 2.1) or for the 21/79 mixture. The poor agreement between the experimental and calculated detonation velocities of the 70/30 mixture suggests that it may be behaving like a nonideal explosive.

For many conventional explosives, the BKW equation of state gives detonation properties that are adequate for engineering purposes. Such calculations are most useful for estimating the effect of small changes in composition or density. They are also useful in determining the interrelationships between the composition of the detonation products and the temperature, pressure, and particle density.

The effect of oxygen balance has been investigated experimentally with nitromethane-tetranitromethane mixtures. One observes that the heat of detonation increases with the approach of oxygen balance. The simple gamma law predictions indicate that the increasing heat of detonation should increase the C-J pressures and detonation velocities. The velocity difference between the $CO-$ and the CO_2- balanced nitromethane-TNM mix-

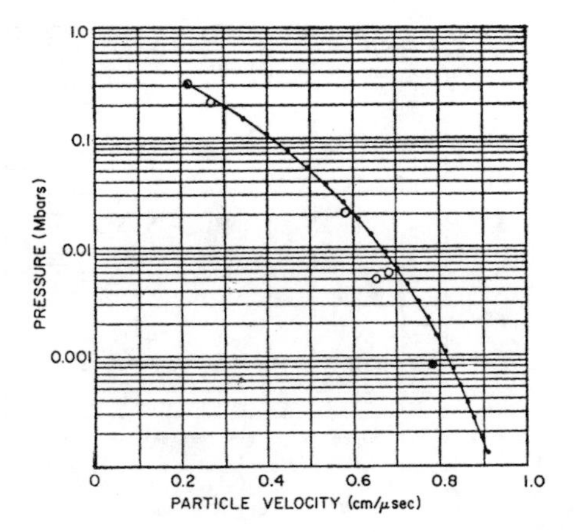

Figure 2.8 BKW isentrope for EDC-11 through C-J point in pressure and particle velocity plane. The experimental C-J and isentrope states obtained using Perspex, argon, and air are shown.

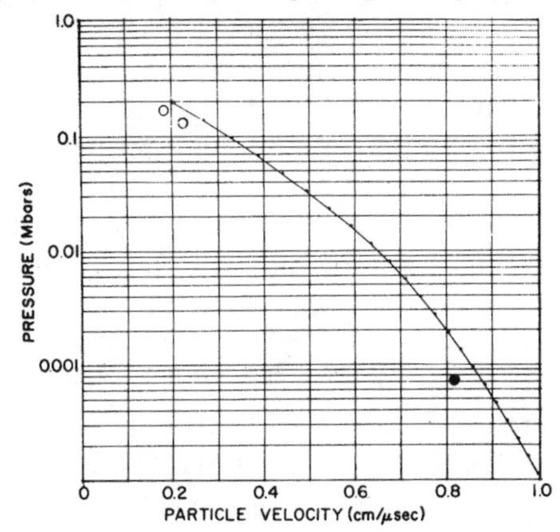

Figure 2.9 BKW isentrope for mixture of 1/3.75 moles of EDB and TNM through C-J point in pressure-particle velocity plane. The experimental C-J and isentrope states obtained using polyethylene and air are shown.

tures may be attributed entirely to the density difference. The temperature increases as the amount of CO_2 increases until excessive oxygen is present, then the temperature decreases. The observed C-J performance may be explained by the lower particle density of the detonation products at the C-J state in systems that produce CO_2 instead of CO. The extra energy in such a system is primarily thermal energy rather than intermolecular potential energy, because the CO_2 decreases the C-J detonation product particle density. Since the intermolecular potential energy primarily determines the C-J pressure, and the temperatures are determined by the thermal energy, the temperatures increase as the amount of CO_2 formed is increased, but the pressures and velocities remain relatively unchanged.

The effect of adding "exotic" elements such as boron, aluminum, or fluorine has been investigated because the heat of detonation is as much as doubled. Usually the observed C-J pressures and detonation velocities are not as high as those of the better CHNO explosives at the same densities. The poor C-J performance of the boron and fluorine explosives relative to CHNO explosives results from formation of complex detonation product molecules such

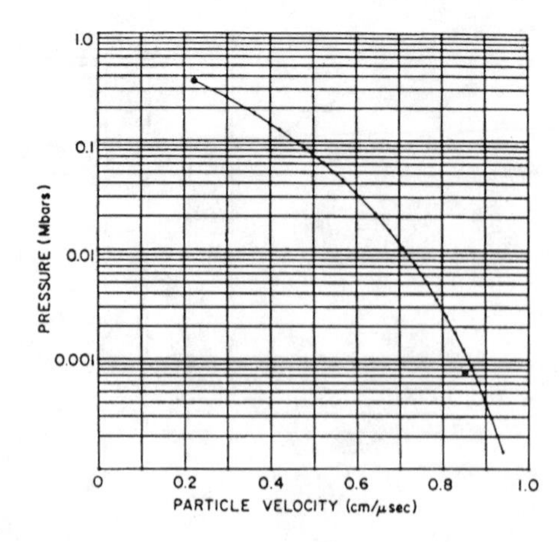

Figure 2.10 BKW isentrope for 9404 through C-J point in pressure-particle velocity plane. The experimental C-J and air isentrope states are shown.

as B_2O_3 and BF_3. The particle density at the C-J point is lower in systems containing these molecules than in systems containing the product molecules CO, CO_2, H_2O, and N_2 only. Again the energy is partitioned unfavorably, the intermolecular potential energy being low and the thermal energy high. At C-J densities the intermolecular potential energy is the primary pressure-determining part of the energy; thus the C-J pressures of the boron- or fluorine-containing explosives are low and the C-J temperatures are high. The high temperatures and total energy will result in less steep expansion isentropes, and the energy becomes available only upon expansion to very low pressures.

The calculated detonation pressures and velocities of the CHNOF systems are very sensitive to the HF, carbon, and CF_4 equilibrium. CF_4 is less desirable than HF because of its large molecular weight and hence detrimental effect on the particle density, which causes the energy to be partitioned so as to give higher temperatures and lower pressures. The covolumes of the CF, CHF, and COF species were scaled (increased by a factor of 1.6 over geometrical covolumes) to improve the agreement between experimental and calculated C-J pressures for several CHNOF explosives. Therefore, calculations for such systems are less reliable than for systems whose detonation product covolumes have been independently calibrated.

For the BCHNO, BCHNOF, CHNOF, and CHNO explosive systems studied, the elements were mixed on a molecular or elementary basis. No such explosive mixture exists for aluminum, so the effect of mixtures of aluminum powder and various explosives must be examined. Decreasing the aluminum powder particle size does not change the observed C-J performance. This suggests that the detonation performance is similar to what one would observe if the aluminum were mixed molecularly with the explosive. The BKW calculations have been used to adequately reproduce the C-J and isentrope state parameters of many mixtures of aluminum and explosives. Apparently one may conclude that most of the reaction occurs near the C-J plane in these mixtures.

The C-J pressures, velocities, and isentropic expansion states of the non-homogeneous systems of explosive and aluminum powder may be reproduced by assuming that the mixture behaves like a homogeneous explosive, and the product molecule Al_2O_3 is formed as an equilibrium C-J product. Again, the computed and experimental C-J pressures and velocities are lower than one might expect from the heats of explosion because of the low particle density of the detonation products. The C-J temperatures are high and the isentropes are less steep than in similar explosive compositions without aluminum. The energy becomes available upon expansion to low pressures.

The BKW calculation permits us to both estimate the probable maximum performance of a proposed chemical compound and to obtain clues as to why a particular explosive behaves in an unusual fashion. It has increased our understanding of the important features of detonation chemistry and energetics. It offers an explanation for the observed failure of the heat of detonation to correlate with the C-J performance. The most important contribution to our understanding of the detonation process has been the discovery that the C-J performance of an explosive is a very sensitive function of the C-J particle density of the detonation products. Systems with high heats of explosion yield detonation product molecules that have large molecular weights and hence low specific C-J particle densities, so the extra energy is present primarily as thermal energy rather than intermolecular potential energy. Since the intermolecular potential energy primarily determines the C-J pressure ($PV/RT \approx 15$ at C-J conditions instead of 1 for an ideal gas), the pressures are affected only slightly whereas the C-J temperature is increased.

To obtain the highest C-J performance explosive, one wishes first to maximize the C-J particle density or number of moles of gas per gram of explosive, and then to maximize the heat of detonation. Increasing the density or the hydrogen content increases the C-J performance even when the other explosive characteristics are held constant. Increasing the hydrogen content and keeping the density and the rest of the composition constant increases the calculated C-J performance spectacularly, as follows.

System	$T_{CJ}(^oK)$	V_{CJ}	$P_{CJ}(kbar)$	$D(m/sec)$
$B_{10}H_{18}C_{5.75}N_{15}O_{30}$	4900	16.5	191	6890
$B_{10}H_{50}C_{5.75}N_{15}O_{30}$	3710	13.4	256	7980
$B_{10}H_{100}C_{5.75}N_{15}O_{30}$	2430	10.3	368	10050

There is little prospect of increasing the hydrogen content significantly and keeping the density constant. A promising approach has been to look for higher density CHNO explosives with compositions similar to HMX. TATB is one explosive whose high density makes its C-J performance 30% greater than that of TNT even though its composition and heat of detonation are similar. Nitroguanidine has a heat of detonation half that of Composition B, but it has the same C-J performance because of the favorable particle density of the detonation products resulting from the high hydrogen content in the explosive and consequent water content in the detonation products. It will be difficult to find an "exotic" explosive (one containing B, Al, or F) that has a better C-J performance than the presently available CHNO systems.

If an "ideal explosive" is one whose performance can be described adequately for engineering purposes by steady-state detonation calculations using the BKW or some other equation of state, then the results of the calculations may be used to alert us to explosives which behave "nonideally" and which may require special treatment and special physics.

2.2 Nonideal Detonations

An ideal explosive would exhibit steady-state or time-independent behavior, but such an explosive probably does not exist. As previously shown, the performance of many explosives can be described adequately for engineering purposes by steady-state theory calibrated with experimental data in the geometry of interest. These may be called nearly ideal explosives. Some explosives behave so nonideally that they cannot be described by steady-state theory. The dividing line between nonideal and ideal explosives is arbitrary.

One may define a nonideal explosive as having a C-J pressure, velocity, or expansion isentrope significantly different from those expected from equilibrium, steady-state calculations such as BKW. Pressure differences of 50 kbar and velocity differences of 500 m/sec are probably nonideal. A significant isentrope difference depends upon the application, but if the experimental and calculated air isentrope values differ by more than 0.1 cm/μsec, the explosive can be nonideal. Nonideal explosives often exhibit other differences, such as larger sensitivity to diameter or confinement.

Adding a nonexplosive component to an explosive mixture does not necessarily result in a nonideal explosive. The addition of powdered aluminum to an explosive will not make it nonideal, as previously shown.

There are several different types of nonideal behavior including that exhibited by inert metal loaded explosives,[26–28] by mixtures of explosives and ammonium salts such as ammonium nitrate[29] or ammonium perchlorate, by mixtures of nonexplosive fuel and ammonium salts, and probably by other types not yet identified. To evaluate the performance of nonideal explosives, much more experimental data are required than for conventional high explosives.

The development of the image intensifier camera which permitted multiple exposures of a detonation wave traveling along a cylindrical charge in water allowed the required data to be generated. The aquarium test was developed to measure simultaneously detonation velocity and pressure, confinement effects, and the release isentrope from the C-J state. The test consists of an explosive cylinder placed in a water tank with two parallel transparent sides, as shown in Figure 2.11. The explosive may be contained within a watertight tube or pipe if varied confinement is desirable. If the explosive is unaffected by water and is rigid, it can be placed directly in the water tank without confinement. The explosive cylinder is detonated from above. As the detonation wave travels downward, an oblique shock wave radiates into the water, and the gaseous products behind the detonation front expand.

Several photographic exposures are taken of the detonation front, the shock waves in the water, and the expanding tube-water interface (or bubble-water interface if the tube is absent). The optical data, shown schematically in Figure 2.12 are used to infer detonation velocity, detonation pressure (C-J states), and the release isentrope of the detonation products. A collection of aquarium test data is available in the data volume entitled *Los Alamos Explosives Performance Data*[39] and on the CD-ROM.

The aquarium test is also particularly attractive for numerical modeling since water behaves as a true fluid and its equation of state is well defined. The shock wave in the water and the interface between the explosive cylinder and the water expand in a smooth manner without distortions, which makes modeling of the flow easy and accurate. Because of the importance of the aquarium test in the determination of the physics and chemistry of ideal and nonideal explosives, it was chosen for the book frontpiece.

Figure 2.11 Water tank containing explosive column for aquarium test. The flood lights furnish the back lighting for the image intensifier camera.

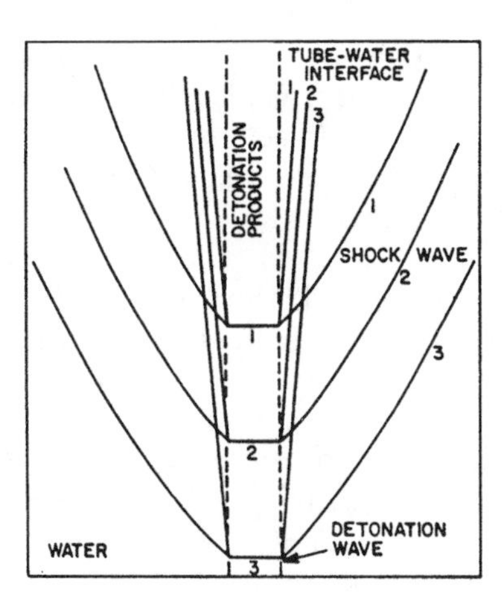

Figure 2.12 Schematic diagram of typical aquarium test result. Three exposures (1, 2, and 3) are shown.

Ammonium Salt-Explosive Mixtures

As Table 2.4 shows, the ammonium nitrate (AN) loaded explosives exhibit differences in observed and calculated performance (assuming either complete reaction or no reaction of the ammonium nitrate with the rest of the detonation products) sufficient to classify them as nonideal. The difference between observed and calculated detonation velocities increases with increased ammonium nitrate concentration, assuming complete ammonium nitrate decomposition.

Hershkowitz and Rigdon[29] demonstrated that the detonation velocity, or cylinder wall velocity, of Amatex 40 and Amatex 20 was insensitive to the particle size (50 to 500 μm) of the ammonium nitrate. Craig[2] found only 101 $\pm$ 40 m/sec difference in the detonation velocities of 2.54-cm-diameter pressed charges of Amatex 20 with ammonium nitrate particle sizes of 15 to 500 μm. This suggests that as much reaction as possible is being obtained between the ammonium nitrate and the detonation products at the observed C-J state. The remaining ammonium nitrate is assumed to be present as an inert. If these assumptions are correct, and if the equation of state of the explosive mixture, assuming that the ammonium nitrate is an inert is valid, then the observed C-J state must lie on a partially reacted Hugoniot between the inert and completely reacted Hugoniots. The experimentally observed detonation velocity may be reproduced if 50% of the ammonium nitrate in the Amatex 40 and 20 is decomposed and 50% is inert. The experimentally observed detonation velocity of Amatol can be reproduced if 19% of the ammonium nitrate decomposes and the rest is an inert. One possible explanation for the different amounts of ammonium nitrate decomposition in Amatex and Amatol is that the Amatex detonation temperatures are higher and more ammonium nitrate decomposition occurs at the higher temperatures.

If ammonium nitrate decomposition depends primarily on the temperature rather than on diffusion and other transport phenomena, then the amount of decomposed ammonium nitrate would not be expected to change significantly along the expansion isentrope from the C-J point, since the temperature is decreasing along the expansion isentrope. One would also expect the explosive to exhibit large effects from side rarefactions as the detonation products cooled. The observed small effect of charge diameter on detonation velocity is consistent with this model. The calculated state parameters for the ammonium nitrate containing explosives are given in Table 2.4. The reactive Hugoniots are given in Figures 2.13-2.15.

In a classic paper, Miron, Watson, and Hay[30] presented experimental evidence that large amounts of the ammonium salts remain in the explosive residue left from exploded ammonium salt-explosive mixtures used as commercial explosives in a 3.7 meter diameter steel sphere with the inside surface metallized with aluminum. Up to 50% of the ammonium nitrate was recovered from the explosion residues of granular, water-gel, and gelatinous explosives while no oil was recovered. Large percentages of sodium chloride, sodium nitrate, calcium carbonate, and even nitrocellulose were recovered from some of the commercial explosives. Up to 96% of the aluminum was recovered in the residue of an aluminum containing water-gel explosive.

Hershkowitz and Rigdon[29] studied the detonation product composition of Amatex 20 and Amatol, both confined and unconfined, which had come to a steady-state inside steel spheres. Their results for confined and unconfined geometries differ considerably, and in no case was undecomposed ammonium nitrate recovered. Other gas-analysis studies[31] after detonation have shown great dependence of the composition upon the experimental

TABLE 2.4 Experimental and Calculated C-J Parameters

Explosive	Experimental				BKW			Description
	ρ_o	P_{CJ}	D_{CJ}	Ref	P_{CJ}	D_{CJ}	T_{CJ}	
Alex 20 44/32/20/4 wt% RDX/TNT/Al/Wax	1.801	230	7530	2	252	7496	5166	Al burns to compressible Al_2O_3 solid
Alex 32 37/28/31/4 wt% RDX/TNT/Al/Wax	1.880	215	7300	2	213	7066	5928	Al burns to compressible Al_2O_3 solid
90/10 vol% RDX/Exon	1.787	320	8404	27	317	8403	2468	
80/10/10 vol% RDX/Exon/Pb	2.700		6734	27	296	6692	2447	Pb treated as compressible inert
70/10/20 vol% RDX/Exon/Pb	3.650		5709	27	285	5748	2352	Pb treated as compressible inert
60/10/30 vol% RDX/Exon/Pb	4.60	150	5012	27	270	5096	2242	Pb treated as compressible inert
55/10/35 vol% HMX/Exon/W	7.90	230	4900	2	328	4807	1755	W treated as compressible inert
Comp B 61/39 wt% RDX/TNT	1.70	287	7900	2	277	8004	2787	

TABLE 2.4 Experimental and Calculated C-J Parameters (continued)

Explosive	Experimental ρ_o	P_{CJ}	D_{CJ}	Ref	BKW P_{CJ}	D_{CJ}	T_{CJ}	Description
Amatex 40 41/38/21 RDX/TNT/AN	1.66		7545	29	265	7963	2472	Ammonium Nitrate (AN) complete reaction
					235	7571	2380	50% AN reacted, rest inert[a]
					206	7154	2271	AN as compressible inert[a]
Amatex 20 20/38/42 RDX/TNT/AN	1.61		7009	29	250	7866	2211	Ammonium Nitrate (AN) complete reaction
					191	7031	1943	50% AN reacted, rest inert[a]
					137	6105	1670	AN as compressible inert[a]
Amatol 40/60 TNT/AN	1.60		5760	29	247	7896	1940	Ammonium Nitrate (AN) complete reaction
					118	5796	1393	19% AN reacted, rest inert[a]
					88	5163	1217	AN as compressible inert[a]

[a] AN inert Hugoniot assumed to be described by $U_s = 0.25 + 1.5U_p$ and $\rho_o = 1.725$.

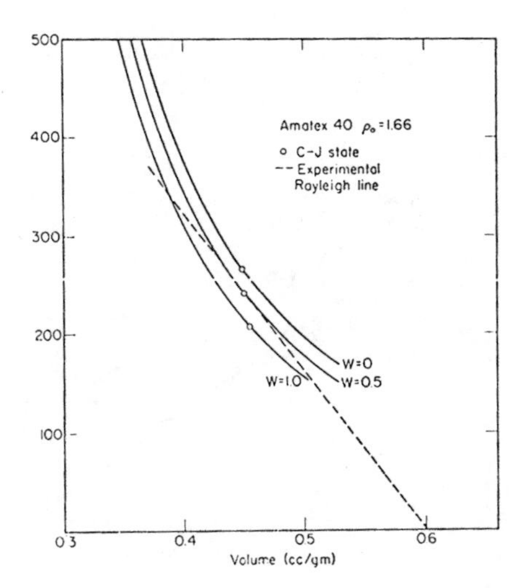

Figure 2.13 BKW calculated Hugoniots for Amatex 40 for $W_{AN} = 0.0, 0.5$, and 1.0.

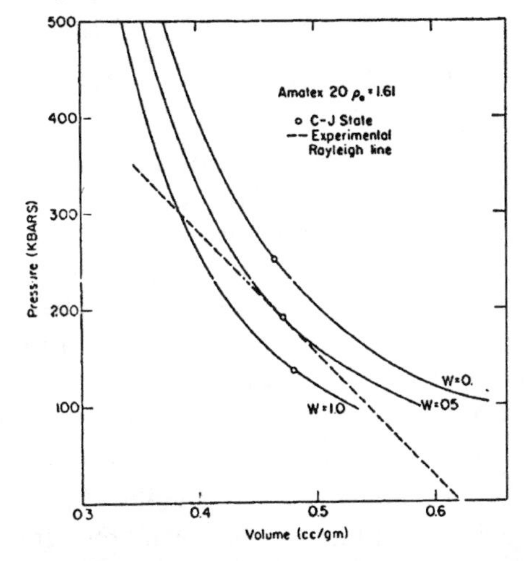

Figure 2.14 BKW calculated Hugoniots for Amatex 20 for $W_{AN} = 0.0, 0.5$, and 1.0.

container, the confinement, and the atmosphere used in the container. The shocks interact with the walls and the explosive confinement and send reflected shocks back into the detonation products, resulting in additional heating and varying amounts of additional chemical reaction. Because the gas-analysis studies give such varied results, not the products along a C-J isentrope but rather the products many steps removed are being measured. The equation of state described in this section is inadequate to account for the additional reaction of the ammonium nitrate and detonation products that apparently occurred in the Hershkowitz experiments.

The model[32] of diffusion-controlled grain burning of the ammonium nitrate is still a candidate; however, the apparent absence of appreciable particle size effects for Amatex explosives in the studies of Hershkowitz and Rigdon[29] or Craig[2] is hard to reconcile with the diffusion model. Walker[32] showed that 5-μm ammonium perchlorate in nitromethane appears to react completely and 200-μm ammonium perchlorate appears to be only about 30% reacted initially. Perhaps the particle size effects depend on the nature of the explosive

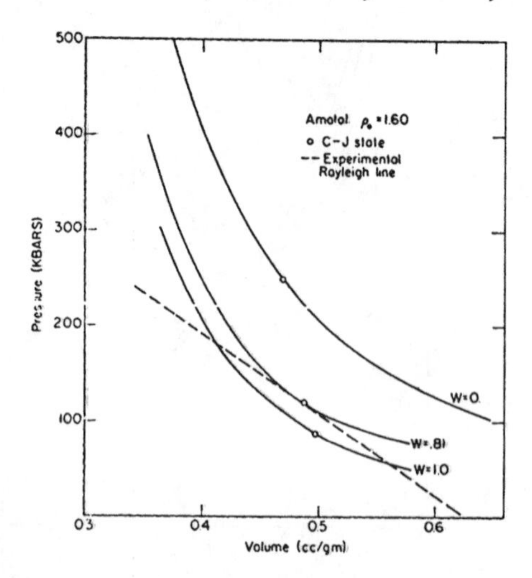

Figure 2.15 BKW calculated Hugoniots for Amatol 20 for $W_{AN} = 0.0, 0.81$, and 1.0.

mixture. No single nonideal model will be satisfactory for different types of nonideal explosives.

Another interesting characteristic that ammonium nitrate containing nonideal explosives exhibit may be inferred from Craig's[2] experimental studies of the over-driving of 7.6 cm of Amatex 20 with 5.0 cm of Composition B initiated by a P-40 lens. The ~295-kbar pressure that Craig observed in the Amatex 20 was much higher than would be calculated (~ 220 kbar) if the usual decay with distance of run were obtained. This indicates that overdriving a nonideal explosive could cause a self-supporting shift in the nonideal behavior. The detonation state achieved by a nonideal explosive probably depends on the magnitude of the initiation pulse. Future experimental studies of nonideal detonation should include variations of the initiating pulse magnitude.

It is interesting to speculate as to why a powdered metal such as aluminum appears to be completely reacted near the C-J plane, whereas a powdered inorganic such as ammonium nitrate does not. One possible explanation is found in the behavior of the detonation product temperature as reaction with the additive occurs. Although both additives increase the heat of detonation as they react, the aluminum raises the temperature of the products, where as the ammonium nitrate *lowers* it. This is readily understood from the principles of partition of energy discussed in section 2.1. The product molecule from burning of the aluminum is solid Al_2O_3, whereas the product molecules from ammonium nitrate decomposition are water and nitrogen. The ammonium nitrate gives products that raise the particle density of the detonation products; the aluminum gives products that lower it. Increased particle density shifts the energy from thermal to intermolecular potential energy. The temperature is lowered by the decomposing ammonium nitrate, and this may determine how much ammonium is decomposed near the C-J plane. The temperature is raised by the burning aluminum, which would increase the rate of aluminum burning until all the aluminum is burned near the C-J plane.

Consequently, a reactive additive probably will result in a nonideal explosive if it lowers C-J temperatures significantly by its decomposition, and the explosive will exhibit more nearly ideal behavior if the C-J temperature is raised.

A water-gel explosive called WGE-1 consists of approximately 46 wt% ammonium nitrate, 24 wt% TNT, 15 wt% sodium nitrate, 13.2 wt% water, 1.2 wt% ethylene glycol, and 0.6 wt% thickener, and has a density of 1.5 g/cc. In a 20 cm diameter charge, the detonation speed is 0.481 cm/μsec as described in Reference 33. This is well below the BKW ideal detonation speed of 0.73 cm/μsec and C-J pressure of 187 kbar. A BKW calculation, under the assumption that no ammonium nitrate reacts, gives a velocity of 0.495 cm/μsec and a pressure of 71 kbar. The aquarium test data could be reproduced for WGE-1 with no ammonium nitrate reacting at the C-J plane and all the ammonium nitrate remaining inert behind the detonation wave.

Detonation performance tests have been performed for many ammonium salts containing propellants and explosives. Some of the propellants and commercial explosives that have been examined and the nature of their nonideality are summarized in Table 2.5. The first two explosives in the table are common commercial explosives. Aquarium tests were performed for both explosives. To reproduce the observed performance, all of the ammonium nitrate and some of the aluminum had to be treated as inert. The aquarium test for the first commercial explosive required no additional reaction of the ammonium nitrate behind the C-J state. The aquarium test for the second commercial explosive required some additional reaction of the aluminum or ammonium nitrate below the C-J state. The ammonium perchlorate containing systems in Table 2.5 are various propellants. As the amount of ammonium perchlorate is increased in the propellants with corresponding decrease in HMX, the degree of nonideality increases. The propellant with 36% ammonium perchlorate also had the air isentrope state determined. The measured air isentrope particle velocity was 0.6 cm/μsec and 0.6 kbar, which is in agreement with the isentrope for all the ammonium nitrate remaining inert.

Ammonium Nitrate-Fuel Oil Mixtures

The most commonly used blasting explosive mixture is ammonium nitrate and fuel oil, commonly called ANFO. It exhibits an interesting combination of both "ideal" and "nonideal" behavior at the C-J plane as a function of charge diameter and confinement. It also exhibits detonation product expansion behavior that requires additional reaction behind the C-J plane in contrast to the ammonium salt-explosive mixtures previously described.

ANFO is usually about 6% fuel oil and 94% ammonium nitrate with a density of 0.8 to 1.0 g/cc. Persson[11] has shown that the detonation velocity of ANFO at 0.88 g/cc increases from about 0.35 cm/μsec in 3.5-cm-diameter charges in rock or steel to the "ideal" velocity of 0.55 cm/μsec in 26.8-cm-diameter charges in rock. Finger et al.[34] have reported detonation velocities for 0.8-g/cc ANFO of 0.325 cm/μsec in 5.1-cm-diameter copper cylinders, 0.389 cm/μsec in 10.2-cm-diameter cylinders, and 0.455 cm/μsec in 29.2-cm-diameter cylinders. They also observed that the cylinder wall velocities scale for 10.2- and 29.2-cm-diameter cylinders but failed to scale for the 5.1-cm-diameter cylinder, which is below the unconfined failure diameter of about 8 cm. This result suggests that the energy release is about the same for 10- and 29-cm diameters even though the C-J velocity changes by more than 15%.

Using the partially reacted Hugoniot tangent to the Rayleigh line determined by the experimental detonation velocity, we can obtain an equation of state in the region of the detonation front. The variation of the C-J state parameters with the detonation velocity of ANFO at 0.954 gm/cc follows.

TABLE 2.5 Explosives and Propellants Containing Ammonium Salts

Explosive	Experimental ρ_o	P_{CJ}	D_{CJ}	Dia	BKW P_{CJ}	D_{CJ}	Description
DBA 25/40/15/15/5 wt%	1.50	–	4800	20	187	7300	All AN decomposed
$TNT/AN/NaNO_3/H_2O/CH_2$					71.5	4955	All AN inert[a]
Stratoblast	1.05	–	3700	20	80	5200	All AN decomposed
40/10/15/25/10 wt%		–			32	3700	All AN inert[a]
$AN/NaNO_3/H_2O/Al/CH_2$							and 25% Al inert
VWC-2 46/ 9/18/19/8 wt%	1.835	–	7940	**	299	8039	All AP decomposed
$HMX/AP/Al/NG/CH_2$							
TD-N1028 44/17/19/15/5 wt%	1.845	–	7800	**	303	8031	All AP decomposed
$HMX/AP/Al/TMETN/CH_2$							
UTP-20930 29/36/18/10/7 wt%	1.838	–	6500	**	305	7976	All AP decomposed
$HMX/AP/Al/TMETN/CH_2$					174	6533	All AP inert[b]
SPISS-44 20/49/21/10 wt%	1.83	–	Failed	7.2	303	7990	All AP decomposed
$HMX/AP/Al/CH_2$							
SPISS-45 12/57/21/10 wt%	1.831	–	Failed	7.2	300	7922	All AP decomposed
$HMX/AP/Al/CH_2$							

[a] AN (Ammonium Nitrate) inert Hugoniot described by $U_s = 0.25 + 1.5U_p$ and $\rho_o = 1.725$.
[b] AP (Ammonium Perchlorate) inert Hugoniot described by $U_s = 0.25 + 1.5U_p$ and $\rho_o = 1.95$.
** - Velocity determined from 2.5 cm thick wedge test.
TMETN - [$CH_3 - C - (CH_2 - O - NO_2)_3$], E_o = -70.
NG - Nitroglycerine

D_{CJ} (m/sec)	% AN Reacted	P_{CJ} (kbar)	D_{CJ}	T_{CJ}
5715	100	86	5715	2169
4600	75	47	4600	1718
3800	50	29	3700	1184
3500	45	24	3500	1087
3200	40	19	3200	992

Craig[33] performed experiments on 10-cm-diameter cylinders of ANFO at 0.954 g/cc confined by thin (0.6-cm) walls of Plexiglas and water. He photographically measured the position of the water shock and the Plexiglas/water interface. The image intensifier photograph for Shot C-4632 with three exposures of the ANFO detonation proceeding up the 10.1-cm-diameter charge in water is shown in Figure 2.16. The charge length was 89.3 cm, and it was initiated by a 2.54 cm thick slab of TNT. The time between exposures were 53 and 46.3 μsec. The detonation velocity was 0.333 cm/μsec.

The shock position in the water and the experimental position of the explosive-water interface moved faster than the calculation without any additional decomposition behind the detonation front, as shown in Figure 2.17. The explosive-water interface moved faster near the detonation front and remained a constant amount ahead of the calculated position. This indicates that additional ammonium nitrate decomposition must be taking place near the detonation front.

Using BKW one can program additional reaction of the remaining ammonium nitrate along the isentrope to approximate the additional energy release that the aquarium data requires. Several rates of additional ammonium nitrate reaction were therefore programmed into BKW and the isentropes were used to calculate the position of the interface and water shock. Several of the rates could reproduce the observations; one such rate is shown in Figure 2.18. The observed detonation velocity of about 0.35 cm/μsec can be reproduced if 55% of the ammonium nitrate is assumed to not react at the C-J point.

The additional 55% ammonium nitrate was permitted to react with the detonation products between 24 and 10 kbar at a linear rate decreasing with pressure. This resulted in the isentrope shown in Figures 2.19 and 2.20. The isentrope was calculated by assuming that the entropy increased with the increased energy released from reaction by $\Delta H_{\text{reaction}}/T$. Also shown is the isentrope with no additional reaction.

The air isentrope state for 10-cm-diameter ANFO was measured[33] and is shown in Figure 2.20, along with three release isentropes for ANFO corresponding to complete reaction, 55% inert ammonium nitrate with no additional burn , and 55% inert ammonium nitrate with additional burn to completion. The air isentrope data are in good agreement with the assumption of 45% reaction at the C-J plane followed by additional burn to completion that was required to describe the aquarium data.

As previously described, the performance of ANFO is strongly dependent upon the diameter of the charge. ANFO must be experimentally characterized for *each* diameter and an equation of state determined for each diameter, as was performed above for 10-cm-diameter.

Aquarium tests were also performed[33] for a 15-cm and a 20-cm-diameter ANFO charge in clay pipe surrounded by water. The measured detonation velocities and partially reacted ammonium nitrate interpretations obtained from the aquarium tests follow.

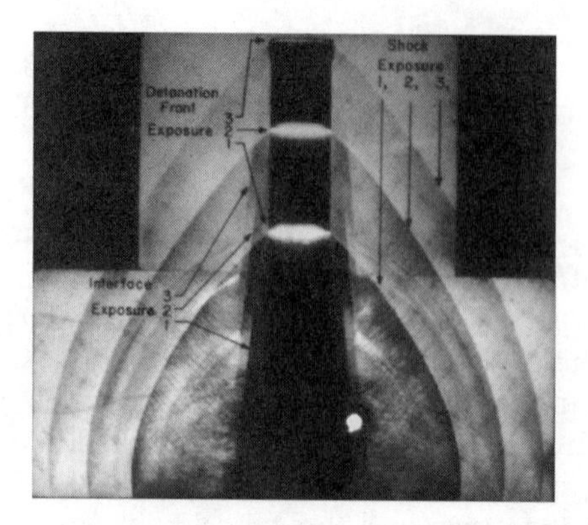

Figure 2.16 The aquarium test image intensifier photograph with three exposures of a detonation proceeding along a 10-cm-diameter cylinder of ANFO confined with plexiglas in water.

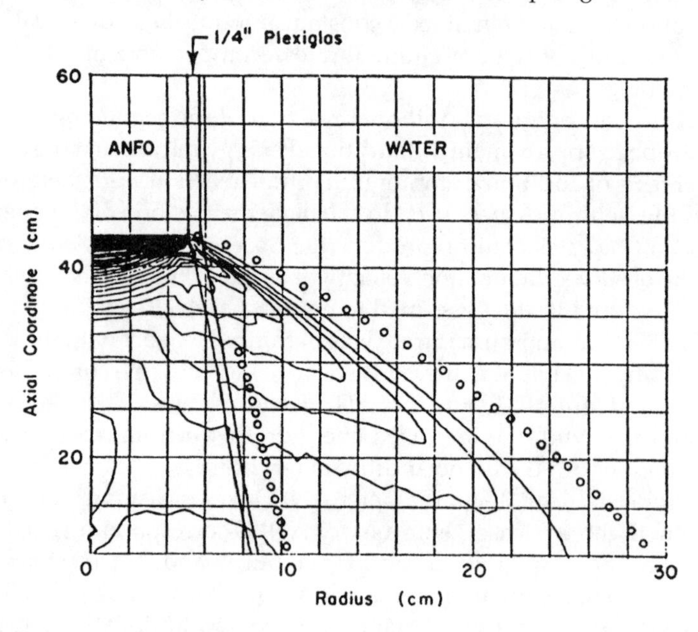

Figure 2.17 Calculated and experimental water shock and explosive-water interface profiles for a 5-cm-radius cylinder of ANFO confined by water. Fifty-five percent of the aluminum nitrate is assumed to remain unreacted. The experimental profile is shown by circles. The isobar interval is 1 kbar.

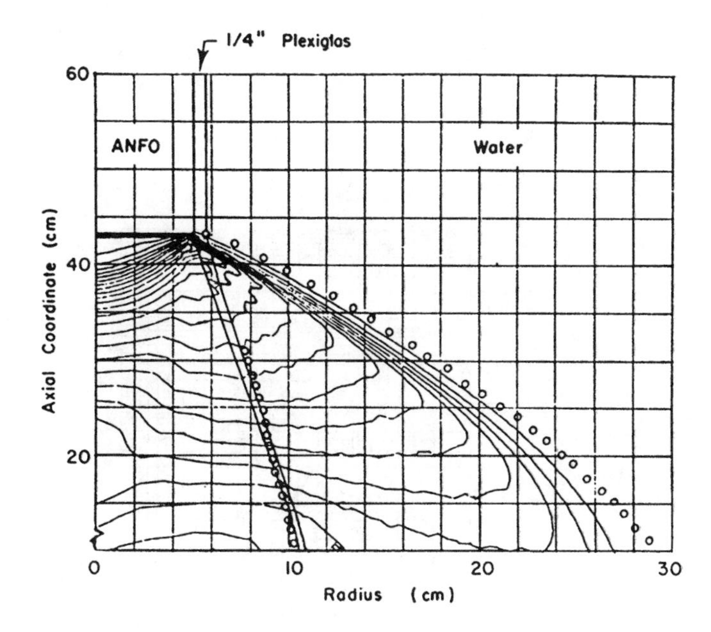

Figure 2.18 Calculated and experimental water shock and explosive-water interface profiles for a 5-cm-radius cylinder of ANFO confined by water. At the C-J plane, 55% of the ammonium is unreacted. The ammonium nitrate is permitted to react with the detonation products at a linear rate with decreasing pressure. The experimental profile is shown by circles. The isobar interval is 1 kbar.

Diameter inches	Density gm/cc	Experimental D_{CJ} (m/sec)	Unreacted AN	Effective P_{CJ} (kbar)
4 (10-cm)	0.954	3500	55%	24
6 (15-cm)	0.954	3780	50%	38
8 (20-cm)	0.900	4147	38%	42
∞	0.880	5440	0%	73

Metal Loaded Explosives

In the early 1950s, an extensive study was undertaken at Los Alamos to measure how explosive performance was affected when both inert and noninert components were added. The experimental data for RDX/Exon/Pb mixtures are shown in Table 2.4. Gamma values as high as 6.7 indicated that these explosive mixtures were of a different class than the common explosives which have gammas around 3.0.

Table 2.4 shows that one could sometimes reproduce the experimental detonation velocity with BKW calculations that have C-J pressures 100 kbar too high. The explosive detonation velocity was ideal, whereas the C-J pressure was obviously nonideal.

The list of inert diluents that caused nonideal behavior grew as other experimenters added inert components to explosive systems. References 26-28 are examples of such work.

In the late 1960s, attention was given to tungsten-loaded explosives. The BKW calculations reproduced the detonation velocity but overestimated the C-J pressure by 100 kbars.

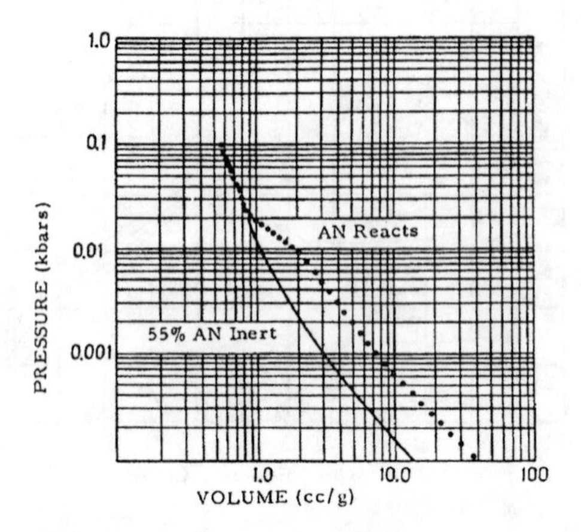

Figure 2.19 Pressure-volume isentrope for 10-cm-diameter ANFO used for calculation shown in Figure 2.18 with additional ammonium nitrate reacted below the C-J state, and isentrope with 55% ammonium nitrate inert used for calculation shown in Figure 2.17.

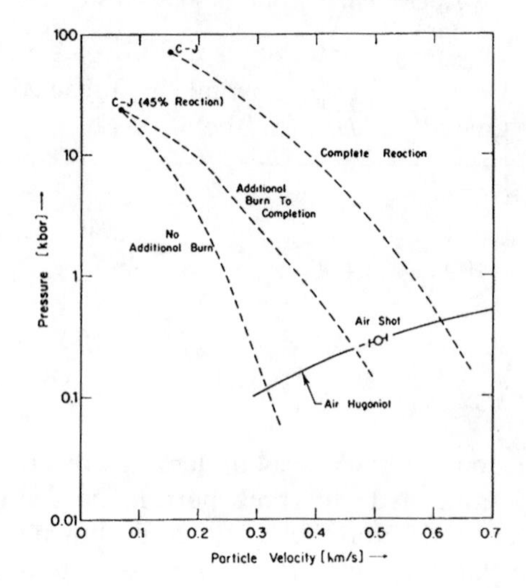

Figure 2.20 Pressure-particle velocity isentropes for 10-cm-diameter ANFO for no ammonium reaction, complete ammonium reaction, and the isentrope used in Figure 2.18 with additional ammonium nitrate reacted below the C-J state. The air isentrope state is shown.

The plate dent test is described in Chapter 5 and in Reference 35. The plate dent test consists of detonating a cylindrical charge of explosive in contact with a heavy steel plate and measuring the depth of the dent produced in the plate. The depth of the dent correlates with the C-J pressure as shown in Figure 2.21 for most explosives. The depth of the dent in the plate dent experiment[35] is much greater for tungsten or lead-loaded explosives than expected for their observed detonation pressure. For example, a 60/30/10 volume percent RDX/Pb/Exon at 4.6 g/cc has a detonation pressure of 150 kbar as determined by metal plate acceleration data and 345 kbar as determined by the plate dent vs. C-J pressure correlation described in Reference 35. The experimentally measured detonation velocity is 0.5012 cm/μs. The BKW-calculated C-J pressure is 270 kbar and velocity is 0.5096 cm/μs. Similar calculated results are obtained whether the lead is considered as compressible or incompressible and whether the lead is in temperature and pressure equilibrium with the detonation products or is in pressure equilibrium alone.

The reactive fluid dynamics of a detonation interaction with a matrix of metal particles was performed in Reference 36 to determine the nature of the observed nonideal behavior.

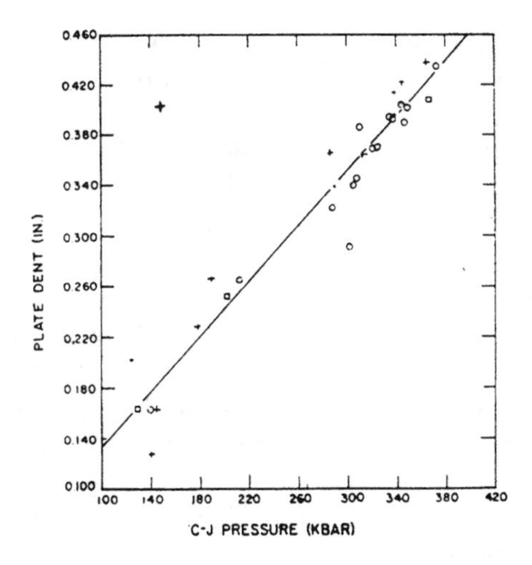

Figure 2.21 The plate dent vs. C-J pressure correlation from Reference 35. The 60/10/30 RDX/Exon/Pb data point is in the upper left side of the graph.

Models Assuming Propagating Detonation Between Particles

Al'tschuler[26] attributed the nonideal behavior of his tungsten-loaded HMX explosives to the explosive detonating at an average velocity determined by the individual detonation wavelets traveling around the metal particles. Dremin[37] attributed the nonideal behavior to a transfer of detonation product energy to the tungsten particles by heat conduction.

A one-dimensional model of layers of explosive and metal with dimensions appropriate to the size of the metal particles was considered. For the explosive system of HMX with tungsten, the C-J pressure of HMX is 395 kbar and velocity is 0.9159 cm/μs. The HMX detonation will interact with a tungsten plate forming a shock of 668 kbar and 830 K, moving at 0.495 cm/μs. A 65/35 volume percent HMX/tungsten system in layers would have a velocity of about 0.77 cm/μs. This velocity is much higher than the experimental value of 0.5 cm/μs.

To examine the Al'tschuler and Dremin proposals, the reactive hydrodynamic code 3DE described in Appendix D was used.

Three-dimensional calculations of a detonation wave in HMX interacting with a matrix of particles showed the detonation wave propagating between the particles at the C-J detonation velocity with the actual velocity determined by the shortest path through the particles. This effective velocity was less than the HMX detonation velocity and greater than the one-dimensional layer model velocity discussed earlier. The explosive shock pressures were C-J or even higher from the shock interaction with the tungsten and colliding detonation waves.

The detailed flow resulting from an HMX detonation wave interacting with a tungsten prism was modeled using the 3DE code. The three-dimensional computational grid contained 19 by 19 by 32 cells, (called I, J, K for the number of cells in the X, Y, and Z directions) each 0.01 mm on a side. The tungsten prism was described with 11 by 11 by 6 cells. The explosive burn was described by an Arrhenius rate law; however, identical results were obtained using the C-J burn technique described in Appendix A. A detonation wave of 395 kbar proceeds at 0.916 cm/μs through the HMX and interacts with the tungsten prism. A reflected shock proceeds back into the detonation products and a shock wave passes through the tungsten. The detonation wave travels around the tungsten prism and collides above the prism with pressures in excess of 800 kbar. The shock wave passing through the tungsten arrives at the upper interface when the detonation wave collision occurs.

An identical calculation was performed permitting heat conduction between the detonation products and the shock-heated tungsten prism. The heat conductivity used for tungsten was 0.1 cal/cm-sec-deg-K and for the detonation products was 0.001 cal/cm-sec-deg-K. An insignificant amount of energy was transferred to the tungsten from the detonation products by heat conduction. The conductivity constants were increased by a factor of 100, and still a negligible amount of energy was transferred.

So the Al'tschuler and Dremin proposals do not describe the observed nonideal behavior of tungsten-loaded explosives because the velocities and pressures are much larger than observed. Models that assume propagating detonation between particles do not describe the observed nonideal behavior.

A well characterized tungsten-loaded HMX explosive is called X0233. The explosive consists of 55/34/11 volume percent of HMX/W/polystyrene and DOP. Its density is 7.41 g/cc. The tungsten particle sizes are 70% coarse (0.02-0.04-mm diameter), 22.4% medium (0.005-0.015 mm), and 7.5% fine (0.0015-0.005 mm). So the tungsten particle size ranges over an order of magnitude. For this study the matrix of tungsten particles was approximately 0.02 mm in diameter. The detonation velocity is 0.464 cm/μs and the detonation pressure from the aquarium test is 160 kbar as measured by S. Goldstein.[38] The steel plate dent depth for a 41.275-mm-diameter cylinder of X0233 is 8.53 mm, as measured by M. Urizar,[2] which corresponds to a detonation pressure of 297 kbar.[35] The BKW-calculated C-J performance is 277 kbar, 0.451 cm/μs, and 2015 K. This explosive exhibits typical nonideal behavior.

Assuming the tungsten particles are located on a hexagonal close-packed lattice (HCP) and have a radius of 0.02 mm, the closest distance between tungsten spheres is 0.0114 mm for 35-volume-percent tungsten in HMX. If the tungsten sphere radius is 0.03 mm, the closest distance between spheres is 0.0170 mm.

For 15-volume-percent tungsten in HMX, the tungsten spheres with a radius of 0.0133 mm are 0.0187 mm apart. If the particle radius is increased by a factor of 4 to 0.0532, the spheres are 0.0748 mm apart.

To determine how detonations propagate in three-dimensional geometries with these dimensions, models that permit failure of propagating detonation must be used.

Models Permitting Failure of Propagating Detonation

As described in Chapter 4, the heterogeneous shock initiation burn model called Forest Fire is used to calculate failure of propagation of detonation as a function of diameter, pulse width of initiating shock, wave curvature, and changes in geometry such as turning corners. These are all dominated by the heterogeneous shock initiation mechanism. The propagation or failure of detonation in a matrix of tungsten particles in HMX is likewise determined by the heterogeneous shock initiation process and described by the Forest Fire model. The Forest Fire rate for PBX-9404 was used to describe the HMX in the matrix of explosive and tungsten.

It was necessary to determine how the failure radius of HMX changes with confinement. Calculations were performed using the 2DL and 2DE codes with Forest Fire to determine the failure radius of HMX and the TATB based shock insensitive explosive PBX 9502. As shown in Table 2.6, the calculated failure radius decreases by a factor of 5 as the density of the confinement increases to that of tungsten. The failure diameter of a cylinder of HMX confined by tungsten is about 0.1 mm. This is larger than the distance between tungsten particles in X0233 so it is possible for detonation to fail to propagate between the tungsten

TABLE 2.6 Failure Radius

Confinement	Calculated Radius (mm)	Experimental (mm)
HMX		
Air	0.6 - 0.4	0.6 ± 0.1
Plexiglas	0.4 - 0.3	
Al	0.3 - 0.2	
W	0.1 - 0.05	
PBX 9502		
Air	5.0 - 4.0	4.5 ± 0.5
Plexiglas	4.0 - 3.0	
Al	3.0 - 2.0	
W	1.5 - 0.5	

particles; however, the actual behavior of the detonation wavelets in the three-dimensional matrix must be calculated.

5% Tungsten/95% HMX

Three-dimensional calculations were performed for a matrix of 5% by volume tungsten in HMX. The computational grid contained 20 by 27 by 51 cells, each 0.00667 on a side. The tungsten spheres had a radius of 0.0133 mm and were described by 4 cells per sphere diameter. Numerical tests with 2 to 6 cells per sphere diameter showed the results were independent of grid size for more than 3 cells per sphere diameter. The closest distance for the HCP matrix between tungsten spheres was 0.0388 mm. The detonation starts to fail as it passes around the tungsten particles but complete decomposition occurs before the shock wave proceeds ahead of the reaction region.

15% Tungsten/85% HMX

Three-dimensional calculations were performed for a matrix of 15% by volume tungsten in HMX for tungsten spheres of 0.0133 mm. The closest distance for the HCP matrix between tungsten spheres was 0.0187 mm. The computational grid contained 4 by 19 by 43 cells, each 0.00667 mm on a side. The tungsten spheres were described by 4 cells per sphere diameter. Some of the detonation wavelets fail to propagate and the shock wave proceeds ahead of the reaction zone in regions of the flow. These partially reacted regions continue to decompose relatively slowly with time.

To demonstrate the effect of particle size, the tungsten sphere radius was increased by a factor of 4, the tungsten spheres being increased to 0.0532 mm. The closest distance for the HCP matrix between tungsten spheres was 0.0748 mm. This is close to the failure diameter of tungsten-confined HMX of 0.2 to 0.1 mm, so most of the HMX in the matrix is large enough to be above the failure diameter. The computational grid contained 14 by 19 by 43 cells, each 0.0267 mm on a side. The tungsten spheres were described by 4 cells per sphere diameter. In contrast to the previous case, most of the detonation wavelets pass around the tungsten particles without exhibiting failure. The shock wave proceeds along with the reaction region.

35% Tungsten/65% HMX

Calculations were performed for a matrix of 35% by volume tungsten in HMX. The computational grid contained 16 by 22 by 46 cells, each 0.00667 mm on a side. The tungsten spheres had a radius of 0.02 mm and were described by 6 cells per sphere diameter. The closest distance for the HCP matrix between tungsten spheres was 0.0114 mm. Many of the detonation wavelets fail to propagate and the shock wave proceeds ahead of the reaction zone. These partially reacted regions continue to decompose with time.

The nonideal behavior of inert metal-loaded explosives can be attributed to failure of the individual detonation wavelets as they pass between the tungsten particles. The shocked, partially decomposed explosive continues to decompose and release energy relatively slowly after shock passage.

The effect of increased concentration of inert metal is to reduce the distance between particles, resulting in more detonation wavelets failing and in lower detonation pressures in the explosive.

The effect of larger particle size of inert metal is to increase the distance between particles, resulting in less detonation wavelets failing and in higher effective pressures in the explosives. This effect of particle size is in agreement with the experimental results of Dremin.[37]

The expansion isentrope of the inert metal-loaded explosive must be less steep than for completely reacted ideal explosives of the same detonation pressure since additional decomposition occurs behind the detonation shock front. The high pressure expansion isentrope would result in larger plate dents and greater aquarium bubble expansions than those characteristic of ideal explosives with the same detonation pressure.

Application to X0233

As described earlier, X0233 consists of 55/34/11 volume percent of HMX/W/polystyrene and DOP at 7.41 g/cc.

The three-dimensional model for the nonideal behavior of tungsten-loaded HMX indicates that the observed low detonation pressure results from failure of some of the individual detonation wavelets. The explosive continues to decompose and release energy after shock passage.

The BKW equation of state represents the ideal behavior of explosive mixtures if all the explosive detonates. The gamma-law equation of state through the experimental detonation pressure and velocity represents the behavior if no additional energy is released behind the detonation front. To a first approximation, the energy actually present during expansion of the detonation products may be accounted for by passing the BKW isentrope through the detonation Hugoniot pressure corresponding to the experimental detonation pressure. If the shock velocity is sufficiently close to the detonation velocity, the resulting BKW isentrope should closely approximate the expansion behavior of the detonation products after complete decomposition has occurred behind the detonation front. For X0233, the shock velocity at 160 kbar on the detonation product Hugoniot is too high. The detonation Hugoniot must include the fact that a significant amount of explosive is not initially decomposed. This was accomplished by increasing (more negative) the heat of formation of the explosive by 47.4 calories per gram of explosive. The energy was returned to the detonation product isentrope in increments of 4.75 calories per gram per isentrope interval of 0.8 times the pressure. The resulting isentrope was insensitive to either the increment or interval details.

The gamma-law and BKW isentrope through the experimental detonation state are shown in Figure 2.22. Calculations of the plate dent using the 2DE code described in Appendix C for the three X0233 equations of state are shown, along with the experimental plate dent, in Table 2.7.

The BKW equation of state through the experimental detonation state reproduces the experimental detonation pressure of 160 kbar, the velocity, and the plate dent that is characteristic of an ideal explosive detonation pressure of 297 kbar. The description of the explosive equation of state was tested by comparing with the experimentally observed behavior of the expansion of the detonation products and shock wave in water.

The aquarium test was performed by S. Goldstein.[38] Calculations of the aquarium test for a 20.63-mm radius cylinder of X0233 in water were performed using the 2DL code described in Appendix B.

The photograph, taken by S. Goldstein with the image intensifier camera, of the shock wave in water and the interface between the detonation products and water is shown in Figure 2.23. The interface between the water and the detonation products is much more ragged than observed previously for any other explosive. Even the ANFO aquarium test shown in Figure 2.16 has a smooth interface between the water and detonation products. This can be interpreted as evidence of irregular decomposition of the explosive, in agreement with our proposed failure of some of the individual detonation wavelets between the tungsten particles and later decomposition of the explosive in these regions.

The calculated and experimental shock wave and detonation product/water interface are shown for the gamma-law equation of state in Figure 2.24, the BKW ideal equation of state in Figure 2.25, and in Figure 2.26 the BKW equation of state through the experimental detonation state with the energy returned to the isentrope. This BKW equation of state

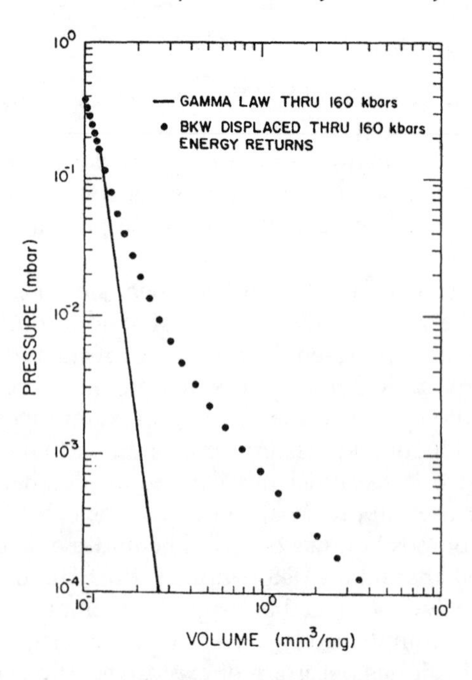

Figure 2.22 The X-0233 gamma-law isentrope through the experimental detonation pressure and velocity, and the BKW isentrope through the experimental detonation pressure and velocity with the energy returned.

reproduces the experimental shock wave and detonation product interface positions. The equation of state constants for X0233 are given in Table 2.8. The constants have the same identity as described in Appendix A and E.

TABLE 2.7 X0233 Plate Dents

Equation of State	Calculated Dent (mm)	Experimental Dent (mm)
BKW ideal	11.0	8.53
Gamma law	5.0	
BKW through experimental detonation state and energy returned to isentrope	9.0	

To model the explosive performance of X0233, it was necessary to displace the BKW detonation product Hugoniot so that it intersected the observed detonation pressure and velocity by decreasing the energy available to the detonation products.

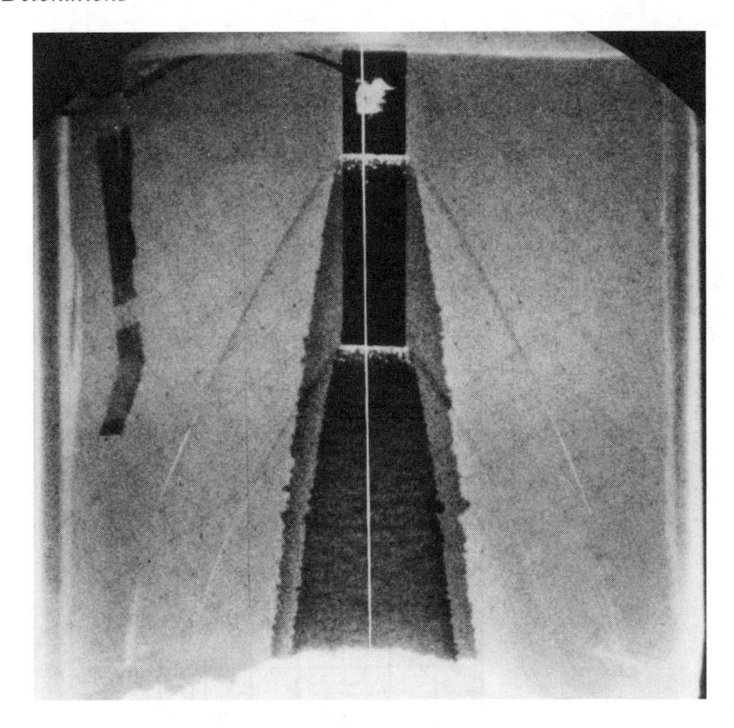

Figure 2.23 The aquarium test for X0233. Two photographic exposures taken with the image intensifier camera. The shock wave in the water and the interface between the detonation products and the water are shown.

The detonation is programmed to travel with the experimental detonation velocity using either the sharp-shock burn or the volume burn described in Appendix A. The velocity or specific volume at the detonation state is larger than the C-J state values for the detonation product Hugoniot passing through the detonation state. This results in a weak-like detonation with the one dimensional pressure-distance profile shown in Figure 2.27. The detonation products behind the detonation front cannot reach the C-J state since the wave is traveling faster than a C-J detonation. A flat top Taylor wave forms behind the detonation front.

To further test the weak detonation model, S. Goldstein measured the water shock velocity in the aquarium test after the detonation wave interacted with the water above the top of the X0233 cylinder. Her experimental water shock velocities, as a function of distance above the top of the explosive cylinder, are shown in Figure 2.28 along with the calculated water shock velocities. They are consistent with a flat top Taylor wave characteristic of a weak detonation and a detonation front pressure of 160 kbars. The initial water shock velocities exhibit behavior characteristic of irregular decomposition of the explosive near the shock front. The 2DL calculated aquarium pressure contours are shown in Figure 2.29.

The displaced BKW isentrope that will describe the observed plate dent, aquarium water shock profiles, explosive interface, and the detonation velocity of X0233 exhibits a weak detonation behavior.

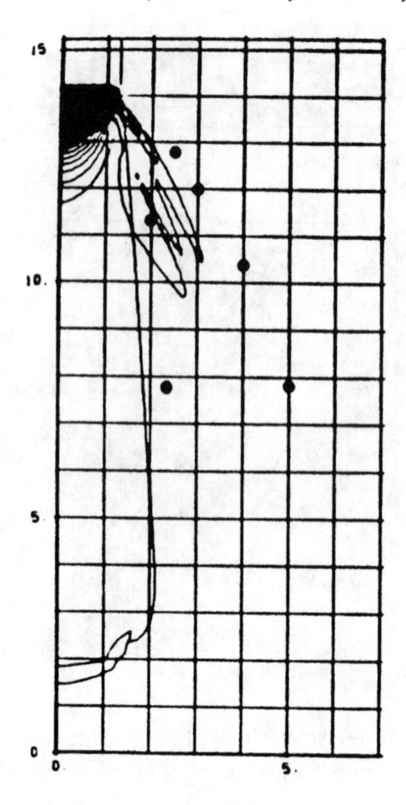

Figure 2.24 The experimental and the calculated X0233 aquarium interfaces for the gamma-law equation of state. The dots are experimental points. The contour interval is 5 kbar.

Application to RDX/Exon/Pb

The explosive mixture 60/10/30 by volume of RDX/Exon/Pb has a density of 4.60 g/cc, a detonation velocity of 0.5 cm/μs, and a detonation pressure determined from aluminum plate push experiments of 150 kbar. The experimental plate dent of 10.23 mm corresponds to a detonation pressure of 346 kbar. The BKW-calculated C-J performance is 270 kbar, 0.5096 cm/μs, and 2242 K. This is one of the most nonideal inert metal-loaded explosives. The very small particle size of the lead powder used results in the failure of many individual detonation wavelets as they pass between the lead particles.

To account for the nonideal behavior, it is insufficient just to displace the BKW isentrope through the detonation product Hugoniot at 150 kbar because the shock velocity of 0.6875 cm/μs is much higher than the experimental velocity of 0.5 cm/μs. The detonation product shock Hugoniot must include the fact that a significant amount of explosive is not initially decomposed. This was accomplished by increasing the heat of formation (more negative) of the explosive mixture until the shock velocity at 150 kbar on the detonation product Hugoniot agreed with the observed detonation velocity. This required increasing the heat of formation by 183 calories per gram of explosive mixture. An isentrope through this Hugoniot state was too weak to reproduce the observed plate dent, so the energy was returned to the detonation product isentrope in increments of 20 calories per gram per isentrope interval of 0.8 times the pressure. The resulting isentrope was insensitive to the

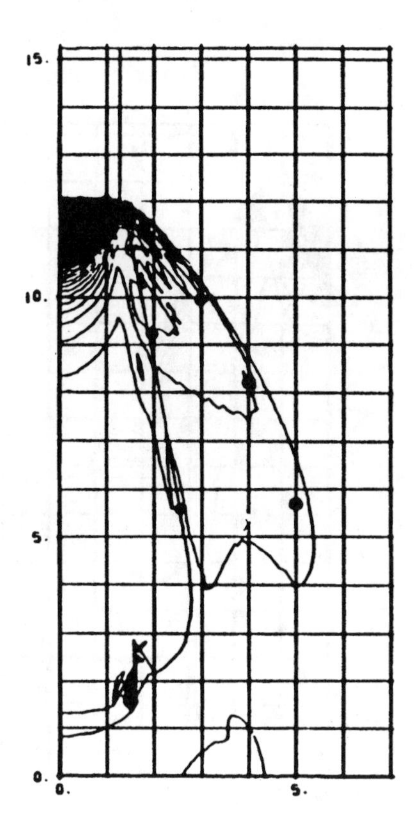

Figure 2.25 The experimental and the calculated X0233 aquarium interfaces for the BKW equation of state. The dots are experimental points. The contour interval is 5 kbar.

increment or to the interval details. The technique is similar to that used to describe ANFO and X0233 earlier in this chapter.

The resulting isentropes, with and without energy returned are shown in Figure 2.30, along with the ideal BKW isentrope and the gamma-law isentrope through the C-J state. The BKW equation of state constants for the isentrope displaced through the experimental detonation pressure and velocity are given in Table 2.8. Calculations of the plate dent, using the 2DE code and for the three equations of state, along with the experimental plate dent are shown in Table 2.9.

The BKW equation of state through the experimental detonation state with the energy returned to the isentrope reproduces the experimental detonation pressure of 150 kbar and velocity of 0.5 cm/μs and the plate dent characteristic of an ideal explosive detonation pressure of 346 kbar. It exhibits weak detonation behavior similar to that observed in X0233.

Metal Loaded Explosive Summary

The nonideal behavior of inert metal-loaded explosives may be attributed to the failure of some of the individual detonation wavelets between the metal particles and the subsequent decomposition of the partially decomposed explosive behind the detonation front. The effect of composition and particle size has been modeled qualitatively.

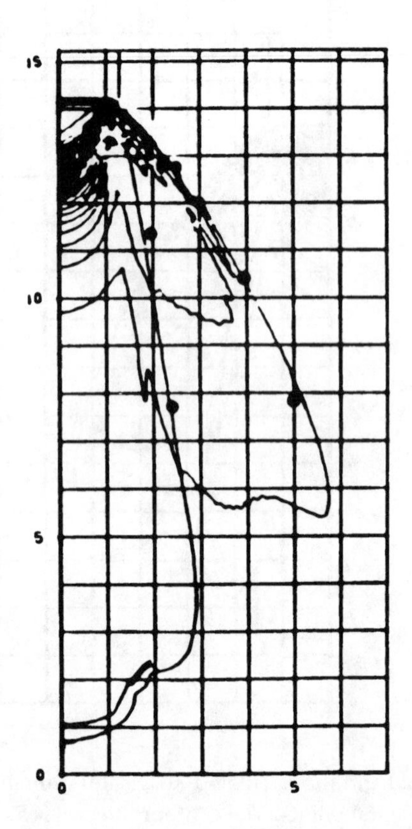

Figure 2.26 The experimental and the calculated X0233 aquarium interfaces for the displaced BKW equation of state with the energy returned to the isentrope. The dots are experimental points. The contour interval is 5 kbar. The weak detonation is shown by the region of no contours near the detonation front.

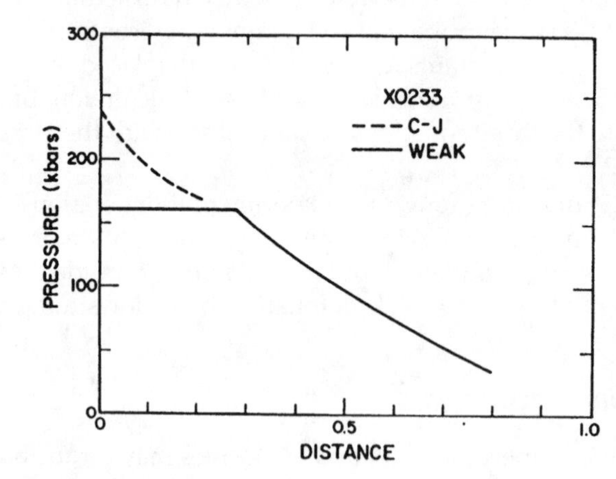

Figure 2.27 The Taylor waves for X0233 assuming a C-J and a weak detonation.

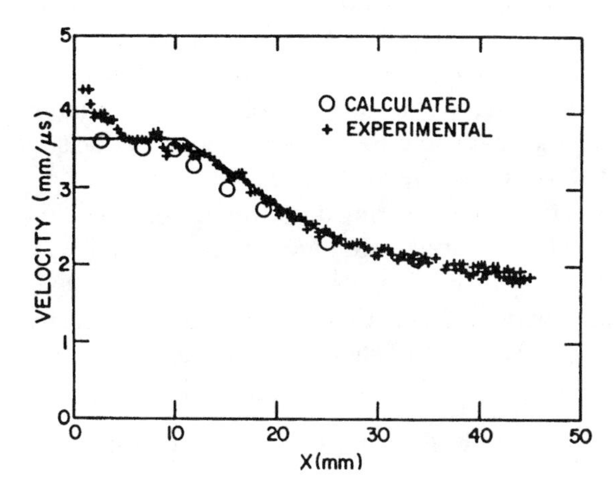

Figure 2.28 The calculated and experimental water shock velocities for X0233 detonation front shocking water.

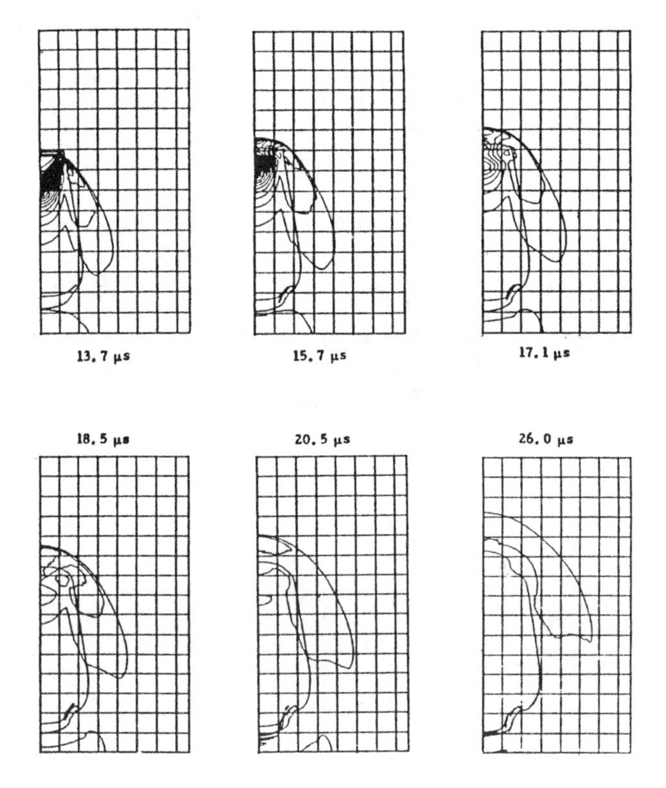

Figure 2.29 The calculated pressure contours for a cylinder of X0233 immersed in water. The contour interval is 5 kbar. The peak detonation pressure is 160 kbar. The weak detonation is shown by the region of no contours near the detonation front.

Characterization of the explosive requires experimental determination of the detonation pressure and velocity. If the experimental state is near the ideal BKW detonation product Hugoniot, the isentrope of the detonation products can be determined by displacing the isentrope through the experimental state. Otherwise, the ideal detonation product Hugoniot must be displaced so that it intersects the observed detonation pressure and velocity by decreasing the energy available to the detonation products. This results in a weak detonation with a flat top Taylor wave.

The technique gives useful engineering descriptions for modeling the nonideal explosive performance, but must be carefully evaluated by comparing with experimentally determined performance data. The isentropic behavior can be modeled at low pressures.

The observed variation of performance with particle size of the inert material reported by Dremin[37] for a few explosive systems should be expected to occur for all inert metal-loaded explosives.

TABLE 2.8 Equation-of-State Constants

	X0223	RDX/Exon/Pb
A	−7.21312551798E+00	−6.25053662637E+00
B	−1.39485557915E+00	−1.95155050436E+00
C	+2.23915871633E−01	−2.90284985132E−01
D	−1.77095802814E−01	−6.32661508840E−02
E	−1.23849736471E−02	−1.07335025294E−03
K	−2.12182312874E+00	−2.07715982237E+00
L	+1.53002477916E−01	+2.36461923017E−01
M	+4.04717285191E−02	+7.23556291125E−02
N	+4.88663567104E−03	+9.99231901764E−03
O	+2.14785041504E−04	+4.86763363442E−04
Q	+6.97901236695E+00	+7.01971398005E+00
R	−5.93317817321E−02	−1.39773676027E−01
S	−8.84798754816E−02	−2.27283819350E−01
T	+7.10719273057E−04	+1.63850564686E−02
U	+2.50870872978E−02	+5.34425309797E−02
C'_V	0.5	0.5
Z	0.1	0.1
P_{CJ}	0.24156	0.21056
D_{CJ}	4.310	4.745
V_{CJ}	0.113	0.17319
P*	0.16000	0.15000
D*	4.688	5.007
V*	0.121698	0.1891

*Displaced Values

It should be possible to significantly change the time history of availability of the explosive energy by variations of the composition and particle size of the inert metal. A systematic experimental study of the explosive performance of an inert metal-loaded explosive with particle size would be a valuable addition to the explosive performance data base.

This study is an example of how a fundamental study of a complicated three-dimensional reactive flow problem can result in sufficient increased understanding of the nature of the problem that a practicable engineering solution can be devised.

TABLE 2.9 60/10/30 RDX/Exon/Pb Plate Dents

Equation of State	Calculated Dent (mm)	Experimental Dent (mm)
BKW ideal	12.0	10.24
Gamma law	5.0	
BKW through experimental detonation state	9.0	
BKW through experimental detonation state and energy returned to isentrope	10.0	

Nonideal Explosive Summary

Three distinct types of nonideal explosive behavior have been found.

The First Type (Amatex)

The addition of ammonium salts to explosives results in increased nonideality with increased ammonium salt concentration. The amount of ammonium salt that reacts defines both the C-J state and the expansion isentrope for the explosive mixture.

The Second Type (ANFO)

The addition of ammonium salts to nonexplosives such as fuel oil results in explosive mixtures that exhibit large changes in performance with diameter and confinement. The amount of ammonium salt that reacts defines the C-J state for the particular diameter only. The ammonium nitrate reaction continues to completion behind the detonation front.

The Third Type (X0233)

The addition of inert metal (or other inerts) particles to explosives results in weak detonations with flat topped Taylor waves. The nonideal behavior is caused by failure of some of the individual detonation wavelets between the metal particles. Subsequent decomposition of the partially decomposed explosive occurs behind the detonation front.

Methods for determining an equation of state for these nonideal explosives for use in numerical hydrodynamic codes have been developed. They depend upon extensive experimental calibration of each explosive in the particular geometry of interest. Aquarium test data are crucial for the evaluation of the nature of the nonideal behavior and for the calibration of the equations of state.

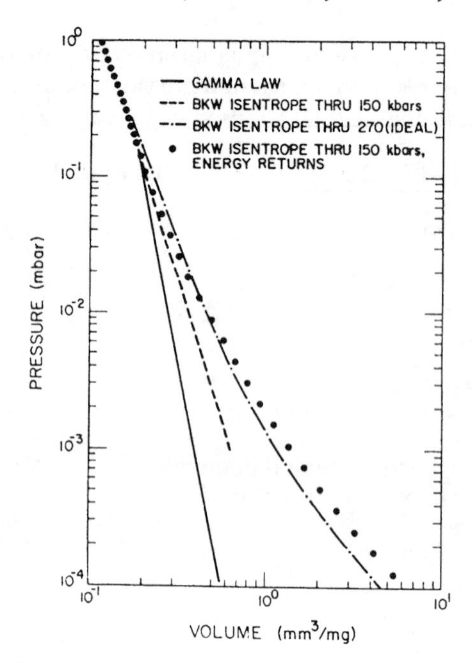

Figure 2.30 The 60/10/30 RDX/Exon/Pb BKW ideal isentrope, the gamma-law isentrope through the experimental detonation pressure and velocity, and the BKW isentrope through the experimental pressure and velocity with and without the energy returned.

2.3 *Nonsteady-State Detonations*

The first observation of build-up of detonation was in the explosive PBX-9404. Build-up of detonation has also been observed in the explosives Composition B and TNT.

In 1965, Craig[40] first discovered the nature of the nonsteady behavior of explosives. He studied the interaction of the explosive 9404 in one-dimensional plane geometry at four charge lengths with Dural, magnesium, and Plexiglas plates. If the 9404 behavior were steady-state, the experimental data should have scaled as a function of plate vs. explosive thickness. The data did not scale and they clearly indicate that the effective C-J pressure (P_{ECJ}) for underinitiated detonation 9404 increases, or builds up, as the detonation wave runs.

The result was not unexpected, because at about the same time, Davis et al.[41] had shown by other experimental studies that the steady-state C-J theory did not accurately describe the behavior of real explosives. The exact nature of the nonsteady-state behavior shown by Craig's data[40] (a 25% change in P_{ECJ} with less than 1% change in detonation velocity) was a surprise to most detonation scientists; however, it permitted one to understand why thin layers of explosives acted so differently from the behavior indicated by the simple calculations calibrated with data from thicker explosive charges.

There are clues, to be discussed later in this chapter, indicating the processes that result in failure of the steady-state detonation model. They are currently of little value to the numerical engineer who wishes to treat an explosive as realistically as practicable. He would like a description of the explosive behavior that would work in divergent and convergent geometry as well as in plane geometry. The study is limited to one-dimensional flow.

The build-up model assumes that a real nonsteady-state detonation can be approximated adequately by a series of steady-state detonations with instantaneous reaction whose "effective C-J pressures" vary with the distance of run. This empirical model depends completely upon experimental data for its calibration. If the magnitude or duration of the initiating pulse is changed, or the explosive is one for which experimental data are not available, new experimental data must be generated and the model must be calibrated for the new system.

It is important to understand that this discussion is about build-up *of* the effective C-J pressure of detonating 9404 initiated at pressures less than infinite-medium effective C-J pressure, but greater than that pressure required for prompt initiation ($\sim$100 kbar for most explosives). This is not to be confused with the build-up *to* detonation, characteristic of the process of shock initiation of heterogeneous explosives initiated by a shock wave of a few tens of kbars described in Chapter 4.

Because the nonsteady-state detonation process requires new concepts, some appropriate definitions are given.

Build-Up TO Detonation – Characteristic of the process of shock initiation of heterogeneous explosives initiated by a shock wave of a few tens of kbars.

Build-Up OF Detonation – The change of the effective C-J pressure of a detonating explosive initiated at pressures less than infinite-medium effective C-J pressure, but greater than that required for relatively prompt initiation.

The Build-up Model – An empirical model for engineering purposes that depends on experimental data for calibration. The model assumes that a real nonsteady-state detonation can be adequately approximated by a series of steady-state detonations with instantaneous reaction and constant detonation velocity whose effective C-J pressures vary with the distance of run for a particular initiation system.

Effective C-J Pressure – The maximum pressure of a completely decomposed explosive in a steady-state detonation, used to approximate the nonsteady-state flow associated with a particular distance of run and initiating system.

Infinite-Medium Effective C-J Pressure – The actual steady-state C-J detonation pressure for an explosive. In practical plane wave systems, it can be achieved only if the explosive is overdriven.

Peak Detonation Pressure – The largest pressure obtained in diverging or converging flow after some distance of travel, using an equation of state with an effective C-J pressure that is characteristic of some distance of run in slab geometry.

The SIN code described in Appendix A was used to compute most of the one-dimensional problems described. For a few problems, the characteristic code RICSHAW,[42] as revised by Rivard, was used to check the results obtained with SIN.

The methods of burning the explosive were studied to determine whether the numerical results were independent of the burn technique. The gamma-law Taylor wave burn technique for slabs, the Arrhenius rate law, the C-J volume burn technique, and the sharp-shock burn technique are described in Appendix A. Another method in common use is the programmed burn, which assumes that the time required to burn an explosive cell can be predetermined from the detonation velocity. Any of the methods is satisfactory for plane geometry, but the gamma-law Taylor wave method is faster and requires fewer cells in the numerical calculation. For divergent geometry, the Arrhenius rate law gives the best results; however, the C-J volume burn technique can be used if, in addition to the usual prescription, the cell burn is completed when expansion of the cell begins. The sharp-shock burn technique of SIN or RICSHAW cannot be used in divergent geometry. In convergent geometry, the C-J volume burn, Arrhenius rate law, or sharp-shock burn of SIN or RICSHAW can be used. The programmed burn technique can be used if the increase in detonation velocity as a function of convergence can be determined. This is possible to a good first approximation as shown later in this section, so the programmed burn technique can be used in convergent geometry.

Most of the plane calculations used 400 cells or space increments in the metal and 350 cells in the explosive. Most of the calculations of spherically diverging detonations were performed with 600 cells in the explosive and 300 cells in the metal. Most of the calculations of spherically converging detonations used 800 cells in the explosive.

The equation of state used was the HOM equation of state described in Appendix A. The Hugoniot of the aluminum was described using experimentally determined linear shock velocity U_s and particle velocity U_p curves, expressed as $U_s = C + SU_p$. The constants used, which are identical to those in Appendix A, are as follows:

Metal	ρ_o	C	S	γ	α	C_v
Al	2.785	0.535	1.35	1.7	2.4x10^{-5}	0.22

Generally the PIC form of the viscosity was used with a constant of 2.0.

The gamma-law equation of state for a steady-state detonation is

$$P_{CJ} V_{CJ}^{\gamma} = C \,,$$

$$\frac{V_{CJ}}{V_o} = \frac{\gamma_{CJ}}{\gamma_{CJ} + 1} \,,$$

$$ln P_i = ln C - \gamma \; ln V$$

$$I_i = \frac{PV}{\gamma - 1} - \frac{P_{CJ} V_{CJ}}{\gamma - 1} + \frac{P_{CJ}}{2}(V_o - V_{CJ}) \,.$$

Given the initial density, ρ_o, detonation velocity, D, and $P_{CJ} = (\rho_o D^2)/(\gamma + 1)$, the isentrope of the detonation products is defined for any given P_{CJ}, γ, or V_{CJ}. Calculations off the isentrope use the constant beta equation of state.

$$P = \frac{1}{\beta V}(I - I_i) + P_i \,.$$

where

$$\beta = \frac{1+\alpha}{\gamma} \text{ and } \alpha = \left[\frac{\gamma+1}{1+\frac{d(lnD)}{d(ln\rho)}} - 2 \right]^{-1}.$$

A value of 0.66 was used for the $d(lnD)/d(ln\rho)$ of 9404, and β varied from 1.30 at P_{ECJ} = 0.365 Mbar to 0.607 at P_{ECJ} = 0.306 Mbar.

The "gamma-law equation of state" means that the gamma-law equation is used to describe the isentrope through the C-J state and the constant beta equation is used to calculate off the isentrope. The "BKW equation of state" means that the Becker-Kistiakowsky-Wilson equation of state (Appendix E) is used to describe the isentrope through the C-J state and a variable beta equation of state is used to calculate off the isentrope, with beta a function of volume as defined by the BKW equation of state.

Build-Up in Plane Geometry

Craig's experimental data show that the effective C-J pressure of underinitiated, detonating 9404 increases or builds up as the detonation wave runs. Craig found that the effective C-J pressures increases with distance as shown in Figure 2.31 for 9404 initiated with a Baratol plane wave lens. Effective C-J pressures were obtained using Dural, magnesium, and Plexiglas plates that were identical within experimental error for the same distance of run. The experimental data used in Figure 2.31 are given in Table 2.10. Craig also observed that the detonation wave velocity remains within 100 m/s of 8800 m/s in all the systems studied. Assuming that real nonsteady-state detonation can be approximated by a series of steady-state detonations whose effective C-J pressures vary with the distance of run,

$$\gamma = \rho_o D^2 / P_{ECJ} - 1$$

(where ρ_o, the initial density, is 1.844 g/cc, D, the detonation velocity, is 0.88 cm/μsec, and P_{ECJ} is the effective C-J pressure in Mbar) is used to calculate gamma as a function of distance of run as shown in Figure 2.32. This may be described by the equation called the "build-up equation"

$$\gamma = 2.68 + 1.39/(\text{distance of run in cm})$$

or

$$\gamma = 2.68 + 1.58/(\text{time of run in } \mu\text{sec}).$$

TABLE 2.10 9404 Experimental Data

Charge Length (cm)	Charge Diameter (cm)	Plate Material	Plate Thickness (mm)	U_{FS} mm/μsec	Lens Diam (cm)	Length/ Diameter Ratio	P_{ECJ} (kbar)
1.27	10.16	2024 ST Dural	3.2	3.12	10.16	0.125	303
			4.5	3.03			
			6.4	2.92			
			12.7	2.57			
1.27	30.48	2024 ST Dural	3.27	3.20	30.48	0.0416	312
			4.55	3.13			
1.27	10.16	Magnesium	3.34	4.18	10.16	0.125	306
			3.72	4.14			
			5.00	4.10			
			7.83	3.95			
2.54	20.32	2024 ST Dural	2.20	3.45	20.32	0.125	335
			2.50	3.45			
			4.50	3.39			
			6.35	3.33			
			12.70	3.075			
			25.40	2.675			
5.08	30.48	2024 ST Dural	2.81	3.67	30.48	0.167	358
			4.59	3.61			
			12.17	3.47			
			21.0	3.27			
10.16	30.48	2024 ST Dural	3.13	3.90	30.48	0.333	375
			25.4	3.45[a]			
			50.9	3.02[a]			
10.16	30.48	Plexiglas	6.0	6.00	30.48	0.333	372
			12.0	5.65[a]			
			24.5	5.34[a]			
			51.0	4.66[a]			

[a] Two-dimensional effects possible.

The fit was used down to 1.53 μsec (1.36 cm); then, the gamma was kept constant at 3.71 (P_{ECJ} = 0.303 Mbar) for shorter times and distances of run. This arbitrary limit was imposed to prevent the fit from giving unreasonable results for short times and run distances. Although this permits good estimates of the effective C-J pressure, the problem remains as to how best to describe the Taylor wave. Calculations were performed for the experimental geometries that Craig studied, assuming that the entire explosive burned with constant gamma for the total distance of run. Calculations were run assuming that the gamma varies

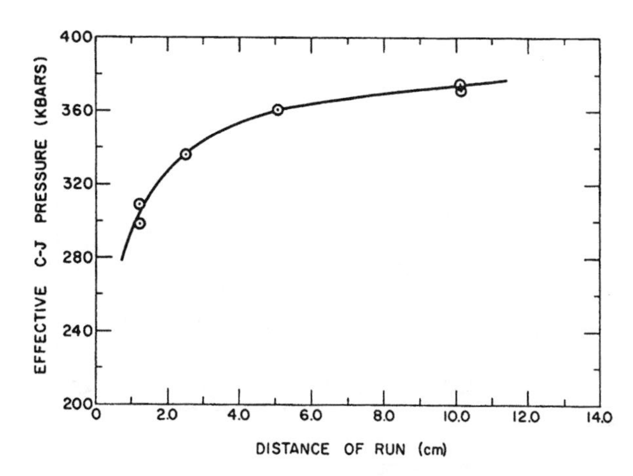

Figure 2.31 Craig's experimental effective C-J pressures of 9404 vs. distance of run for Baratol plane-wave initiation of 9404 slabs.

through the flow according to the build-up equation described above. For the latter case, the gamma of each cell was held constant at its value when the detonation wave first passed through it. The Taylor waves are shown in Figures 2.33 and 2.34, and their effect on the Dural plate motion is shown in Figure 2.35. To a very good first approximation, a constant gamma for the total distance of run can be used to describe the Taylor wave.

Note that from 4 to 10 cm of run, the effective C-J pressure varies only from 350 to 375 kbar. Most plane wave experimental studies are in this range of run. The usual effective C-J pressure stated for 9404 is 365 kbar. The actual infinite-medium effective C-J pressure is approximately 400 kbar. Deal[20] determined the 365-kbar value some 40 years ago, using a charge that was 14.2 cm long and 14.2 cm in diameter. This system is two-dimensional. The effect of the side rarefactions is shown in Figure 2.36, where the experimental free-surface velocities are markedly lower than the calculated velocities as the metal is affected by more of the Taylor wave. It is thus a matter of luck that the two-dimensional effects lowered the extrapolated pressure into the range where most plane wave experiments are performed. This is why the BKW equation-of-state calibrations, using Deal's effective C-J pressures, worked so well in describing most plane wave experimental data. The build-up model will not work for slab systems whose length-to-diameter ratio is greater than 0.25 and which are not supported by a plane wave. The reason such a small value for the length-to-diameter ratio is chosen is that most plane wave systems are plane and uniform enough only over the inside half or less of their diameter, so only the inner half of the explosive charge is properly supported. The small length-to-diameter ratio is also chosen to eliminate the effect of side rarefactions while the shock travels through the metal plate.

The effect of changing the magnitude of the initiating pulse can be estimated by shifting the build-up curve shown in Figure 2.31 so that it intersects the pressure axis at whatever the initiating pressure in the explosive is for the particular driving system. Thus, if the explosive is driven with a flying plate that delivers 400 kbars to 9404, the build-up curve will be flat and no build-up will occur. If the initial pressure is greater than the infinite-medium effective C-J pressure, a delay occurs until the infinite-medium effective C-J pressure is achieved. The build-up curve shown in Figure 2.31 does not address the problem of strongly overdriven detonations.

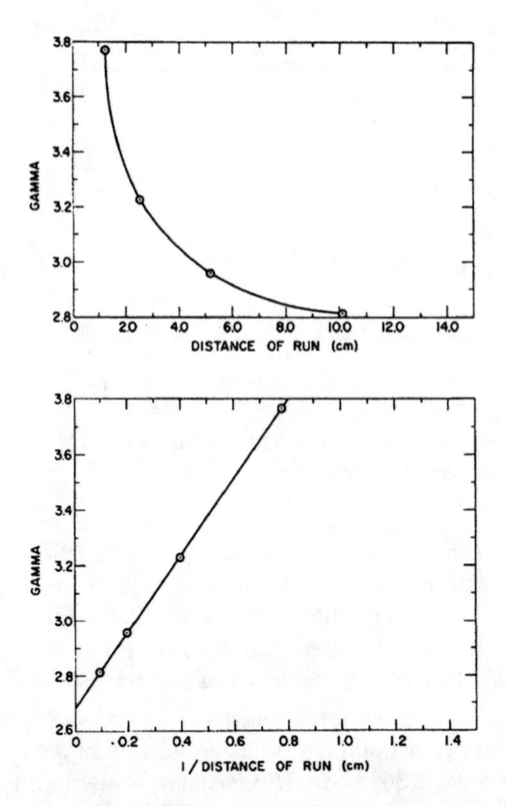

Figure 2.32 Gamma-law equation of state gamma calculated using Craig's experimental effective C-J pressures of 9404 vs. distance of run and vs. reciprocal of distance of run.

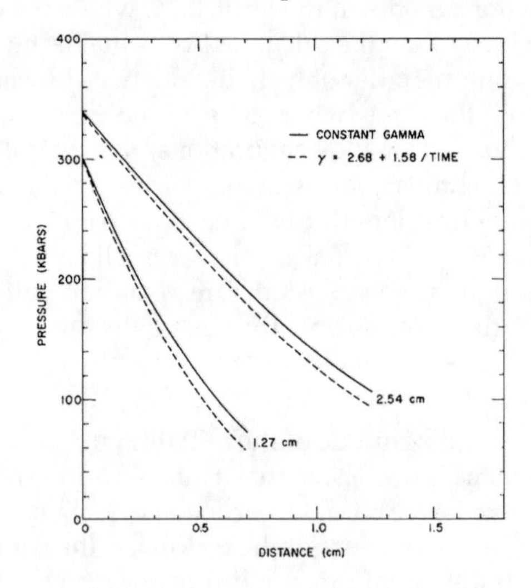

Figure 2.33 Calculated Taylor waves for 1.27 and 2.54 cm of 9404 with constant gammas of 3.77 and 3.227, respectively, and with a variable gamma.

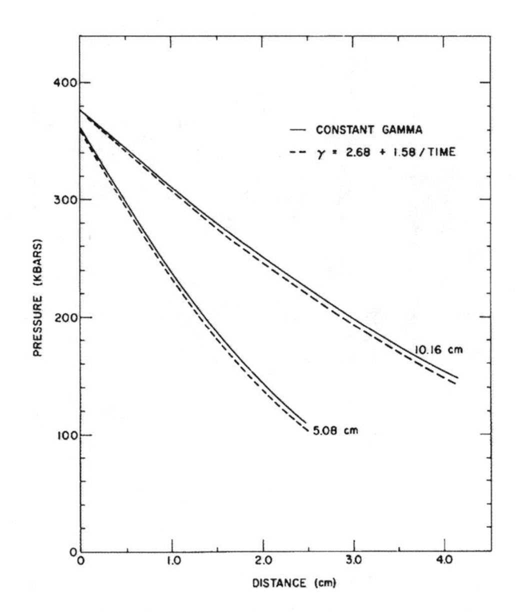

Figure 2.34 Calculated Taylor waves for 5.08 and 10.16 cm of 9404 with constant gammas of 2.9536 and 2.817, respectively, and with a variable gamma.

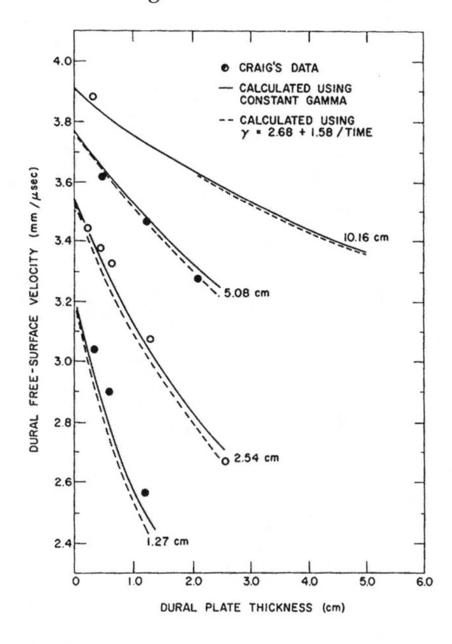

Figure 2.35 Dural free-surface velocities vs. Dural plate thicknesses for 1.27, 2.54, 5.08, and 10.16 cm of 9404. Curves show Craig's data and the calculated results for constant gamma Taylor waves (3.77, 3.227, 2.9536, and 2.817, respectively) and $\gamma = 2.68 + 1.58/\text{Time}$ Taylor wave.

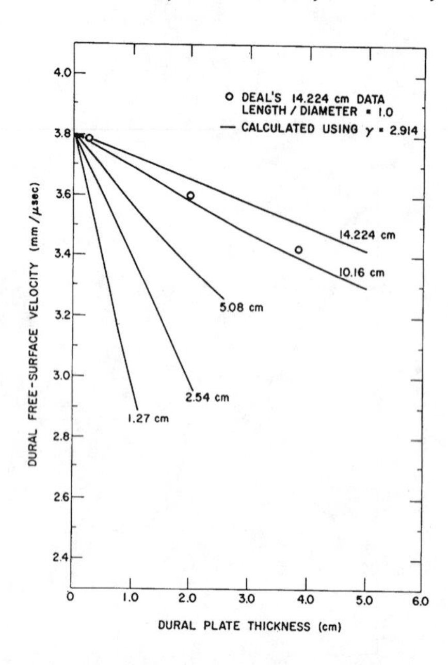

Figure 2.36 Deal's experimental 9404 data for 14.224 cm of run with a two-dimensional system whose length/diameter was 1.0. The calculated results for a constant gamma (2.914) Taylor wave are shown for 1.27, 2.54, 5.08, 10.16, and 14.224 cm of run. These results scale as Dural/HE thickness, but each thickness is shown for comparison with Figure 2.35.

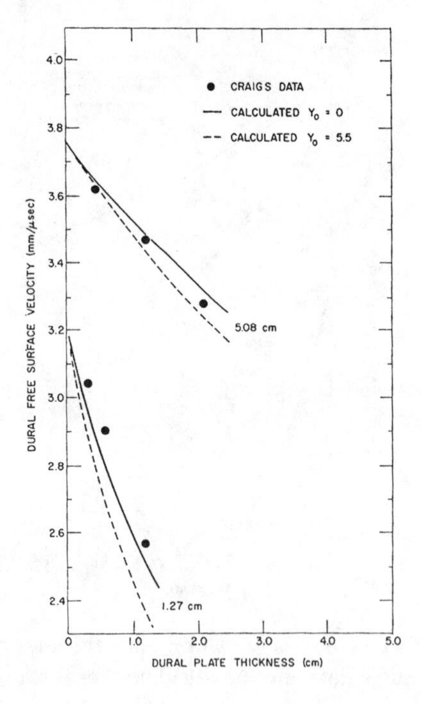

Figure 2.37 Dural free-surface velocities vs. Dural plate thickness for 1.27 and 5.08 cm of 9404. The calculated results for gammas of 3.77 and 2.9536 are shown for a yield of zero (as in Figure 2.36) and for a 5.5-kbar yield in aluminum.

The experimental free-surface velocities for Dural include an effect of the elastic component of the rarefaction. Because this effect is nonscaling if the explosive is nonscaling, its effect on Craig's explosive build-up data must be determined. The yield strength of aluminum at 300-400 kbar is uncertain, but a yield of 5.5 kbar successfully reproduced the Stanford Research Institute data at 330 kbar[43] and those of Isbell et al.[44] at 250 kbar. Calculations were performed for the 9404-aluminum systems tested by Craig, using a 5.5-kbar yield. The largest effect of the elastic component was to decrease the free-surface velocity by 0.01 cm/μsec for the 1.27-cm-thick 9404 and 1.27-cm-thick aluminum system. Thinner aluminum plates and longer explosive distances of run resulted in correspondingly smaller effects as shown in Figure 2.37. The elastic effect was small compared with the build-up effect and, for most of Craig's data, smaller than the experimental error.

Venable, using techniques similar to those described in Reference 45 and the PHERMEX radiographic facility, determined the behavior of a P-081 Baratol lens initiating 10.16 cm of 9404 with 0.00127-cm-thick tantalum foils embedded every 0.638 cm. The effective C-J pressure for 5 cm of 9404 initiated by a P-081 Baratol lens is $\sim$ 365 kbar. The tantalum foils are expected to interrupt the build-up process, and the effective C-J pressure for 9404 with tantalum foils would be less than 365 kbar (the pressure with no tantalum foils), but greater than 300 kbar (the effective C-J pressure for 0.638 cm of 9404 initiated by a 0.1-cm-thick tuballoy plate driven by 9404).

The experimental foil displacement as a function of the foil position is shown in Figure 2.38. Also shown is the calculated displacement assuming the effective C-J pressure of 365 and 300 kbar. An effective C-J pressure of 320 kbar fits the experimental data and is additional confirmation of the importance of the build-up process. Even systems with thin foils will interrupt the build-up behavior and markedly decrease the observed performance.

It is important to realize that an explosive's effective C-J pressure depends on the magnitude of the initiating pulse, the distance of run, and if side rarefactions can affect the flow, the confinement. Craig measured the initial free-surface velocities of Dural plates driven by plane wave nitromethane detonations. The experimental geometries were

Lens	*TNT* (*cm*)	Teflon (cm)	Nitromethane (cm)	Plates
P-040	1.27	0.01524	2.54	Dural
P-081	2.54	0.01524	5.08	Dural

These systems result in nitromethane detonation waves that are slightly overdriven initially and which then decay to the infinite-medium effective C-J pressure. Although the experiments seem to scale, the 5.08-cm thickness deviates from the 2.54-cm thickness by about 1.5% at the largest scaled (Dural/nitromethane) thickness. This deviation is larger than the experimental error (0.5%) and is probably a result of overdrive or of two-dimensional effects.

Figure 2.39 shows the experimental data and the calculated velocities as a function of scaled Dural thicknesses for yields of 0 and 5.5 kbar. In contrast to the underdriven 9404 data, the nitromethane data scale and do not exhibit build-up. Because initiation of a homogeneous explosive begins in the previously shocked but undetonated explosive and proceeds through the compressed explosive at a velocity and pressure greater than the infinite-medium velocity and effective C-J pressure, initiation of a homogeneous explosive results in an overdriven detonation that then decays toward the infinite-medium effective

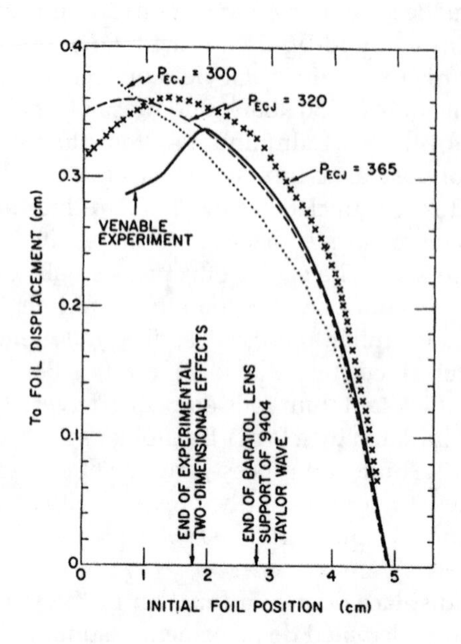

Figure 2.38 Experimental tantalum foil displacements in 4.985 cm of 9404 with 0.00127-cm-thick foils embedded every 0.638 cm. The calculated displacements with effective C-J pressures of 365, 320, and 300 kbar also are shown.

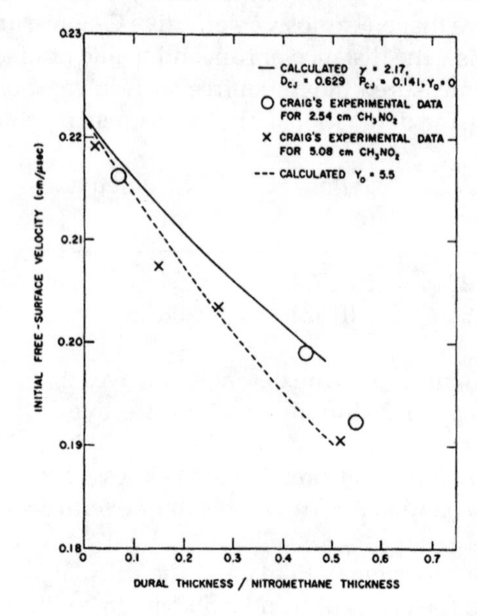

Figure 2.39 Dural free-surface velocities vs. scaled Dural/nitromethane thickness for 2.54 and 5.08 cm of nitromethane. The calculated curves use the experimental infinite-medium velocity of 0.629 cm/μsec, effective C-J pressure of 141 kbar, gamma of 2.17, and Dural yields of 0 and 5.5 kbar.

C-J pressure. Also, because initiation of homogeneous explosives results in overdriven detonations in the practical case, they will not exhibit build-up. The nitromethane experimental data seem to scale and to be adequately described by a steady-state model. The nitromethane overdriven detonation may decay to a steady-state detonation or may decay to a flow that continues to be time dependent (oscillates), perhaps requiring greater experimental resolution to detect.

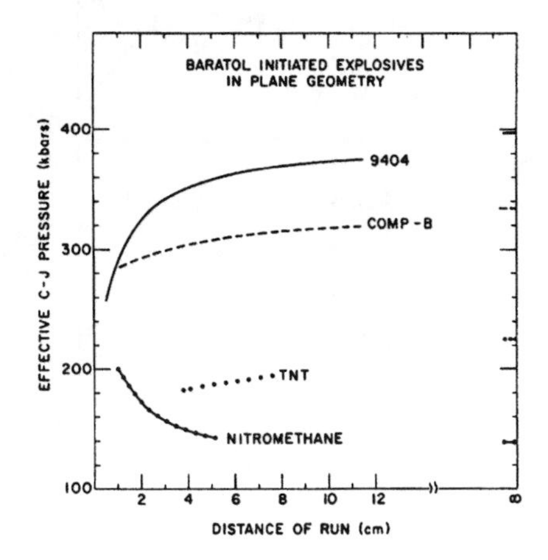

Figure 2.40 Experimental effective C-J pressures of 9404, Composition B, TNT, and nitromethane initiated by a plane-wave Baratol lens vs. distance of run. The infinite-medium C-J pressures are shown on the right-hand side of the figure for each explosive. The nitromethane is self-overdriven and the other explosives are underdriven.

Since the initial discovery of build-up in 9404, additional studies have been made by Craig for TNT and by Davis[23] for Composition B. Figure 2.40 shows that these explosives also exhibit a change in effective detonation pressure as a function of distance of run, and also a detonation velocity that remains essentially constant within the experimental measurement error. A curve is also shown for self-overdriven nitromethane for comparison with the other explosives that are underdriven by the Baratol plane wave initiation. The infinite-medium C-J pressures are shown on the right hand side of the figure. The build-up behavior is apparently different for the various explosives, with a lesser difference between 9404 and Composition B as distances of run decrease. It seems possible that explosive build-up curves could cross and that the ordering of the performance characteristics of explosives could change, depending on the distance of run and perhaps also on the initiating system.

Some explosives do not exhibit appreciable amounts of build-up. PETN has minimal build-up behavior as a function of distance of run. The performance of PBX 9502 (95/5 wt% TATB/Kel F at 1.894 g/cc) is reported in Reference 46. The initial free surface of Dural plates driven by 1.25, 2.5, and 5.0 cm thick slabs of PBX 9502 *scale* as a function of Dural plate thickness to PBX-9502 thickness. No evidence of build-up of detonation pressure as a function of run was observed. All the metal acceleration and aquarium test data can be described using a BKW or a gamma-law equation of state with a single effective C-J pressure (285 kbars), detonation velocity (0.7707 cm/μsec), and gamma (2.95).

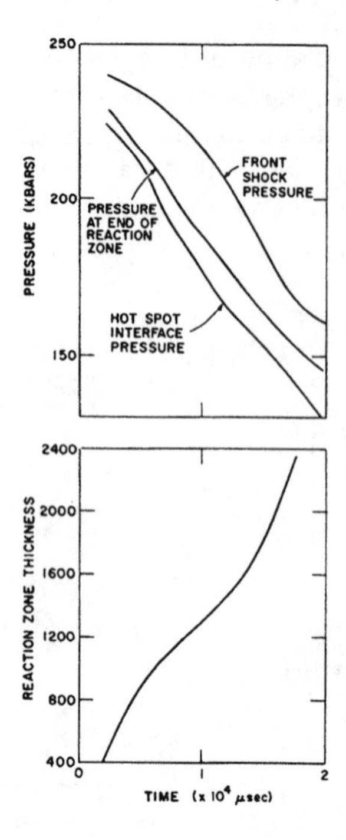

Figure 2.41 Pressure and reaction zone thickness vs. time for a 2.5 μm radius hot spot driving spherically divergent detonation in nitromethane. The calculation was performed using 100-A mesh.

Build-Up in Diverging Geometry

There is no useful theoretical treatment of spherically expanding detonation waves. The Taylor self-similar solution for diverging detonations has been used widely for lack of a better treatment; however, Courant and Friedrichs[47] show that it is incorrect. The Taylor self-similar solution does not permit the pressure at the end of the reaction zone to change with the flow divergence.

In Chapter 1, the results were described of studies done on the time-dependent reaction zones of an ideal gas, nitromethane, and liquid TNT, using one-dimensional numerical hydrodynamics with the Arrhenius rate law. A similar study was made of the reaction zone in spherically diverging geometry. The reaction zone in nitromethane that was overdriven enough to be stable in the plane case, and in an ideal gas that was stable in the plane case at C-J velocity has been modeled. Figures 2.41 and 2.42 show that the detonation process depends upon the rarefaction process that follows it. The pressure at the front and back of the reaction zone quickly drops below the plane-wave C-J values.

One-dimensional numerical hydrodynamic calculations using Arrhenius kinetics and unresolved reaction zones have been performed. The important features of the flow do not depend upon the mesh size or detailed kinetics. Similar results can be obtained using the C-J volume burn. For an overdriven detonation, the pressure decreases until it is considerably less than the effective C-J value and then slowly increases toward the effective C-J value. For an underdriven detonation, the pressure increases slowly toward the effective C-J value.

Craig determined the free-surface velocity of Dural plates in contact with 1.27, 5.08, and 7.62 cm radius spheres, or effective spheres, of 9404 initiated by detonations varying in size from a "point" of 0.1 mm radius of high density PETN to an effective sphere of 0.76 cm radius. The 0.76 cm radius detonator had low-density PETN (1.0 g/cc) for the inner 0.2 cm and high density PETN (1.6 g/cc) for the outer 0.56 cm. The center of the 9404 detonation wave was observed to be within 0.01 cm of the center of the detonator. This is because the high-density PETN produces pressure profiles that are similar to 9404 at short distances of run.

Craig's experimental data are shown in Figure 2.43. Although the data appear to scale, they do not, because the smaller spheres generally give lower velocities at the same scaled thickness. It is important to realize that although scaling is necessary for a Taylor self-similar model, it is not sufficient to show that the Taylor model is correct. The model used below almost scales but is not Taylor self-similar.

Calculations were performed for 1.27, 5.08, 7.62, and 10.16 cm radius spheres of 9404 and detonator, with the detonator described by a 0.2 cm radius PETN hot spot, which initiated the surrounding 0.56 cm thick layer of high-density PETN. The BKW or gamma law equation of state for 9404 with an effective C-J pressure of 365 kbar and a gamma of 2.914 yielded the lower dashed curve in Figure 2.43 for a 10.16 cm radius sphere. The free-surface velocity of the 1.27 cm radius sphere was about 1% less than that of the 10.16 cm thick sphere at the same scaled thickness.

The calculated peak detonation pressure in the 9404 as a function of radius is shown in Figure 2.44. For a 10 cm radius sphere, the calculation using Arrhenius burn and the BKW equation of state gives a detonation velocity of 0.8868 cm/μsec and peak detonation pressure of 312 kbar, compared with the BKW plane wave value of 0.8880 cm/μsec and effective C-J pressure of 365 kbar. Differences of 53 kbar in pressure and only 12 m/s in velocity occur between plane and spherically diverging geometry. The experimental detonation velocities for these systems are within 100 meters/sec of the plane wave values.

The experimental values shown in Figure 2.43 lie below the curve calculated using 365 kbar for the effective C-J pressure and a gamma of 2.914. The 1.27 cm radius values were generally below the 7.62 cm radius points at the same scaled thickness. To obtain the maximum expected build-up value, the 9404 build-up curve was shifted to give the PETN effective C-J pressure of 300 kbar to the 9404 at zero thickness. This gives an effective C-J pressure of 338 kbar and a gamma of 3.227 for the expected build-up equation of state of 9404 initiated with high density PETN at the minimum (1.27 cm) run distance studied experimentally. The bottom line in Figure 2.43 shows the spherically diverging calculation with the maximum amount of build-up that can be reasonably expected. This curve is just below the experimental 1.27 cm radius values.

Craig also determined the free-surface velocity of Dural plates in contact with 1.27, 2.54, and 5.08 cm radius cylinders or effective cylinders, of 9404 initiated with line generators of high density PETN. Figure 2.43 shows the experimental and calculated results for 9404 cylinders with a gamma of 2.914 and for the maximum amount of build-up gamma of 3.227. Also shown are parts of the plane results described previously. The build-up effect decreases with increasing flow divergence. The build-up effect is also small because the initiating pressures in diverging geometry are high.

Although the above agreement between the experimental data and calculations in diverging geometry is impressive, it does not suffice to show that the flow is not self-similar. It might be possible to devise an equation of state with a steep enough isentrope to compensate for keeping the pressure at the infinite-medium effective C-J pressure as assumed in the Taylor self-similar solution. What is needed is a direct measurement of the Taylor wave pressure or density behind plane and spherically diverging detonation waves in an

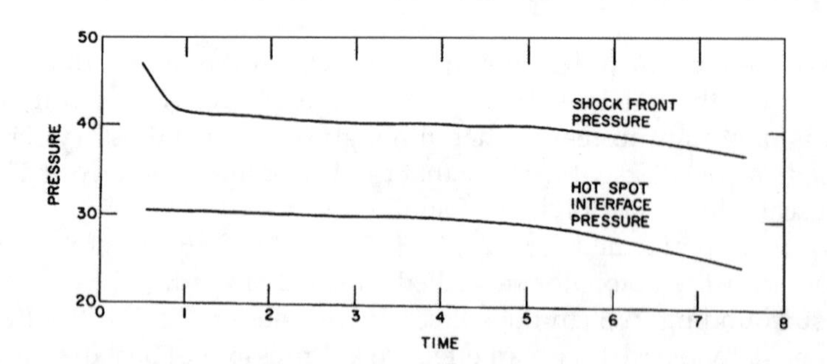

Figure 2.42 Pressure vs. time for hot spot driving spherically divergent detonation in ideal gas described by $E^* = 10$, $f = 1.0$, $P_{W=1} = 42.2$, and $P_{W=0} = 21.5$ where $f = D^2/D_{CJ}^2$ and $P = P/P_o$.

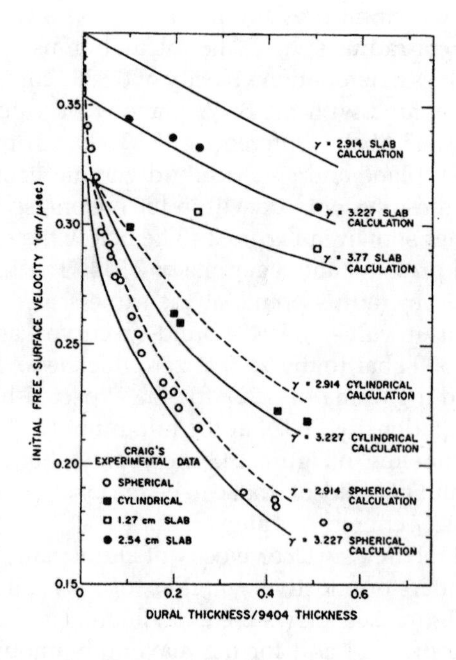

Figure 2.43 Initial free-surface velocities vs. scaled Dural/9404 thickness for plane, cylindrically, and spherically diverging 9404 detonations. Calculated results are shown for gammas of 2.914 (identical to results obtained using the BKW equation of state) and 3.227 (the maximum gammas expected from build-up).

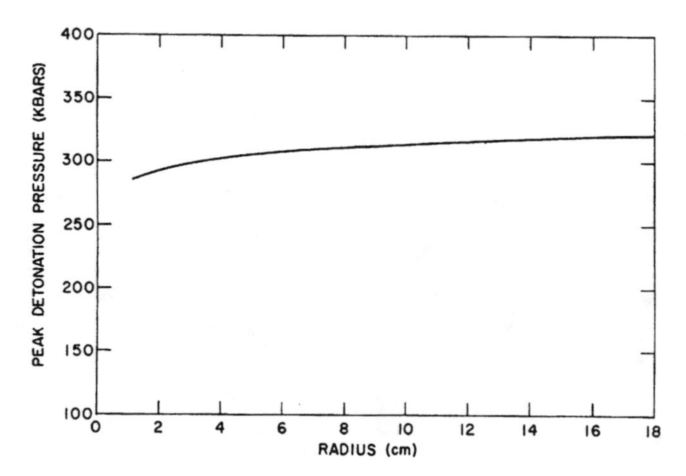

Figure 2.44 Calculated peak detonation pressures of 9404 in spherically diverging geometry using high-density PETN detonator vs. detonation wave front radius.

explosive. If the experimentally observed Taylor waves for the two systems do not give the same maximum density and pressure values, the flow is not Taylor self-similar. Radiographic studies of the density profiles have been performed and the results show that the flow is not Taylor self-similar.

Venable[2] took PHERMEX radiographs of the detonation wave of Composition B-3, with embedded tantalum foils, to determine the particle velocity and density throughout the Taylor wave in plane and spherically diverging geometry. Using a gamma-law or BKW equation of state, excellent agreement is obtained between the calculated and experimental Taylor wave densities for the front quarter of the wave, as Figure 2.45 shows. The experimental slab effective C-J density is 2.4 ± 0.05 g/cc and the experimental diverging peak density is 2.2 ± 0.05 g/cc. This is conclusive evidence that the Taylor self-similar solution is incorrect and that the calculated flow in diverging geometry adequately reproduces the actual flow.

The one-dimensional flow of spherically and cylindrically diverging detonation waves may be closely approximated numerically if it is not forced to be Taylor self-similar. The build-up effect in diverging geometry is masked by the larger divergence effect and higher pressure initiation systems.

Build-Up in Converging Geometry

Converging detonations present no special numerical difficulties. Similar results are achieved for converging detonations using the Arrhenius rate law, the C-J volume burn, or the sharp-shock method of burning the explosives. Figure 2.46 shows the calculated peak detonation pressure as a function of scaled radius for spheres and cylinders of 9404 and nitromethane using either the BKW equation of state or the gamma-law equation of state with the same effective C-J pressure of 365 kbar for 9404 and the BKW equation of state for nitromethane. Figure 2.47 shows the detonation velocity as a function of scaled radius for a sphere of 9404. If build-up did not occur, the above results would describe the convergence effect for any size of explosive sphere, because the results scale. However, because there is build-up, different initial effective C-J pressures must be used, depending upon the initial

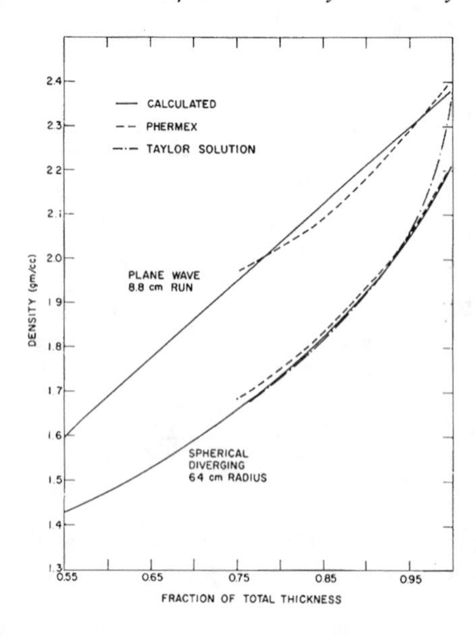

Figure 2.45 Calculated and PHERMEX Taylor wave densities of 1.73 g/cc Composition B-3 explosive.

radius of the sphere and the magnitude of the initiating pulse. The results in Figure 2.46 are about correct for a 10 cm radius sphere of 9404; however, if an exact treatment is needed the build-up equation must be used.

One requirement for using a programmed burn is a description of the change in detonation velocity with convergence. With Figure 2.47 it is possible to describe the velocity for 9404 as a function of convergence for one equation of state. In the search for a more general description, converging detonation calculations were performed for 9404, Composition B, and nitromethane using different effective C-J pressures and both the SIN and RICSHAW computer codes. As shown in Figure 2.48, the results could be scaled as the peak detonation pressure divided by the effective C-J pressure as a function of the scaled radius. Figure 2.49 shows a similarly scaled plot for the detonation velocity.

As a good first approximation, the detonation velocity, D, can be described as a function of the detonation front radius divided by the initial radius, R/R_o, in spherical geometry by

$$\frac{D}{D_{CJ}} = 0.22049\left(\frac{0.2}{R/R_o}\right)^{1.65} + 0.9845\,,$$

$$\text{for } 1.0 > \frac{R}{R_o} > 0.25\,,$$

and in cylindrical geometry by

$$\frac{D}{D_{CJ}} = 0.1255\left(\frac{0.1}{R/R_o}\right)^{1.00} + 0.9875\,,$$

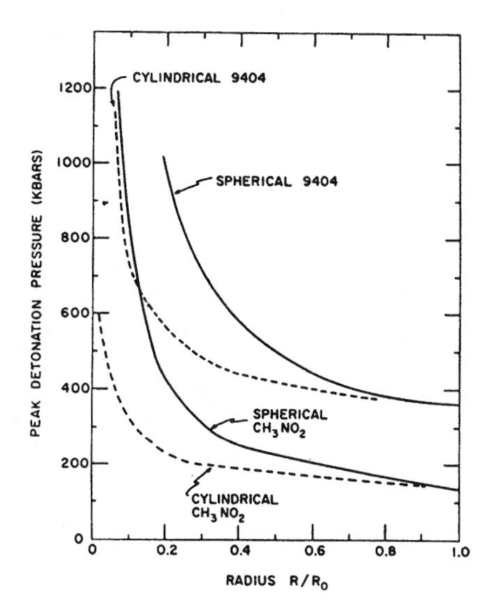

Figure 2.46 Calculated peak detonation pressure vs. scaled radius for spherically and cylindrically converging detonations of 9404 and nitromethane.

$$\text{for } 1.0 > \frac{R}{R_o} > 0.15 .$$

These equations permit converging detonation calculations using the programmed velocity burn down to R/R_o of at least 0.5. They must be used with care because they are approximate, and other equations of state could give different convergence effects.

Morales and Venable[2] have used the PHERMEX facility to collect experimental evidence that the calculated converging detonation wave profiles are realistic. They detonated a 9404 sphere 12.319 cm in outside radius and studied the shock wave formed when the converging detonation wave interacted with a 3.048 cm radius aluminum sphere. Embedded foils permitted determination of the density profile before and after the shock wave converged at the center of the sphere. The explosive was described with an effective plane C-J pressure of 365 kbar. The peak detonation pressure upon arrival of the detonation front at the surface of the aluminum sphere was 820 kbar. The calculated timing of the detonation front arrival agreed with the experimental observations. The 1100 aluminum had an initial density of 2.710 g/cc, and the equation of state used was $U_s = 0.5222 + 1.428U_p$ with a gamma of 1.7. Above 5 Mbar, the Barnes equation of state was used. The coefficients used were A_b = 1.62047, B_r = 2.15566, and B_a = 3.78754. Figure 2.50 shows the calculated and experimental positions of the shock waves and interface, and Figure 2.51 shows the foil position through the shock wave in aluminum at 1.63 μsec after the shock arrived at the explosive-aluminum interface.

Since the convergence and build-up interact, a more realistic treatment is obtained if the gamma is varied through the flow according to the build-up equation. When the build-up equation was used, the initial effective C-J pressure was 303 kbar (γ = 3.70) and the final effective C-J pressure was 373 kbar (γ = 2.83). The peak detonation pressure upon

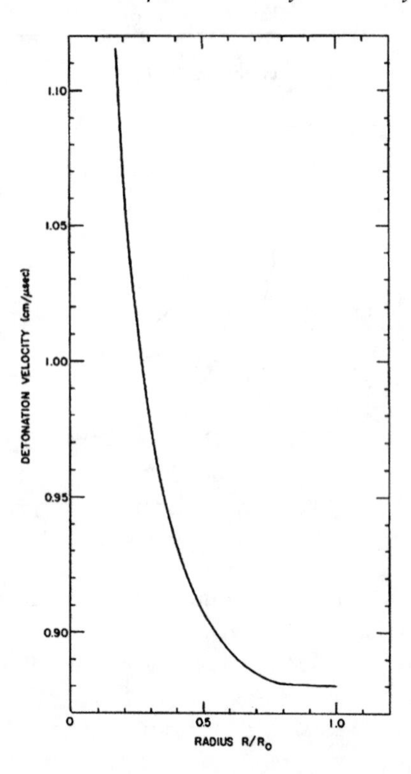

Figure 2.47 Calculated detonation velocities vs. scaled radius for spherically converging 9404 detonation.

arrival of the detonation front at the surface of the aluminum sphere was 740 kbar. The calculated experimental positions of the shock were the same, within experimental error, as in Figures 2.50 and 2.51.

These results indicate that the calculated convergent detonation profiles are realistic, but that the experimental results are not very sensitive to the peak detonation pressure. This is because the convergent detonation Taylor wave steepens near the front while converging toward the center of the sphere.

Chemistry of Build-Up

What is the possible chemistry of the build-up process? One of the most unexpected and puzzling results of recent experimental studies has been the observed constancy of the detonation velocity of an explosive, and the associated large variations of the effective C-J pressures and Taylor waves. The observed lack of appreciable curvature at the front of unconfined and confined explosive charges is further evidence that the detonation velocity cannot be related to the other state parameters in any simple manner. The lack of appreciable front curvature in charges with length-to-diameter ratios of one-half or more is difficult to understand, considering the observed steeper Taylor waves (Figure 2.36) in such charges compared with charges where two-dimensional effects are not present.

Many explosives, which have solid carbon as a detonation product, exhibit behavior that is not described adequately without including some time-dependent phenomenon, such

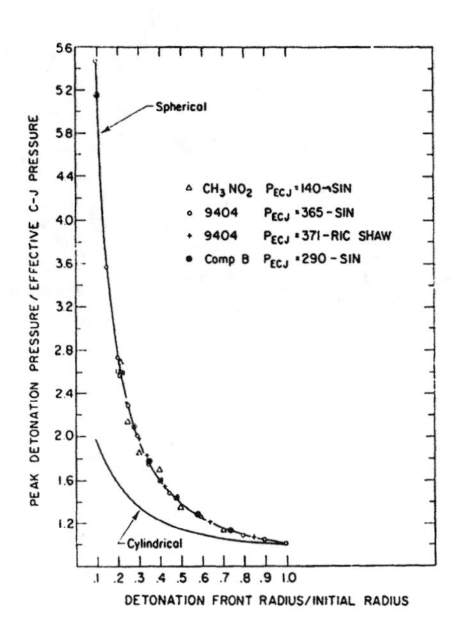

Figure 2.48 Scaled peak detonation pressures vs. scaled radius for spherically and cylindrically converging detonations of 9404, nitromethane, and Composition B.

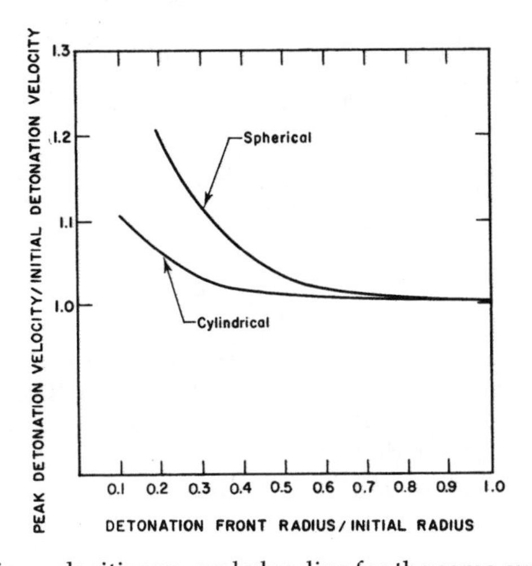

Figure 2.49 Scaled detonation velocities vs. scaled radius for the same explosives as in Figure 2.48.

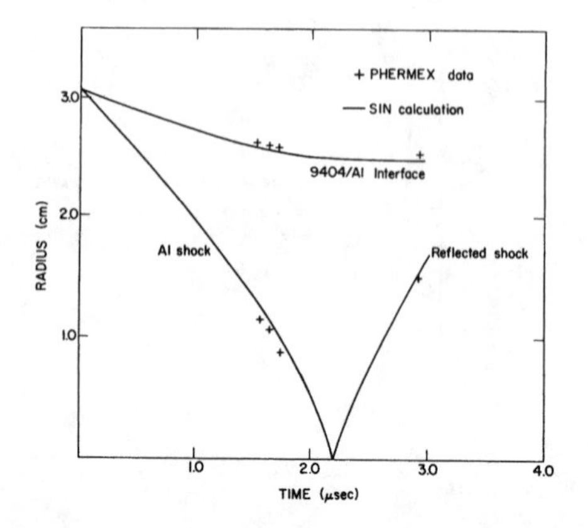

Figure 2.50 Radius vs. time profiles of shock waves and interfaces inside a 3.048 cm radius sphere of aluminum shocked by a surrounding sphere of 9.271-cm-thick 9404. PHERMEX radiographic data are shown at four different times.

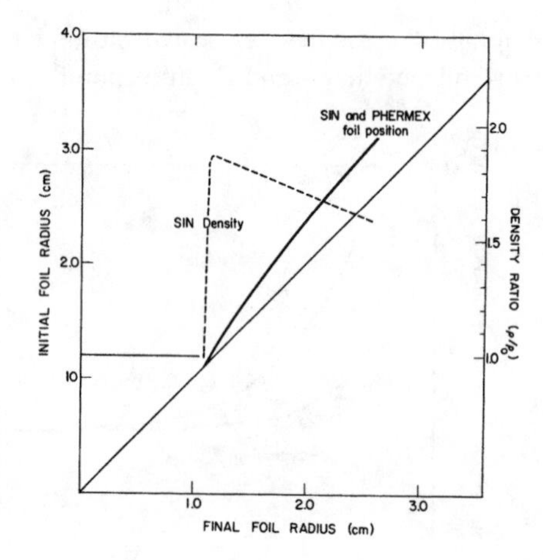

Figure 2.51 Initial and final foil radii and shock wave density for the system described in Figure 2.50 at 1.63 μsec after detonation wave arrived at surface of aluminum sphere.

as diffusion-controlled carbon deposition or some other kinetic behavior of the detonation products. A time-dependent carbon deposition is the only process known that could account for the large energy deficits required by the build-up model. The observed velocity constancy and large C-J pressure variations can be reproduced by the time-dependent carbon deposition mechanism.

As the run distance increases and the Taylor wave becomes less steep, the process of carbon coagulation ($C_{gas} \rightarrow C_2 \rightarrow C_5 \rightarrow C_n \rightarrow C_{graphite}$) has more time to proceed before the product expansion decreases the carbon species interactions. The increasing energy from near the C-J plane increases the C-J pressures and densities. It is not obvious what the detonation velocity would do, but it is reasonable to expect it would increase along with the other state parameters. To determine the effect on C-J state values, calculations were performed using the BKW equation of state and holding the amount of carbon and other detonation products fixed at the usual equilibrium value but forcing part of the carbon to remain as C, C_2, or C_5.

Table 2.11 shows the C-J state parameters calculated using BKW and changing the $C_{gas}/C_{graphite}$ ratio of carbon in the reaction of 9404 at 1.844 g/cc of

$$C_{4.4236}H_{8.6597}N_{8.075}O_{8.467}Cl_{0.0993}P_{0.0331}$$

$$\rightarrow 4.3298H_2O + 2.07CO_2 + 4.0375N_2 + 0.033POCl_3 + 2.35C\ .$$

The velocity is found to be nearly constant while the pressure is decreasing by 20% if C, C_2, C_5 gas is formed instead of graphite.

Since BKW is calibrated for explosive charges about 10 cm long, to obtain the 390 kbar infinite-medium C-J pressure for 9404, the constants were recalibrated. The new set of constants, called the "Infinite Geometry BKW Parameter Set," are alpha = 0.71, beta = 0.05, kappa = 92, and theta = 400. The calculated C-J state parameters for 9404 with various carbon-gas/graphite ratios are shown in Table 2.12. The calculated pressures vary from 390 to 320 kbar while the detonation velocity changes by less than 3%. The calculated temperatures are too low and this set of BKW parameters should be used with caution. The calculated mass fraction of carbon gas and distance of run as a function of pressure are shown in Figure 2.52. Also shown is the amount of carbon present as graphite as a function of pressure from Table 2.12.

Since little build-up has been observed in PETN, similar calculations were performed for PETN at 1.77 g/cc. The calculated pressures and velocities are changed only slightly by changing the carbon-gas/graphite reaction.

The constant detonation product composition used for TNT at 1.64 g/cc was

$$C_7H_5N_3O_6 \rightarrow 2.5H_2O + 1.75CO_2 + 1.5N_2 + 5.25C\ .$$

TABLE 2.11 9404 BKW Fixed Composition

Moles	D(m/sec)	P(kbar)	T(^oK)	γ	Q(kcal/g)
$C_{graphite}/C_{gas}$					
2.35/0.0	8897	363	2468	3.02	1.03
2.00/0.35	8921	357	2017	3.10	0.83
1.70/0.65	8940	351	1660	3.19	0.65
1.40/0.95	8953	342	1328	3.32	0.48
1.15/1.20	8951	332	1075	3.45	0.34
0.95/1.40	8891	322	889	3.53	0.23
0.85/1.50	8917	315	804	3.64	0.17
0.60/1.75	8841	297	613	3.85	0.025
$C_{graphite}/C_{2\ gas}$					
2.35/0.0	8897	363	2468	3.02	1.03
1.85/0.25	8902	356	2032	3.10	0.84
1.35/0.50	8909	347	1626	3.22	0.51
0.85/0.75	8911	334	1249	3.38	0.47
0.35/1.00	8891	318	912	3.58	0.29
0.15/1.10	8869	309	793	3.68	0.21
$C_{graphite}/C_{5\ gas}$					
2.35/0.0	8897	363	2468	3.02	1.03
1.50/0.17	8855	354	2165	3.08	0.85
1.00/0.27	8831	349	1995	3.12	0.82
0.50/0.37	8808	342	1828	3.18	0.75
0.00/0.47	8785	336	1667	3.23	0.68

With all the carbon present in the TNT detonation products, it is reasonable to expect a larger build-up for TNT than 9404. However, if only 1.70 moles of the 5.25 moles of carbon are gaseous, the heat of detonation becomes positive. The TNT pressures and temperatures drop rapidly and the velocity changes slowly with increasing carbon gas.

Since TATB has the same elemental composition as TNT, one would expect it to exhibit similar build-up behavior. As described previously, PBX-9502 which contains 95% TATB does not exhibit build-up. The density and therefore the C-J pressure of TATB is much higher than TNT so perhaps the carbon condensation rate is increased sufficiently that graphitic carbon forms too rapidly for build-up to occur.

The carbon coagulation mechanism is compatible with the observed large changes in effective C-J pressures associated with small changes in detonation velocity characteristic of build-up to detonation; however, some explosives with large amounts of solid carbon in their detonation products do not exhibit build-up of detonation.

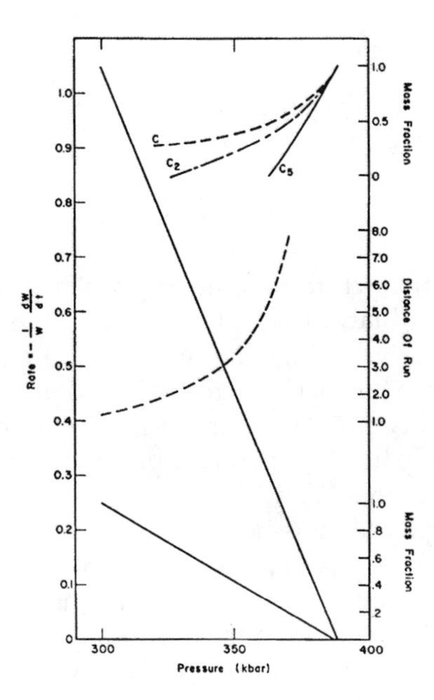

Figure 2.52 Build-up of 9404 detonation from 300 to 388 kbar as a function of pressure. The percent of carbon present as graphite for various carbon-gas/graphite compositions from Table 2.12 is shown as a function of pressure.

TABLE 2.12 9404 Infinite Geometry BKW Parameters

Moles	D(m/sec)	P(kbar)	$T(^oK)$	γ
$C_{graphite}/C_{gas}$				
2.35/0.0	8878	389	1415	2.73
2.00/0.35	8977	385	1065	2.86
1.40/0.95	9113	370	631	3.14
1.15/1.20	9131	359	509	3.28
0.85/1.50	9099	341	392	3.48
0.60/1.75	9013	320	315	3.68
$C_{graphite}/C_{2\ gas}$				
2.35/0.0	8878	389	1415	2.73
1.85/0.25	8953	382	1077	2.86
0.85/0.75	9077	361	592	3.21
0.35/1.00	9070	343	439	3.42
0.00/1.175	9013	326	355	3.59
$C_{graphite}/C_{5\ gas}$				
2.35/0.0	8878	389	1415	2.73
1.50/0.17	8887	380	1176	2.83
1.00/0.27	8895	374	1049	2.89
0.50/0.37	8902	369	933	2.96
0.00/0.47	8908	363	827	3.03

2.4 Nitrogen Oxide

Nitrogen Oxide has a very simple elemental and initial molecular constitution. Its almost irreversible decomposition reaction

$$2\,NO \rightarrow O_2 + N_2$$

generates a mixture of homonuclear diatomic molecular products that are miscible with each other and also with residual Nitrogen Oxide.

Schott[48] has measured the shock Hugoniots of liquid Nitrogen Oxide, Nitrogen, Oxygen, and synthetic equal molar mixture of Oxygen and Nitrogen dissolved in each other as a homogeneous liquid phase. This equivalent composition represents complete decomposition of Nitrogen Oxide to its main high temperature products. The details of the experimental study are described in References 48 and 49.

Measurements of the separate first-shock Hugoniots obtainable from liquid Nitrogen Oxide and from its elemental equivalent an equal molar Oxygen and Nitrogen mixture, demonstrated uniqueness, within experimental tolerances, of the product state at their common point. Reaching this point from the chemically "opposite" starting conditions of "pure reactant" and "pure product" satisfied the fundamentally necessary condition for the shock-state data to be representative of a single equation of state. The shock Hugoniots in the pressure-volume or pressure-energy planes cross each other near 205 kbar and demonstrate attainment of the same equilibrium states from Nitrogen Oxide in overdriven detonations as from equal molar Oxygen and Nitrogen mixtures in endothermically reacting shocks. This is the most positive test of attainment of chemical equilibrium yet accomplished in a condensed-phase substance behind a shock wave.

Liquid Nitrogen Oxide was first studied by Ramsay and Chiles.[19] It was observed to detonate with a detonation pressure of approximately 100 kbar and a velocity of 0.562 cm/μsec. The BKW calculated C-J parameters[50] given in Table 2.1 are 106 kbar, 0.5607 cm/μsec, and 1850 K. The calculated detonation products are Nitrogen and Oxygen with 0.003 moles of Nitrogen Oxide gas and 0.00001 moles of Nitrogen Dioxide gas.

The experimental Hugoniot values and the BKW calculated Hugoniots are shown in Figure 2.53 for Nitrogen at 0.77 g/cc, equal molar Nitrogen-Oxygen mixtures at 0.947 g/cc, Oxygen at 1.202 g/cc, and Nitrogen Oxide at 1.26 g/cc. The experimental Hugoniots are well described by the BKW equation of state.

The experimental Hugoniot and reflected shock states off Aluminum for Nitrogen and the BKW calculated Hugoniot and reflected curves are shown in Figure 2.54.

The experimental Hugoniot and reflected shock states off Aluminum for equal molar mixtures of Nitrogen and Oxygen at 0.947 g/cc and the BKW calculated Hugoniot and reflected curves are shown in Figure 2.55.

The experimental Hugoniot and reflected shock states off Magnesium, Aluminum, and Copper for Nitrogen Oxide at 1.26 g/cc and the BKW calculated Hugoniot and reflected curves are shown in Figure 2.56

The experimental data for single and multiple shocked Nitrogen, Nitrogen plus Oxygen, and Nitrogen Oxide are in agreement with those predicted by the BKW equation of state.

Schott also investigated the shock Hugoniot of Carbon Dioxide.[51] Shocked states of Carbon Dioxide at pressures from 100 to 300 kbar were measured using high explosives in systems with cryogenically liquefied Carbon Dioxide confined between parallel metal layers. The Hugoniot states of shocked liquid Carbon Dioxide are shown in Figure 2.57 along

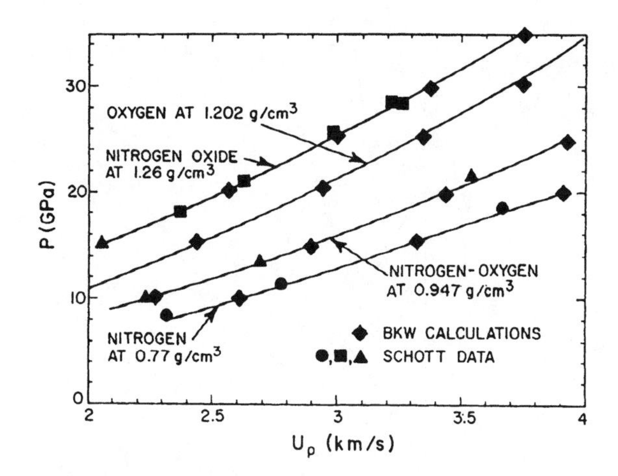

Figure 2.53 The experimental Hugoniot values and the BKW calculated Hugoniots for Nitrogen at 0.77 g/cc, equal molar Nitrogen-Oxygen mixtures at 0.947 g/cc, Oxygen at 1.202 g/cc, and Nitrogen Oxide at 1.26 g/cc.

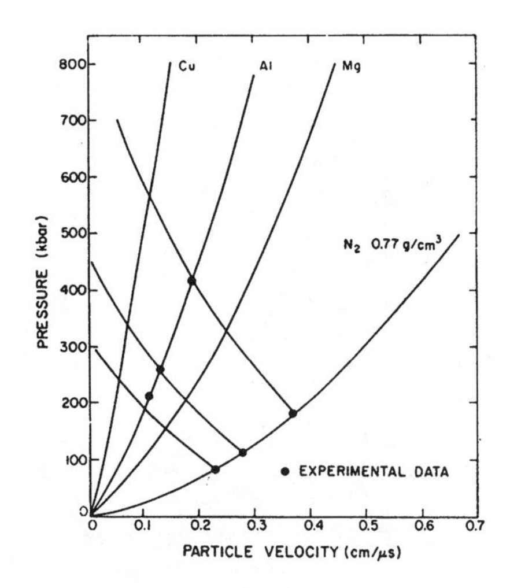

Figure 2.54 The experimental Hugoniot and reflected shock states off Aluminum for Nitrogen and the BKW calculated Hugoniot and reflected curves.

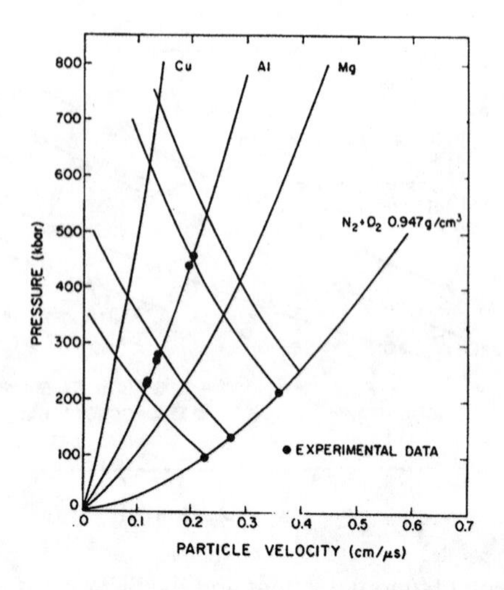

Figure 2.55 The experimental Hugoniot and reflected shock states off Aluminum for equal molar mixtures of Nitrogen and Oxygen at 0.947 g/cc and the BKW calculated Hugoniot and reflected curves.

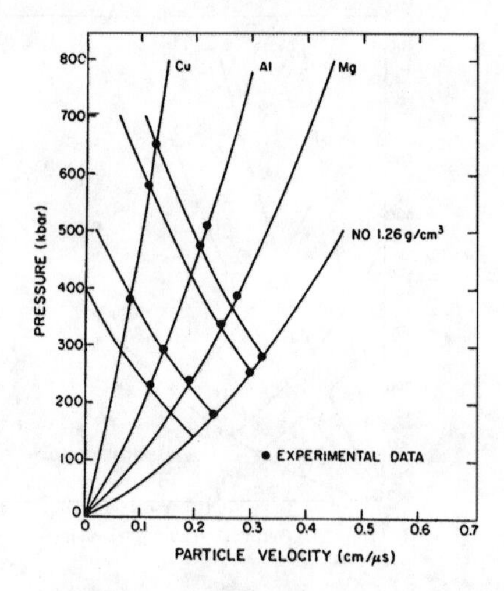

Figure 2.56 The experimental Hugoniot and reflected shock states off Magnesium, Aluminum, and Copper for Nitrogen Oxide at 1.26 g/cc and the BKW calculated Hugoniot and reflected curves.

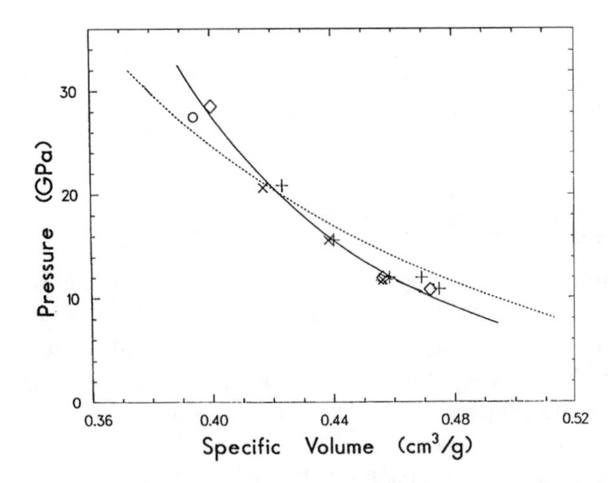

Figure 2.57 Hugoniot states of liquid Carbon Dioxide. The solid line is the ESP equation of state. The dashed line is the BKW equation of state Hugoniot using the RDX parameter set.

with the BKW calculated Hugoniot using the RDX parameters and Carbon Dioxide covolume of 600 and the ESP Hugoniot of Shaw.[52]

Nellis, Mitchell, Ree, Ross, Holmes, Trainor, and Erskine[53] at Lawrence Livermore National Laboratory, have measured the equation of state of shocked liquid Carbon Dioxide in the pressure range of 280 to 710 kbar. The shock Hugoniot data show an unexpected slope change at 400 kbar. Neither the ESP or BKW equation of state for Carbon Dioxide exhibits such a slope change. The slope change is probably an experimental artifact of having only four data points.

The four measured Hugoniot states and the calculated BKW Hugoniot states are listed below.

Source	Pressure kbar	Volume cc/gm	Shock Vel. cm/μs	Particle Vel. cm/μs	Temp. K	CO_2 Mole Fraction
BKW	700.	0.2995	0.959	0.622	6732	0.934
Exp	708.	0.30	0.965	0.626		
BKW	550.	0.3223	0.868	0.540	5714	0.96
Exp	545.	0.322	0.865	0.538		
BKW	425.	0.345	0.781	0.464	4057	0.993
Exp	422.	0.363	0.791	0.455		
BKW	275.	0.388	0.656	0.352	2022	0.999
Exp	275.	0.395	0.661	0.356		

The BKW equation of state for Carbon Dioxide gives a good fit to the Hugoniot data for solid and liquid Carbon Dioxide. This is in contrast to the ESP equation of state which is harder than the Nellis high pressure data points. The 420 kbar Nellis point is in poorest agreement with the BKW Hugoniot. The sharp slope change is probably a result of fitting two straight lines to data that are smoothly changing in slope and weighting the 420 kbar data point too highly.

2.5 Carbon Condensation

The major uncertainty in the description of the detonation process is the carbon condensation that occurs in explosives with excess carbon.

The classic example of observed explosive properties that cannot be described by the usual steady-state, chemical equilibrium models with any equation of state was discussed at the beginning of this chapter. The detonation velocity of TNT as a function of density reported by Urizar, James, and Smith[1] has a sharp change of slope from 3163 to 1700 m/sec/g/cc at 1.55 g/cc. The slope change between 1.55 and 1.64 g/cc, requires a change in condensing carbon mechanism with an energy of +6 to 10 kcalories/mole of carbon. Another possible explanation for the break is that at 1.55 g/cc all the carbon is solid graphite, and at 1.64 g/cc, 0.25 moles of the carbon remains gaseous.

Real explosives, which have solid carbon as a detonation product, exhibit behavior that is not described adequately without including some time-dependent phenomenon, such as diffusion-controlled carbon deposition or some other kinetic behavior of the detonation products. A time-dependent carbon deposition is the only process known that could account for the large energy deficits required by the build-up process.

The TNT slope change in detonation velocity has also been reported by Pershin.[54] This important paper proposes that the slope change in TNT detonation velocity as a function of density is a result of carbon being graphite at 1.55 g/cc and being diamond at 1.64 g/cc. He proposes that multiple slope changes must be occurring for all explosives with large amounts of carbon in the detonation products at around 200 kbar pressure. He measured the detonation velocity as a function of density for 75/25 and 50/50 TNT/RDX mixtures. He found two slope changes at densities between 1.5 and 1.7 g/cc for these explosives.

Considerable discussion continues in the detonation community regarding whether one should treat the carbon as graphite or diamond. For many years it was thought that, since the carbon equation of state was fit to shock Hugoniot data, it described carbon (whether graphite, diamond, or something else) as well as possible in the detonation products. It is now clear that the shocking of a piece of graphite into the diamond phase is quite a different process than the condensing of carbon atoms into small pieces of graphite or diamond during the formation of detonation products. Apparently, the transition occurs much more quickly and in a different and smaller parameter space in detonation products.

While the observation of multiple slope changes is a fact, what about the proposal that the changes are due to a graphite-diamond effect? To test this proposal, BKW calculations were performed with a diamond equation and a pure graphite equation of state using RDX parameters for the detonation products. The carbon equation of state pressure as a function of specific volume is shown in Figure 2.58. Each model for carbon gave a detonation velocity as a function of density slope of 3000-3200 meters/sec/g/cc. The two curves and experimental points are shown in Figure 2.59.

The diamond form for the carbon gave results that reproduced the observed C-J performance of 1.64 g/cc TNT and the graphite form for carbon gave results that reproduced the observed C-J performance of 1.55 g/cc TNT as shown in the following table.

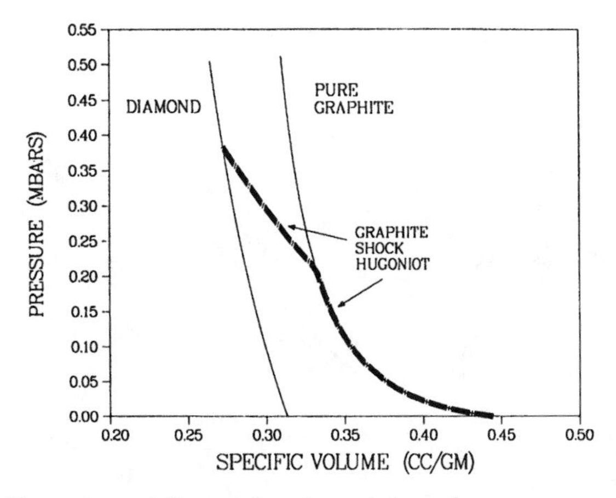

Figure 2.58 The shock Hugoniots of diamond and graphite carbon.

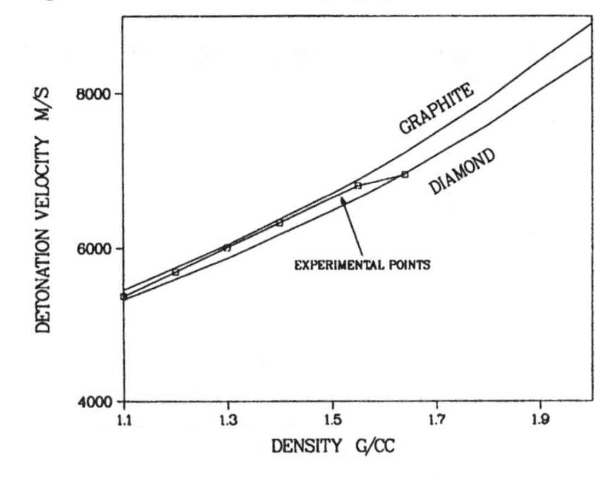

Figure 2.59 The detonation velocity of TNT as a function of density with the carbon detonation product treated as diamond and as graphite. The BKW equation of state RDX parameters were used.

TNT	1.55 g/cc			1.64 g/cc		
	Graphite	Diamond	Exp	Graphite	Diamond	Exp
Detonation Vel(m/s)	6888	6650	6812	7237	6969	6950
C-J Pressure(kbar)	180	170		205	193	190

The calculations support the proposal that the carbon is mostly graphite at 1.55 g/cc and mostly diamond at 1.64 g/cc. This surprising result also supports the proposal that all explosives producing large amounts of solid carbon may exhibit similar slope changes in their detonation velocity-density curves. It also suggests that the carbon present in high density RDX is similar to diamond. As shown in the following table, the BKW calculations for RDX with the diamond equation of state gave about the same C-J states as with the

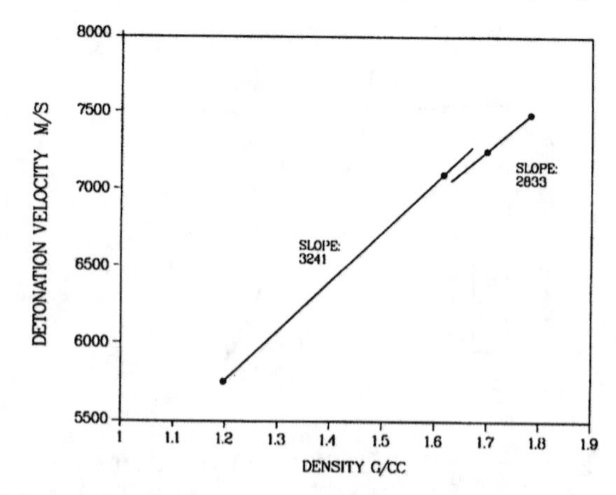

Figure 2.60 The experimental detonation velocity of DATB as a function of density.

"standard" Cowan carbon equation of state that was fit to the graphite shock Hugoniot data. This is not surprising since at high pressure, the graphite shock Hugoniot data overlap the diamond Hugoniot data as shown in Figure 2.58.

RDX	1.80 g/cc Graphite	Diamond	Exp	1.0 g/cc Graphite	Diamond	Exp
Detonation Vel(m/s)	8754	8740	8754	6128	6165	5981
C-J Pressure(kbar)	347	337	347	108	110	

The detonation velocity of DATB (diamino trinitrobenzene) as a function of density is shown in Figure 2.60. The data can be fit with two lines whose slopes are 3241 and 2833 m/sec/g/cc. Craig[55] performed eight detonation velocity shots at four different densities. Only three of these data points are reported in the Los Alamos Explosive Performance data volume.[39] Unfortunately, sometime after 1964 it was decided (not by Craig) that the data that were not on a straight line were in error and should not be included in the data volume. The data point missing from the data volume is the 1.616 g/cc density point which has a velocity of 0.7108 cm/μsec. Craig concluded after reporting his DATB data that "The present evidence supports the hypothesis that a bend in the density vs. detonation velocity curve is characteristic of carbon rich explosives. It is proposed that the investigation be extended to obtain sufficient data for publication in the open literature." It is unfortunate that the extension to the investigation did not occur. It would be valuable to have more DATB data, as it is difficult to establish a break with only four points.

The BKW calculations using RDX parameters for DATB with the carbon treated as graphite and as diamond are shown in Figure 2.61. The velocity difference between the calculated and experimental curves is similar, but the calculated and experimental velocities are quite different at higher densities. This is in marked contrast to the good agreement obtained for TNT.

The BKW calculations using RDX parameters for TATB with the carbon treated as graphite and as diamond are shown in Figure 2.62. The high density experimental TATB detonation

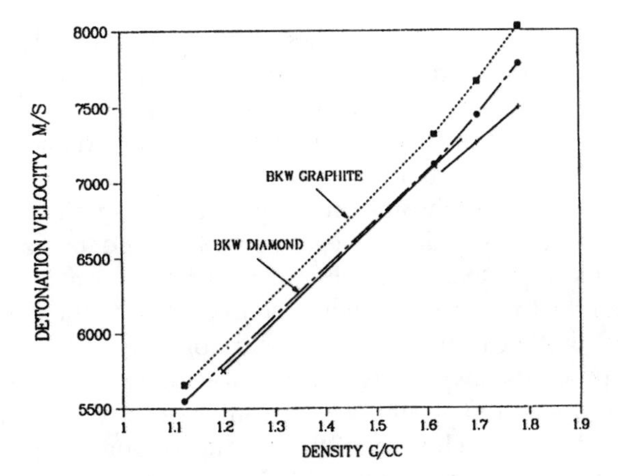

Figure 2.61 The detonation velocity of DATB as a function of density with the carbon detonation product treated as diamond and as graphite. RDX parameters were used for the BKW calculations.

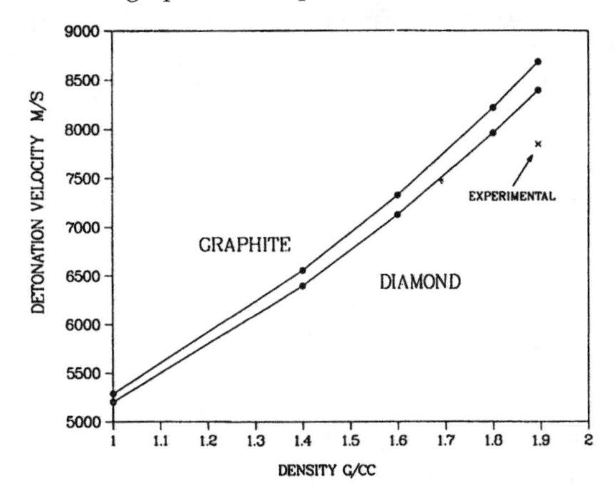

Figure 2.62 The detonation velocity of TATB as a function of density with the carbon detonation product treated as diamond and as graphite. RDX parameters were used for the BKW calculations.

velocity is not well described by the calculations with the carbon treated as graphite or diamond. More detonation velocity-density data are needed for TATB to determine if a break occurs. Even though TATB has a considerable amount of carbon in its detonation products, it does not exhibit build-up behavior characteristic of other carbon containing explosives such as HMX, RDX/TNT, and TNT as described earlier in this chapter.

It was not possible to reproduce the detonation performance of RDX and TNT without two sets of BKW parameters – one for explosives like RDX with small amounts of carbon and the second one for explosives with large amounts of carbon like TNT. Another method would be to use the BKW RDX parameters for all explosives and change the form of the carbon equation of state (diamond, graphite, or something else) according to the pressure, explosive density, or some other set of conditions. Since high density TATB and DATB are not well described using RDX parameters and the diamond equation of state for carbon, the correct method of describing the carbon is unknown.

Unfortunately, our detonation data base for evaluating the presence of breaks in the velocity-density plane is sparse. Most studies of detonation velocity as a function of density in the 1960-1990 time frame were concentrated on the "end points" to get the best straight line fit. This design resulted in our missing the most important feature that the data could exhibit – that of deviation from straight line behavior. Such data would be very useful in evaluating models for graphite-diamond transitions and in helping to explain the build-up behavior of explosives. Hopefully other investigators will examine their original detonation velocity as a function of density data base to see if valuable data were not reported because they did not fit on the expected straight line as was the case for DATB.

Since the effect of the form of the carbon is so important to the performance of an explosive, it is important that we examine the evidence for diamond and other forms of carbon in the detonation products. Experiments where the detonation products are recovered at ambient conditions do not represent the composition of the detonation products at the C-J state and do not furnish us the history of the products along the expansion isentrope. The recovered products will have had a complicated history of expansion, being reshocked after interaction with the container walls and associated non-equilibrium composition changes. If diamond is seen, the diamond phase can probably be associated with the higher pressures near the C-J state. Graphite can be formed near the C-J state and during the release process.

Russian scientists have extensively investigated the carbon detonation products from explosives, as described in References 56 thru 60. They exploded mixtures of TNT and RDX in explosive chambers with inert gas. They collected the solid products, which were 8 to 9 percent of the initial mass of the explosive. They have been able to obtain up to 80% of the solid product as diamond powder with an average particle diameter of 4 nanometers. Larger diamond particles were obtained which were made up of "welded" 4 nanometer particles. From electrical conductivity measurements they found that the carbon production was completed in 0.2 to 0.5 microseconds.[56] They studied the formation of carbon using isotopic carbon methods[59] and found that the condensed carbon isotopic ratio was the same as in the initial explosive. They reported producing about a ton of industrial diamonds a year from explosives.

Diamonds have also been found by U.S. and German scientists. Greiner, Phillips, Johnson, and Volk[61] used the Franhofer Institute 1500 liter explosive chamber filled with argon to detonate explosive compositions of 40/60 TNT/RDX, 50/50 TNT/TATB, and 50/50 TNT/Nitroguanidine. The solid detonation products collected had two distinct powder forms. One form was compact spheroids of about 7 nanometer diameter identified as diamonds and the other form was curved ribbons of about 4 nanometers in thickness, identified as graphite. The structures were identified using convergent beam electron diffraction. The solid products collected were about 25% diamond.

The clustering of carbon in the detonation regime has been studied theoretically by Shaw and Johnson.[62,63] They assume a diffusion-limited clustering process. With the size dependence of the cluster energy treated with a surface term, for any given time it takes 1000 times as long to release the next 90% of the carbon energy. This leads to a very slow condensation rate which can couple to the reaction zone to produce nonideal time-dependent detonations on a scale of microseconds and centimeters. Most of the energy release occurs before cluster sizes reach 10,000 atoms. Using transmission electron microscopy, the diamond cluster sizes observed were compact spheres of about 10,000 atoms.[61] The clusters collide and stick with an annealing process driven by the hot high pressure gases and the energy released by the formation of bonds between the initial clusters.

Which combination of mechanisms causes the termination of annealing is not known, but all have the same dependence on the fraction of surface atoms. Some suggested mechanisms are: 1. the heat from bond formation becomes too small a fraction of the total energy of a cluster; 2. the effective melting temperature for a finite cluster shifts with cluster size; 3. the kernel in the coagulation equations changes character as a function of cluster size; and 4. the thermal fluctuations in a finite cluster are smaller on a percentage basis in larger clusters than in smaller clusters.

The morphology of carbon is of interest because as we have shown previously the modeling of carbon in the detonation environment with bulk equations of state – either graphite or diamond – is not adequate to describe the observed detonation behavior. The carbon found in the detonation products is not bulk carbon, but made up of clusters of either diamond or graphite form, and the cluster sizes are small enough to have significant surface contributions. At early times during cluster growth, the surface contributions must be even larger. The diamond form found in detonation products differs from diamond produced by shocking graphite, which is dirty but bulk polycrystalline diamond. This confirms our previous conclusion that one cannot use the equation of state of shocked bulk graphite to describe the carbon formation in detonation products.

Chirat and Baute[64,66] have found it necessary to assume either graphite or diamond for the solid carbon phase in their application of WCA4 equation of state to explosives. In their calculations they have found that carbon in the detonation products can be diamond as predicted by its phase diagram, or in some graphite-like phase even if its phase diagram would imply diamond formation.[65,66]

An attempt to include the observed carbon cluster structure for carbon into the detonation equation of state has been made by Van Thiel and Ree.[67] They replaced their diamond equation of state with an estimated diamond cluster treatment and found a larger compressibility, heat of formation, and initial volume for the cluster compared to diamond. They concluded that a more accurate description of the surface layer will be required.

The major problem of modeling the formation of carbon in the detonation products remains unsolved. Until a realistic model for the carbon formation is developed, the improvement of our detonation performance models and our understanding of the nature of the time-dependent features of detonations will remain at its present unsatisfactory state.

Since the experimental evidence suggests that the carbon formation may take 0.5 microseconds[56] and the best carbon clustering models suggest that carbon clustering occurs for centimeters and microseconds behind the detonation front,[62,63] it seems unlikely that any equation of state treatment that assumes that the detonation process is a steady-state, equilibrium process regardless of its carbon equation of state treatment will describe in detail the detonation performance of explosives with excess carbon. What will be required is treating the explosive as a non-steady-state, non-equilibrium process. A combination of a reactive hydrodynamic code with complete mixture equations of state and associated rates for the explosive decomposing to gaseous detonation products and for carbon clustering will be the minimum treatment required.

The reaction zone of heterogeneous explosives has been shown to occupy a region of varying dimensions and pressures, since the density discontinuities in heterogeneous explosives are of the same order of size as the reaction zone in the homogeneous explosive as described in Chapter 1. Heterogeneity and the associated pressure variations in the reaction regions will have a significant effect on the rates of carbon condensation. This effect will also need to be included in any realistic description of the carbon clustering.

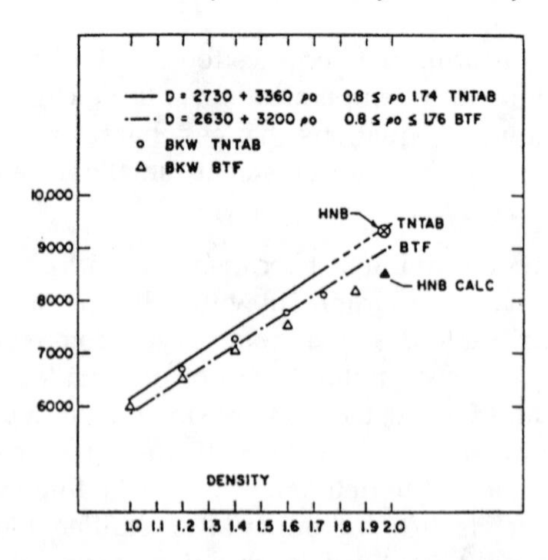

Figure 2.63 The measured and BKW calculated detonation velocity as a function of density for the CNO explosives TNATB, BTF, and Hexanitro benzene.

2.6 CNO Explosives

For CNO systems there are three groups of data. The solid and the liquid Carbon Dioxide Hugoniot data and the detonation velocity data for high density CNO (Carbon-Nitrogen-Oxygen) explosives, and, in particular, Hexanitro benzene. BKW reproduces the solid and the liquid Carbon Dioxide Hugoniot data, but gives velocities too low for high density CNO explosives. ESP reproduces the solid Carbon Dioxide Hugoniot data and the high density Hexanitro benzene explosive detonation velocity, but is too hard to reproduce the higher pressure liquid Carbon Dioxide Hugoniot data. ESP gives over 5000 for the detonation velocity-density slope for Hexanitro benzene while BKW gives about 3200. The detonation velocity-density data fits for other CNO explosives such as TNTAB (1,3,5-Triazido-2,4,6-trinitrobenzene) and BTF (1,2,5-oxadiazole-1-oxide) which do not have as high a crystal density as Hexanitro benzene (1.973) are listed below.

TNTAB	0.8 to 1.74 g/cc	$D = 2730 + 3360\ \rho_0$
BTF	0.8 to 1.76 g/cc	$D = 2630 + 3200\ \rho_0$

These are plotted along with the calculated BKW detonation velocities in Figure 2.63. It is important to note that the Hexanitro benzene detonation velocity is on the extrapolated TNTAB curve. It is difficult to believe that the detonation velocity of Hexanitro benzene could change as quickly with density as suggested by the ESP calculations.

To obtain the observed detonation velocities for high density CNO explosives using BKW, it was necessary to increase the Carbon Dioxide covolume from 600 to 680. This results in calculated solid and liquid Carbon Dioxide Hugoniots that are too hard and calculated detonation velocity-density slopes that are too steep.

Mixtures of Carbon Dioxide and Nitrogen at high densities probably do not mix as modeled in ESP or BKW and additive mixing is unrealistic for such mixtures. Russell Pack[68] offered the following insights. His ab-inito calculations show that Carbon Dioxide prefers (is less repulsive) to line up in a parallel arrangement rather than a random or end-on arrangement. The least repulsive arrangement is

$$O - C - O \qquad O - C - O \qquad O - C - O$$

$$O - C - O \qquad O - C - O$$

On the other hand, Nitrogen prefers to line up in a diagonal cross-like manner.

Pack expects that Nitrogen interferes with the stacking of Carbon Dioxide and makes it less compressible. Thus, spherical potentials for Carbon Dioxide are too repulsive at high pressures and ideal mixing is incorrect for mixtures of Carbon Dioxide and Nitrogen. Spherical potentials are more realistic for Water than Carbon Dioxide because the Water molecule under compression resembles a sphere with small Hydrogen bumps. Mixtures of Water, Nitrogen, and Carbon Dioxide would mix more ideally than mixtures of Nitrogen and Carbon Dioxide alone. Ab-inito quantum chemical calculations of intermolecular potentials of mixtures of Carbon Dioxide and Nitrogen and Carbon Dioxide, Nitrogen and Water in the 100 to 500 kilobar and 2000 to 3000 K range offer the promise of determining the source of the CNO problem.

2.7 *Density*

Density is the primary physical parameter in detonation performance and shock sensitivity. To obtain the highest C-J performance explosive, one wishes first to maximize the C-J particle density or number of moles of gas per gram of explosive and then to maximize the heat of detonation. The most productive method of maximizing the C-J particle density is to start with as high an initial density explosive as possible. The performances of HMX and RDX at the same density are essentially identical. Because HMX has an initial density of 1.9 g/cc and RDX has an initial density of 1.8 g/cc, HMX explosives typically have C-J pressures 10% higher than RDX explosives. The performance of TATB is slightly less than that of TNT at the same density. The crystal density of TATB is 1.9 g/cc and of TNT is 1.64 g/cc, which makes TATB the explosive of choice for shock insensitive explosives as its C-J pressure is more than 30% greater than that of TNT in typical explosive applications. The search for explosives with higher C-J performance becomes a search primarily for explosives with higher initial densities. Packing more atoms into smaller spaces occurs in cubane-type structures. These materials are being experimentally and theoretically studied.

As the density and C-J pressure and detonation velocity increase, the intermolecular potential energy increases at the expense of the thermal energy since the total available energy per unit of mass remains constant. Increasing the density results in decreasing the C-J temperature since the available thermal energy determines the temperature. In principal, it should be possible to increase the density of an explosive enough that the temperature decreases to below a temperature where chemical decomposition of the explosive can occur on the time scale of a propagating detonation. An explosive that could exhibit such a property is Krypton Floride.

Solid Krypton Floride decomposes to Krypton and Fluorine gas and the only energy available is from its positive heat of formation. This is similar to a Nitrogen Oxide detonation. The heat of explosion is less than 0.1 kilocalorie/gram.

The BKW calculations were performed with Krypton gas entropy fits obtained using the TDF code described in Appendix F and data from NBS Circular 467. The Krypton term value is 112.04 with a weight of 4 and an initial weight of 2. The Krypton Floride heat of formation from its elements at zero K is 8.114 kcalories/mole. The covolume used for Krypton was 400. It was estimated using the corresponding states model and critical constants of Argon and Krypton as described in Reference 52 as were the Krypton potential values for the ESP equation of state calculations.[49]

The BKW and ESP calculated detonation states for Krypton Floride at various densities are given in the following tables:

KRYPTON FLORIDE
BKW CALCULATED C-J STATES

Density (g/cc)	Press (kbar)	Velocity (m/s)	Temperature (K)
1.0	9.	1800	941
1.5	19.	2200	764
2.0	32.	2700	588
2.5	51.	3200	438
3.0	76.	3700	325

KRYPTON FLORIDE
ESP CALCULATED C-J STATES

Density (g/cc)	Press (kbar)	Velocity (m/s)	Temperature (K)
1.0	4.3	1270	1138
1.5	9.9	1690	1164
2.0	20.5	2220	1148
2.5	38.9	2820	1072
3.0	68.0	3470	917
3.24(crystal)	86.4	3800	811
3.5	109.9	4150	674
4.0	165.0	4800	354

The ESP equation of state predicts higher temperatures and a slower decrease in temperature with increasing density than the BKW equation of state. The partition of intermolecular repulsive and thermal energy is also significantly different between the two equations of state.

2.8 Propellant Performance

As shown in the section on nonideal explosives, propellants containing HMX and ammonium perchlorate can exhibit nearly ideal detonation performance with small amounts of ammonium perchlorate. With increasing percentages of ammonium perchlorate, the propellants detonate at detonation velocities and C-J pressures which are less than the ideal explosive performance. As the HMX concentration is decreased and the ammonium perchlorate concentration is increased, the propellants will fail to support propagating detonation.

The BKW equation of state describes gases from highly compressed to the ideal gas state. Since the BKW equation of state describes the equation of state of mixtures of gases and solid decomposition products whether generated by a detonation or by burning, it is useful for evaluating the propellant performance in addition to the explosive performance of energetic materials.

The version of BKW that calculates the propellant performance is called ISPBKW and the executable code is included on the CD-ROM. Calculations can be performed for both the performance of an energetic material as an explosive and as a propellant using the same input parameters for the composition, density, and heat of formation of the energetic material and the same equation of state and thermodynamic properties for the decomposition products. The nonideal gas behavior of the mixture of products is included in the modeling of the propellant performance as well as in the modeling of the explosive performance. While degree of the nonideal gas behavior of propellants is small, the uncertainty regarding the effect is removed when using ISPBKW. The solid products from burning propellants have had various treatments and are one of the largest uncertainties in the modeling of propellant performance. The use of the ISPBKW code permits the detailed solid product models that have been developed for explosives to be used to calculate propellant performance. The modeling of the performance of aluminum containing propellants requires a treatment for liquid and solid Al_2O_3 products since it melts at 2307 K. Details of the treatment used in ISPBKW are given in Appendix E.

The specific impulse rigorously defines the amount of useful energy which may be obtained from the combustion and expansion of a fuel and oxidizer. In addition to its thermodynamic significance, specific impulse also possesses an inherent ballistic importance which will be described in the ballistic section.

The specific impulse, I_{SP}, in pounds of thrust per pound of propellant per second is defined as

$$I_{SP} = 9.33\sqrt{(H_c - H_e)}$$

where H_c is the energy in calories/gram of the combustion products at 1000 pounds per square inch (68.0457 atmospheres) or 68.94733 bars. H_c is called the chamber energy for the decomposition products inside the rocket chamber as the propellant is burning. H_e is the energy of the expanded products as they exit the rocket nozzle at 1 atmosphere pressure or 1.01325 bars.

The ISPBKW code calculates the equilibrium product composition for the propellant at 68.94733 bars and its associated temperature, density, energy, and entropy. The products are expanded at constant entropy to 1.0 bar and the associated exhaust temperature, density, and energy, E, are calculated. Since $\Delta H = \Delta E + P\Delta V$, the enthalpies H_c and H_e can be calculated and used to calculate the specific impulse.

The burnout velocity, the vertical range, the horizontal range, the payload weight, and the physical size of a rocket are all influenced by the specific impulse of the propellant.

$$Burnout\ Velocity = I_{SP} + g + ln\frac{W_i}{W_f}$$

$$Vertical\ Range = H_{bo} + K'\ I_{SP} + ln\left[\frac{W_i}{W_f}\right]^2$$

$$Horizontal\ Range = 2S_{bo} + K'\, I_{SP} + ln\left[\frac{W_i}{W_f}\right]^2$$

$$Payload = f\,(I_{SP})(W_e)(W_s)$$

$$Physical\ Size = f\,(I_{SP})(\rho_{bulk})$$

where

I_{SP}= Specific Impulse
W_i= Total Initial Mass of Rocket with Propellant
W_f= Total Final Mass of Rocket after Burnout
g= Acceleration of Gravity
H_{bo}= Height of Missile at Burnout
S_{bo}= Horizontal Distance of Missile from Launch Point at Burnout
K'= Physical Constant
ln= Natural Logarithm to base e

Frequently, the measured performance of a rocket engine exceeds 95% of the ideal, theoretical values. The accepted practice for designing rocket engines is to utilize ideal rocket parameters and modify them using empirical corrections.

The propellant properties of several explosives and aluminized (20 wt%) explosive mixtures were calculated using ISPBKW and the standard Naval Weapons Center propellant performance code.[69]

	HMX		TATB		HMX-Al		TATB-Al	
	BKW	Navy	BKW	Navy	BKW	Navy	BKW	Navy
Chamber Temperature	3293	3277	1737	1666	3905	3769	3834	3711
Exhaust Temperature	1633	1557	997	984	2307	2313	2285	2313
I_{SP}	266.5	265.2	203.6	198.7	275.4	271.4	265	261.1

Comparisons of specific impulse calculations from different sources may not be as useful as the difference between propellants from the same source. The effect of using a nonideal gas equation of state and thermodynamic functions fit to different temperature ranges is small and within the uncertainties associated with assuming chemical equilibrium for the products.

The propellant performances calculated using ISPBKW are shown in Table 2.13 for various high performance propellants, including those of Table 2.5 whose explosive performance has been determined.

TABLE 2.13 Propellant Performance

Propellant	ρ_o	Chamber Temperature	Exhaust Temperature	I_{SP}
FKM 23/ 8/18/32/15/4 wt% $HMX/AP/Al/NG/NC/CH_2$	1.808	3914	2293	276
VRP 43/ 8/19/21/9 wt% $HMX/AP/Al/NG/CH_2$	1.836	3200	1693	261
VTG 40.5/10/19.5/21/9 wt% $HMX/AP/Al/NG/CH_2$	1.839	3168	1662	260
VWC-2 46/ 9/18/19/8 wt% $HMX/AP/Al/NG/CH_2$	1.835	3347	1854	267
TD-N1028 44/17/19/15/5 wt% $HMX/AP/Al/TMETN/CH_2$	1.845	3561	2144	272
UTP-20930 29/36/18/10/7 wt% $HMX/AP/Al/TMETN/CH_2$	1.838	3653	2220	270
SPISS-44 20/49/21/10 wt% $HMX/AP/Al/CH_2$	1.83	3550	2249	269
SPISS-45 12/57/21/10 wt% $HMX/AP/Al/CH_2$	1.831	3623	2250	272

TMETN - $C_5H_9N_3O_9$, E_o = -70.
NG - Nitroglycerine
NC - Nitrocellulose

Conclusions

The conclusions of the chapter on equations of state for detonation in the previous edition of this book is 20 years old but seem even more appropriate.

Some of the practical consequences of nonsteady-state behavior of explosives have been elucidated. Our previous attempts to describe an explosive with a single effective C-J pressure, regardless of run distance or the initiating system, were as certain of failure as if we had used a single C-J pressure and velocity to describe all explosives. Determination of the nature of the carbon coagulation rate as a function of pressure, density, temperature, and the gradients of these state values will be necessary for more detailed modeling and understanding of the nonsteady-state detonation process. Our objective should be to understand and model build-up to detonation and build-up of detonation as one continuous time-dependent process. In multidimensional flow, the two processes interact in interesting and important ways which we can only imagine at the present state of understanding of detonation physics and chemistry.

References

1. M. J. Urizar, E. James, Jr., and L. C. Smith, "Detonation Velocity of Pressed TNT," Physics of Fluids 4, 262 (1961).
2. Experimental data generated at Los Alamos National Laboratory.
3. H. C. Hornig, E. L. Lee, and M. Finger, "Equation of State of Detonation Products," Fifth International Symposium on Detonation, ACR-184, 503 (1970).
4. S. Fujiwara, M. Kusakabe, and K. Shuno, "Homogeneous Liquid Explosives Containing Ureaperchlorate," Sixth International Symposium on Detonation, ARC-221, 450 (1976).
5. Prince Rouse, Jr., "Enthalpies of Formation and Calculated Detonation Properties of Some Thermally Stable Explosives," Journal of Chemical and Engineering Data 21, 16 (1976).
6. "Properties of Explosives of Military Interest," Army Materiel Command pamphlet AMPC 706-177 (1967).
7. Unpublished experimental data generated at Atomics Weapons Research Establishment.
8. Unpublished experimental data generated at Lawrence Livermore National Laboratory.
9. Private communication.
10. Unpublished experimental data generated at Commissariat a L'Energie Atomique.
11. P. A. Persson, "Swedish Methods for Mechanized Blasthole Charging," Proceedings of 104th AIME Meeting, February (1975), and P. A. Persson, R. Holmberg, J. Lee, "Rock Blasting and Explosive Engineering", CRC Press, Boca Raton, Florida (1994).
12. M. Kusakabe and S. Fujiwara, "Effects of Liquid Diluents on Detonation Propagation in Nitromethane," Sixth International Symposium on Detonation, ACR-221, 133 (1976).
13. A. W. Campbell, C. L. Mader, J. B. Ramsay, and L. C. Smith, "Comments on Paper by Kusakabe and Fujiwara," Sixth International Symposium on Detonation, ACR-221, 142 (1976).
14. M. Finger, E. Lee, F. Helm, H. Cheung, and D. Miller, "Participation of LLL in JSEXP," Lawrence Livermore Laboratory Report UCID-16439 (1973).
15. P. J. Humphris and L. C. R. Thompson, "Flexible Explosives with Low Velocities of Detonation," Australian Defence Standards Laboratories Technical Memorandum 27 (1969).
16. A. Y. Apin, Y. A. Lebeder, and O. I. Nefedova, "The Reactions of Nitrogen in Explosives," Journal of Physical Chemistry (Russian) 32, 819 (1958).
17. F. E. Walker, "Participation of LLL in JSEXP," Lawrence Livermore Laboratory Report UCID-16557 (1974).

18. M. A. Cook, A. S. Filler, R. T. Keys, W. S. Partridge and W. O. Ursenbach, "Aluminized Explosives," Journal of Physical Chemistry 61, 194 (1957).

19. John B. Ramsay and W. C. Chiles, "Detonation Characteristics of Liquid Nitrogen Oxide," Sixth International Symposium on Detonation, ACR-221, 723 (1976).

20. W. E. Deal, "Measurement of Chapman Jouguet Pressure for Explosives," Journal of Chemical Physics 27, 796 (1957).

21. Stanley P. Marsh, "LASL Shock Hugoniot Data," University of California Press (1980) and "Selected Hugoniots," Los Alamos Scientific Laboratory Report LA-4167-MS (1969).

22. W. E. Deal, "Measurement of the Reflected Shock Hugoniot and Isentrope for Explosive Reaction Products," Physics of Fluids 1, 523 (1958).

23. W. C. Davis, private communication.

24. W. E. Deal, "Low Pressure Points on the Isentropes of Several Explosives," Third Symposium on Detonation, ACR-52 (1960).

25. J. W. S. Allan and B. D. Lambourn, "An Equation of State of Detonation Products at Pressures below 30 Kilobars," Fourth International Symposium on Detonation, ACR-126, 52 (1965).

26. L. V. Al'tschuler, V. T. Ryazonov, and M. P. Speranskava, "The Effect of Heavy Admixture on the Detonation Conditions of Condensed Explosive Substances," Zhurnal Prikladnov Mekhaniki i Tekhnicheskoy Fisiki 1, 122 (1972).

27. I. M. Voskoboinikov, A. A. Kotomin, and N. F. Voskoboinikova, "The Effect of Inert Additives on the Velocity of Pushing Plates by Composite Explosives," Physics of Combustion and Explosion 18, 108 (1982).

28. Charles L. Mader, "Theoretical Estimates of the Detonation Velocities of Explosives Containing Inert Diluents," Australian Defence Standards Technical Memorandum 29 (1969).

29. J. Hershkowitz and J. Rigdon, "Evaluation of a Modified Cylinder Test of Metal Acceleration of Nonideal Explosives Containing Ammonium Nitrate," Picatinny Arsenal Technical report 4611 (1974).

30. Yael Miron, Richard W. Watson, and Edmund Hay, "Nonideal Detonation Behavior of Suspended Explosives as Observed from Unreacted Residues," Proceedings of the International Symposium on the Analysis and Detection of Explosives, FBI Academy, March (1984).

31. Brigitta M. Dobratz, "Properties of Chemical Explosives and Explosive Simulants," Lawrence Livermore Laboratory report UCRL-51319, pp. 8-9 (1972).

32. F. E. Walker, "Performance of Nonideal HE," Lawrence Livermore Laboratory report UCID-16557 (1974); also UCID-16439 (1973).

33. B. G. Craig, J. N. Johnson, C. L. Mader, and G. F. Lederman "Characterization of Two Commercial Explosives" Los Alamos Scientific Laboratory report LA-7140 (1978).

34. M. Finger, F. Helm, E. Lee, R. Boat, H. Cheung, J. Walton, B. Hayes, and L. Penn, "Characterization of Commercial, Composite Explosives," Sixth International Symposium on Detonation, ARC-221, 729 (1976).

35. Louis C. Smith, "On Brisance, and a Plate-Denting Test for the Estimating of Detonation Pressure," Explosivstoffe 5, 6 (1967).

36. Charles L. Mader, James D. Kershner, and George H. Pimbley, "Three-Dimensional Modeling of Inert Metal-Loaded Explosives," Journal of Energetic Materials 1, 293-324 (1983).

37. K. K. Dhvedov, A. I. Aiskin, A. A. Il'in, and A. N. Dremin, "Detonation of High Dilute Porous Explosives, II Influence of Inert Additive on the Structure of the Front," Combustion, Explosion and Shock Waves 18, 64 (1982).

38. J. N. Johnson, C. L. Mader, and S. Goldstein, "Performance Properties of Commercial Explosives," Propellants, Explosives, and Pyrotechnics 8, 8-18 (1983).

39. J. N. Johnson, Charles Mader, and Sharon Crane, "Los Alamos Explosive Performance Data," University of California Press (1983).

40. Charles L. Mader and B. G. Craig, "Nonsteady-State Detonations in One-Dimensional, Plane, Diverging and Converging Geometries," Los Alamos Scientific Laboratory report LA-5865 (1975).

41. W. C. Davis, B. G. Craig, and J. B. Ramsay, "Failure of the Chapman Jouguet Theory for Liquid and Solid Explosives," Physics of Fluids 8, 2169 (1965).

42. B. D. Lambourn and N. E. Hoskins, "The Computation of General Problems in One-Dimensional Unsteady Flow by the Method of Characteristics," Fifth International Symposium on Detonation, ACR-221,501 (1970); also Methods of Computational Physics. 3, 265 (1964).

43. Charles L. Mader, "One-Dimensional Elastic-Plastic Calculations for Aluminum," Los Alamos Scientific Laboratory report LA-3678 (1967).

44. W. M. Isbell, F. H. Shipman, and A. H. Jones, "Use of a Light Gas Gun in Studying Material Behavior at Megabar Pressures," Behavior of Dense Media Under High Dynamic Pressures, Gordon and Breach Science Publishers, New York, p. 179 (1968).

45. W. C. Rivard, D. Venable, W. Fickett, and W. C. Davis, "Flash X-Ray Observation of Marked Mass Points in Explosive Products," Fifth International Symposium on Detonation, ACR-184, 3 (1970).

46. Charles L. Mader, M. S. Shaw, and John B. Ramsay, "PBX 9502 Performance," Los Alamos National Laboratory Report LA 9053-MS (1981).

47. R. Courant and K. O. Friedrichs, Supersonic Flow and Shock Waves, Interscience Publishers, New York, p. 430 (1948).

48. Gary L. Schott, M. S. Shaw, and J. D. Johnson, "Shocked States from Initially Liquid Oxygen-Nitrogen Systems," Journal of Chemical Physics, 82, 4264-4275 (1985).

49. Charles L. Mader, "Detonation Performance," Chapter 3, pages 165-247, Organic Energetic Compounds, edited by Paul L. Marinkas, Nova Science Publishers, Inc., Commack, New York (1996).

50. Charles L. Mader, "Numerical Modeling of Detonations," U. C. Press (1979).

51. Gary L. Schott, "Shock-Compressed Carbon Dioxide: Liquid Measurements and Comparisons with Selected Models," High Pressure Research 1991, 6, 187-200 (1991).

52. M. S. Shaw and J. D. Johnson, "The Theory of Dense Molecular Fluid Equations of State with Application to Detonation Products," Eighth Symposium on Detonation, pages 531-539 (1985).

53. W. J. Nellis, A. C. Mitchell, F. H. Ree, M. Ross, N. C. Holmes, R. J. Trainor, and D. J. Erskine, "Equation of State of Shock-Compressed Liquids: Carbon Dioxide and Air", Journal of Chemical Physics, 95, 5268-5272 (1991).

54. S. V. Pershin, "Formation of Diamonds During Detonation of TNT," IV All Union Conference on Detonation (1987).

55. B. G. Craig, Los Alamos reports GMX-8-MR-64-12, GMX-8-MR-59-6, GMX-8-MR-59-5, and private communication.

56. A. M. Staver, A. P. Ershov, and A. I. Lyamkin, "Study of Detonations in Condensed Explosives by Conduction Methods," Fizika Goreniya i Vzryva, 20, 79-83 (1984).

57. A. M. Staver, N. V. Gubareva, A. I. Lyamkin, and E. A. Petrov, "Ultradispersed Diamond Powders Obtained with the use of Explosive Energy," Fizika Goreniya i Vzryva, 20, 100-104 (1984).

58. A. P. Ershov and A. L. Kupershtokh, "Temperature of Detonation Products with Explosion in a Chamber," Fizika Goreniya i Vzryva, 22, 118-122 (1986).

59. V. F. Anisichkin, B. G. Derendiaev, V. A. Koptiug, N. Iu. Malkov, I. F. Salakhutdinov, and V. M. Titov, "Study of the Disintegration Process in a Detonation Wave using the Isotopic Method," Fizika Goreniya i Vzryva, 24, 121-122 (1988).

60. A. I. Lyamkin, E.A. Petrov, A. P. Ershov, G. V. Sakovich, A. M. Staver, and V. M. Titov, "Production of Diamonds from Explosives," Doklady, 302, 611-613 (1988).

61. N. Roy Greiner, D. S. Phillips, J. D. Johnson, and Fred Volk, "Diamonds in Detonation Soot," Nature, 333, 440-442 (1988).

62. M. S. Shaw and J. D. Johnson, "Carbon Clustering in Detonations," Journal of Applied Physics, 62, 2080-2085 (1987).

63. M. S. Shaw and J. D. Johnson, "A Slow Reaction Rate in Detonations due to Carbon Clustering," Shock Waves in Condensed Matter 1987, Elsevier Science Publishers, pages 503-506 (1988).

64. R. Chirat and J. Baute, "An Extensive Application of WCA4 Equation of State for Explosives," Ninth Symposium on Detonation, pages 751-761 (1991).

65. J. Baute and R. Chirat, "Which Equation of State for Carbon in the Detonation Products?" Ninth Symposium on Detonation, pages 521-529 (1991).

66. Roger Cheret, "Detonation of Condensed Explosives," Springer-Verlag, New York (1994).

67. M. Van Thiel and F. H. Ree, "Nonequilibrium Effects of Slow Diffusion Controlled Reactions on the Properties of Explosives," Ninth Symposium on Detonation, pages 743-749 (1991).

68. Russel T. Pack, private communication and P. T. Hay, R. T. Pack, and R. L. Martin "Electron Correlation Effects on the $N_2 - N_2$ Interaction." Journal of Chemical Physics, 81, 1630-1372 (1984).

69. 69. Thomas L. Boggs, Naval Weapons Center, China Lake private communication (1978).

chapter three

Initiation of Detonation

An explosive may be initiated by various methods of delivering energy to it. Bulk heating or "thermal initiation" can be treated satisfactorily for engineering purposes by solving the Frank-Kamenetskii equation with Arrhenius kinetics.

Whether an explosive is dropped, burned, scraped, or shocked to furnish the first impulse that starts the initiation process, a shock is formed. Initiation of an explosive always goes through a stage in which a shock wave is an important feature. Most theoretical studies at Los Alamos of the initiation properties of explosives have been primarily studies of initiation by shock waves. Modeling of simple pressure-dependent burning combined with shock decomposition of the explosive has resulted in numerical simulation of the burning to detonation process.

One may initiate propagating detonation in an explosive by sending a strong enough shock wave into it. There is a minimum shock pressure below which propagating detonation does not occur in an explosive of a particular density and geometry.

If one introduces gas bubbles or grit into a homogeneous explosive, such as a liquid or a single crystal, thereby producing a heterogeneous explosive (one containing a density discontinuity), the minimum shock pressure necessary to initiate propagating detonation can be decreased by an order of magnitude.

Explosive	Physical State	Initiation State	Type of Initiation
HMX	1.905 g/cc Single Crystal	358 kbar in 0.3 μsec	Homogeneous
HMX	1.891 g/cc Pressed Powder	100 kbar in 0.3 μsec	Heterogeneous
PETN	1.774 g/cc Single Crystal	100 kbar	Homogeneous
PETN	1.000 g/cc Pressed Powder	2 kbar	Heterogeneous
CH_3NO_2	1.128 g/cc liquid	85 kbar in 2 μsec	Homogeneous
CH_3NO_2	1.128 g/cc with grit	23 kbar in 2 μsec	Heterogeneous
CH_3NO_2	1.128 g/cc with bubble	85 kbar in 0.1 μsec	Heterogeneous

Experimental and theoretical studies suggest that shock initiation of homogeneous explosives results from simple thermal explosion caused by shock heating the bulk of the material. The voids or density discontinuities in a heterogeneous explosive cause irregularities of the mass flow when shocked. The heterogeneous explosive is initiated by local hot spots formed in it by shock interactions with density discontinuities. The hot spot mechanism is important in the propagation and failure of the detonation wave. As shown in Chapter 1, the density discontinuities are also a dominant feature of the heterogeneous explosive reaction zone.

An increased understanding of the initiation properties of explosives is necessary for development of safer explosives and propellants and for safer handling of those in use today. During the first year that the author worked in explosive research, he was part of a team studying what was thought to be a safer explosive for detonators than low-density PETN. It could be initiated by bursting bridgewires at lower energy inputs. The explosive, Thallous Azide, had such a low energy release that the standard impact sensitivity tests indicated that it took a large impulse to initiate it. Actually, it was taking a large impulse to initiate *enough* of it to be recorded as a positive event in the impact test. Only after this explosive had claimed the life of a member of the team, was it found that the explosive was more sensitive to impact than Lead Azide. Being the first to arrive at the scene of the accident, the author found a sight that often returns when he is tempted to be overconfident about his understanding of the initiation mechanisms of explosives. Since that incident, six others have died at Los Alamos in explosive accidents. The need for increased understanding of explosive initiation is obvious if we are to avoid or reduce the chance of future accidents. As shown in Chapter 2 and Appendix E, many common materials such as ammonium nitrate, ammonium perchlorate, hydrazine, liquid nitrogen oxide, and even liquid carbon monoxide are potentially hazardous as explosives, but little is known about what might increase or decrease the hazard they present. Most solid propellants are also high explosives. An energetic material designed to be a propellant is often assumed to somehow be safer than if it was called an explosive. The difference between an explosive and a propellant is primarily in the application. Many current propellants are better explosives than anything available in World War II. To understand the process well enough to evaluate explosive hazards accurately is the challenge facing the experimental and theoretical scientist studying detonation initiation.

3.1 Thermal Initiation

For those cases of explosive initiation in which the important mechanism of energy transfer is heat conduction, the usual method of numerical solution is by a finite difference technique as is described in Appendix A for the SIN code. The technique for two-dimensional geometries is described in Appendixes B and C. If one is interested only in heat conduction, one can use special purpose codes such as TEPLO described in Reference 1.

Gray and Lee[2] give an elegant review of the theory of thermal explosions. The first numerical solution of the nonlinear heat conduction equation with zero-order Arrhenius kinetics was obtained by Zinn and Mader.[3] They did not use a finite difference scheme, and their results have been used as a standard for comparison with the faster and more general treatments like those in the TEPLO or SIN codes.

The Frank-Kamenetskii[4] equation for a single zero-order rate process is

$$-\lambda\nabla^2 T + \rho C(\partial T/\partial t) = \rho Q Z e^{-E/RT} .$$

Here T is temperature in $^\circ K$, λ is thermal conductivity (cal/deg/cm/sec), ρ is density (g/cc), C is heat capacity (cal/g/deg), Q is heat of decomposition (cal/g), Z is collision number or frequency factor (sec^{-1}), E is activation energy (cal/mole), and R is the gas constant (1.987 cal/mole/deg). It will be assumed that the above constants are independent of temperature.

The Laplacian operator ∇^2, in the special cases of spheres, infinitely long cylinders, or infinite slabs, reduces in one dimension to

$$(\partial^2/\partial x^2) + (m\partial/x\partial x) ,$$

if the heating is uniform over the surface (m= 0 for slabs, 1 for cylinders, and 2 for spheres).

Where the reaction-heating term $(\rho QZ)exp(-E/RT)$ is zero, the heating equation, in the one-dimensional case (infinite slab), is

$$\partial^2 T/\partial x^2 = \rho C \partial T/\lambda \partial t .$$

Under the boundary conditions $T = f_o(x)$, when t = 0 and $T = T_1$, at x = 0 and x = 2a, the solution is

$$T(x,t) = T_1 + \frac{1}{a}\sum_{n=1}^{\infty} exp\left(\frac{-n^2\pi^2\lambda t}{4\rho C a^2}\right) sin\frac{n\pi x}{2a}\int_o^{2a} [f_o(x) - T_1] sin\frac{n\pi x}{2a} dx .$$

If at any time the temperature $f_o(x)$ is known at enough points, the integrals can be evaluated numerically; thus the temperature distribution at later times can be calculated.

When the self-heating term is nonzero, then for the infinite slab,

$$-\lambda(\partial^2 T/\partial x^2) + \rho C(\partial T/\partial t) = \rho Q Z e^{-E/RT} ,$$

which has not been solved analytically. However, numerical solutions can be obtained in the following manner.

Suppose the time scale to be divided into equal intervals each τ sec long. Suppose that the heat of reaction is not liberated continuously, but is added to the system batchwise at the end of each interval. To calculate the amount of reaction heat liberated at any point in an interval τ, it is convenient to assume that the temperature remains constant during τ, an approximation which improves as the intervals are made shorter. Thus, if in the absence of reaction during τ the temperature at some point should be T, and the reaction that has occurred in the interval should actually raise the temperature at the point to $T + (QZ\tau/C)exp(-E/RT)$.

Suppose the slab is initially at a uniform low temperature T_o. At time zero the two faces at x = 0 and x = 2a are suddenly raised to a higher temperature T_1. Treating the problem for the first small interval τ_1 as though the slab were inert, an approximation is obtained to the temperature distribution $T(x, \tau_1)$ after τ_1 seconds, as given by the $T(x,t)$ equation where $f_o(x)$ is set equal to T_o and $t = \tau_1$. Owing to the chemical reaction which will have occurred, this first estimate of the temperature should be revised to $T(x,\tau_1) + (QZ\tau_1/C)exp[-E/RT(x,\tau_1)]$. During the next interval τ_2, the slab is again treated as an inert and a first approximation to the temperature is again given by the $T(x, t)$ equation, where $f_o(x)$ is set equal to $T(x, \tau_1) + (QZ\tau_1/C)exp[-E/RT(x, \tau_1)]$ and t equal to τ_2. Again the temperature rise due to reaction is added on. This process is repeated until a steady state is attained, or until the temperature at some point in the slab reaches some very high value which may be said to define an explosion.

An exactly analogous technique has been applied to cylinders, where the inert heat equation takes the form,

$$\partial^2 T/\partial r^2 + \partial T/r\partial r = \rho C \partial T/\lambda \partial t \ .$$

Under the boundary conditions $T = f_o(r)$ when $t = 0$ and $T = T_1$ for $r = a$, the solution is

$$T(r, t) = T_1 + \frac{2}{a^2}\sum_{n=1}^{\infty}\frac{exp\left(-g_n^2\lambda t/\rho Ca^2\right)}{J_1^2(g_n)}J_o\left(\frac{g_n r}{a}\right)\int_o^a r\left[f_o(r) - T_1\right]J_o\left(\frac{g_n r}{a}\right)dr \ ,$$

where the g_n are the roots of the zero-order Bessel function J_o.

The procedure has also been applied to spheres. Here the inert heat equation is

$$\partial^2 T/\partial r^2 + 2\partial T/r\partial r = \rho C \partial T/\lambda \partial t \ .$$

Under the boundary conditions $T = f_o(r)$ when $t = 0$ and $T = T_1$ for $r = a$, the solution is

$$T(r, t) = T_1 + \frac{2}{ar}\sum_{n=1}^{\infty} exp\left(\frac{-n^2\pi^2\lambda t}{\rho Ca^2}\right) sin\frac{n\pi r}{a}\int_o^a r\left[f_o(r) - T_1\right]sin\frac{n\pi r}{a}dr \ .$$

The calculations were performed numerically as early as 1956. The gross results of the calculations are not at all unexpected and may be summarized as follows:

If the surface of a given explosive in a charge of a given size and shape is maintained at a temperature higher than a certain critical value, which is designated by T_m, the explosive eventually will explode. If the surface temperature is kept below T_m, no thermal explosion will occur. This critical temperature decreases as the dimensions of the charge are increased.

If the charge is initially at room temperature, and its surface temperature is raised to T_1, where T_1 is above T_m, the explosion will occur after a certain induction time which depends on the explosive, the geometry of the charge, and T_1. If T_1 is only slightly higher than the critical temperature T_m, the induction time is relatively long and the explosion develops at the center of the charge. With increasingly higher values of T_1 the induction time becomes progressively shorter and the explosion commences progressively closer to the surface as shown in Figures 3.1 and 3.2. For a given value of T_1 the induction times for small charges are much shorter than for large ones.

Figure 3.1 shows the calculated variation of temperature with time at two locations within a 1-inch sphere of RDX for various values of T_1. The RDX constants used were $\rho = 1.8$ g/cc, $C = 0.5$ cal/g/°C, $Q = 500$ cal/g, $Z = 10^{18.5} sec^{-1}$, and $E = 47.5$ kcal/mole. For relatively low surface temperatures the hottest region prior to explosion is at the center. Points near the surface remain relatively cool until the end of the induction period, suffering a discontinuous jump at the time of explosion. Somewhat the reverse situation applies when

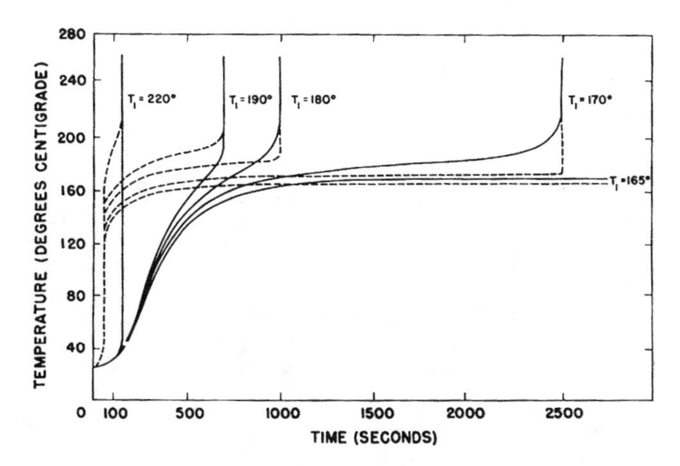

Figure 3.1 Temperature vs. time at two locations within a 1-inch-diameter sphere of RDX initially at 25°C for various values of T_1. Solid curve: r/a = 0 (center of sphere); dashed curve: r/a = 0.9.

T_1 is high. The same facts are evident from Figure 3.2, where the temperature vs. position profiles are shown for various times near the end of the induction period.

For charges of a given size, shape, and composition, it is convenient to plot logarithms of the calculated explosion times vs. the reciprocal of the surface temperature T_1, if the charges all start out at a uniform low temperature T_o. For slabs, cylinders, and spheres, these plots are linear over quite a large region but bend upward sharply near the critical temperature T_m. At T_m they become vertical, indicating an infinitely long induction period.

Frank-Kamenetskii solved his equation under the steady-state condition $\partial T/\partial t = 0$ and obtained the following expression for critical temperature in terms of the related physical parameters.

$$T_m = \frac{E}{2.303\ R\ log(\rho\ a^2\ QZE/\lambda RT_m^2\delta)}\ ,$$

where δ = 0.88 for slabs, 2.00 for cylinders, and 3.32 for spheres. This equation can be solved quickly by interation, and it gives values for T_m which agree with those obtained from the calculated log t_{exp} vs. $1/T_1$ curves.

Note from the method of calculation and the form of the equations that for a given shape only three parameters, $\rho a^2 QZ/\lambda$, E, and $\lambda/\rho Ca^2$, instead of seven independent ones, are used to describe the explosive charge. Thus, one can make the following generalizations: (1) changing Q by a multiplicative factor is entirely equivalent to changing Z; (2) in the same way, ρ, a^2, and $1/\lambda$ are equivalent; (3) E and C are independent of the others.

The t_{exp} vs. $1/T_1$ relationship is of the form

$$t_{exp} = \frac{\rho Ca^2}{\lambda}\ F(E/T_m - E/T_1)\ ,$$

where F is a function depending only on the type of geometry (spheres, cylinders, or slabs) and the initial temperature T_o. Values of F as a function of the argument $E/T_m - E/T_1$ are plotted in Figure 3.3 for spheres, cylinders, and slabs all initially at 25°C. The previous

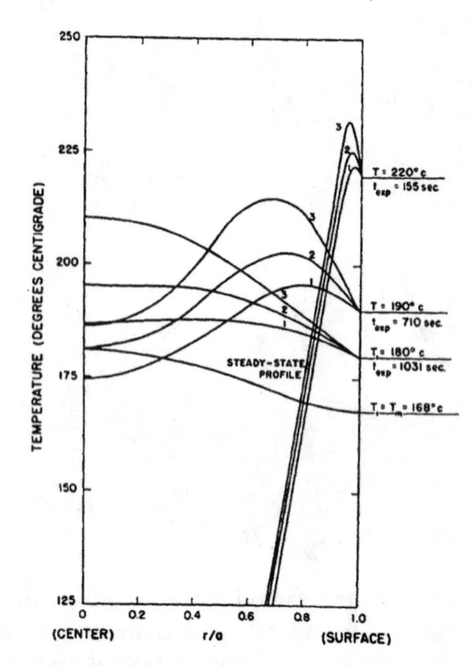

Figure 3.2 Temperature profiles for times near the end of the induction period, as calculated for 1-inch spheres of RDX initially at 25°C. Curve 1: t = 0.9 t_{exp}; curve 2: t = 0.95 t_{exp}; curve 3: t = 0.98 t_{exp}. Also shown is the steady-state profile at the critical temperature, T_m.

two equations together with Figure 3.3 make possible the calculation of explosion times for spheres, cylinders, or slabs of any arbitrary explosive (when ρ, C, a, λ, Z, Q, and E are known).

Longwell[5] has used these results to form nomographs for calculation of explosion times.

Figure 3.4 shows the calculated explosion times vs. temperature for three particular cases, 1-inch RDX spheres, cylinders, and slabs. The results can be explained qualitatively as follows: The critical temperature T_m below which no explosions are obtained are related to the rate at which heat can escape to the surface, which in turn is related to the surface-to-volume ratio. The surface-to-volume ratios of spheres, cylinders, and slabs are, respectively, $3/a$, $2/a$, and $1/a$, if a represents radius or half-thickness. Thus, for a given value of a, the slab should have the lowest critical temperature. At relatively low temperatures explosion times relate to surface-to-volume ratios in the opposite manner. The sphere, having the smallest volume to be heated per unit surface, will explode the fastest.

Although at relatively low temperatures geometry has an important influence on induction times, its effect becomes less pronounced as temperature is increased. This is associated with the fact that for high external temperatures, explosions begin near the surface of the explosive while the interior is still cool. Above 270°C the log t_{exp} vs. $1/T_1$ plots for 1-inch RDX spheres, cylinders, and slabs are coincident. Moreover they are coincident with the plots obtained by G. B. Cook[6] for a semi-infinite solid heated at its single surface. Cook's line, adjusted to correspond to our own choice of physical parameters, is included in Figure 3.4.

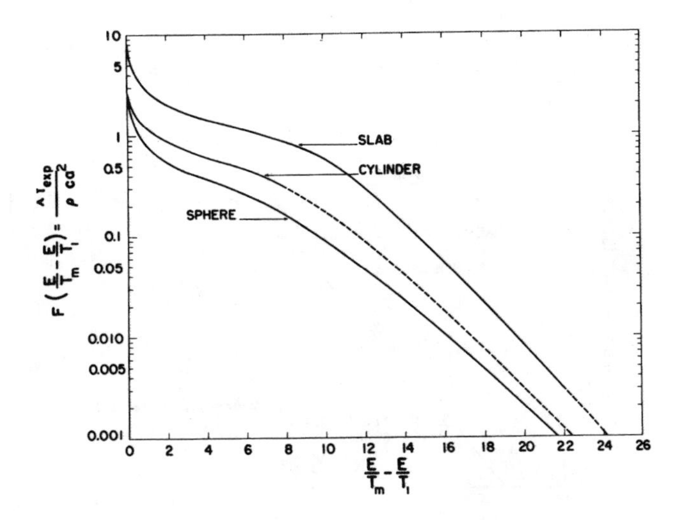

Figure 3.3 Graphs of $\lambda t_{exp}/\rho Ca^2$ vs. $E/T_m - E/T_1$, for spheres, cylinders, and slabs, all initially at 25°C.

Also plotted in Figure 3.4 are "adiabatic" explosion times, defined by the relation

$$t_{exp} = \int_{T_1}^{T_{exp}} \frac{C}{QZ} exp(E/RT)\, dT \cong \frac{CRT_1^2}{QZE} exp(E/RT_1)\ .$$

Physically this represents the time it would take for an infinite mass of explosive to ignite if it were initially at the uniform temperature T_1. The adiabatic explosion time is necessarily shorter than explosion times obtained by raising only the surface of the explosive to T_1.

The Arrhenius constants to describe the thermal decomposition are a subject of much debate. Rogers[7] has used his differential thermal calorimetry technique to obtain the constants, and the experimental and calculated critical temperatures for several explosives are compared in Table 3.1.

TABLE 3.1 Experimental and Calculated Critical Temperatures

	T_m(°C)		Values Used					
Explosive	Experimental	Calculated	a (cm)	ρ	Q(cal/g)	Z (sec^{-1})	E	λ x 10^4
HMX (liq.)	253-255	253	0.033	1.81	500	5.000 x 10^{19}	52.7	7.0
RDX (liq.)	215-217	217	0.035	1.72	500	2.015 x 10^{18}	47.1	2.5
TNT	287-289	291	0.038	1.57	300	2.510 x 10^{11}	34.4	5.0
PETN	200-203	196	0.034	1.74	300	6.300 x 10^{19}	47.0	6.0
TATB	331-332	334	0.033	1.84	600	3.180 x 10^{19}	59.9	10.0
DATB	320-323	323	0.035	1.74	300	1.170 x 10^{15}	46.3	6.0
BTF	248-251	275	0.033	1.81	600	4.110 x 10^{12}	37.2	5.0
NQ	200-204	204	0.039	1.63	500	2.840 x 10^{7}	20.9	5.0
PATO	280-282	288	0.037	1.70	500	1.510 x 10^{10}	32.2	3.0
HNS	320-321	316	0.037	1.65	500	1.530 x 10^{9}	30.3	5.0

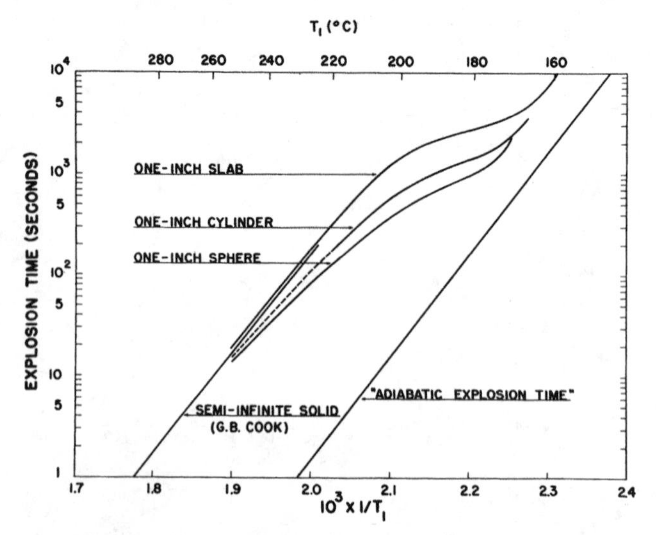

Figure 3.4 Explosion times vs. $1/T_1$ for RDX in various geometries. Initial temperature is 25°C.

Numerical solution of the Frank-Kamenetskii equation with first order Arrhenius kinetics was first performed by Zinn and Rogers.[8]

While agreement such as Rogers obtained between the experimental and calculated critical temperatures is essential for the thermal initiation theory to be correct, it is not necessary that the details of the flow are correct or that one would obtain temperature-time profiles such as shown in Figure 3.2. John Zinn investigated the problem in 1959. He measured the temperature-time profiles inside a 10-cm-diameter sphere of 9404 while the surface of the sphere was maintained at 168°C. Ignition was observed to occur in 112 minutes.

The heater consisted of a 76-cm-cube box lined with asbestos and aluminum foil. Eight 250-watt heating lamps mounted at the corners were directed toward the center. The air inside was circulated by hair dryers. The explosive sphere, blackened with India ink, was suspended on a wire tripod at the center of the box. The controlling element was connected to a thermocouple (30-gauge iron constantan) mounted flush with the explosive surface. The surface could be maintained within 5 degrees of the desired temperatures with a period of 15 seconds. Six additional theromocouples were mounted in the equatorial section of the sphere. Three were flush with the spherical surface, one was at the center, one was at one-half, and one at three-quarters radius from the center. As shown in Figure 3.5, during the earlier portion of the run the temperature at the interior points was found to rise in a manner characteristic of heat flow in an inert solid. After about 100 minutes, all three interior thermocouples indicated temperatures in excess of the surface temperature and continued to rise. The temperature at the center proceeded upward to 215°C, at which time the explosive ignited, melting the heating lamps and wire tripod.

The kinetics of 9404 is sufficiently complicated that it is not an ideal explosive for numerical simulation; however, the experiment does show temperature-time profiles similar to those expected. The effect of self-heating is negligible during the early portion of the run, so $\lambda/\rho C$ can be obtained from Gurney-Lurie charts. The value in the vicinity of 50 minutes is 0.0015 cm^2/sec.

If the explosion time is taken as 112 minutes, then $\lambda t_{exp}/\rho C a^2 = 0.39$. Making use of the reduced explosion time relation shown in Figure 3.3, one obtains a value of 3.6 for

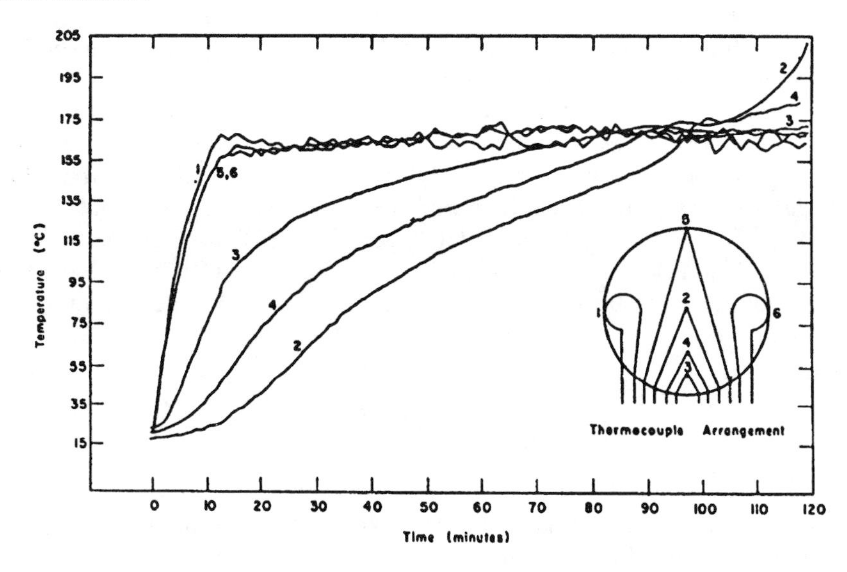

Figure 3.5 Zinn's temperature-time profiles for a 10-cm-diameter sphere of 9404 heated on the surface to 168°C.

the quantity $E(1/T_m - 1/T_1)$. Rogers gave an activation energy of 31.3 and frequency factor of 4.3 x 10^{12} for 9404 when nitrocellulose was dominating the kinetics. The critical temperature, T_m, is 419°K. Using the HMX kinetics in Table 3.1 gives a critical temperature of 428°K.

It would be interesting to repeat the experiment with several pure explosives and to make a detailed comparison with thermal initiation calculations.

The SIN code described in Appendix A was used to simulate the experiments. The heat capacity in calories/gram was described by 0.025 + 6.88 x $10^{-4}T(°K)$. A coefficient of thermal conductivity of 0.00091 cal/°C/cm/sec was calculated from the previous estimate of $\lambda/\rho C$ using 0.328 calories/gram at 441°K for the heat capacity. The temperature profiles shown in Figure 3.5 were reproduced by the numerical calculation without any chemical reaction to 70 minutes. The HMX kinetics in Table 3.1 was too slow, and the RDX kinetics (an HMX impurity) was too fast unless it was adjusted to account for the amount of RDX present (~ 9%) and a contribution from the 3% nitrocellulose was included. To reproduce the profiles in Figure 3.5, the nitrocellulose activation energy used was 32.5. The limited nitrocellulose contribution first occurs near the surface of the sphere and sweeps toward the center, causing the temperature humps observed between 80 and 100 seconds. The higher temperature at the center of the charge then significantly decomposes the RDX and causes explosion of the charge from the center.

Current use of heat conduction as an energy transfer mechanism in initiation calculations is usually in conjuction with hydrodynamic calculations. For example, the one-dimensional reactive hydrodynamic code SIN described in Appendix A was used to obtain a numerical description of the underwater experiments[9] designed to investigate the mechanism of initiation from the shock compressions of various layers of gases in contact with explosives.

To describe the experiment, the numerical model consisted of a 1.27-cm-radius sphere of 9404 explosive containing a 0.1587-cm PETN initiator at the center surrounded by a 0.47628-cm-thick shell of PETN. The explosive sphere was surrounded by 9.68 cm of water, 0.15 cm of Plexiglas, 0.1 cm of air, 2.54 cm of TNT, and finally 6 cm of water. Calculations were also performed without the air. The calculation could reproduce the shock behavior as it decayed in the water as shown in Chapter 5.

The air was described using the BKW equation of state and an ideal gas equation of state. The TNT and Plexiglas equations of state were determined using the experimentally measured single-shock Hugoniots and the HOM equation of state described in Appendix A.

For Plexiglas the Hugoniot was described by $U_s = 0.243 + 1.5785U_p$ and $\rho_o = 1.18$ g/cc, $\gamma = 1.0$, $\lambda = 1.57 \times 10^{-14}$ Mbar/cc/°K/μsec. For TNT the Hugoniot was described by $U_s = 0.3033 + 1.366U_p$ and $\rho_o = 1.625$ g/cc, $\lambda = 2.59 \times 10^{-14}$ Mbar/cc/°K/μsec, E^* = 41.4 kcal/mole, $Z = 1 \times 10^7$ μsec^{-1}. The air heat conductivity constant was described by the equation

$$\lambda = 2.6 \times 10^{-15}\left[\frac{T(^\circ K)}{300}\right]^{0.5}.$$

The SIN code can be used to run with both hydrodynamic and conductive energy transfer, but it is more economical to perform two separate calculations with the hydrodynamic state values at maximum compression being used as the starting conditions for the heat conduction and reaction calculations.

The calculation without air between the Plexiglas and TNT gave a maximum TNT shock pressure of 3.7 kbar and a temperature rise of approximately 5°K. It is not surprising that no reaction was experimentally observed in the system without an air gap. The pressure at the TNT-Plexiglas interface had decreased to 1.8 kbar by the time the shock wave had traveled through the TNT explosive. The pressure at the front of the shock wave was 2.5 kbar upon arrival at the TNT-water interface. The time for the shock to travel through the TNT was about 8 μsec. The calculated water shock pressure upon arrival of the shock at the Plexiglas-water interface was 2 kbar, approximately linearly decreasing to 1 kbar at 2 cm behind the Plexiglas-water interface.

The calculation with 0.1 cm of air required 6.5 μsec for the gap to close. The maximum air pressure was 3.4 kbar; the density was 0.41 g/cc; and the particle velocity was 0.000675 cm/μsec. The gap was compressed to 1/320 of its original thickness of 0.00031 cm. The maximum temperature of the gas was 2700°K.

These results were used as the starting conditions for the heat transfer calculation. They gave a maximum temperature of 820°K in the TNT and 910°K in the Plexiglas. The temperature in the gas was lowered to a maximum of 2000°K within 3 μsec, and insufficient reaction had occurred to make the first cell decompose. At later times the explosive surface cooled as heat was being conducted away faster than it was furnished by the hot gas and explosive decomposition.

The conclusions are not sensitive to the gas temperature calculated or the constants used. The initial compressed gas temperature could be more than 1000°K higher than calculated and still be insufficient to result in any appreciable decomposition at the TNT surface.

Identical calculations with the air described by an ideal gas equation of state with a γ of 1.28 and C_v of 0.25 cal/g/°K gave a maximum air pressure of 3.54 kbar, density of 0.76 g/cc, particle velocity of 0.007 cm/μsec, and maximum gas temperature of 1900°K. The gas temperature using the ideal gas equation of state was about 800°K lower than calculated using the more realistic BKW air equation of state.

These calculations eliminated the commonly accepted mechanism of plane surface heat conduction for energy transfer. The experimental observations are that approximately the same amount of decomposition was observed for 0.1 cm of air, Krypton, methane, or vacuum. Our most favorable calculations indicate that heat conduction alone is insufficient to cause appreciable reaction in air or vacuum and that the methane filled gap should give TNT temperatures at least 300°K lower than the Krypton-filled gap. Therefore, it would appear that some phenomenon other than plane surface heat conduction dominates the initiation process of explosives when gaps are present. Some mechanism is required for heat in the gas to be concentrated in local areas of the explosive surface, or some other source of initiation energy is required such as shock interactions or internal void compression. The concentration mechanism appears to be relatively independent of the gas temperature and essentially independent of the gas.

It is disturbing that the mechanism of initiation that is probably important in the accidental premature initiation of explosives in shells is unknown. Our calculations confirm what the experimental evidence indicates, that some phenomenon other than plane surface heat conduction and adiabatic gas compression is dominating the initiation process.

3.2 *Shock Initiation of Homogeneous Explosives*

The classic experimental studies by Campbell, Davis, and Travis[10] of the shock initiation of homogeneous explosives showed that the detonation of the heated, compressed explosive begins at the interface, where the explosive has been hot the longest, after an induction time which is a decreasing function of the shock strength. The detonation proceeds through the compressed explosive at a velocity greater than the steady-state velocity in uncompressed explosive, overtaking the initial shock and overdriving the detonation in the unshocked explosive.

The initial and boundary conditions for the calculation of plane wave shock initiation of nitromethane, liquid TNT, and single-crystal PETN were chosen to match the following experimental arrangement. A homogeneous explosive is adjacent to a plane wall which, at initial time, accelerates impulsively to a prescribed velocity, thereby producing a shock. This velocity is retained until the adjacent explosive detonates, at which time the wall again comes to rest. If the wall velocity remains constant the essential features of the calculation, such as induction time, C-J pressure, and velocity in the shocked explosive remain unchanged. In the experiments of greatest interest, the wall velocity is such as to produce shock heating just subcritical for immediate detonation. As a result, the shock travels some distance before initiation occurs at the wall (where the explosive has been hot the longest). The delay time between initial shocking and later initiation (the "induction time"), and the velocity and C-J pressure of the detonation in the shocked explosive are the primary results of the experiments.

A typical sequence of pressure profiles for the initiation of nitromethane is shown in Figure 3.6. The shock travels into the explosive; shock heating occurs and results in chemical decomposition. Explosion occurs at the rear boundary, and a detonation develops with the C-J pressure and velocity characteristic of the explosive at the shocked pressure and density. Because these pressures and velocities are often much larger than the explosive normal maximum C-J performance, these detonations in the shocked explosive are called "super detonations." The detonation wave overtakes the shock wave and then decays to the normal density C-J pressure and velocity. An animation of the shock initiation of nitromethane is on the CD-ROM.

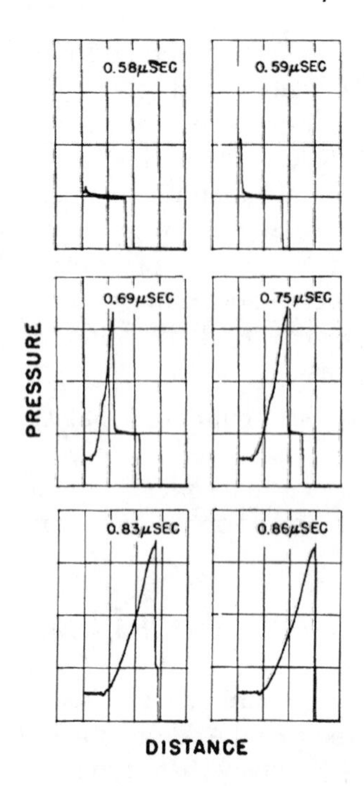

Figure 3.6 Pressure-distance profiles at various times for shock initiation of nitromethane by a 90 kbar shock.

The experimental data and shock initiation computed using the SIN code in Appendix A for nitromethane, liquid TNT, and single crystal PETN were described in Reference 11. Since that time new experimental measurements of the equations of state and Arrhenius rate constants have been performed and Craig has repeated the PETN single crystal experiments. The shock initiation experiments have been re-examined, and the calculations have been performed with the new equation of state and Arrhenius rate constants. The activation energy and frequency factors for nitromethane are those of Reference 12 given in Table 1.1; those for liquid TNT and PETN are given in Table 3.1. Rogers' liquid TNT values differ from those given in Table 1.1, but give about the same adiabatic explosion times for the shock initiation experiments. The Hugoniot data for liquid TNT were reported by Garn[13] and the Hugoniot data for nitromethane and PETN were measured by Marsh and Craig. The equation of state parameters for nitromethane and liquid TNT are given in Table 1.1; those for PETN, in Table 3.3. The coefficients to fits of calculated Hugoniot temperatures for nitromethane, liquid TNT, and PETN using the Walsh and Christian technique described in Appendix E are given in Tables 1.1 and 3.3. The experimental and computed results for the shock initiation experiments are shown in Table 3.2. The computed explosion times are within a factor of 2 of adiabatic explosion times computed using Table 3.1 parameters. The difference is primarily a result of the higher heat of explosion at the higher pressures and rates in the shock initiation model.

TABLE 3.2 Shock Initiation

Parameter	Input	Calculated	Experimental
Nitromethane			
Particle Velocity (cm/μsec)	0.171		0.171 ± 0.01
Initial Temperature (°K)	300		300
Shock Pressure (Mbar)		0.086	0.086 ± 0.005
Shock Temperature (°K)		1180.	
Shock Velocity (cm/μsec)		0.445	0.45
Induction Time (μsec)		0.64	2.26 ± 1.
Detonation Velocity + Particle Velocity[a] (cm/μsec)		1.00	1.02± 0.03
Detonation Velocity (Steady-State)[a] (cm/μsec)		0.83	0.85± 0.02
Pressure[a] (Mbar)		0.35	0.300
Liquid TNT			
Particle Velocity (cm/μsec)	0.176		
Initial Temperature (°K)	358.1		358.1
Shock Pressure (Mbar)		0.125	0.125
Shock Temperature (°K)		1180.	
Shock Velocity (cm/μsec)		0.490	0.49 ± 0.01
Induction Time (μsec)		0.30	0.70
Detonation Velocity + Particle Velocity[a] (cm/μsec)		1.01	1.10± 0.1
Detonation Velocity (Steady-State)[a] (cm/μsec)		0.83	0.92± 0.1
Pressure[a] (Mbar)		0.44	
PETN Single Crystal			
Particle Velocity (cm/μsec)	0.124		0.125 ± 0.01
Initial Temperature (°K)	298.17		298.17
Shock Pressure (Mbar)		0.112	0.110 ± 0.01
Shock Temperature (°K)		660.	
Shock Velocity (cm/μsec)		0.51	0.50 ± 0.01
Induction Time (μsec)		0.35	0.30 ± 1.
Detonation Velocity + Particle Velocity[a] (cm/μsec)		1.12	1.09± 0.1
Detonation Velocity (Steady-State)[a] (cm/μsec)		1.00	0.97± 0.1
Pressure[a] (Mbar)		0.61	

[a] In Compressed Explosive (Super Detonation State)

TABLE 3.3 PETN Equation of State and Rate Parameters

	HOM SOLID	EXP		HOM GAS	EXP
C	+2.33	−001	A	−3.47697078840	+000
S	+2.74	+000	B	−2.47673506638	+000
F_s	−3.88361737598	+000	C	+2.56165371703	−001
G_s	−5.20933072942	+001	D	−3.99581663941	−003
H_s	−1.05066847323	+002	E	−3.88233591806	−003
I_s	−9.06683024970	+001	K	−1.60841591960	+000
J_s	−2.62417534059	+001	L	+4.87554479239	−001
C_v	+2.6	−001	M	+5.48253018197	−002
α	+7.35	−005	N	+2.60937128503	−003
V_o	+5.6338028169	−001	O	+2.72202328270	−005
γ_s	+7.7	−001	Q	+7.50597305073	+000
T_o	+3.0	+002	R	−4.84640053341	−001
C1	+2.95	−001	S	+5.28235558979	−002
S1	+1.71	+000	T	+3.45536575786	−002
VSW	+4.775	−001	U	−1.06733104740	−002
			C'_v	+5.0	−001
			Z	+1.0	−001
		Reaction			
E*	+4.70	+004			
Z	+6.3	+013			
		BKW CJ Parameters			
P_{CJ}	+3.19	−001			
D_{CJ}	+8.422	−001			
T_{CJ}	+2.832	+003			

Explosion time data for nitromethane are available for pressures of 67 to 104 kbar. While some of the data are not one-dimensional, it is interesting to determine whether the data are consistent with an adiabatic explosion model with the same kinetics and temperature calculations as used in the numerical hydrodynamic calculations. To perform the adiabatic explosion time calculations the same nitromethane constants and a heat of decomposition of 1000 cal/g were used.

A summary of the available data and calculated temperatures follows.

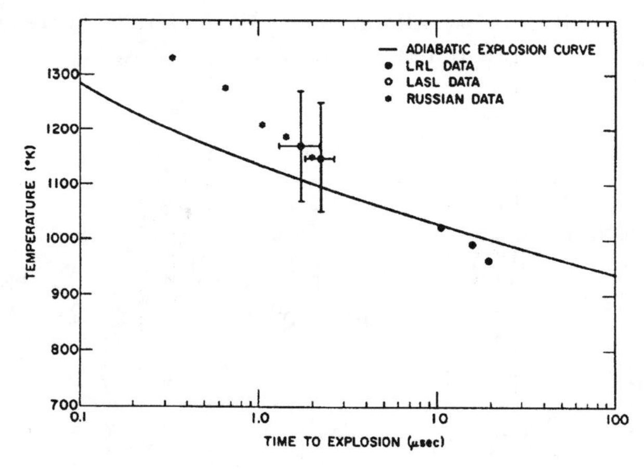

Figure 3.7 Nitromethane time to adiabatic explosion vs. temperature profile and experimental data.

Experimental Shock Velocity (cm/μsec)	Hugoniot Pressure (kbar)	Experimental Explosion Time (μsec)	Walsh and Christian Temperature (°K)	Data Source
0.445	86	2.26± 0.5	1150± 100	LANL[10]
0.451	89	1.74± 0.5	1180± 100	LANL[10]
	86	2.54	1150	Russian[15]
	90	1.42	1190	Russian[15]
	93	1.05	1215	Russian[15]
	99	0.67	1280	Russian[15]
	104	0.33	1337	Russian[15]
0.420	74	10.4	1027	LLNL[14]
0.411	70	16.0	990	LLNL[14]
0.404	67	20.0	960	LLNL[14]

The data and the adiabatic explosion curve are plotted in Figure 3.7. The data are consistent with the adiabatic explosion model within any reasonable estimate of the experimental error and the error in the calculated temperature.

The best method of obtaining Hugoniot temperatures is to rely on experimental data, as is done to obtain Hugoniot pressures and volumes. How well does the technique used to calculate the temperature agree with the experimental data? The Walsh and Christian technique described in Appendix E, with the C_p at standard pressure and temperature set equal to C_v and with $(dP/dT)_v/C_v)$ held constant, will reproduce experimentally observed temperatures of carbon tetrachloride, water, and nitromethane shown in Table 3.4 along with

TABLE 3.4 Experimental and Walsh and Christian Calculated Hugoniot Temperatures

	Experimental Values	Computed Values
Carbon Tetrachloride		
$U_s = 0.1558 + 1.4726U_p$	0.170 Mbar	0.170 Mbar
ρ_o=1.5772	2400°± 50	2412°K
K=90.7		
$\alpha = 4.12 \times 10^{-4}$		
$C_p = 0.198$		
Water		
$U_s = 0.1504 + 1.786U_p$	0.190 Mbar	0.190 Mbar
ρ_o=1.0	1080°± 100	1010°K
K=50		
$\alpha = 6. \times 10^{-5}$		
$C_p = 1.0$		
Nitromethane		
$U_s = 0.1647 + 1.637U_p$	0.085 Mbar	0.085 Mbar
ρ_o=1.128	1200°± 100	1150°K
K=22.56		
$\alpha = 3. \times 10^{-4}$		
$C_p = 0.414$		

the parameters used in the calculations. The experimental temperatures were measured by John Ramsay at Los Alamos using photomultipliers to measure the experimental brightness. The observed agreement between the calculated and experimental temperatures implies a fortunate compensation of errors resulting from the variation of γ and C_v over the temperature and pressure range of interest. The heat capacity, isothermal compressibility, and alpha (the coefficient of thermal expansion) decrease with increasing pressure, so the Walsh and Christian $b = 3\alpha/KC_v$ remains nearly constant. With decreasing specific volume and increasing pressure, gamma decreases from $\gamma = 3\alpha V/KC_v$. The Hugoniot temperature calculation is performed using the STM code which is on the CD-ROM.

The equation of state used in the calculations yields pressures, volumes, and temperatures in agreement with the available experimental data and is as realistic as possible without additional experimental data. One can use Arrhenius kinetics to describe the gross features of the explosive thermal decomposition of nitromethane over a surprising range of temperatures, times, and pressures. Since it is well known that the dominant decomposition mechanism changes with temperature and pressure (which is grossly taken into account by changing the heat of decomposition by a factor of 2 from low-temperature thermal initiation calculations [~500 cal/g] to high-temperature and high-pressure [~1000 cal/g] detonation

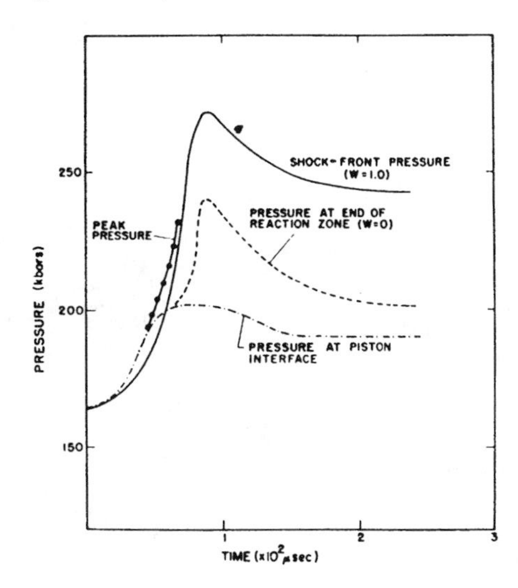

Figure 3.8 Shock initiation of nitromethane with a 0.25-cm/μsec constant velocity piston. $E^* = 40$, $Z = 1.27 \times 10^6$, $\gamma = 0.68$, mesh is 4×10^{-5} cm, and reaction zone is resolved.

calculations), it would be useful to investigate the compensating mechanisms which permit the use of such a simple description of the kinetics.

As shown in Chapter 1, it is possible to perform shock initiation calculations with a resolved reaction zone. Since the nitromethane detonation wave is stable at the C-J detonation velocity for an activation energy of 40 kcal/mole, a gamma of 0.68, and a frequency factor of 1.27×10^6, it is possible to study how the reaction zone is formed by constant velocity pistons that initially shock the unreacted nitromethane to less than the C-J pressure.

The pressures at various positions of the flow as a function of time are shown in Figures 3.8 and 3.9 for 0.25- and 0.23-cm/μsec velocity pistons, respectively, and in Figure 1.12 for a 0.21-cm/μsec piston. Figure 3.10 shows a similar profile for the unresolved calculation shown in Figure 3.6.

The shocked nitromethane first completely decomposes at or near the piston and results in a detonation that builds up toward the C-J pressure of the high-density shocked nitromethane. The detonation wave overtakes the shock wave and then decays toward the piston pressure. These calculations with a resolved reaction zone show the details of the process of shock initiation of nitromethane and are basically the same as computations with an unresolved reaction zone at lower pressures.

Hot Spot Initiation of Homogeneous Explosives

The initial conditions of the calculations were a hot spot assumed to be a spherical disturbed region whose density and temperature were uniform throughout. The surrounding shocked nitromethane was likewise uniform in density and temperature, and all initial values were zero.

Two types of hot spots were considered: the "temperature hot spot" in which the temperature was higher and the density of the hot spot was the same as in the surrounding nitromethane, and the "pressure hot spot" in which the density was also increased. In the pressure hot spot, the density and temperature chosen were values on the single-shock

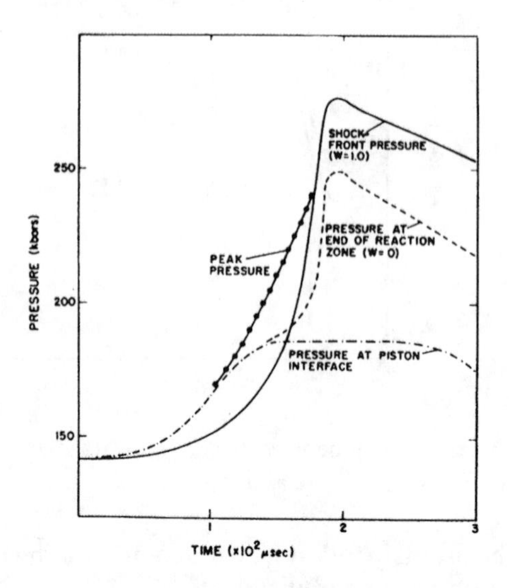

Figure 3.9 Shock initiation of nitromethane with a 0.23-cm/μsec constant velocity piston. $E^* = 40$, $Z = 1.27 \times 10^6$, $\gamma = 0.68$, mesh is 4×10^{-5} cm, and reaction zone is resolved.

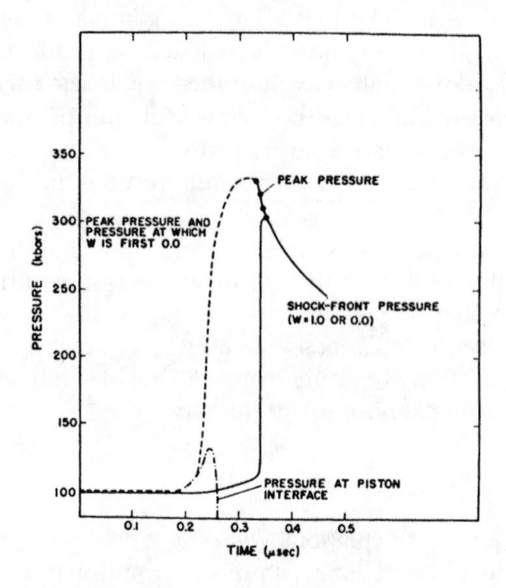

Figure 3.10 Shock initiation of nitromethane with a 0.18/0.0 step piston. $E^* = 53.6$, $Z = 4 \times 10^8$, $\gamma = 0.68$, mesh is 2.5×10^{-2} cm, and reaction zone is unresolved.

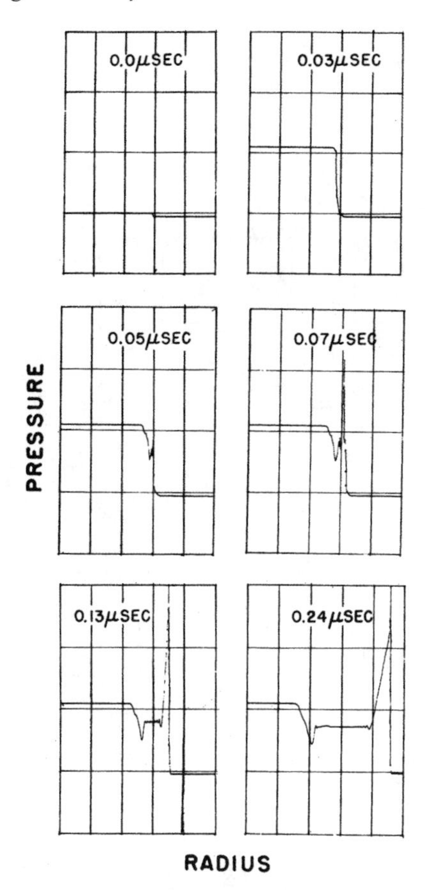

Figure 3.11 Pressure-radius profiles at various times for development of detonation in shocked nitromethane (94.7 kbar, 1230°K) from spherical temperature hot spot (1404°K) of 0.292-cm radius. The ordinate is pressure in scale divisions of 100 kbar, and the abscissa is radius in scale divisions of 0.1 cm.

Hugoniot. In both cases the temperature was chosen so that the hot spot would decompose almost immediately relative to the surrounding methane.

For the temperature hot spot, the pressure was only slightly higher than in the surrounding medium. Figure 3.11 shows the pressure-radius profiles for a 0.292-cm-radius temperature hot spot (1404°K) in nitromethane which had been shocked to 94.7 kbar. Initially, at the boundary between the hot spot and the shocked nitromethane, a small shock is sent into the outer nitromethane and a rarefaction is sent back into the hot spot. At ~ 0.03 μsec, part of the hot spot explodes and initiates the remainder of the hot spot which has been cooled by the rarefaction. A strong shock goes into the undetonated nitromethane and a rarefaction goes back into the detonation products. The strongly shocked nitromethane at the hot spot boundary explodes at about 0.06 μsec and initiates propagating detonation through the rest of the nitromethane. The detonation propagates at a velocity of 0.856 cm/μsec, which is the computed equilibrium detonation velocity of the shocked nitromethane.

Figure 3.12 shows in considerable detail the mechanism of initiation of detonation from a 0.06-cm-radius temperature hot spot (1404°K). Initially, at the hot spot boundary, a small

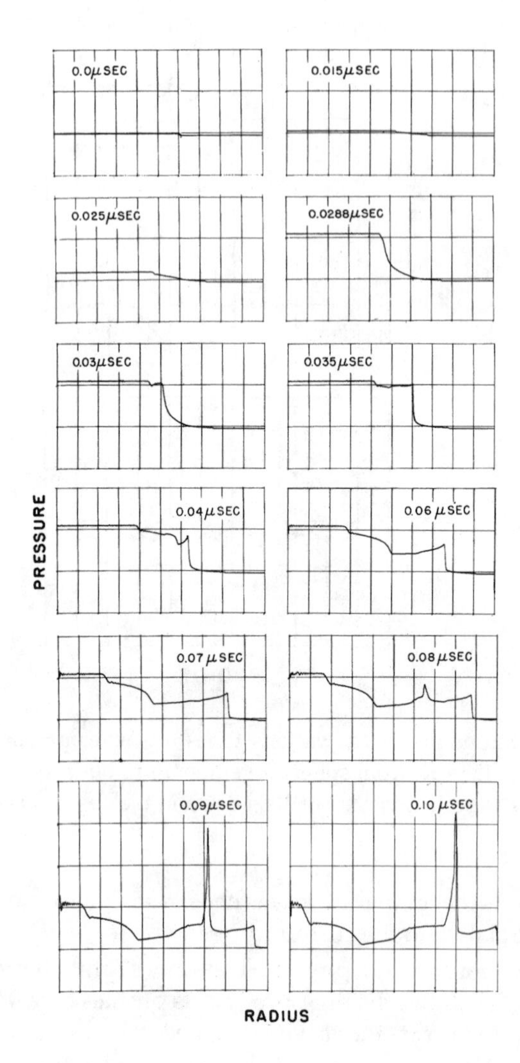

Figure 3.12 Pressure-radius profiles at various times for development of detonation in shocked nitromethane (94.7 kbar, 1230°K) from spherical temperature hot spot (1404°K) of 0.06-cm radius. The ordinate is pressure in scale divisions of 100 kbar, and the abscissa is radius in scale divisions of 0.01 cm.

shock is sent into the shocked nitromethane and a rarefaction is sent back into the hot spot. The pressure of the hot spot increases as a result of chemical reaction. At 0.0288 μsec, about 0.045 cm of the original hot spot explodes. At 0.035 μsec the entire hot spot has exploded. A shock is sent into the undetonated explosive and a rarefaction is sent into the detonation products. The undetonated explosive at the hot spot boundary does not explode until 0.08 μsec, or after an induction time of 0.045 μsec. In 0.1 μsec the detonation is propagating at full velocity and pressure.

Figure 3.13 shows the pressure-radius profiles for a 0.0292-cm-radius temperature hot spot (1404°K) in shocked nitromethane. The behavior computed is essentially the same as for the larger hot spots except for the fact that shock divergence is sufficient to cool the undetonated explosive at the hot spot interface to the point that detonation is not initiated.

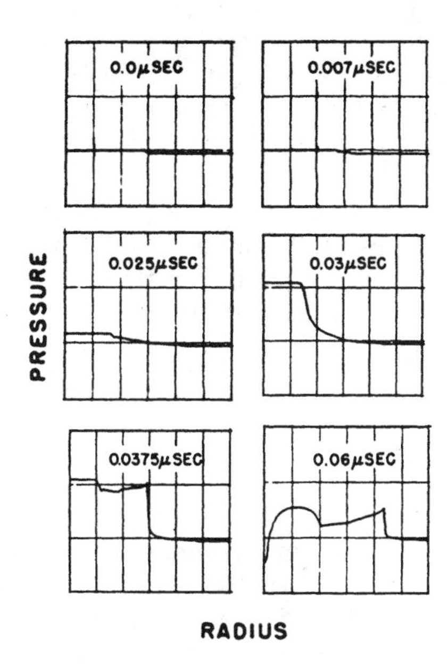

Figure 3.13 Pressure-radius profiles at various times for failure of a 0.0292-cm-radius temperature hot spot (1404°K) to initiate propagating detonation in shocked nitromethane (94.7 kbar, 1230°K). The ordinate is pressure in scale divisions of 100 kbar; the abscissa is radius in scale divisions of 0.01 cm. The nitromethane at the boundary of the hot spot is cooling (1393°K) at 0.06 μsec. At 0.08 μsec the boundary temperature has decreased to 1370°.

The model for a pressure hot spot is a sphere whose density and energy have been increased along the Hugoniot to values greater than the density and energy of the surrounding explosive. For a given temperature the pressure is therefore considerably higher than in the corresponding temperature hot spot.

Figure 3.14 shows the pressure-radius profiles for a 0.03 cm-radius pressure hot spot (112 kbar, 1395°K) in shocked nitromethane (94.7 kbar, 1230°K). The rarefaction travels into the hot spot until at 0.032 μsec about 0.013 cm of the hot spot explodes. The rarefaction has cooled the outer part of the hot spot sufficiently that it does not explode almost instantaneously, as it does in the case of the temperature hot spot. A shock is sent into the remainder of the undecomposed hot spot and, at 0.034 μsec, the shocked hot spot explosive decomposes and builds up to a C-J detonation pressure and velocity characteristic of the high density nitromethane. Upon arriving at the hot spot boundary, the shock pressure is so high that the detonation front continues unchanged.

Figure 3.15 shows the pressure-radius profiles for a 0.0175-cm-radius pressure hot spot in shocked nitromethane. At 0.032 μsec the inner 0.013 cm explodes; however, the converging rarefaction is strong enough so that the hot spot fails to initiate propagating detonation.

Over a considerable range of hot spot temperatures the critical radius of a temperature hot spot in shocked ($\sim$ 90 kbar) nitromethane is 0.03 to 0.06 cm, whereas for a pressure hot spot it is 0.015 to 0.03 cm.

The numerical solutions show that two competing processes occur just outside the hot spot. The first process is the expansion from spherical divergence of the heated fluid, which tends to cool it. The second is chemical decomposition of the fluid, which tends to heat it. Propagating detonation occurs only if the second process dominates.

Many calculations of hydrodynamic hot spots in slab geometry have been performed. The critical half width found for slab temperature hot spots in shocked ($\sim$ 90 kbar) nitromethane was 0.005-0.01 cm; that for slab pressure hot spots in shocked nitromethane was 0.01-0.02 cm.

The critical radius of spherical temperature hot spots in shocked nitromethane is 0.03-0.06 cm and that of spherical pressure hot spots is 0.015-0.03 cm. As anticipated, the critical size of slab hot spots is somewhat smaller than that of spherical hot spots.

A few calculations have included heat conduction as an additional mechanism of energy transfer. Its effect on the critical sizes and times of hydrodynamic hot spots was found to be negligible.

Hot Spot Formation and Initiation

Travis[16] has determined experimentally the shock induction times for nitromethane-filled corners of various types. The corresponding problems have been modeled for the corners formed by the intersection of a Plexiglas, Aluminum, or Gold plate with a Plexiglas plate. The computed results reproduced Travis' experimental results.

The problems are exemplified by a plane, piston supported, flat topped shock normally incident upon one side (which is taken as the bottom) of a cubical, liquid filled box of semi-infinite extent. The corner was formed by the bottom and one vertical slice of the box. The problem thus became a two-dimensional one described in the Cartesian coordinates X and Z.

As long as the extensions of the boundaries are infinite, a description of the hot spot at a given time may be scaled as X/t and Z/t to any other time, t, until a signal returns from a boundary. Therefore, increased numerical resolution of the hot spot can be obtained by allowing the problem to be run until a boundary is reached and then scaling the result as desired. For the hot spots described, about 1000 cells were in the region of the nitromethane hot spot.

The 2DL code described in Appendix B was used, and the HOM equation of state constants are given in Table 1.1 and Reference 20.

Plexiglas Corner

Figure 3.16 shows the interface positions and the isobars for the interaction of a 96 kbar shock (shock velocity $U_s = 0.5$ cm/μsec, particle velocity $U_p = 0.163$ cm/μsec) with a nitromethane

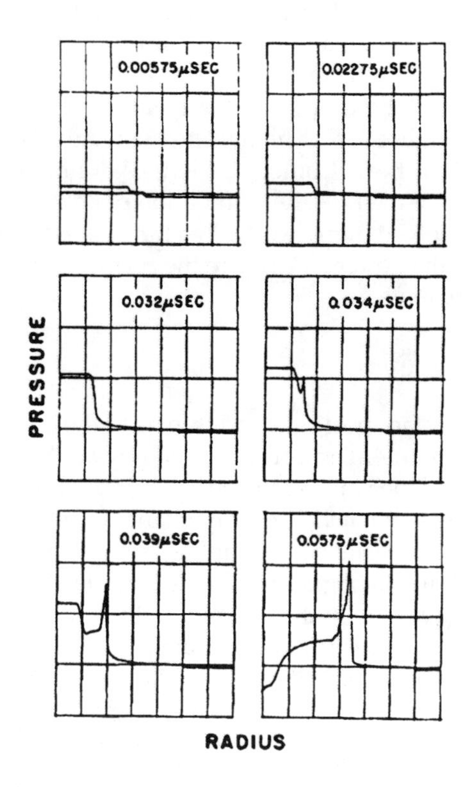

Figure 3.14 Pressure-radius profiles at various times for development of detonation in shocked nitromethane (94.7 kbar, 1230°K) from spherical pressure hot spot (112 kbar, 1395°K) of 0.03-cm radius. The ordinate is pressure in scale divisions of 100 kbar, the abscissa is radius in scale divisions of 0.01 cm.

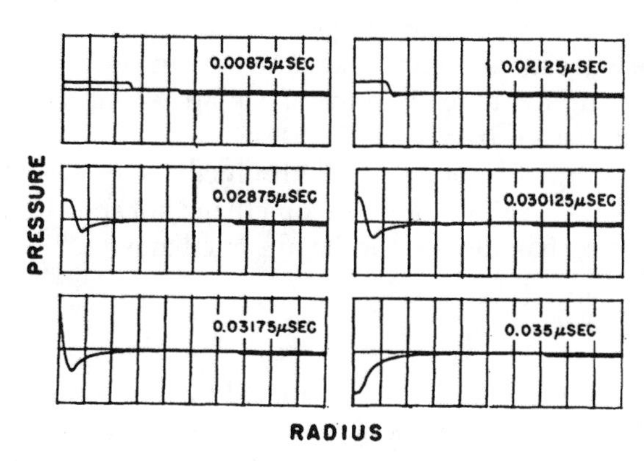

Figure 3.15 Pressure-radius profiles at various times for failure of 0.0175-cm radius pressure hot spot (112 kbar, 1395°K) to develop into detonation in shocked nitromethane (94.7 kbar, 1230°K). The ordinate is pressure in scale divisions of 100 kbar, the abscissa is radius in scale divisions of 0.005 cm.

filled corner of Plexiglas. The plane wave matched nitromethane pressure is 87.6 kbar (U_s = 0.448, U_p = 0.174, T_H = 1180°K). If the distance units in Figure 3.16 are cm, the time since the shock arrived at the nitromethane/Plexiglas interface is 0.85 μsec. The time since the piston started to move at full velocity is 3.35 μsec. The shock moves faster in the Plexiglas than in the nitromethane, and multiple shocking of the nitromethane occurs. The hot spot isotherms are shown in Figure 3.17. The nitromethane hot spot has small regions that are 70° hotter than the bulk of the nitromethane. Since the hottest region is near the shock front and is soon cooled by the flow, determination of an induction time is a complicated problem which must be solved numerically. The hot spot isobars are shown in Figure 3.18. The nitromethane hot spot had regions where the pressure was 7 kbars higher than that in the bulk of the nitromethane.

Aluminum-Plexiglas Corner

Figure 3.19 shows the interface positions and isobars of the interaction of a 96 kbar shock in Plexiglas with a nitromethane-filled corner formed by an Aluminum plate perpendicular to a Plexiglas plate. The plane wave matched Aluminum pressure is 162 kbar (U_s = 0.665, U_p = 0.089). The plane wave matched nitromethane pressure is 87.6 kbar (U_s = 0.448, U_p = 0.174, T_H = 1180°K). If the distance units in Figure 3.19 are cm, the time since the shock arrived at the nitromethane/Plexiglas or Aluminum/Plexiglas interface is 0.85 μsec. The shock moved considerably faster in the Aluminum than in the nitromethane, and multiple shocking of the nitromethane occurred. The hot spot isotherms are shown in Figure 3.20. The nitromethane hot spot had small regions that were 260° hotter than the bulk of the nitromethane. The hot spot isobars are shown in Figure 3.21. The nitromethane hot spot has regions where the pressure is 27 kbar higher than that in the bulk of the nitromethane.

Gold-Plexiglas Corner

Figure 3.22 shows the interface positions and isobars of the interaction of a 96 kbar shock in Plexiglas with a nitromethane filled corner formed by a gold plate perpendicular to a Plexiglas plate. The plane wave matched gold pressure is 261 kbar (U_s = 0.365, U_p = 0.037). As before, the plane wave matched nitromethane pressure is 87.6 kbar. If the distance units in Figure 3.22 are in cm, the time since the shock arrived at the nitromethane/Plexiglas or gold/Plexiglas interface is 0.85 μsec. The shock moves slower in the gold than in the nitromethane, and multiple shocking of the nitromethane occurs. The hot spot isotherms are shown in Figure 3.23. Regions of the nitromethane hot spot are 170° hotter, and have a pressure 20 kbar higher, than the bulk of the nitromethane.

Hot Spot Initiation

The induction times for explosion were calculated by repeating the previously described calculations with chemical reaction permitted. Travis' computed and normalized experimental induction times[16] (obtained by dividing the induction time for detonation at the discontinuity by that for detonation in the surrounding plane shocked region) are shown in the following table. Since the computed induction time of nitromethane plane shocked to 87.6 kbar was 1 ± 0.1 μsec, the computed induction times obtained at discontinuities are automatically normalized, and the computed and experimental times may be compared directly.

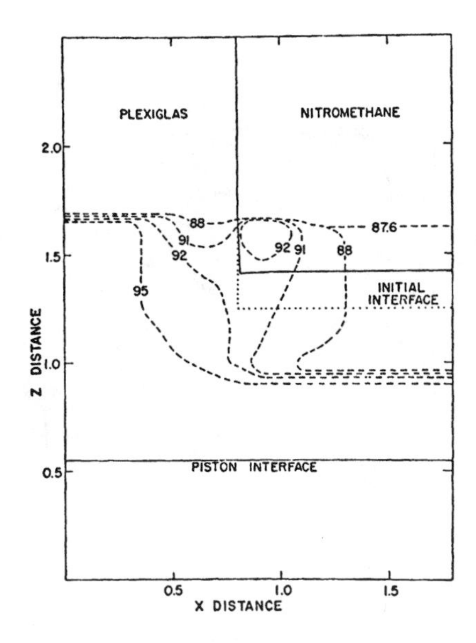

Figure 3.16 Interface positions and isobars for interaction of 96 kbar shock in Plexiglas with nitromethane-filled Plexiglas corner. The interface position is shown by a solid line; the initial interface by a dotted line. The isobars are shown by dashed lines, and the units are kbars.

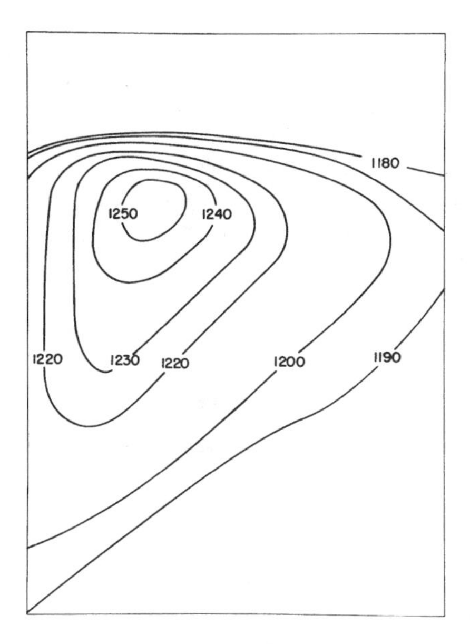

Figure 3.17 Hot spot isotherms for hot spot formed upon interaction of 96 kbar shock with a nitromethane-filled Plexiglas corner. The units are °K.

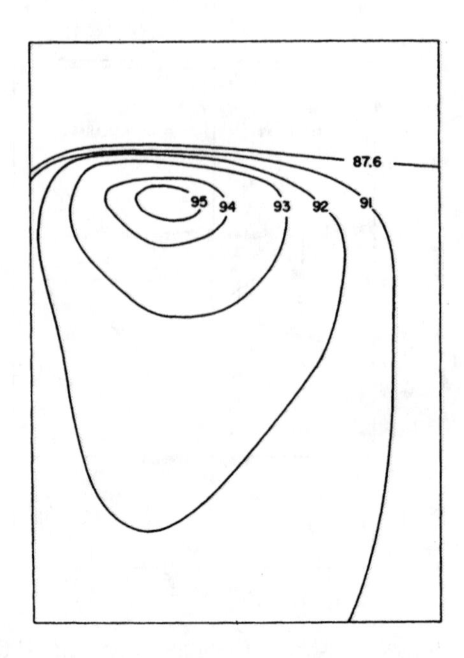

Figure 3.18 Hot spot isobars for hot spot formed upon interaction of 96 kbar shock with nitromethane-filled Plexiglas corner. The units are kbar.

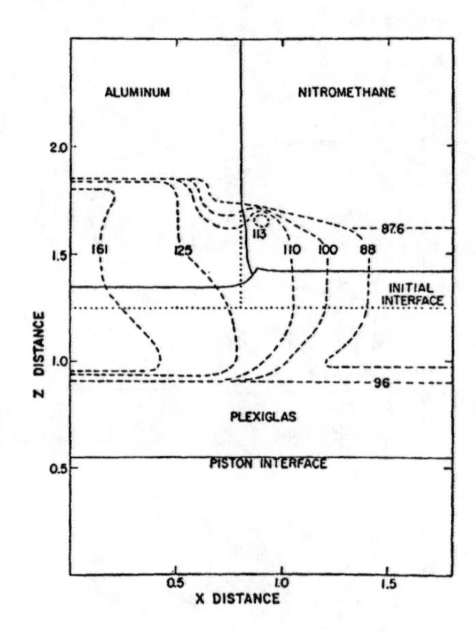

Figure 3.19 Interface positions and isobars for interaction of 96 kbar shock in Plexiglas with nitromethane-filled corner formed by an aluminum plate perpendicular to a Plexiglas plate. The interface position is shown by a solid line; the initial interface by a dotted line. The isobars are shown by dashed lines, and the units are kbars.

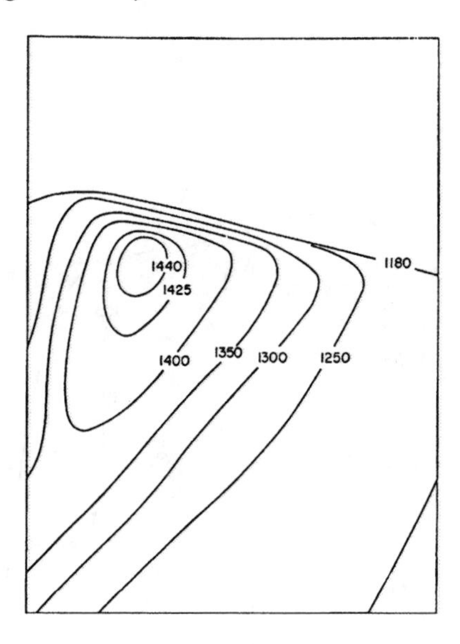

Figure 3.20 Hot spot isotherms for hot spot formed upon interaction of 96 kbar shock in Plexiglas with a nitromethane-filled corner formed by an aluminum plate perpendicular to a Plexiglas plate. The units are °K.

System	Calculated Induction Time (μsec)	Normalized Experimental Induction Time (μsec)
Plexiglas Corner	0.6 ± 0.1	0.46
Aluminum-Plexiglas Corner	0.057 ± 0.01	0.05
Gold-Plexiglas Corner	0.12 ± 0.02	0.16

The remarkable agreement of the experimental and theoretical induction times increases the confidence in the accuracy of the theoretical description of hot spot formation and the numerical description of the thermal decomposition resulting from the hot spot.

The shock interactions formed in nitromethane by corners of Plexiglas, aluminum, and gold, the resulting formation of a hot spot, and the build up to propagating detonation have been computed. The accuracy of the results was demonstrated by their agreement with the experimental induction times.

Since the interaction of shocks with density discontinuities as simple as corners produces a very complicated fluid flow, it appears that detailed numerical studies are essential for an understanding of the experimental results of more complicated systems such as heterogeneous explosives.

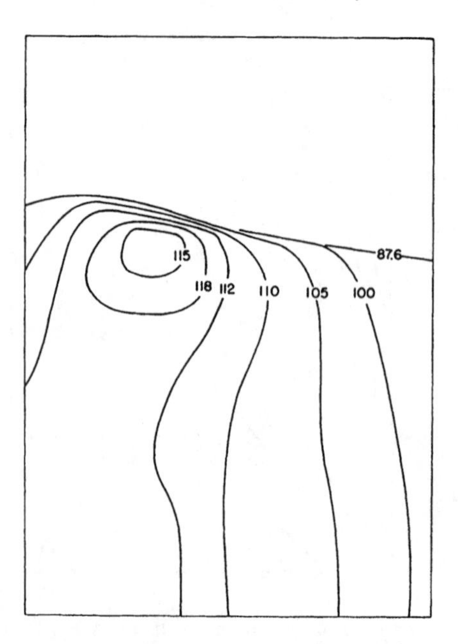

Figure 3.21 Hot spot isobars for hot spot formed upon interaction of a 96 kbar shock in Plexiglas with a nitromethane-filled corner formed by an aluminum plate perpendicular to a Plexiglas plate. The isobar units are kbars.

3.3 *Shock Initiation of Heterogeneous Explosives*

Campbell, Davis, Ramsay, and Travis[17] have studied experimentally the mechanism of shock initiation of detonation in heterogeneous explosives. Their experiments indicate that initiation of inhomogeneous explosives cannot result only from uniform shock heating of the explosives, as previously found adequate for homogeneous explosives. It is necessary to consider the result of shock interactions with density inhomogeneities which cause the formation of hot spots. The hot spots decompose and liberate energy which strengthens the shock so that, when it interacts with additional inhomogeneities, hotter hot spots are formed and more of the explosive is decomposed. The shock wave grows stronger and stronger, releasing more and more energy, until it becomes strong enough for all of the explosive to react and detonation begins.

Hydrodynamic Hot Spot Model

To increase the understanding of the basic processes involved in shock initiation of inhomogeneous explosives, numerical two and three-dimensional hydrodynamics have been used to study the formation of hot spots from shocks interacting with single and multiple voids, air holes, and other density discontinuities.

The process of heterogeneous shock initiation is described by the *Hydrodynamic Hot Spot Model*,[18–25] which models the hot spot formation from the shock interactions that occur at density discontinuities and describes the decomposition using the Arrhenius rate law and the temperature from the HOM equation of state described in Appendix A.

The studies of the one-dimensional hot spot generation and propagation were extended to two dimensions in the 1960s. In Reference 18, the basic processes in the shock initiation of

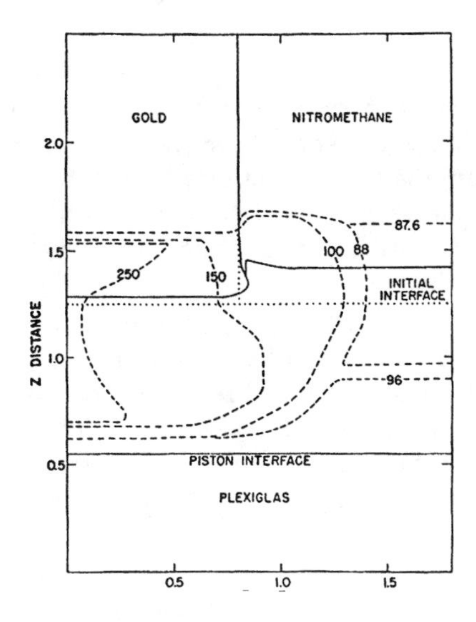

Figure 3.22 Interface positions and isobars for interaction of 96 kbar shock in Plexiglas with nitromethane-filled corner formed by a gold plate perpendicular to a Plexiglas plate. The interface position is shown by a solid line; the initial interface by a dotted line. The isobars are shown by dashed lines, and the units are kbars.

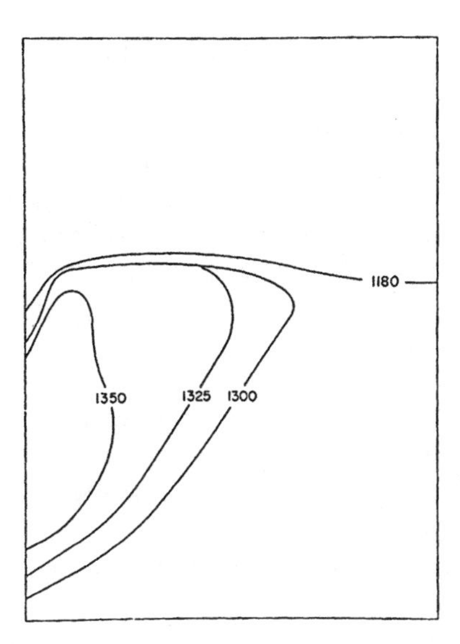

Figure 3.23 Hot spot isotherms for hot spot formed upon interaction of 96 kbar shock with a nitromethane-filled corner formed by a gold plate perpendicular to a Plexiglas plate. The units are °K.

inhomogeneous explosives were investigated using the model of a cylinder of nitromethane containing a spherical void. The interaction of a shock with a spherical void, the resulting formation of a hot spot, and build up to propagating detonation were modeled. In Reference 19, the study was extended to a cylinder of nitromethane containing either a cylindrical or biconical void, or a sphere or cylinder of aluminum. In Reference 20, the shock interactions that occur at corners described earlier in this chapter was modeled. In Reference 21, the formation of hot spots from the interaction of a shock in nitromethane with a cylindrical or rectangular void, and the failure of the hot spot to initiate propagating detonation as a result of the rarefactions interacting with the reaction zone was modeled using the 2DE code described in Appendix C. The interaction of the hot spots formed from several voids was also modeled.

When the shock wave interacts with a hole, a hot spot with temperatures several hundred degrees hotter than the surrounding explosive is formed in the region above the hole after it is collapsed by the shock wave. The hot region decomposes and contributes energy to the shock wave, which has been degraded by the hole interaction. Whether this energy is sufficient to compensate for the loss from the hole interaction depends on the magnitude of the initial shock wave, the hole size, and the interaction with the flow from its nearest neighbor hot spots.

The development of the three-dimensional Eulerian code, 3DE, described in Appendix D, allowed the *Hydrodynamic Hot Spot Model* of heterogeneous shock initiation to be used to investigate the shock interaction with a matrix of holes and the resulting formation of hot spots, their interaction, and build up toward a propagating detonation. The *Hydrodynamic Hot Spot Model* has been used to evaluate the relative effect of explosive shock sensitivity as a function of composition, pressure, temperature, density, and particle size. It has also been used to understand the desensitization of explosives by preshocking.

The process of shock initiation of heterogeneous explosives has been analyzed numerically[22] by studying the interaction of shock waves with a cube of nitromethane containing 91 holes. An 85 kbar shock interacting with a single 0.0002-cm hole did not build toward detonation. When the shock wave interacted with a matrix of 0.0002-cm holes, it became strong enough to build toward detonation. When the size of the holes was reduced to 0.00004 cm, a marginal amount of the explosive decomposed to compensate for the energy loss to the flow caused by the shock wave interacting with the holes. The shock wave slowly grew stronger, but it did not build to detonation in the time of the calculation. A 55 kbar shock wave interacting with a matrix of 0.0002-cm holes resulted in insufficient heating of the resulting hot spots to cause significant decomposition. Desensitization by preshocking resulted when holes were closed by the low-pressure 55 kbar shock wave. A higher pressure 85 kbar shock wave that arrived later had no holes with which to interact and behaved like a shock wave in a homogeneous explosive until it caught up with the lower pressure preshock wave.

The interaction of a shock wave with a single air hole and a matrix of air holes in PETN, HMX, TATB, and Nitroguanidine was modeled in References 23 and 24. The basic differences between shock-sensitive explosives (PETN or HMX) and shock insensitive explosives (TATB or Nitroguanidine) were described by the *Hydrodynamic Hot Spot Model*.

Shock Sensitivity and Composition

The *Hydrodynamic Hot Spot Model* has been used to describe the experimentally observed sensitivity to shock sensitivity of the heterogeneous explosives PETN, HMX, TATB, and

Nitroguanidine, with PETN being the most sensitive and Nitroguanidine the least sensitive.

To initiate PBX-9404 (HMX-based explosive) or PBX-9502 (TATB-based explosive) at maximum pressed density within 0.4 cm of shock run requires a shock wave in PBX-9404 of 50 kbar and in PBX-9502 of 160 kbars as determined from the experimental Pop plots.[26]

To initiate PETN at 1.75 g/cc (crystal density is 1.778) within 0.4 cm of shock run requires a pressure of only 20 kbar, while to initiate Nitroguanidine at 1.723 g/cc (crystal density is 1.774) within 0.4 cm of shock run requires a pressure of 250 kbar.[26]

The hole size present in such pressed explosives varies from holes of 20 to 600 A in TATB crystals to holes as large as 0.05 cm in the explosive-binder matrix. Most of the holes vary in size from 0.005 to 0.0005 cm in diameter, so holes in that range of diameter were modeled.

The hot spot formed when a shock wave interacts with a spherical hole scales with the radius of the hole as long as no chemical reaction occurs. Using hot spot temperatures in the calculated range of 700 to 1300°K and calculating the adiabatic explosion is shown below. The ordering is identical to that observed experimentally.

Explosive	Hot Spot 700 K	Temperature 1000 K	1300 K
Nitroguanidine	5504 μsec	124 μsec	18.47 μsec
TATB	1290 μsec	$6\text{x}10^{-3}$ μsec	$1\text{x}10^{-5}$ μsec
HMX	5.26 μsec	$1\text{x}10^{-4}$ μsec	$5\text{x}10^{-7}$ μsec
PETN	0.08 μsec	$7\text{x}10^{-6}$ μsec	$5\text{x}10^{-8}$ μsec

The interaction of shock waves of various pressures with single cubical air holes of various sizes in PETN, HMX, TATB, and Nitroguanidine (NQ) was investigated. The calculations model the hot spot explosion and failure to propagate because of rarefaction cooling of the reactive wave. If the reaction comes too fast to numerically resolve the cooling by rarefactions, the flow builds toward a detonation too quickly.

A summary of the results of the study is shown in Table 3.5. The ordering of shock sensitivity of the explosives correlates well with the observed Pop plot data.[26] To evaluate the sensitivity to shock more realistically, a matrix of holes in the explosives HMX, TATB, and PETN was modeled.

The computational grid contained 24 x 22 x 36 cells, each 1×10^{-4} cm on a side. The 36 air holes were described by 4 cells per sphere diameter. Numerical tests with 2 to 6 cells per sphere diameter showed the results were independent of grid size for more than 3 cells per sphere diameter. The air holes were located on a hexagonal close-packed lattice (HCP). The closest distance for the HCP matrix between holes was 3.8×10^{-4} cm. The time step was 1×10^{-5} μsec.

The interaction of a 50 kbar shock wave in HMX with a matrix of spherical holes of 4×10^{-4} cm diameter was modeled. The void fraction was 10%. While a single hole fails to build toward detonation as shown in Figure 3.24, the matrix of holes builds toward a detonation as shown in Figure 3.25. The experimental run to detonation for a 50 kbar shock wave in 1.71 g/cc HMX is 0.17 cm. While a propagating detonation would not be expected to occur experimentally in this geometry, the enhancement of the shock wave would occur.

TABLE 3.5 Single Cubical Air Hole Study

Explosive	Air Cube Size (cm)	Pressure (kbar)	Result
HMX	0.0005	25	Fails to Build Toward a Detonation
	0.0005	50	Fails to Build Toward a Detonation
	0.0005	75	Builds Toward a Detonation
	0.005	25	Fails to Build Toward a Detonation
	0.005	50	Marginal
	0.005	75	Builds Toward a Detonation
TATB	0.005	50	Fails
TATB	0.005	125	Fails
TATB	0.005	150	Builds Toward a Detonation
TATB	0.05	75	Fails
PETN	0.0005	25	Builds Toward a Detonation
PETN	0.0005	75	Builds Toward a Detonation
NQ	0.005	250	Fails
NQ	0.5	250	Builds Toward a Detonation

The interaction of a 125 kbar shock wave in TATB with a single spherical hole of 4×10^{-3} cm diameter is shown in Figure 3.26. It fails to build toward a detonation. The interaction of a 125 kbar shock wave in TATB with a matrix of spherical holes of 4×10^{-3} cm diameter with a void fraction of 10% is shown in Figure 3.27. The flow builds toward a detonation. The experimental run to detonation for a 125 kbar shock wave in 1.71 g/cc TATB is 0.3 cm. The computed detonation occurs too quickly because of insufficient numerical resolution when the shock wave is enhanced to high enough pressures and temperatures by the interacting hot spots.

The interaction of a 20 kbar shock wave in PETN with a single spherical hole of 4×10^{-4} cm diameter is shown in Figure 3.28. Build up toward a detonation did not occur. The interaction of a 20 kbar shock wave in PETN with a matrix of spherical holes of 4×10^{-4} cm diameter with a void fraction of 10% is shown in Figure 3.29. The flow builds towards a detonation after the hot spots interact. The computed detonation in this geometry is a result of insufficient numerical resolution at high decomposition rates. The experimental run to detonation for a 20 kbar shock wave in 1.60 g/cc PETN is 0.2 cm.

The experimental run to detonation values are about the same for a 125 kbar shock wave interacting with TATB with 10% voids, for a 50 kbar shock wave interacting with HMX with 10% voids, and for a 20 kbar shock wave interacting with PETN with 10% voids.

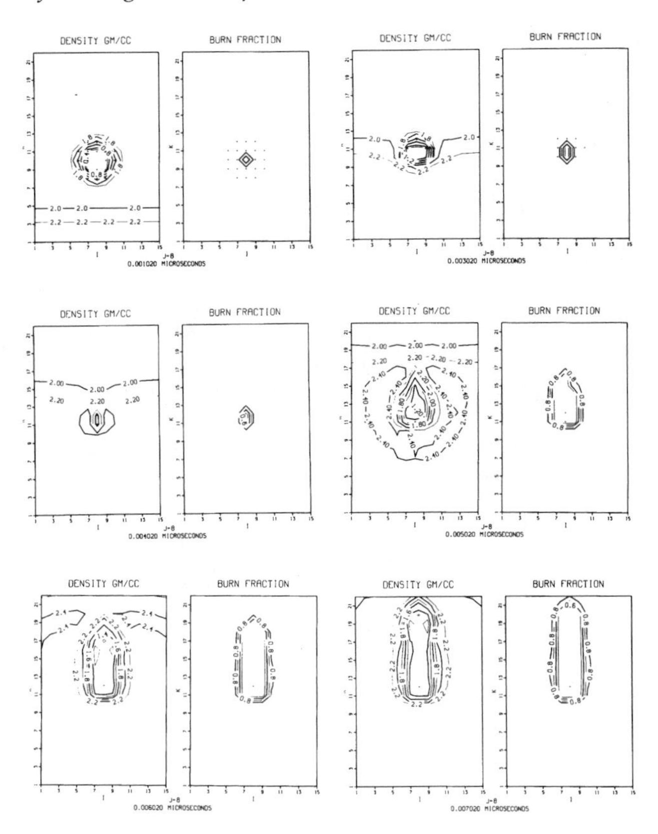

Figure 3.24 A 4 x 10^{-4} cm diameter spherical air hole in HMX. The initial shock pressure is 50 kbar. The density and burn fraction cross sections through the center of the hole are shown at various times. The flow does not build toward detonation.

The *Hydrodynamic Hot Spot Model* describes the basic differences between shock sensitive and shock insensitive explosives. The interaction of a shock wave with air holes in PETN, HMX, TATB, and Nitroguanidine, the resulting hot spot formation, interaction, and the build up toward detonation or failure have been modeled. Increased hole size results in larger hot spots that decompose more of the explosive and add their energy to the shock wave and result in increased sensitivity of the explosive to shock. Increasing the number of holes results in more hot spots that decompose more explosive and increase the sensitivity of the explosion to shock. The interaction between hole size and number of holes is complicated and requires numerical modeling for adequate evaluation of specific cases. The hole size can become sufficiently small (the critical hole size) that the hot spot is cooled by side rarefactions before appreciable decomposition can occur. Since increasing the number of holes while holding the percentage of voids present constant results in smaller holes, there are competing processes that may result in either a more or less shock sensitive explosive. If the hole size is below the critical hole size, then the explosive will become less sensitive with increasing number of holes of decreasing diameter.

To evaluate the potential shock sensitivity of an explosive for engineering purposes, one needs to determine experimentally the Arrhenius constants. Then one calculates the adiabatic explosion times for several assumed hot spot temperatures to determine the relative

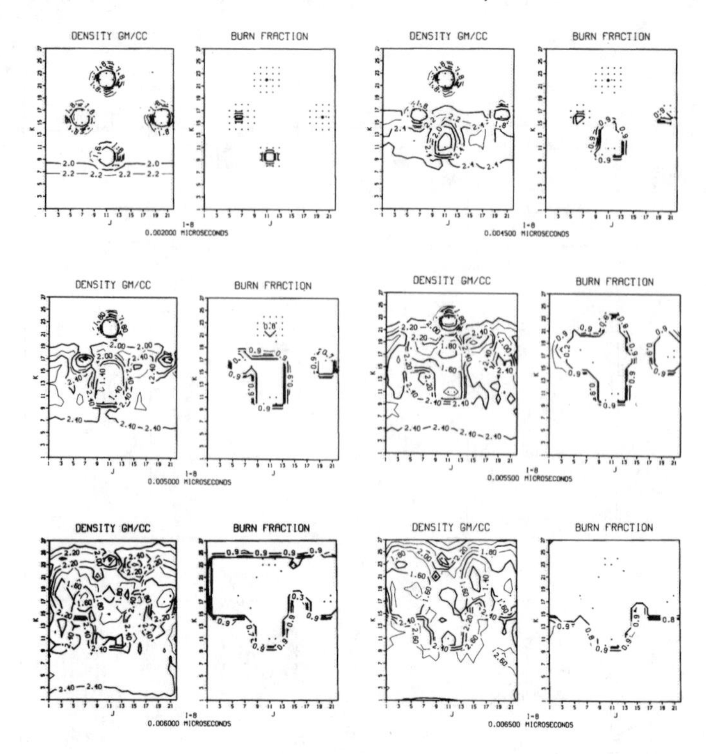

Figure 3.25 A matrix of 10% air holes in HMX. The spherical air holes have a diameter of 4×10^{-4} cm. The initial shock pressure is 50 kbar. The density and mass fraction contours are shown for a cross section through the center of the matrix. The flow builds toward a propagating detonation.

sensitivity of the explosive compared with explosives of known sensitivity. A more detailed evaluation can be obtained from calculations using the *Hydrodynamic Hot Spot Model*.

Particle Size and Temperature Effects on Shock Sensitivity

The effects of particle size and the resulting hole or void size on the shock initiation properties of cast and pressed heterogeneous explosives have been studied using the wedge test and the gap test for many years. Studies have also been performed investigating the effect of temperature on shock initiation properties of explosives.

Cast TNT has been observed to be less shock-sensitive than pressed TNT at the same density. Cast TNT has fewer and larger holes than pressed TNT, resulting in fewer hot spots that are too far apart to be as effective in supporting the reactive shock wave as pressed charges with more, although smaller, hot spots. Campbell, Davis, Ramsay, and Travis[17] have used the wedge test to measure the distance of run to detonation as a function of shock pressure (the Pop plot) for pressed-TNT charges of coarse TNT (200-250 micrometers) and found that they are less shock-sensitive than pressed charges of finer TNT (20-25 micrometers).

Price[27] has reviewed the effect of particle size on the shock sensitivity of pure porous explosives considering both gap test and wedge test data. She showed that in many gap tests, coarse porous explosive seems more shock-sensitive than fine, whereas in most wedge tests, the reverse is true. She concludes that the apparent reversal is actually a result of

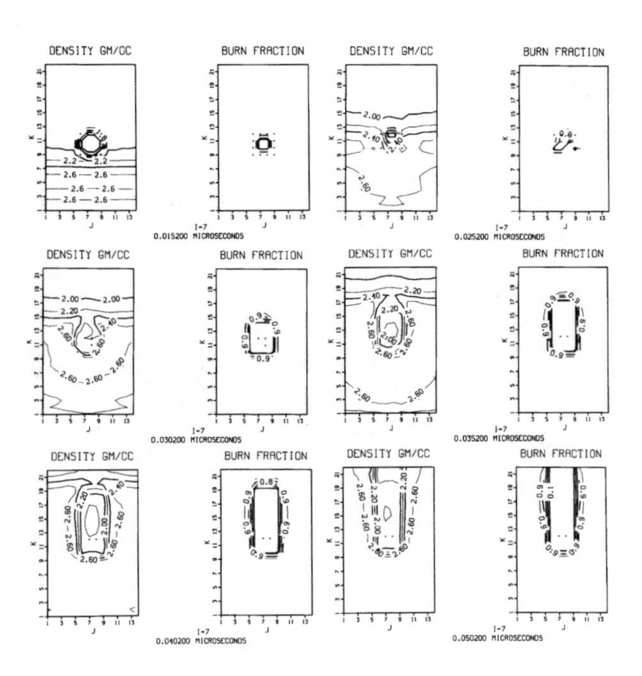

Figure 3.26 A 4 x 10^{-3} cm diameter spherical hole in TATB. The initial pressure is 125 kbar. The density and burn fraction cross sections through the center of the hole are shown at various times. The flow does not build toward a detonation.

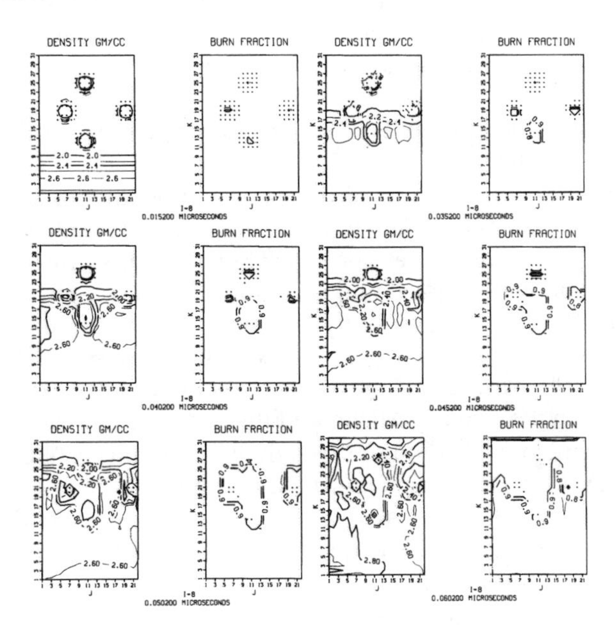

Figure 3.27 A matrix of 10% air holes in TATB. The spherical air holes have a diameter of 4 x 10^{-3} cm. The initial shock pressure is 125 kbar. The density and mass fraction contours are shown for a cross section through the center of the matrix. The flow builds toward a detonation.

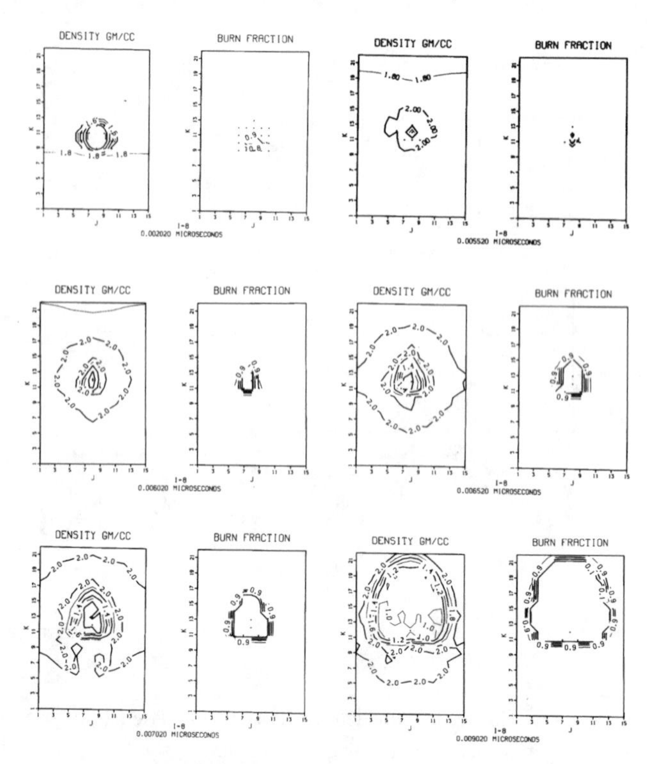

Figure 3.28 A 4 x 10^{-4} cm diameter spherical air hole in PETN. The initial shock pressure is 20 kbar. The density and burn fraction cross sections through the center of the hole are shown at various times. The flow does not build toward a detonation.

crossing Pop plots and that the gap tests sense a lower pressure region than those usually available in wedge test data.

Moulard, Kury, and Delcos[28] studied two special monomodal RDX polyurethane cast PBX formulations of 70 wt% RDX (either 6 micrometers or 134- micrometers medium particle size) and 30 wt% polyurethane. The shock initiation properties of these formulations were measured in thin plate impact, projectile impact, and wedge tests. The formulations containing the fine RDX were significantly less sensitive than those with coarse RDX at the same density of 1.44 g/cc, at least up to 100 kbar shock pressure. From these studies it was concluded that at the same density, the most shock-sensitive explosives are those with particle sizes between the coarse particles and the very fine particles.

Waschl and Richardson[29] studied the sensitivity of hexanitrostibene (HNS) with variations in specific surface area. Maximum sensitivity was found to occur for specific surface areas of 10 to 20 m^2/g for HNS pressed to 90 percent of maximum density. The sensitivity was also found to decrease as the density decreased and to approach homogeneous explosive shock initiation behavior at very high densities.

Urizar[30] studied the impact sensitivity of RDX, HMX, and PETN using a Type-12 impact tool and showed a decrease in the 50% point from 27 cm at 20°C to 15 cm at 160°C for RDX. Similar behavior was observed for HMX and PETN. He found that the violence of the explosion increased with increasing temperature.

Schwarz[31] showed that a less energetic flyer plate is required to initiate a pellet of TATB

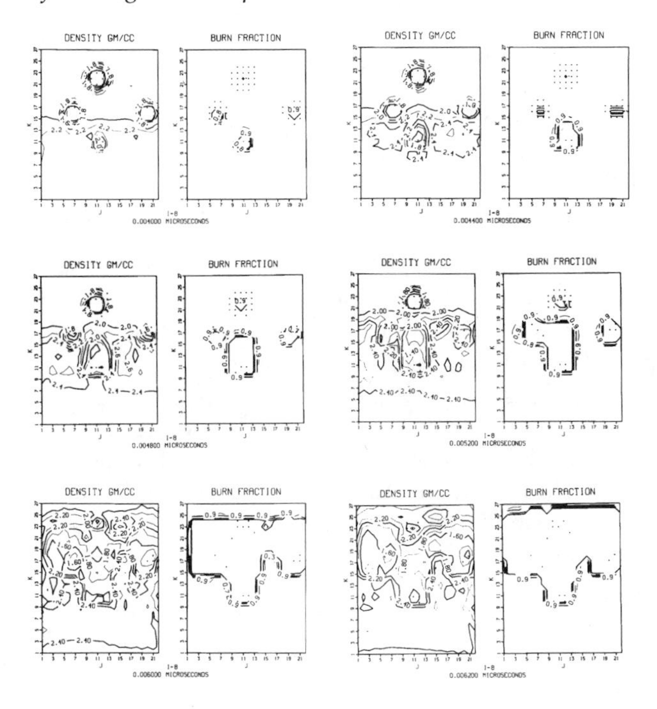

Figure 3.29 A matrix of 10% air holes in PETN. The spherical air holes have a diameter of 4 x 10^{-4} cm. The initial shock pressure is 20 kbar. The density and mass fraction contours are shown for a cross section through the center of the matrix. The flow builds toward a detonation.

when the initial temperature is 260°C compared to that required when the temperature is 25°C. The failure diameter of explosives has been found to decrease with increasing temperatures[32] and since the failure diameter is directly related to the Pop Plot, the explosives are more shock-sensitive with increasing temperature.

Ramsay, Craig, and Dick[33] performed wedge tests for PBX-9501 (95/2.5/2.5 HMX/Estane /nitroplasticizer), DATB (95/5 DATB/Viton), and TNT at temperatures between 25° and 150°C. A decrease in the distance-to-detonation of approximately 30% for the same input pressure was found for DATB and PBX-9501 at 150°C. A small decrease in distance-to-detonation was found for cast TNT between 24° and 73°C.

The shock initiation of LX-17 (92.5/7.5 TATB/Kel-F) at temperatures between -54°C and 88°C was studied using embedded gauges in Reference 34. The shock sensitivity increased with increasing temperature.

Dallman[35] performed wedge tests for PBX-9502 (95/5 TATB/Kel-F) explosive at 75° and 250°C. The PBX-9502 shock sensitivity increased with increasing temperature until at 250°C it was as shock sensitive as PBX-9501 (95/5 HMX/Estane) at pressures greater than 50 kbar.

It is known that low temperatures decrease the reliability of marginal boosters to initiate detonation in explosives such as Composition B, PBX-9404, and TNT. The reliability increases at elevated temperatures.

From the *Hydrodynamic Hot Spot Model*, one can postulate that the coarse particle explosives have fewer holes or voids per unit volume than the fine-particle explosives, resulting in fewer but larger hot spots. As the explosive particles become finer, the number of hot spots formed by a shock wave increases while the hot spot size decreases. When the explosive becomes very fine, the hot spot size can become so small that the hot spots cool from side rarefactions before appreciable decomposition can occur. This results in a less shock-sensitive explosive if the explosive has very fine particle sizes. A three-dimensional study[25] of the effect of particle size and the resulting void or hole size on the shock initiation of heterogeneous charges of TNT showed that at the same density, the most shock-sensitive explosive was one with particle sizes between coarse and extremely fine material.

Another result of the *Hydrodynamic Hot Spot Model* is that explosives with faster Arrhenius kinetics form hot spots that decompose faster and are less affected by side rarefactions before appreciable decomposition occurs. Explosives with faster Arrhenius kinetics exhibit increasing shock sensitivity with decreasing particle size for smaller particle sizes than explosives with slower kinetics. The effect of increasing the initial temperature of an explosive in the hydrodynamic hot spot model is to increase the temperature of the hot spots resulting from shock interactions with voids. The hotter hot spots decompose more and result in faster build up to detonation. Thus increasing the initial temperature of an explosive without significantly changing the density or density discontinuities results in a more shock-sensitive explosive.

To model the effect of particle size and temperature, the Arrhenius reactive rate law was used with the constants determined experimentally by Raymond N. Rogers and described in Table 3.1. The HOM equation-of-state constants used for HMX, TATB, and TNT are described in References 24 and 25. The equation of state for the detonation products were described using the BKW equation of state.

In the model a constant-velocity piston was applied to the bottom of a cube of explosive, shocking the explosive initially to the desired pressure. When the shock wave interacts with a hole, a hot spot with temperatures several hundred degrees hotter than the surrounding explosive is formed in the region above the hole after it is collapsed by the shock wave, which has been degraded by the hole interaction. Whether this energy is sufficient to compensate for the loss from the hole interaction depends on the magnitude of the initial shock wave, the hole size, and the interaction with the flow from its nearest neighbor hot spots.

Single Hole Study

The interaction of a shock wave with a single rectangular air hole in TNT, TATB, and HMX was modeled using the 3DE code. The calculations were performed with 15 x 15 x 22 (i, j, k) cells. The air hole was 5 x 5 x 5 cells located in the center of the i-j plane and the bottom of the hole was 5 cells above the piston. The results of the calculations were classified according to the amount of burn caused by the hot spot. If the initial burned region was not expanding at the end of the calculation, it was classified as failed. If the initial burn region was expanding along with the front of the shock wave, the hot spot was classified as propagating. Intermediate cases where the burn region was expanding, but was behind the shock front were classified as marginal. A summary of the 54 cases modeled is presented in Table 3.6.

By comparing the initiation characteristics as a function of particle size at fixed pressures, one can determine approximate shock sensitivity equivalency. For example, from Table 3.6

the following can be concluded:

TATB at 250°C and 50 kbar ≅
TATB at 75°C and 100 kbar ≅
TATB at 27°C and 125 kbar ≅
HMX at 27°C and 75 kbar

or

TATB at 27°C and 150 kbar ≅
TATB at 75°C and 125 kbar ≅
TATB at 250°C and 75 kbar ≅
HMX at 27°C and between
75 and 100 kbar

TNT is less shock sensitive than TATB and, since TNT melts at 82°C, it becomes even less shock sensitive than TATB at higher temperatures since it becomes a homogeneous explosive.

From these results, conditions were chosen for the more realistic multiple hole model. Pressures modeled were for cases where the single hole hot spots decomposed but failed to expand appreciably. For HMX the pressure chosen for the multiple hole model study was 50 kbar and for TATB it was 75 kbar. The pressure chosen for TNT was 125 kbar.

Multiple Hole Study

The interaction of a shock wave with a matrix of spherical air holes in HMX, TATB, and TNT was modeled with the void fraction held at 0.5% and the hole size varied from 0.5 to 0.000005 cm radius.

Calculations were performed using 22 x 20 x 111 (i, j, k) or 48,840 cells with a hole diameter of at least two cells. The spherical air holes were placed on a hexagonal close-packed lattice. Five layers of cells above the piston did not contain any holes. With 2 cells per sphere diameter, the matrix contains a maximum of 60 holes. The densities used for HMX were 1.90 at 27°C, 1.8864 at 75°C, and 1.8385 at 250°C. The densities used for TATB were 1.937 at 27°C, 1.9245 at 75°C, and 1.8756 at 250°C. A summary of the multiple air hole calculations is given in Table 3.7.

The density and mass fraction of undecomposed explosive cross sections through i=11 are shown at various times for TATB with a matrix of 0.005-cm-radius holes at 27°C in Figure 3.30, and at 75°C in Figure 3.31. The amount of decomposition increases with temperature with propagation of detonation occurring at 75°C. The detonation occurs more quickly at 250°C than at 75°C.

The density and mass fraction of undecomposed explosive cross sections through i=11 are shown at various times for HMX with a matrix of 0.005 cm air holes at 27°C in Figure 3.32, which is marginal, and for 0.0005 cm air holes in Figure 3.33 which propagates. The air holes are 0.000005 cm radius at 27°C in Figure 3.34 and at 250°C in Figure 3.35. The matrix of holes fails to form a propagating detonation at 27°C and propagates at 250°C.

TABLE 3.6 Single Air Hole

Explosive	Hole Size cm	Temperature Deg K	Pressure kbar	Result
TATB	0.05	300	75.	failed
TATB	0.05	300	100.	marginal
TATB	0.05	300	125.	propagated
TATB	0.005	300	125.	marginal
TATB	0.0005	300	125.	failed
TATB	0.00005	300	125.	failed
TATB	0.05	300	150.	propagated
TATB	0.005	300	150.	propagated
TATB	0.0005	300	150.	marginal
TATB	0.00005	300	150.	failed
TATB	0.05	348	75.	marginal
TATB	0.005	348	75.	failed
TATB	0.0005	348	75.	failed
TATB	0.00005	348	75.	failed
TATB	0.05	348	100.	propagated
TATB	0.005	348	100.	marginal
TATB	0.0005	348	100.	failed
TATB	0.00005	348	100.	failed
TATB	0.05	348	125.	propagated
TATB	0.005	348	125.	propagated
TATB	0.0005	348	125.	marginal
TATB	0.00005	348	125.	failed
TATB	0.05	523	50.	propagated
TATB	0.005	523	50.	marginal
TATB	0.0005	523	50.	failed
TATB	0.00005	523	50.	failed
TATB	0.05	523	75.	propagated
TATB	0.005	523	75.	propagated
TATB	0.0005	523	75.	marginal
TATB	0.00005	523	75.	failed
HMX	0.05	300	50.	failed
HMX	0.05	300	75.	propagated
HMX	0.005	300	75.	marginal
HMX	0.0005	300	75.	failed
HMX	0.00005	300	75.	failed
HMX	0.05	300	100.	propagated
HMX	0.005	300	100.	propagated
HMX	0.0005	300	100.	propagated
HMX	0.00005	300	100.	marginal

TABLE 3.6 Single Air Hole (continued)

Explosive	Hole Size cm	Temperature Deg K	Pressure kbar	Result
HMX	0.05	300	125.	propagated
HMX	0.005	300	125.	propagated
HMX	0.0005	300	125.	propagated
HMX	0.00005	300	125.	marginal
HMX	0.05	348	50.	failed
HMX	0.005	348	50.	failed
HMX	0.0005	348	50.	failed
HMX	0.00005	348	50.	failed
HMX	0.05	348	75.	propagated
HMX	0.005	348	75.	propagated
HMX	0.0005	348	75.	marginal
HMX	0.00005	348	75.	failed
TNT	0.05	300	150.	propagated
TNT	0.05	300	125.	marginal
TNT	0.005	300	150.	failed

TABLE 3.7 Multiple Air Holes

Explosive	Hole Size cm	Temperature Deg K	Pressure kbar	Result
TATB	0.5	300	75.	propagated
TATB	0.05	300	75.	propagated
TATB	0.005	300	75.	marginal
TATB	0.0005	300	75.	marginal
TATB	0.005	348	75.	propagated
TATB	0.005	523	75.	propagated
TATB	0.005	523	50.	propagated
TNT	0.5	300	125.	propagated
TNT	0.05	300	125.	propagated
TNT	0.005	300	125.	marginal
TNT	0.0005	300	125.	fails
HMX	0.005	300	50.	marginal
HMX	0.0005	300	50.	propagated
HMX	0.00005	300	50.	marginal
HMX	0.000005	300	50.	fails
HMX	0.000005	348	50.	fails
HMX	0.000005	523	50.	propagated

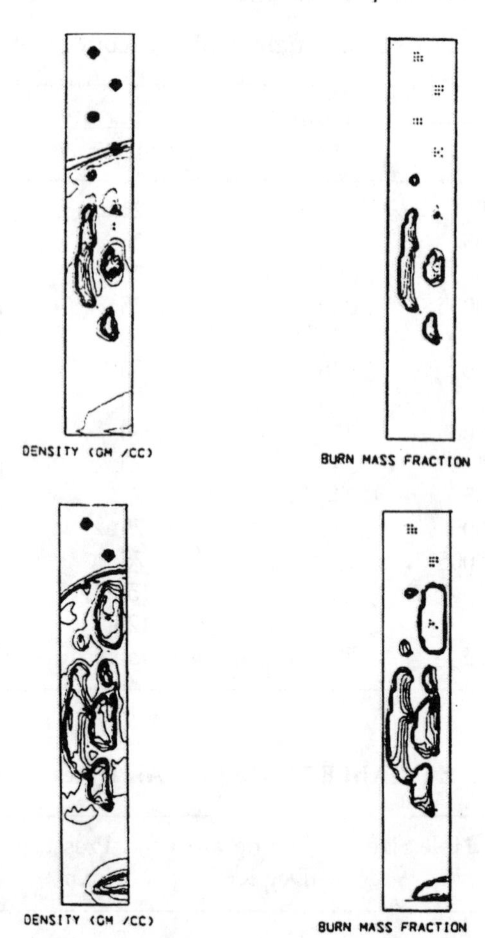

Figure 3.30 The density and mass fraction of undecomposed explosive cross sections through i=11 are shown for various times for TATB with a matrix of 0.005-cm-radius holes at 27°C. The density contour is 0.2 g/cc and the mass fraction contour interval is 0.2. The shock pressure is 75 kbar. The initiation is marginal.

The most sensitive HMX explosive is the one with particle sizes between coarse and very fine explosive. The hole size for a matrix of HMX with maximum sensitivity is two orders of magnitude smaller than for TATB or TNT as are the hole size for failure to occur.

The *Hydrodynamic Hot Spot Model* describes the basic differences between the shock-sensitive and shock-insensitive explosives. The interaction of a shock wave with air holes in HMX, TATB, and TNT, the resulting hot spot formation, interaction, and the build up toward detonation or failure have been modeled. An increase in hole size results in larger hot spots that decompose more of the explosive, add their energy to the shock wave, and result in increased sensitivity to shock. An increase in number of holes also results in more hots spots that decompose more explosive and increase the sensitivity of the explosive to shock. The interaction between hole size and number of holes is complicated and requires numerical modeling for adequate evaluation of specific cases. The hole size can become sufficiently small that the hot spot is cooled by side rarefaction before appreciable decomposition can

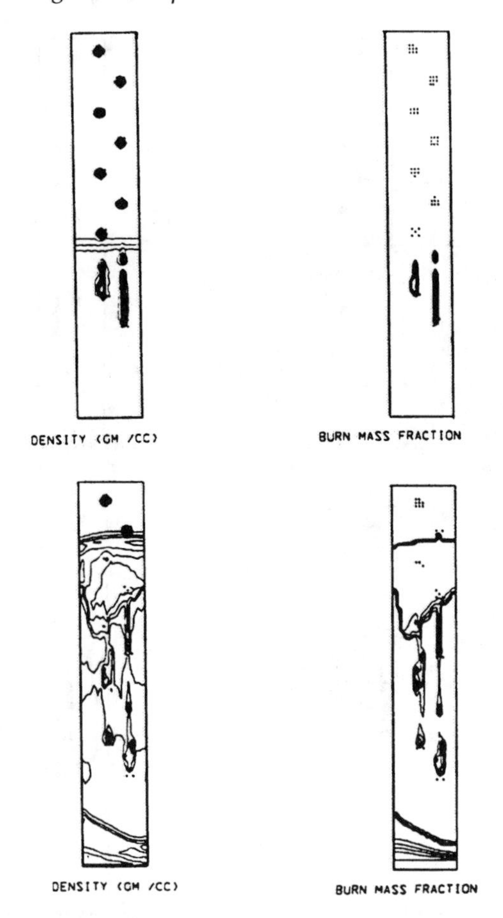

Figure 3.31 The density and mass fraction of undecomposed explosive cross sections through i=11 are shown at various times for TATB with a matrix of 0.005-cm-radius holes at 75°C. The density contour interval is 0.2 gm/cc and the mass fraction contour interval is 0.2. The shock pressure is 75 kbar. The initiation propagates a detonation.

occur. Since increasing the number of holes, while holding the percentage of voids present constant, results in smaller holes, competing processes may result in either a more or less shock sensitive explosive. If the hole size is below the critical hole size, then the explosive will become less sensitive with an increasing number of holes of decreasing diameter.

Increasing the temperature of an explosive causes a small decrease in density which results in slightly lower pressure hot spots; however, the resulting hot spot is hotter for the higher temperature explosives. More of the hot spot explosive decomposes resulting in an increasing shock sensitivity with increasing temperature. The *Hydrodynamic Hot Spot Model* indicates that the effect of temperature is greatest for the explosives with slower Arrhenius decomposition rates. The model gives about the same shock sensitivity for TATB at 250°C as for HMX at 25°, which is in agreement with the experimental observations.[35]

The *Hydrodynamic Hot Spot Model* can be used to evaluate the relative effect of explosive shock sensitivity as a function of composition, pressure, temperature, and density (as represented by the number and sizes of the holes present for hot spot generation). A higher

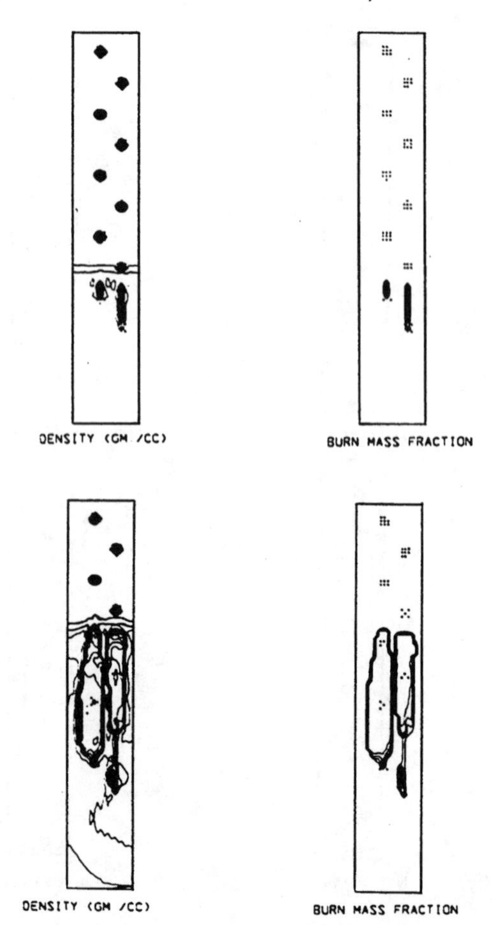

Figure 3.32 The density and mass fraction of undecomposed explosive cross sections through i=11 are shown at various times for HMX with a matrix of 0.005-cm-radius holes at 27°C. The density contour interval is 0.2 gm/cc and the mass fraction contour interval is 0.2. The shock pressure is 50 kbar. The initiation is marginal.

decomposition rate results in explosives that exhibit a decrease in sensitivity with smaller particle sizes than for explosives with slower rates. The faster decomposition rate results is less time for the hot spot to be cooled before complete decomposition has occurred.

Desensitization of Explosives by Preshocking

Preshocking a heterogeneous explosive with a shock pressure too low to cause propagating detonation in the time of interest can cause a propagating detonation in unshocked explosive to fail to continue propagating when the detonation front arrives at the previously shocked explosive. Initiation of explosives by jets that first penetrate barriers of inert materials in contact with the explosive has resulted in decreased sensitivity of the explosive to jet impact. The bow shock wave that travels ahead of the jet through the barrier can desensitize the explosive sufficiently that it cannot be initiated by the higher pressure generated near the jet interface.

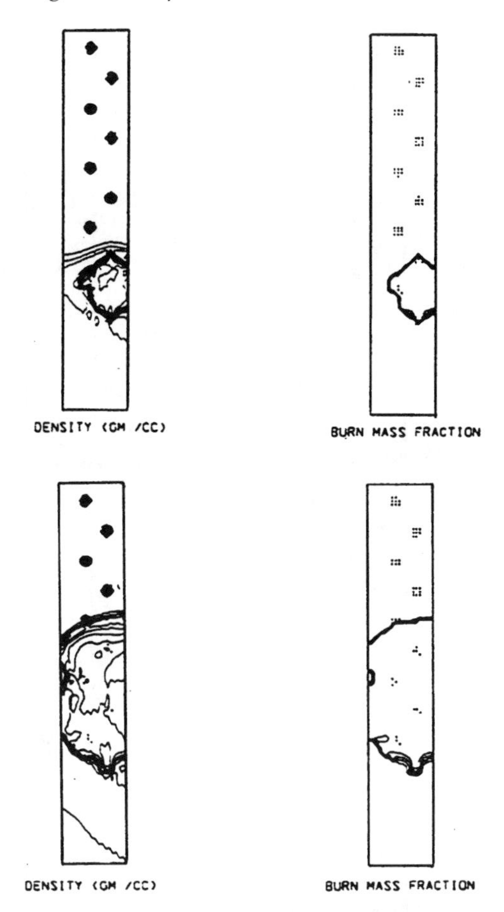

Figure 3.33 The density and mass fraction of undecomposed explosive cross sections through i=11 are shown at various times for HMX with a matrix of 0.0005-cm-radius holes at 27°C. The density contour interval is 0.2 gm/cc and the mass fraction contour interval is 0.2. The shock pressure is 50 kbar. The initiation propagates.

The mechanism of explosive desensitization by preshocking has been studied using a three-dimensional reactive hydrodynamic model of the process. With the mechanism determined, it was possible to modify the Forest Fire heterogeneous shock initiation decomposition rate to include both the desensitization and failure to desensitize effects as will be described in Chapter 4.

The study to determine the mechanism of the explosive desensitization by preshocking used the three-dimensional reactive hydrodynamic code, 3DE, described in Appendix D. A constant velocity piston was applied to the bottom of a TATB explosive cube shocking the explosive to the desired pressure. When a higher pressure second shock was to be introduced, the piston velocity was increased and other piston state values changed appropriately for a multiple shock of the required pressure. The Arrhenius reactive rate law was used for TATB with the constants determined for solid TATB by Raymond N. Rogers shown in Table 3.1 of an activation energy of 59.9 kcal/mole and a frequency factor of 3.18 x $10^{13} \mu\text{sec}^{-1}$.

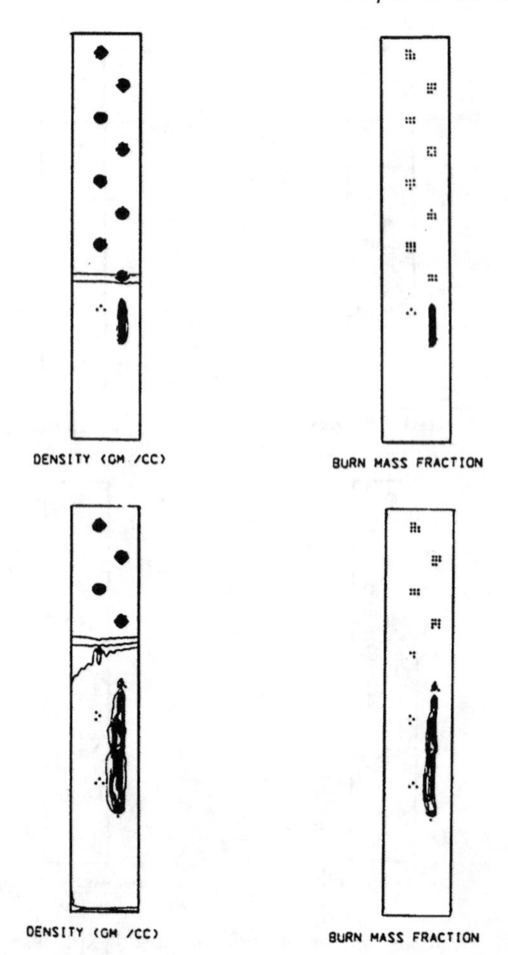

Figure 3.34 The density and mass fraction of undecomposed explosive cross sections through i=11 are shown at various times for HMX with a matrix of 0.000005-cm-radius holes at 27°C. The density contour interval is 0.2 gm/cc and the mass fraction contour interval is 0.2. The shock pressure is 50 kbar. The hot spots fail to initiate the explosive.

A single-shock pressure of 290 kbar in TATB has a density of 2.8388 g/cc, particle velocity of 0.21798 cm/μsec, and temperature of 1396° K. The adiabatic explosion time is 1.0 x $10^{-6}\mu$sec. A second shock pressure of 290 kbar in TATB initially shocked to 40 kbar has a density of 2.878 gm/cc, particle velocity of 0.204 cm/μsec, and a temperature of 804° K. The time to explosion is 3.85 μsec. TATB shocked to 40 kbar has a temperature of 362° K and the time to explosion is 1.0 x $10^{+19}\mu$sec. Thus, the multiply shocked explosive requires 3.85 μsec for a propagating detonation to form and only 1.0 x $10^{-6}\mu$sec if singly shocked to 290 kbar. If the initial 40 kbar shock converts the heterogeneous explosive into a more homogeneous explosive, for which Arrhenius kinetics are appropriate, then a detonation would not be expected to propagate through the preshocked TATB. When a shock wave interacts with a hole or some other density discontinuity in an explosive, a hot spot with temperatures hotter than the surrounding explosive is formed in the region above the hole after it is collapsed by the shock wave. The hot region decomposes and contributes energy

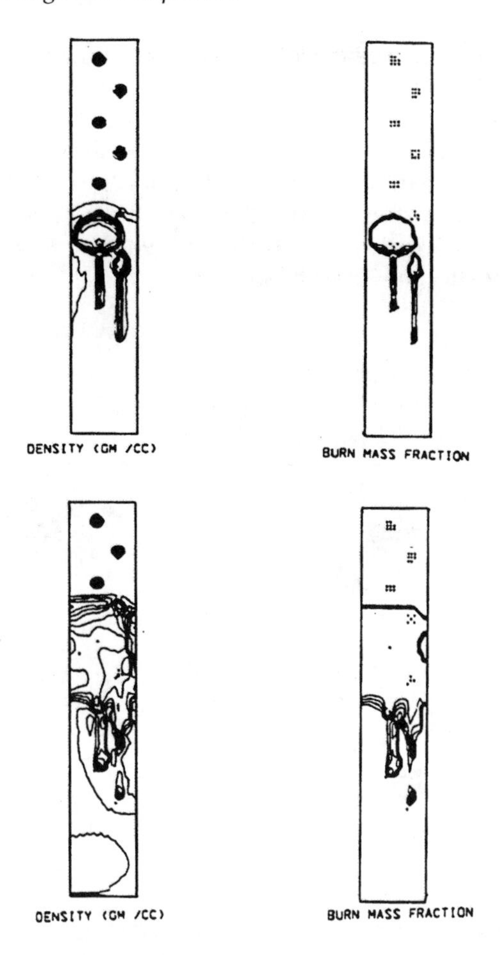

Figure 3.35 The density and mass fraction of undecomposed explosive cross sections through i=11 are shown at various times for HMX with a matrix of 0.000005-cm-radius holes at 250°C. The density contour interval is 0.2 gm/cc and the mass fraction contour interval is 0.2. The shock pressure is 50 kbar. The initiation propagates.

to the shock wave, which has been degraded by the hole interaction. Whether this energy is sufficient to compensate for the loss from the hole interaction depends on the magnitude of the initial shock wave, the hole size, and the interaction with the flow from nearest neighbor hot spots.

The hole size present in pressed explosives varies from holes of 20 to 600 A in the TATB crystals to holes as large as .05 cm in the explosive-binder matrix. Most of the holes vary in size from 0.005 to 0.0005 cm in diameter, so holes in that range of diameters were examined.

The interaction of a 40 kbar shock with a single 0.004 cm diameter air hole in TATB was modeled. The three-dimensional computational grid contained 15 by 15 by 36 cells, each 0.001 cm on a side. The time step was 0.002 μsec. After 0.025 μsec, the 40 kbar shock had collapsed the hole and a 290 kbar shock wave was introduced that passed through the 40 kbar preshocked region and overtook the 40 kbar shock wave.

The density and burn fraction surface contours are shown in Figures 3.36 and 3.37, and the cross sections through the center of the hole are shown in Figure 3.38. The 40 kbar shock wave collapsed the hole and formed a small, weak hot spot that was not hot enough to result in appreciable decomposition of the TATB. The 290 kbar shock wave temperature was not hot enough to cause explosion during the time studied in the bulk of the explosive previously shocked to 40 kbar; however, the additional heat present in the hot spot formed by the 40 kbar shock wave after it interacted with the hole was sufficient to decompose some of the explosive after it was shocked by the 290 kbar wave. Propagating detonation occurred immediately after the 290 kbar shock wave caught up with the 40 kbar preshock.

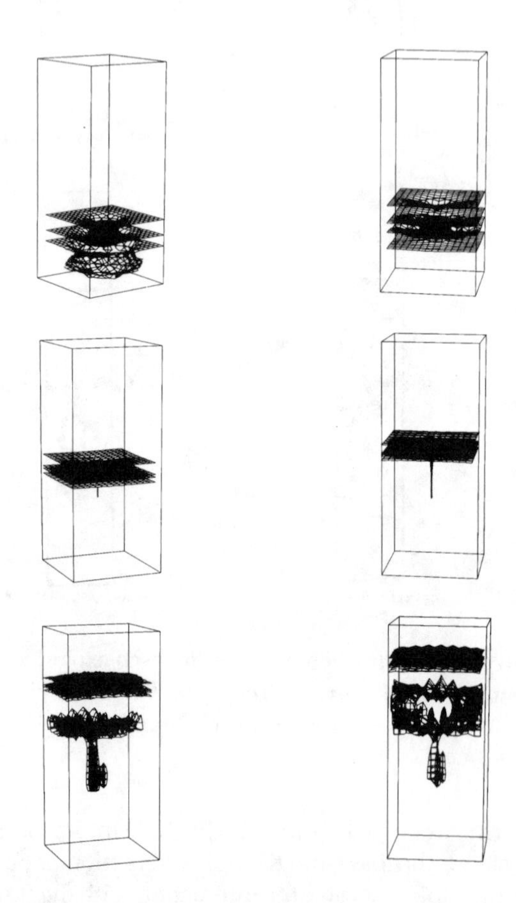

Figure 3.36 The density surface contours for a 40 kbar shock interacting with a single 0.004 cm diameter air hole in TATB, followed after 0.025 μsec by a 290 kbar shock wave.

To investigate the effect of the interaction of a matrix of holes with a multiple shock profile, a matrix of 10% air holes located on a hexagonal close-packed lattice in TATB was modeled. The spherical air holes had a diameter of 0.004 cm. The initial configuration is shown in Figure 3.39. The three-dimensional computational grid contained 16 by 22 by 36 cells each 0.001 cm on a side. The time step was 0.0002 μsec. Figure 3.40 shows the density and mass fraction cross sections for a 40 kbar shock wave followed after 0.045 μsec by a 290 kbar shock wave interacting with a matrix of 10% air holes of 0.004 cm diameter in TATB.

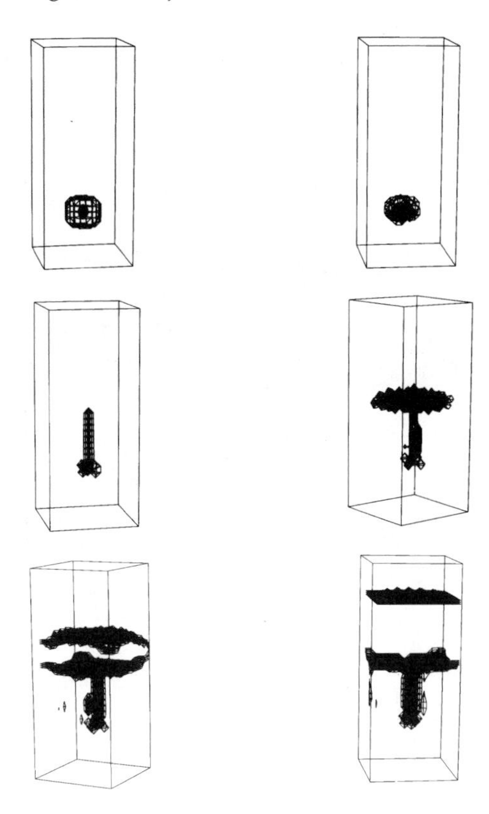

Figure 3.37 The burn fraction contours for a 40 kbar shock interacting with a single 0.004 cm diameter air hole in TATB, followed after 0.025 μsec by a 290 kbar shock wave.

The preshock desensitized the explosive by closing the voids and making it more homogeneous. The higher pressure second shock wave proceeded through the preshocked explosive until it caught up with the preshock.

The three dimensional modeling study demonstrated that the desensitization occurs because the preshock interacts with the holes and eliminates the density discontinuities. The subsequent higher pressure shock waves interact with a more homogeneous explosive. The multiple shock temperature is lower than the single shock temperature at the same pressure, which is the cause of the observed failure of a detonation wave to propagate in preshocked explosive for some ranges of preshock pressure.

The modification indicated by the three-dimensional study to the Forest Fire heterogeneous shock initiation decomposition rate was to limit the heterogeneous explosive decomposition to that characteristic of the initial shock pressure and then to add the homogeneous explosive decomposition using the Arrhenius rate law. The resulting multiple shock Forest Fire (MSFF) will be described in Chapter 4.

Conclusions

Numerical modeling using reactive hydrodynamic codes have given us increased understanding of the effect of heterogeneities or density discontinuities on explosive initiation.

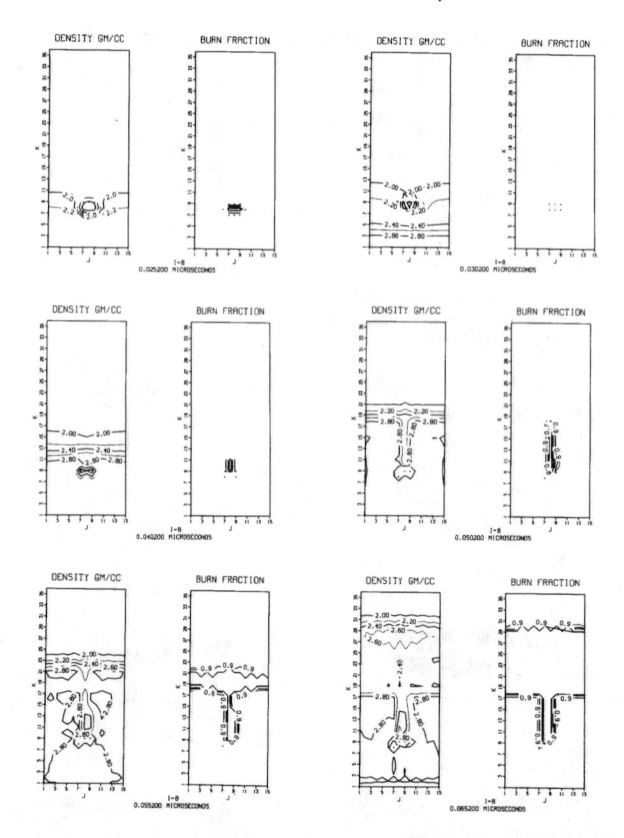

Figure 3.38 The density and burn fraction cross sections through the center of the hole for a 40 kbar shock interacting with a single 0.004 cm diameter air hole in TATB, followed after 0.025 μsec by a 290 kbar shock wave.

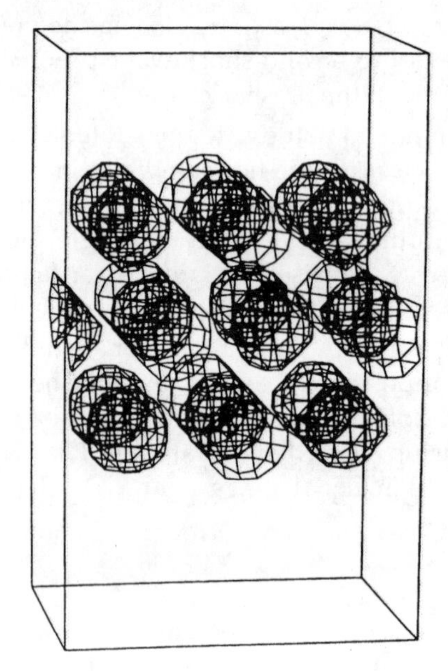

Figure 3.39 The initial configuration of a matrix of 10 % air holes in TATB.

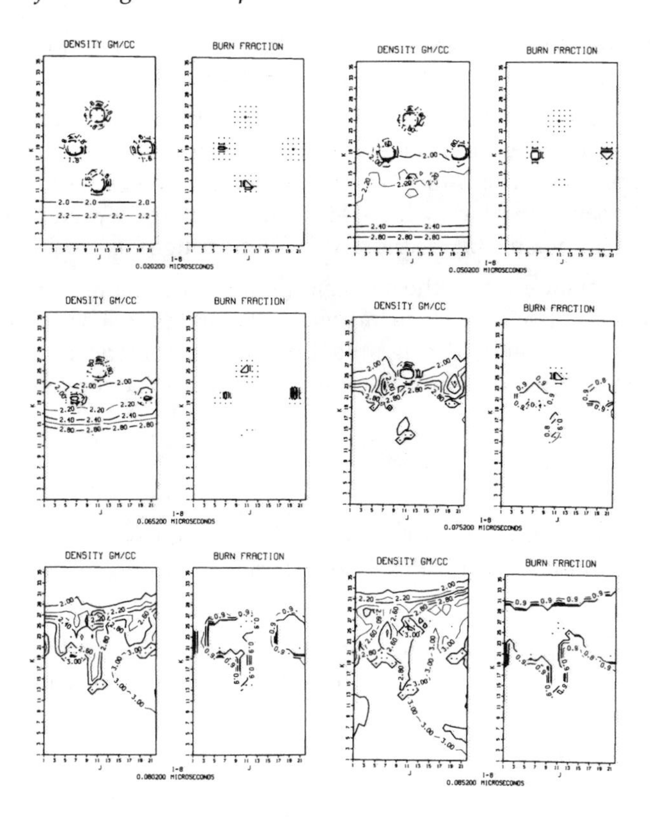

Figure 3.40 The density and mass fraction cross sections are shown for a 40 kbar shock wave, followed after 0.045 μsec by a 290 kbar shock wave interacting with a matrix of 10% holes of 0.004 cm diameter in TATB.

The *Hydrodynamic Hot Spot Model* has resulted in an increased understanding of the effect of explosive composition, hole size, number of holes, explosive Arrhenius decomposition rate, initial temperature, and shock pressure on shock initiation of heterogeneous explosives. The *Hydrodynamic Hot Spot Model* reproduces the observed shock initiation behavior. This indicates that the dominate features are shock heating by hydrodynamic hot spot formation, cooling by rarefactions, and the Arrhenius decomposition rate as a function of temperature.

The *Hydrodynamic Hot Spot Model* describes the basic differences between shock sensitive and shock insensitive explosives. The interaction of a shock wave with air holes in PETN, HMX, TATB, and Nitroguanidine, the resulting hot spot formation, interaction, and build up toward detonation or failure have been modeled.

The effects of particle size (and the remaining void or hole size) and initial temperature on the shock initiation of heterogeneous explosive charges have been modeled. Investigations have included shocks of various pressures interacting with TNT, HMX, and TATB with 0.5 % void fraction and hole sizes from 0.5- to 0.000005-cm radius. At the same density, the most shock-sensitive explosive is the one with particle sizes between coarse and fine material. The shock sensitivity of HMX continues to increase with decreasing hole sizes for hole sizes where TNT or TATB fail. The shock sensitivity of TNT, TATB, and HMX increases with initial temperature. TATB at 250°C is calculated to be as shock sensitive as HMX at

25°C. This is in agreement with experimental observations. The shock sensitivity of HMX is less dependent on temperature than that of TATB or TNT.

Desensitization occurs because the preshock interacts with the holes and eliminates the density discontinuities. The subsequent higher pressure shock waves interact with a more homogeneous explosive. The multiple shock temperature is lower than the single shock temperature at the same pressure, which is the cause of the observed failure of a detonation wave to propagate in preshocked explosive for some ranges of preshock pressure.

The detailed understanding of the shock initiation process in heterogeneous explosives that has been obtained from the *Hydrodynamic Hot Spot Model* has been used to develop a technique for performing engineering modeling of many practical explosive vulnerability problems. The technique is called Forest Fire and will be described in Chapter 4.

References

1. Charles A. Anderson, "TEPLO: A Heat Conduction Code for Studying Thermal Explosions in Laminar Composites," Los Alamos Scientific Laboratory report LA-4511 (1970).
2. P. Gray and P. R. Lee, "Thermal Explosion Theory," in Oxidation and Combustion Reviews, edited by C. F. H. Tipper, Elsevier Publishing Company, New York 2, 1-185 (1967).
3. John Zinn and Charles L. Mader, "Thermal Initiation of Explosives," Journal of Applied Physics 31, 323 (1960).
4. D. A. Frank Kamenetskii, "Calculation of Thermal Explosion Limits," Acta Physicochimica U.R.S.S. 10, 365 (1939).
5. P. A. Longwell, "Calculation of Critical Temperatures and Time-to-Explosion for Propellants and Explosives," U. S. Naval Ordnance Test Station, China Lake, report NAVWEPS 7646 (1961).
6. G. B. Cook, "The Initiation of Explosion in Solid Secondary Explosives," Proceedings of the Royal Society (London) 246, 154 (1958).
7. Raymond N. Rogers, "Thermochemistry of Explosives," Thermochimica Acta 11, 131 (1975).
8. John Zinn and Raymond N. Rogers, "Thermal Initiation of Explosives," Journal of Physical Chemistry 66, 2646 (1962).
9. A. Popolato, Editor, JSEP Program, Los Alamos Scientific Laboratory report LA-5521-PR, pp. 9-18 (1973).
10. A. W. Campbell, W. C. Davis, and J. R. Travis, "Shock Initiation of Detonation in Liquid Explosives," Physics of Fluids 4, 498 (1961).
11. Charles L. Mader, "Shock and Hot Spot Initiation of Homogeneous Explosives," Physics of Fluids 6, 375 (1963).
12. T. I. Cottrell, T. E. Graham, and T. Y. Reid, "The Thermal Decomposition of Nitromethane," Transactions of the Faraday Society 47, 584 (1951).

13. W. B. Garn, "Determination of the Unreacted Hugoniot for Liquid TNT," Journal of Chemical Physics 30, 819 (1959).

14. F. E. Walker and R. J. Wasley, "Initiation of Nitromethane with Relatively Long-Duration, Low Amplitude Shock Waves," Combustion and Flame 15, 233 (1970).

15. I. M. Voskoboinikov, V. M. Bogomolov, A. D. Margolin, and A. Ya. Apin, "Determination of the Decomposition Times of Explosive Materials in Shock Waves," Doklady Akademii Nauk. SSSR 167, 610 (1966).

16. J. R. Travis, "Experimental Observations of Initiation of Nitromethane by Shock Interactions at Discontinuities," Fourth International Symposium on Detonation, ACR-126, 386 (1965).

17. A. W. Campbell, W. C. Davis, J. B. Ramsay, and J. R. Travis, "Shock Initiation of Solid Explosives," Physics of Fluids 4, 511 (1961).

18. Charles L. Mader, "The Two-Dimensional Hydrodynamic Hot Spot," Los Alamos Scientific Laboratory report LA-3077 (1964).

19. Charles L. Mader, "The Two-Dimensional Hydrodynamic Hot Spot - Volume II," Los Alamos Scientific Laboratory report LA-3235 (1965).

20. Charles L. Mader, "The Two-Dimensional Hydrodynamic Hot Spot - Volume III," Los Alamos Scientific Laboratory report LA-3450 (1966).

21. Charles L. Mader, "The Two-Dimensional Hydrodynamic Hot Spot - Volume IV," Los Alamos Scientific Laboratory report LA-3771 (1967).

22. Charles L. Mader and James D. Kershner, "Three-Dimensional Modeling of Shock Initiation of Heterogeneous Explosives," Nineteenth Symposium (International) on Combustion, William and Wilkins, Eds., 685-690, (1982).

23. Charles L. Mader and James D. Kershner, "The Three-Dimensional Hydrodynamic Hot Spot Model," Eighth Symposium (International) on Detonation, NSWC-MP 86-194, 42-51 (1985).

24. Charles L. Mader and James D. Kershner, "The Three-Dimensional Hydrodynamic Hot-Spot Model Applied to PETN, HMX, TATB, and NQ," Los Alamos National Laboratory report LA-10203-MS (1984).

25. Charles L. Mader and James D. Kershner, "Numerical Modeling of the Effect of Particle Size of Explosives on Shock Initiation Properties," GTPS 4th Congress International de Pyrotechnie du Groupe de Travail de Pyrotechnie, 45-54 (1989).

26. Terry R. Gibbs and Alphonse Popolato,"LASL Explosive Property Data," University of California Press (1980).

27. Donna Price, "Effect of Particle Size on the Sensitivity of Pure Porous HE," NSWC-TR-86-336 (1986).

28. H. Moulard, J. Kury, and A. Delcos, "The Effect of RDX Particle Size on the Shock Sensitivity of Cast PBX Formulations," Eighth Symposium (International) on Detonation, NSWC-MP 86-194, 902-913 (1985).

29. J. Waschl and D. Richardson, "Effects of Specific Surface Area on the Sensitivity of Hexanitrostilbene to Flyer Plate Impact," Journal of Energetic Materials 9, 269-282 (1991).

30. M. J. Urizar, Los Alamos Scientific Laboratory unpublished data, (1956).

31. A. C. Schwarz, "Flyer Plate Performance and the Initiation of Insensitive Explosives by Flyer Plate Impact," Sandia Laboratory Report SAND 75-0461 (1975).

32. C. H. Johansson and P. A. Persson, "Detonics of High Explosives," Academic Press (1970).

33. J. B. Ramsay, B. G. Craig, and J. J. Dick, "High-Temperature Shock Initiation of Explosives," Los Alamos National Laboratory report LA-7158 (1978).

34. P. A. Urtiew, L. M. Erickson, D. F. Aldis, and C. M. Tarver, "Shock Initiation of LX-17 as a Function of its Initial Temperature," Proceedings Ninth Symposium (International) on Detonation, OCNR 113291-7, Vol 1, 112-122 (1989).

35. John C. Dallman, "Temperature-Dependent Shock Sensitivity of PBX-9502," Private Communication (1990).

chapter four

Modeling Initiation of Heterogeneous Explosives

As shown in Chapter 3, the basic mechanism of heterogeneous explosive shock initiation is shock interaction at density discontinuities which produce local hot spots that decompose and add their energy to the flow. Since it is not possible to model in detail all the density discontinuities of a heterogeneous explosive or even a representative sample of the discontinuities, the problem is to model the gross features of the shock initiation of a heterogeneous explosive or propellant.

The behavior of the shock front as it proceeds in a heterogeneous explosive was first clearly confirmed by Campbell, Davis, Ramsay, and Travis.[1] They concluded that a heterogeneous explosive is initiated by local hot spots formed in it by shock interactions with density discontinuities. When a shock wave interacts with the density discontinuities, producing numerous local hot spots that explode but do not propagate, energy is released which strengthens the shock so that, when it interacts with additional inhomogeneities, higher temperature hot spots are formed and more of the explosive is decomposed. The shock wave grows stronger and stronger, releasing more and more energy, until it becomes strong enough to produce propagating detonation.

The behavior behind the accelerating front is also important to the modeling of the performance of explosives and propellants. Craig and Marshall[2] studied the contribution of shocked, but not detonating, 9404 explosive to the velocity of metal plates. It was apparent as early as 1961 that considerable decomposition of 9404 occurred, releasing energy that was available to push plates. The first experimentalists to report observations of the nature of this decomposition were Dremin and Koldunov.[3] Using electromagnetic techniques to study the particle velocities as a function of time, they observed that the particle velocity increased with time behind, as well as at, the shock front. They observed velocity humps that moved so slowly that they did not catch up with the shock front until after complete decomposition had occurred.

Craig and Marshall[2] and Kennedy[4] have provided experimental studies of the process. Attempts to describe a rate of decomposition by calibrating some previously defined rate law are unsuccessful at describing the observed flow. Programming the flow[5] could be used to describe some experimental observations, but the model did not respond realistically to local state variables.

A model, called Forest Fire after its originator, Charles Forest[6] has been developed for describing the decomposition rates as a function of the experimentally measured distance of run to detonation vs. shock pressure (the Pop plot named after its originator, A. Popolato[7]) and the reactive and nonreactive Hugoniots. The model can be used to describe the decomposition from shocks formed either by external drivers or by internal pressure gradients formed by the propagation of a burning front.

The Forest Fire rate of heterogeneous explosive or propellant shock initiation has been used in the modeling of the vulnerability of systems containing explosives or propellants. Failure or propagation of an explosion as a function of diameter, shock pulse width, wave curvature, corner turning, side rarefactions, and density are all dominated by the heterogeneous initiation mechanism.

4.1 The Forest Fire Model

The similarity among overlapping portions of the experimentally measured shock distance and time coordinated from experiments having different shock pressures observed by Lindstrom[8] for RDX/Exon and by Craig[2] for 9404, TNT, and TATB supports the assumption first made in Reference 9 that the explosive will pass through the same pressure, distance, and decomposition states at the shock front regardless of the initial conditions. The *Single Curve Build-Up Principle* is the assumption that a reactive shock wave grows to detonation along a unique line in distance, time, and state space. Applying the *Single Curve Build-Up Principle* to the Pop plot means that the Pop plot is a description of the entire shock initiation and detonation front propagation process.

The HOM equation of state described in Appendix A was used to calculate the Hugoniots for partially reacted explosive. The parameters used for 9404 are given in Table 4.1 and on the CD-ROM. Ramsay's[7] unreacted equation of state, $U_s = 0.2423 + 1.883U_p$, was used to describe the unreacted 9404 explosive. The BKW equation of state described in Appendix E was used to describe the 9404 detonation products.

The computed partially reacted HOM Hugoniots and Ramsay's experimental reactive Hugoniot, $U_s = 0.246 + 2.53U_p$, are shown in Figure 4.1. For each state point on the experimental reactive Hugoniot there is a corresponding point on a partially reactive HOM Hugoniot for some degree of reaction.

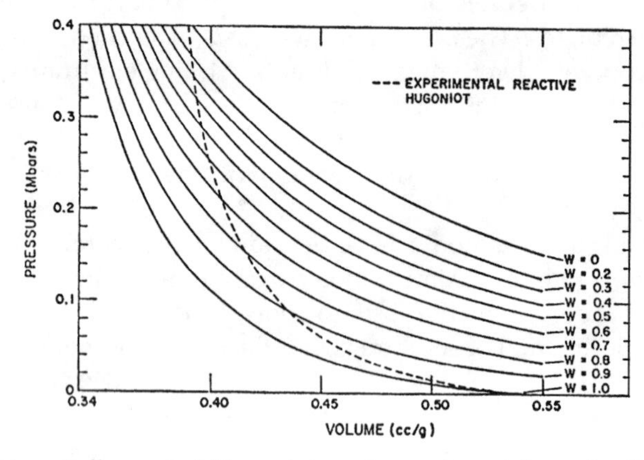

Figure 4.1 HOM 9404 partially reacted Hugoniots and experimental reactive Hugoniot.

TABLE 4.1 9404 Equation of State Parameters

	HOM SOLID	EXP		HOM GAS	EXP
C	+2.423	−001	A	−3.53906259946	+000
S	+1.883	+000	B	−2.57737590393	+000
F_s	−9.04187222042	+000	C	+2.60075423332	−001
G_s	−7.13185252435	+001	D	+1.39083578508	−002
H_s	−1.25204979360	+002	E	−1.13963024075	−002
I_s	−9.20424177603	+001	K	−1.61913041133	+000
J_s	−2.21893825727	+001	L	+5.21518534192	−001
C_v	+4.0	−001	M	+6.77506594107	−002
α	+5.00	−005	N	+4.26524264691	−003
V_o	+5.42299349241	−001	O	+1.04679999902	−004
γ_s	+6.75	−001	Q	+7.36422919790	+000
T_o	+3.0	+002	R	−4.93658222389	−001
Y_o	+1.20	−002	S	+2.92353060961	−002
μ	+4.78	−002	T	+3.30277402219	−002
PLAP	+5.000	−002	U	−1.14532498206	−002
			C'_v	+5.0	−001
			Z	+1.0	−001

The Pop plot for 9404 is shown in Figure 4.2 and may be expressed by

$$ln\ X = -5.499637 - 1.568639\ ln\ P\ ,$$

where X is in cm and P is in Mbar. Applying the *Single Curve Build-Up Principle*, the shock initiation of PBX-9404 can be described from the Pop plot and the experimental partial reacted Hugoniot. The resulting "experimental" description of heterogeneous shock initiation of 9404 is shown in Figures 4.1–4.4.

More information is needed to obtain a burn rate function that is consistent with the Pop plot and the reactive Hugoniot data line. The assumption that the pressure gradient at the front is zero was found to be adequate for many purposes. The assumption that the pressure gradient at the front is zero is about equal to assuming a growing square wave if the rate derived at the front is used throughout the flow. When the shock front approaches the detonation state, the growing square wave is inappropriate and the Forest Fire model ceases to approximate a square wave.

The following equations present a general derivation of the Forest Fire model. The derivation is then restricted to the growing square wave. The Forest Fire executable code for solving the Forest Fire heterogeneous shock initiation rates is on the CD-ROM.

Equation Nomenclature

P = Pressure
U_s = Shock Velocity

U	= Particle Velocity
V	= Specific Volume
t	= Time
ρ	= Density
I	= Internal Energy
W	= Mass Fraction
x	= Distance

Notation: the Lagrangian "mass coordinates" are

$$\frac{\partial}{\partial m} = \frac{1}{\rho}\frac{\partial}{\partial x}$$

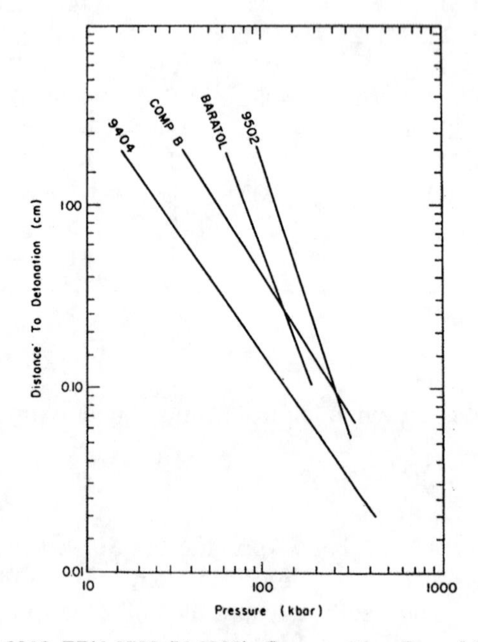

Figure 4.2 Pop Plots for X-0219, PBX-9502 (X-0290), Composition B, and 9404.

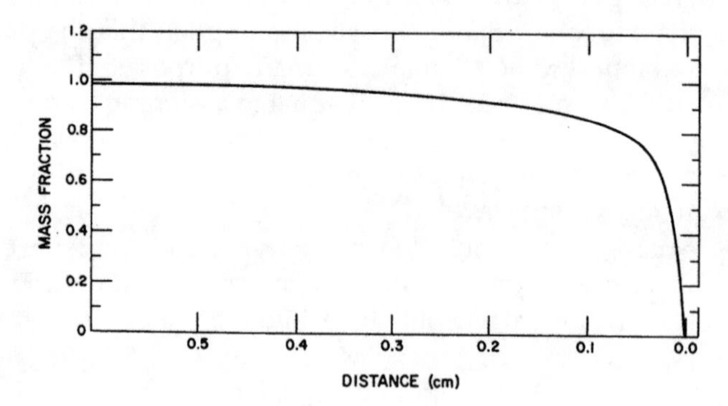

Figure 4.3 Mass fraction of undecomposed 9404 explosive as a function of distance.

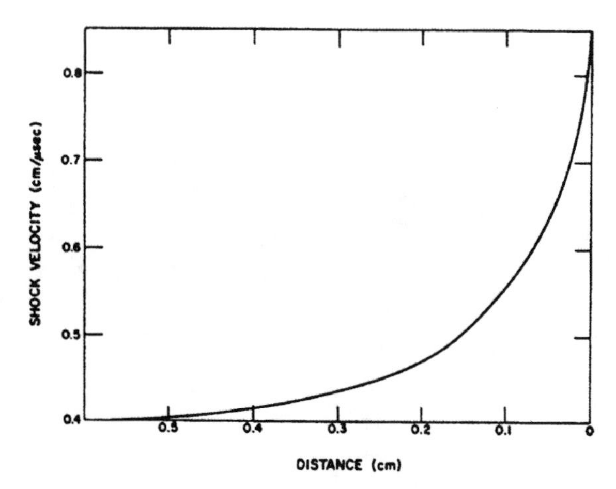

Figure 4.4 Shock velocity of 9404 reactive shock front.

and

$$\frac{\partial}{\partial r} = \frac{\partial}{\partial t} + U\frac{\partial}{\partial x} .$$

The fluid flow equations are

$$U_\tau = -P_m, \; where \; U_\tau = \frac{\partial U}{\partial \tau} \; and \; P_m = \frac{\partial P}{\partial m} ,$$

$$V_\tau = U_m .$$

and

$$I_\tau = -P \; V_\tau ,$$

The solution is sketched as follows. Using the Pop plot and the shock jump equations, solve for P, V, I and then using the flow equations behind the shock, solve for P_τ, V_τ, and I_τ. Then solve for W and W_τ from

$$P = H(V, I, W)$$

and

$$P_\tau = H_V \; V_\tau + H_I \; I_\tau + H_W \; W_\tau .$$

Notation:

Let $\widehat{P}$, $\widehat{V}$, $\widehat{I}$, and $\widehat{U}$ be shock front functions.

Let $m_s(t)$ = mass position of the shock so that $m_s(0) = 0$.

Let $t_s(m)$ = time of shock arrival at mass point m.

Note:

$$\frac{dm_s(\tau)}{d\tau} = \rho_o U_s(\tau)$$

and

$$\frac{dt_s(m)}{dm} = \frac{1}{\rho_o U_s(m)} \, .$$

Assumption: Let

$$\frac{\partial P}{\partial m} \equiv P_m = f(\tau) \, .$$

Then the shock pressure wave looks like

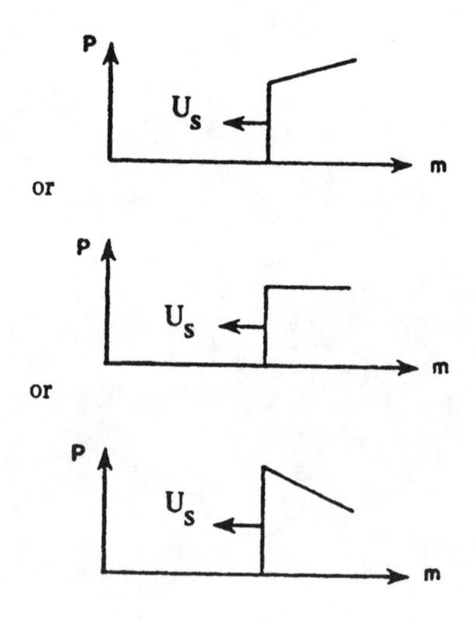

The solution of the flow equations is

$$P(m, \tau) = \widehat{P}[m_s(\tau)] + P_m(\tau)[m - m_s(\tau)] \, ,$$

$$P_\tau(m, \tau) = \left[\frac{d\widehat{P}}{dm} - P_m(\tau)\right]\frac{dm_s(\tau)}{d\tau} + \frac{dP_m(\tau)}{d\tau}[m - m_s(\tau)] \, ,$$

$$U(m, \tau) = \widehat{U}(m) - \int_{t_s(m)}^{\tau} P_m(t') \, dt' \, ,$$

$$U_m(m, \tau) = \frac{d\widehat{U}}{dm} + P_m[t_s(m)]\frac{dt_s(m)}{dm} \quad a \ function \ of \ m \ only,$$

$$V(m,\tau) = \widehat{V}(m) + U_m(m)\int_{t_s(m)}^{\tau} dt' ,$$

$$I(m,\tau) = \widehat{I}(m) - U_m(m)\int_{t_s(m)}^{\tau} P(m,t')dt' ,$$

$$V_\tau = U_m(m) ,$$

and

$$I_\tau = -P(m,\tau)\, U_m(m) .$$

The reactive shock Hugoniot and shock jump relations are

$$U_s = C + S\widehat{U},$$

$$\widehat{P} = \rho_o U_s \widehat{U},$$

$$\widehat{V} = V_o(U_s - \widehat{U})/U_s,$$

and

$$\widehat{I} = \widehat{U}^2/2 .$$

Then

$$\widehat{U} = -[-C + (C^2 + 4V_o S\widehat{P})^{1/2}]/2S ,$$

$$U_s = -[C + (C^2 + 4V_o S\widehat{P})^{1/2}]/2,$$

$$d\widehat{P} = \rho_o(S\widehat{U} + U_S)d\widehat{U} .$$

The Pop plot is

$$ln\ (run) = \alpha_1 + \alpha_2\, ln(\widehat{P} - P_o),$$

where run is distance to detonation, and

$$\frac{dP}{d\,run} = \frac{(\widehat{P} - P_o)}{\alpha_2\, run\ (\widehat{P})} .$$

Then,

$$\frac{d\widehat{P}}{dm} = \frac{V_o(\widehat{P} - P_o)}{\alpha_2\, run\ (\widehat{P})} ,$$

and

$$\frac{d\widehat{U}}{dm} = \frac{1}{\rho_o(S\widehat{U} + U_s)}\frac{d\widehat{P}}{dm} .$$

The solution for t_s is

$$\frac{dt_s}{d\,run} = \frac{1}{(d\,run/dt_s)} .$$

Thus integrate

$$\frac{dt_s}{d\,run} = \frac{-2}{C + \left[C^2 + 4V_oS\left(e^{-\alpha_1/\alpha_2}run^{1/\alpha_2} + P_o\right)\right]^{1/2}} ,$$

with initial conditions

$$t_s(o) = 0 .$$

Remember that $m_s = \rho_o run$.

Special integral evaluations are given for $\widehat{P}_m \equiv 0$. Here $\widehat{P}$ may be used as the independent variable $dt' = d\widehat{P}/(d\widehat{P}/dt)$. Thus,

$$\int_{t_s(m)}^{\tau} dt' = \int_{\widehat{P}(m)}^{\widehat{P}(\tau)} \frac{-\alpha_2\, run\,(P)}{U_s(P)(P - P_o)} dP \tag{1}$$

and

$$\int_{t_s(m)}^{\tau} P(t')dt' = \int_{\widehat{P}(m)}^{\widehat{P}(\tau)} \frac{-P\alpha_2\, run\,(P)}{U_s(P)(P - P_o)} dP ,$$

where

$$\frac{run\,(P)}{U_s(P)} = \frac{2\,exp\,[\alpha_1 + \alpha_2\, ln(P - P_o)]}{C + (C^2 + 4V_oSP)^{1/2}} .$$

The integral (1) can also be used to calculate the time to detonation, t_{DET}, if the upper limit is set to P_{DET}, where

$$P_{DET} = \rho_o D_{CJ}(D_{CJ} - C)/S ,$$

D_{CJ} is the C-J detonation velocity and P_{DET} is the reactive Hugoniot pressure on the Rayleigh line.

Restriction of the square-wave solution ($\widehat{P}_m = 0$) to the shock front only gives further simplification. In summary, using $\widehat{P}$ as the independent variable,

$$ln\ (run) = \alpha_1 + \alpha_2\ ln(\widehat{P} - P_o),$$

$$\widehat{U} = [-C + (C^2 + 4V_oS\widehat{P})^{1/2}]/(2S),$$

$$U_s = C + S\widehat{U},$$

$$V = V_o(U_s - \widehat{U})/U_s,$$

and

$$I = \widehat{U}^2/2\,.$$

W is solved from $\widehat{P} = H(V, I, W)$,

$$P_\tau = (P - P_o)\ U_s/(\alpha_2\ run),$$

$$U_m = \frac{-V_o^2(\widehat{P} - P_o)}{\alpha_2\ run\ (U_s + S\widehat{U})},$$

$$V_\tau = U_m,$$

and

$$I_\tau = -PV_\tau\,.$$

Finally, W is solved from

$$P_\tau = H_V\ V_\tau + H_I\ I_\tau + H_W\ W_\tau.$$

Temperature is calculated as an additional output of the HOM equation of state.

Figure 4.2 shows the Pop plots for PBX-9404, Composition B(60/40 RDX/TNT), PBX-9502 (X-0290) (95/5 TATB/Kel-F), and X-0219 (90/10 TATB/Kel-F). The partially reactive Hugoniots for each explosive are shown in Figure 4.5. If experimental data are not available for the reactive Hugoniot, it can be estimated by using the known sound speed and the C-J state since the reactive Hugoniot will pass through the C-J state.

Figure 4.6 shows the decomposition rate calculated, using the Forest Fire model as a function of pressure. Forest Fire was incorporated in the SIN code, as shown in Appendix A, and the rate is fit as a function of pressure [$ln\ (rate) = A + BP + CP^2 \ldots XP^n$]. The rate normally is set to zero if the pressure is less than the minimum pressure used in the fit, and the burn is completed if the pressure is greater than C-J pressure. If the mass fraction of unburned explosive (W) is less than 0.05, it is often set to zero.

The Forest Fire results for PBX-9404, Composition B, PBX-9502 (X-0290), X-0219 and other explosives are presented on the CD-ROM along with the executable code FIRE, which can be used to calculate the Forest Fire heterogeneous shock initiation rates for other explosives.

Pressure and mass fraction profiles for 2 cm of PBX-9404 initially shocked to 22.5 kbar are shown in Figure 4.7. The calculations closely reproduce the experimentally observed shock front profiles. To determine if the calculated state values were valid behind the shock front, Forest Fire was used to model experiments designed to characterize the flow.

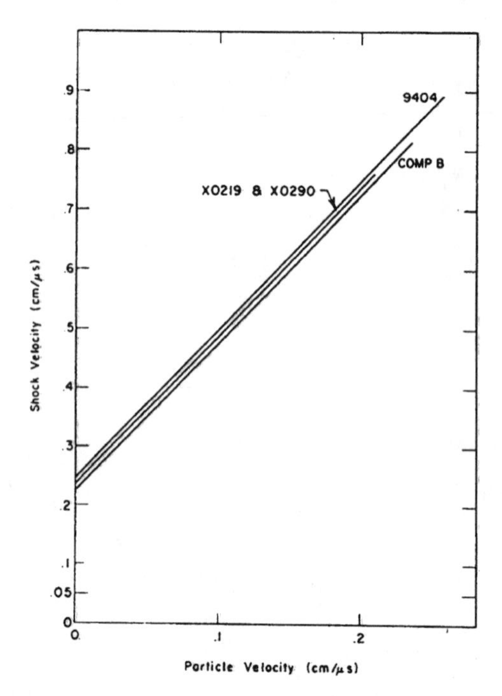

Figure 4.5 Reactive Hugoniots.

Craig and Marshall[2] performed a series of experiments shocking various thicknesses of PBX-9404 to various pressures and measuring the time histories of the velocity of free surfaces of Lucite plates in contact with the explosive. The PBX-9404 HOM equation of state constants are given in Table 4.1.

The calculated and experimental velocity vs. time profiles for a 63 kbar shock and for a 30 kbar shock initiating explosion in PBX-9404 are shown in Figures 4.8 and 4.9, respectively. The calculated and experimental initial free surface velocity profiles for various thicknesses of PBX-9404 in contact with a 0.508 cm thick Lucite plate are shown in Figure 4.10. Figure 4.11 shows the velocity for various thicknesses of Lucite plates in contact with 0.254 cm of PBX-9404 initially shocked to 63 kbar by an Aluminum driver. The energy available from shocked and decomposing explosive that has not detonated is important for vulnerability modeling. As shown in Figure 4.11, decomposing shocked explosives can push plates and deliver their energy more effectively than can detonating explosives of the same thickness.

Calculations were also performed to determine the Forest Fire model's response to driver pulse width. Gittings[10] reported the excess transit time for PBX-9404 shocked by 0.0127 to 0.040 cm thick aluminum foils traveling 0.14 to 0.20 cm/μsec, resulting in shock pressures of 135 to 85 kbars. Trott and Jung[11] have studied the effect of driver pulse width on the initiation of PBX-9404 from 35 to 65 kbars. They measured the effect of driver pulse width on Composition B in the 50 to 85 kbar range.

Forest Fire modeling described the effect of pulse width upon detonation propagation or failure within the 135 to 85 pressure range covered by Gittings' experiments and the Trott and Jung experiments down to 35 kbars.

The calculated pressure and mass fraction are shown in Figure 4.12 for PBX-9404 shocked to 50 and 40 kbar by a 0.1 cm thick Aluminum plate going at velocities of 0.1 to 0.08 cm/μsec.

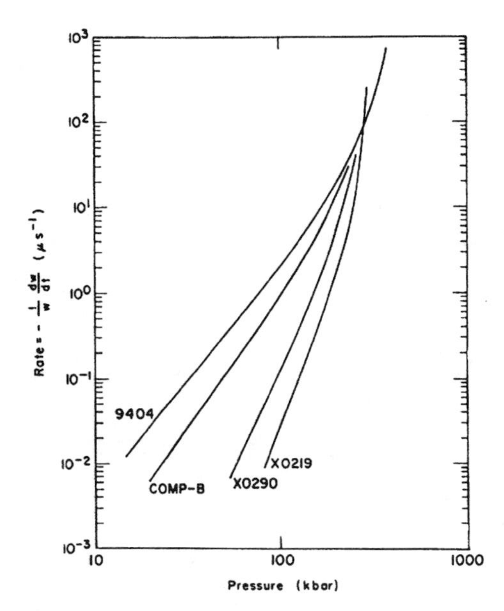

Figure 4.6 Forest Fire rate vs. Pressure.

The 50 kbar shock grows and detonation occurs at 0.390 cm in 0.83 μsec. A long duration pulse would result in detonation at 0.386 cm in 0.736 μsec. The 40 kbar shock does not grow to detonation as the rear rarefactions remove energy from the shock front as fast as it is generated by the hot spots.

Modeling using Forest Fire reproduced the observed quantitative behavior of the shock initiation by short pulses of PBX-9404 and Composition B. The energy available from shocked and decomposing explosive that has not detonated is important for vulnerability modeling.

Forest Fire has been successfully applied to many explosives and shock-sensitive propellants. The results are sensitive to the Pop plot, and small experimental errors at low pressures can be magnified in the rate sufficiently to cause large errors in the calculated time histories of a building shock wave. For example, the PBX-9404 that Kennedy[12] used for some of his experiments was slightly different from the type described in this chapter. Unfortunately, different batches of the same explosive can have different void distributions even at the same density and, hence, have different Pop plots.

As shown in Reference 29, if the Pop plot for an explosive is known, it is possible to estimate the Pop plot for other densities. The Pop plot for an explosive can be estimated if gap sensitivity data[7] are available for the explosive by using the Pop plot for an explosive with the same gap sensitivity.

4.2 Heterogeneous Detonations

Heterogeneous explosives, such as PBX-9404 or Composition B, behave differently than homogeneous explosives when propagating along confining surfaces. A heterogeneous explosive can turn sharp corners and propagate outward and, depending upon its sensitivity, it may show either very little or much detonation front curvature. The mechanism of initiation for heterogeneous explosives is different than the Arrhenius kinetic model found

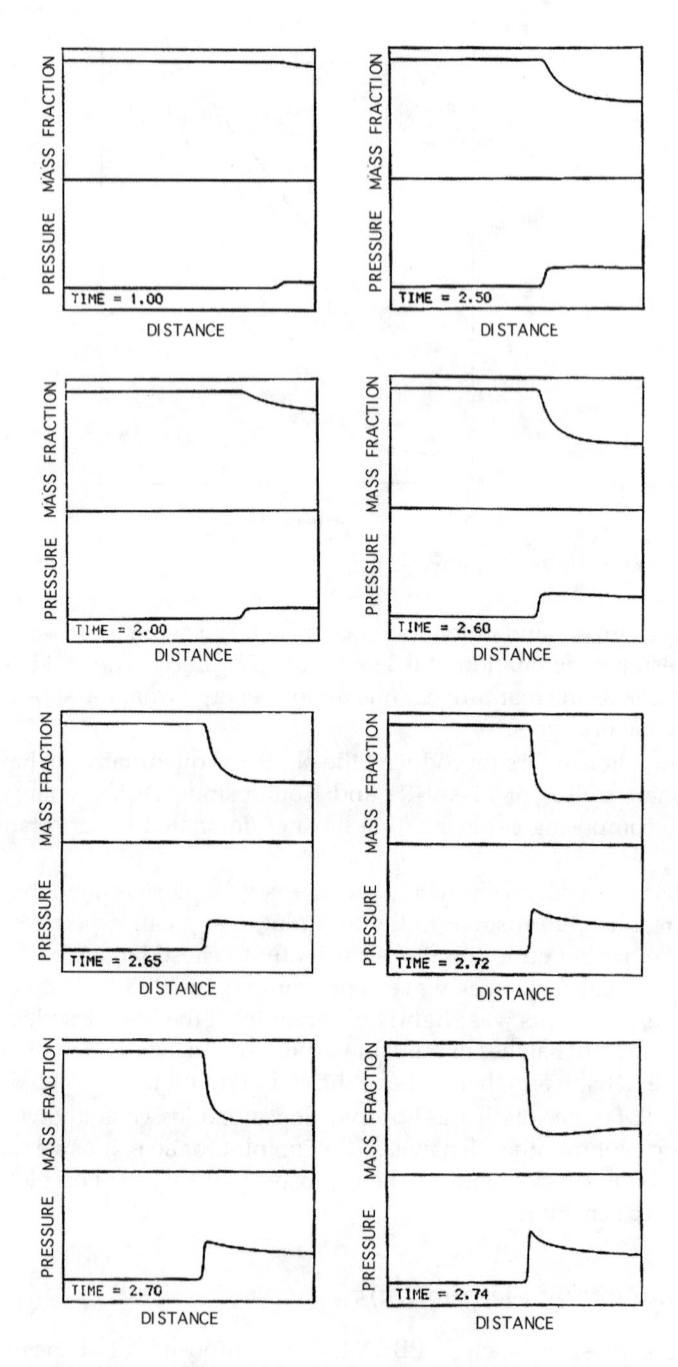

Figure 4.7 Pressure and mass fraction profiles for 2 cm of 9404 initially shocked to 22.5 kbar. The pressure scale is from -100 to 400 kbar and the mass fraction scale is 0 to 1.1.

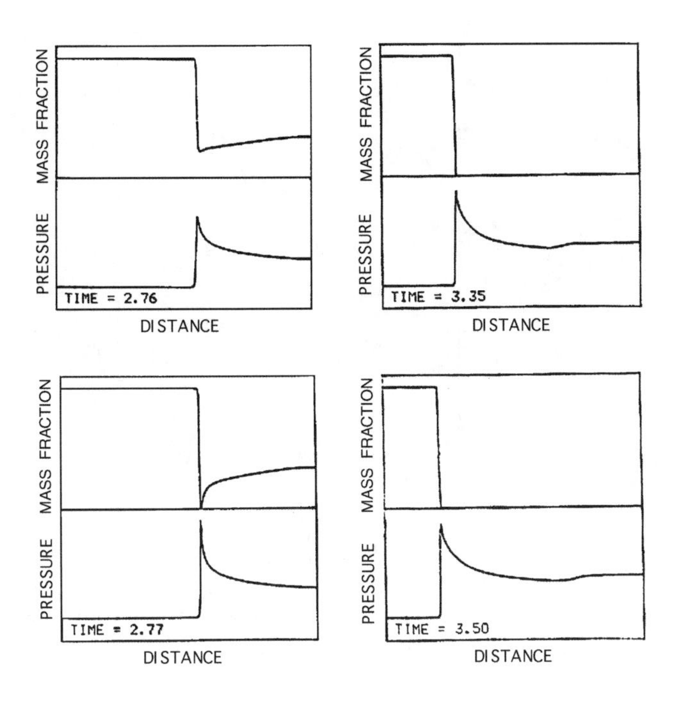

Figure 4.7 *(Continued)* Pressure and mass fraction profiles for 2 cm of 9404 initially shocked to 22.5 kbar. The pressure scale is from -100 to 400 kbar and the mass fraction scale is 0 to 1.1.

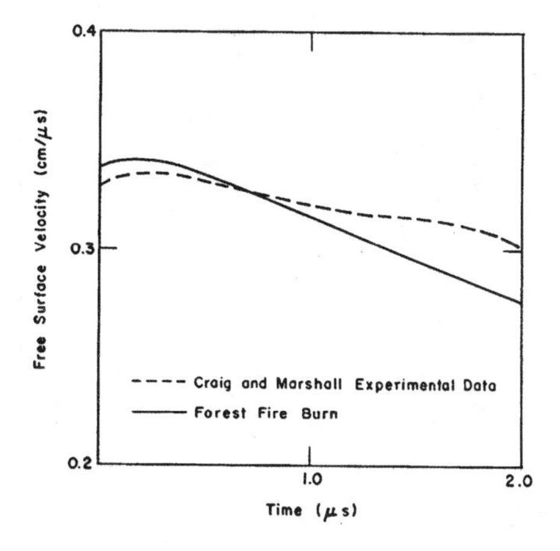

Figure 4.8 Free-surface velocity of 0.5 cm thick Plexiglas plate in contact with 0.254 cm of 9404 initially shocked to 63 kbar.

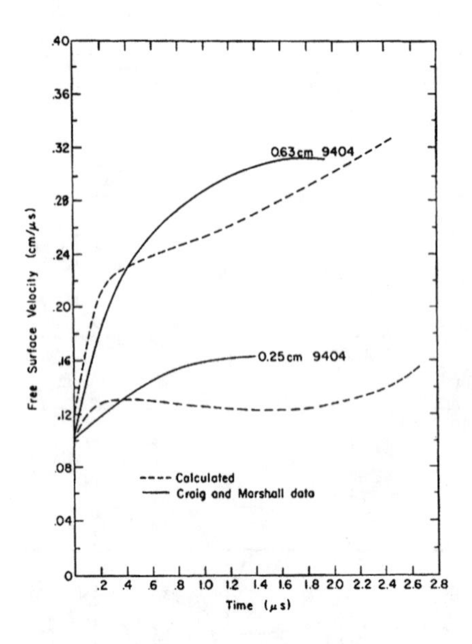

Figure 4.9 Free-surface velocity of 0.2 cm thick Plexiglas plates in contact with 0.25 and 0.63 cm thick pieces of 9404 initially shocked to 30 kbar.

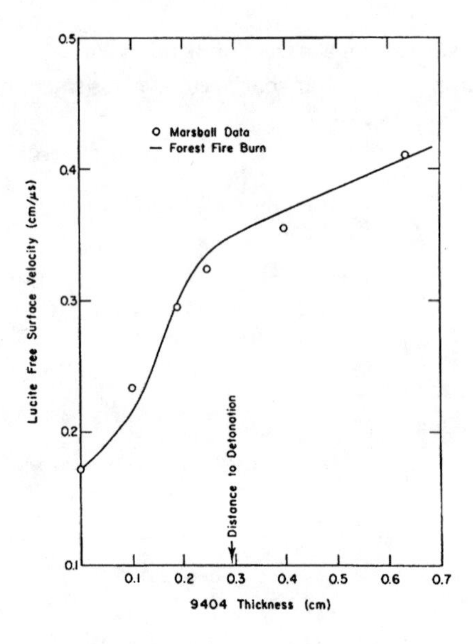

Figure 4.10 Initial free-surface velocity of 0.508 cm thick Lucite plate in contact with various thicknesses of 9404 initially driven by an Aluminum plate to 63 kbar.

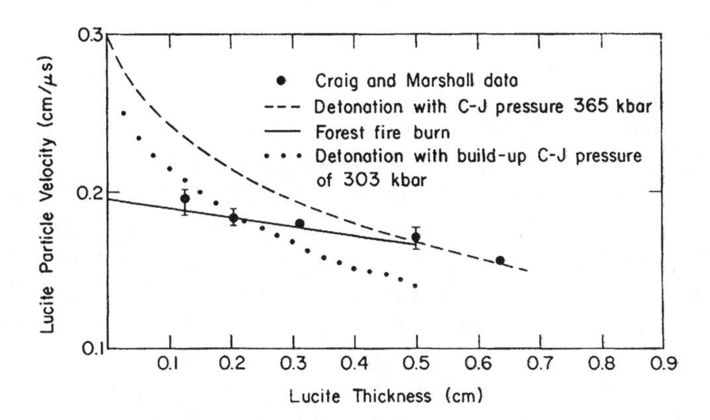

Figure 4.11 Initial free-surface velocity of various thicknesses of Lucite plates in contact with 0.254 cm of 9404 initially shocked to 63 kbar by an Aluminum driver.

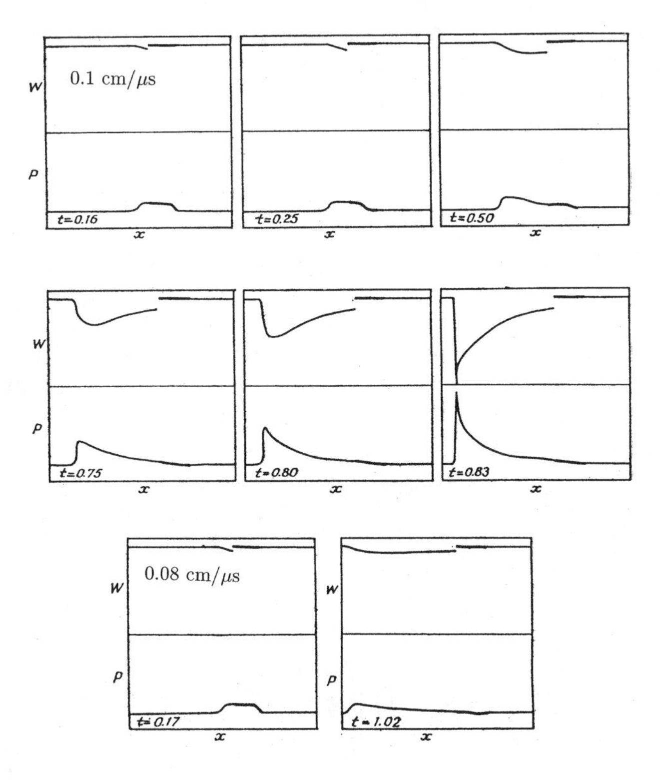

Figure 4.12 Pressure and mass fraction profiles of 0.4 cm of 9404 initially driven by a 0.1 cm thick Aluminum plate going 0.1 and 0.08 cm/μsec. The pressure scale is from -100 to 400 kbar and the mass fraction scale is from 0 to 1.1. The Aluminum flyer going 0.1 cm/μsec initiates propagating detonation while the flyer going 0.08 cm/μsec fails to initiate detonation.

adequate for homogeneous explosives. Heterogeneous explosives are initiated and may propagate by the process of shock interaction with density discontinuities such as voids. These interactions result in hot regions that decompose and give increasing pressures that cause more and hotter decomposing regions. Some heterogeneous explosives may require hot spots for propagation of the detonation wave.

Forest Fire may be used to reproduce the explosive behavior in many one-, two-, and three-dimensional situations where data are available. Forest Fire gives the rate of explosive decomposition as a function of local pressure, or any other state variable, in the explosive. The Forest Fire description of heterogeneous shock initiation is also useful for describing heterogeneous detonation propagation along surfaces, around corners, the failure of a propagating detonation if the thickness or diameter becomes less than the critical value, and the failure of a propagating detonation to propagate into precompressed explosive regions.

The Los Alamos National Laboratory radiographic facility, PHERMEX, [13,14,18] has been used to study detonation wave profiles in heterogeneous explosives as they proceed up metal surfaces.[15,18] PHERMEX has also been used to study the density profiles when a detonation wave turns a corner.

As described in Reference 15, a radiographic study was made of a 10.16 cm cube of Composition B, with and without Tantalum foils, initiated by a plane wave lens confined by 2.54 cm thick Aluminum plates. The radiographs show a remarkably flat detonation front followed by a large decrease in density originating near the front of the wave as it intersects the metal plate. Figure 4.13 shows the explosive without the foils and Figure 4.14 shows the same system with Tantalum foils embedded in the system. Figure 4.15 is a sketch of the prominent features of the radiograph. The shock wave in the Aluminum and rarefaction from the free surface are shown. Also shown are a remarkably flat detonation front and a small displacement of the foils across it, followed by a large decrease in density originating near the front of the wave as it intersects the metal plate, and a large displacement of the foils.

The Forest Fire model of heterogeneous shock initiation results in a calculated flow that reproduces that observed experimentally, as shown in Figure 4.16. This result suggests that the observed detonation behavior is a consequence of the heterogeneous shock initiation process. The more shock insensitive explosives should give greater wave curvature and have larger failure diameters. Explosives initiated and burned with the heterogeneous shock initiation model do not show scaling behavior; therefore, failure depends upon the pressure magnitude and how long it can be maintained. Energy continues to be delivered by shocked but not detonated explosive even though a propagating detonation was not generated.

Corner Turning

Venable performed a radiographic study of a Baratol plane wave initiated Composition B detonation proceeding perpendicular to an Aluminum block, up a 45° wedge, and around a 90° block. Calculations using the Forest Fire model reproduced the radiographic features.

Dick performed a radiographic study of a detonation wave proceeding up a block of a very shock insensitive TATB based explosive, X-0219, and its failure to propagate around a corner. The radiographs are shown in Figure 4.17. Dick's experimental profiles and the calculated profiles using the Forest Fire model in the 2DL code are shown in Figure 4.18. The amount of explosive that remains undecomposed after passage of the shock wave depends primarily upon the curvature of the detonation wave before it turns the corner. If the wave

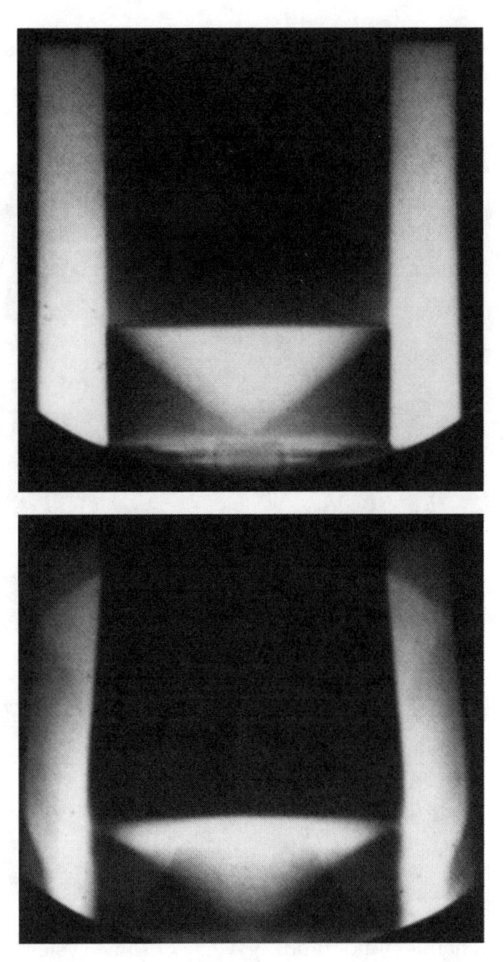

Figure 4.13 Static and dynamic radiographs NW-423 of a 10.16 cm cube of Composition B explosive initiated by a plane wave lens and confined by 2.54 cm thick plates of Aluminum.

is sufficiently curved, the detonation proceeds like a diverging detonation wave and little or no explosive remains undecomposed. If the wave is flat, or nearly so, when it arrives at the corner, then much more partially decomposed explosive will remain after shock passage. Because the actual experiment was performed with air in the corner, the Lagrangian calculation required that some low-density material corner (Plexiglas was used). The calculation slightly underestimates the amount of explosive that remains undecomposed.

To study this system in a more realistic geometry, the Eulerian code, 2DE, described in Appendix C, was used because it can describe large distortion problems such as an explosive-air interface. The results calculated using the Forest Fire burn are shown in Figure 4.19. Again, the results depend upon the detonation wave profile before it reaches the corner. If the wave started out flat, the explosive region near the explosive-air interface remained partially decomposed and the detonation wave never completely burned across the front until the wave became sufficiently curved at the front and near the interface. The failure process of a heterogeneous explosive is a complicated interaction of the effective reaction zone thickness which determines how flat the wave should be and the curvature

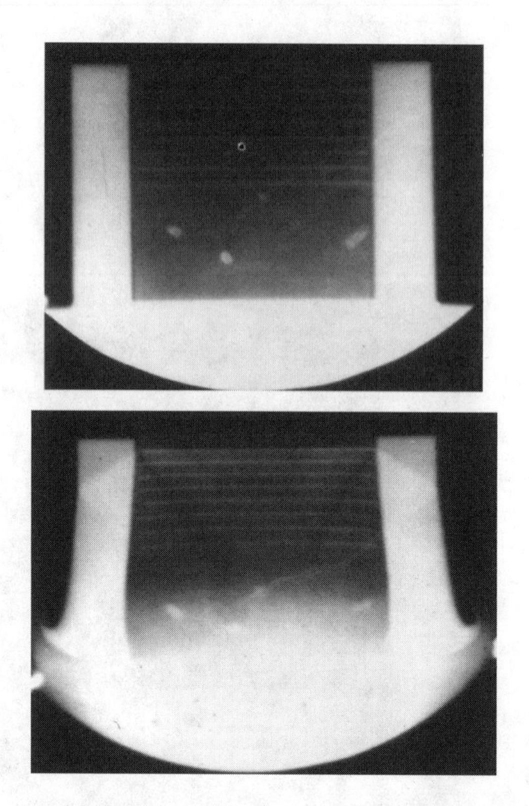

Figure 4.14 Static and dynamic radiographs NW-459 of same system as Figure 4.13 with addition of 0.00125 cm thick Tantalum foils spaced every 0.635 cm.

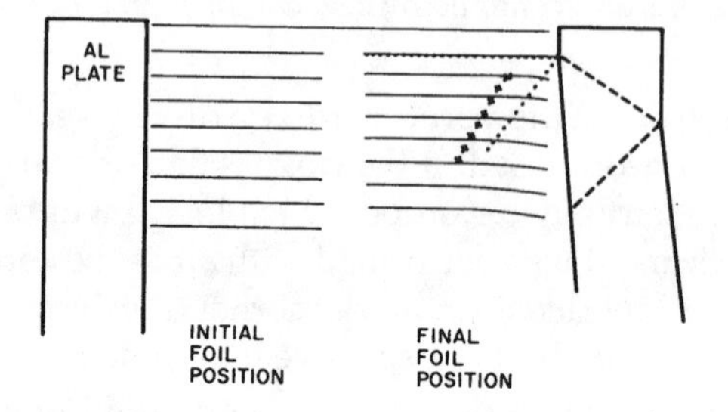

Figure 4.15 Prominent features of radiograph shown in Figure 4.14. The initial and final Tantalum foil positions, the detonation front, the Aluminum shock wave and rarefaction, the position of the Aluminum shock wave and rarefaction, the position of the Aluminum plate, and approximate positions of the rarefactions in the detonation products are shown.

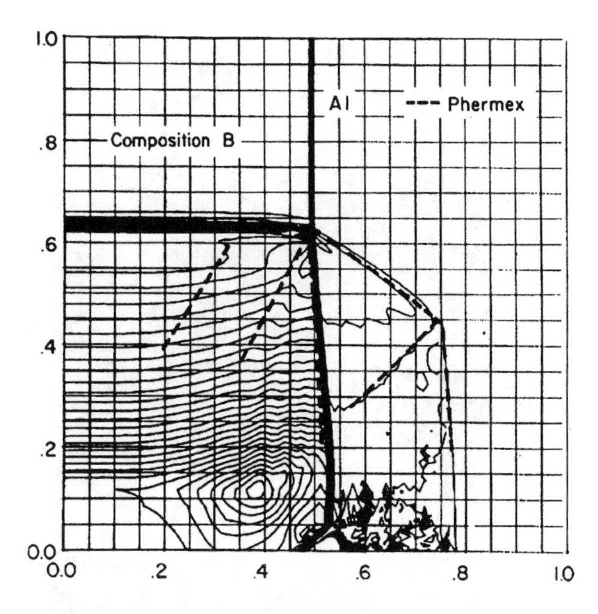

Figure 4.16 Constant density profiles at 8 μsec for a 5 cm half thickness slab of Composition B detonation proceeding perpendicular to a 2.5 cm thick Aluminum plate. The Forest Fire burn technique was used. The prominent features of the radiograph of the flow are shown by dashed lines.

required for decomposition to occur near the surface of the charge. As shown in Chapter 1, the heterogeneous explosive reaction zone is dependent upon the void distribution and the resulting hot spot size and decomposition rate. Because details of the hot spot reaction zone are missing from the calculations, detailed realistic modeling of corner turning or failure diameter has yet to be achieved.

Failure Diameter

Calculations were performed using the Forest Fire burn in 2DL for 0.7 and 1.3 cm radius cylinders of X-0219 confined by Plexiglas and for half thickness slabs of 1.3 and 2.6 cm. The thinner charges developed greater curvature and the 0.7 cm radius cylinder failed to propagate. Calculations were also performed using the Forest Fire burn in the 2DE code for 0.65 and 1.3 cm radius cylinders of X-0219 confined by air. The 0.65 cm radius cylinder failed to propagate, as shown in Figure 4.20. The experimentally determined failure radius is 0.75 cm. Similar calculations were performed for PBX-9404, Composition B, X-0290, and the very shock insensitive, nitroguanidine based, X0288. Results are compared with Campbell and Engelke's experimental failure radii[17] in Table 4.2.

The failure thickness of slabs of PBX-9502 confined by air, Plexiglas, Aluminum, Copper, and Tungsten were modeled. Ramsay measured the failure thickness of PBX-9502 confined by air, Plexiglas, Aluminum, and Copper. The results of the experiments and the modeling are shown in Table 4.3.

The dominant feature of failure in heterogeneous explosives is the same hot spot decomposition that determines the shock initiation behavior.

The wave curvatures that Campbell and Engelke observed in X-0219[15] were reproduced by the calculations. Donguy and Legard[16] have shown that a Forest Fire type rate can be used to model the effect of diameter on the detonation velocity.

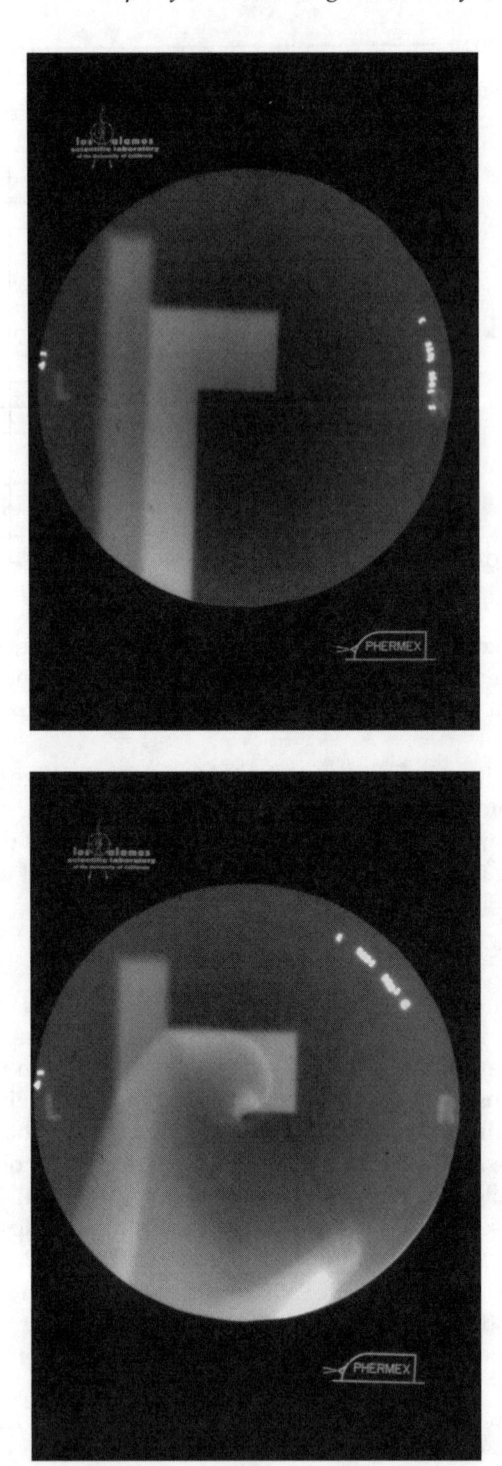

Figure 4.17 PHERMEX static and dynamic radiograph of shot NW-1942. A corner of X-0219 with a Plexiglas plate initiated by a plane wave lens. The corner region of undecomposed X-0219 is shown.

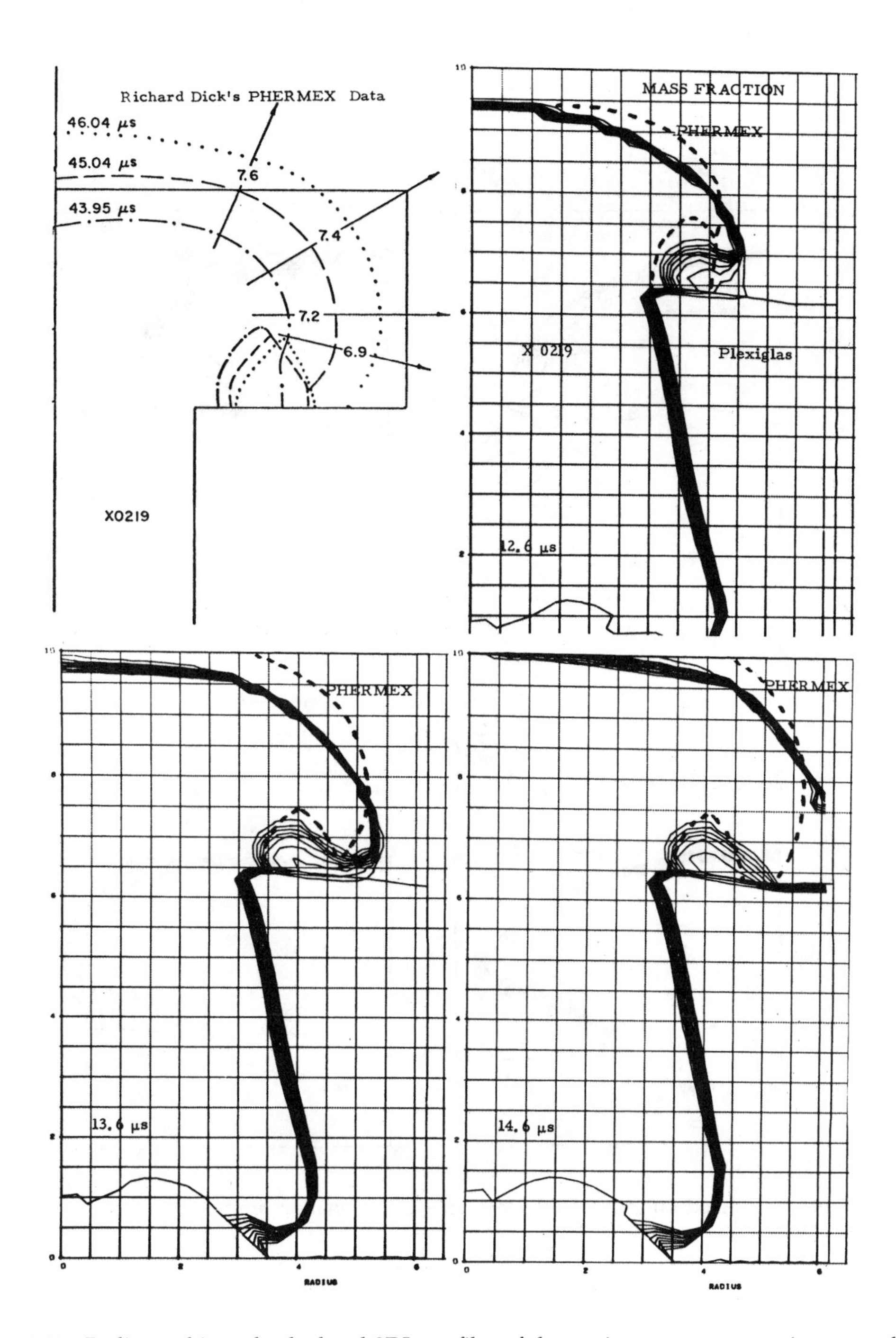

Figure 4.18 Radiographic and calculated 2DL profiles of detonation wave propagating around a corner of X-0219. The corner was filled with air in the experiment and with Plexiglas in the calculation.

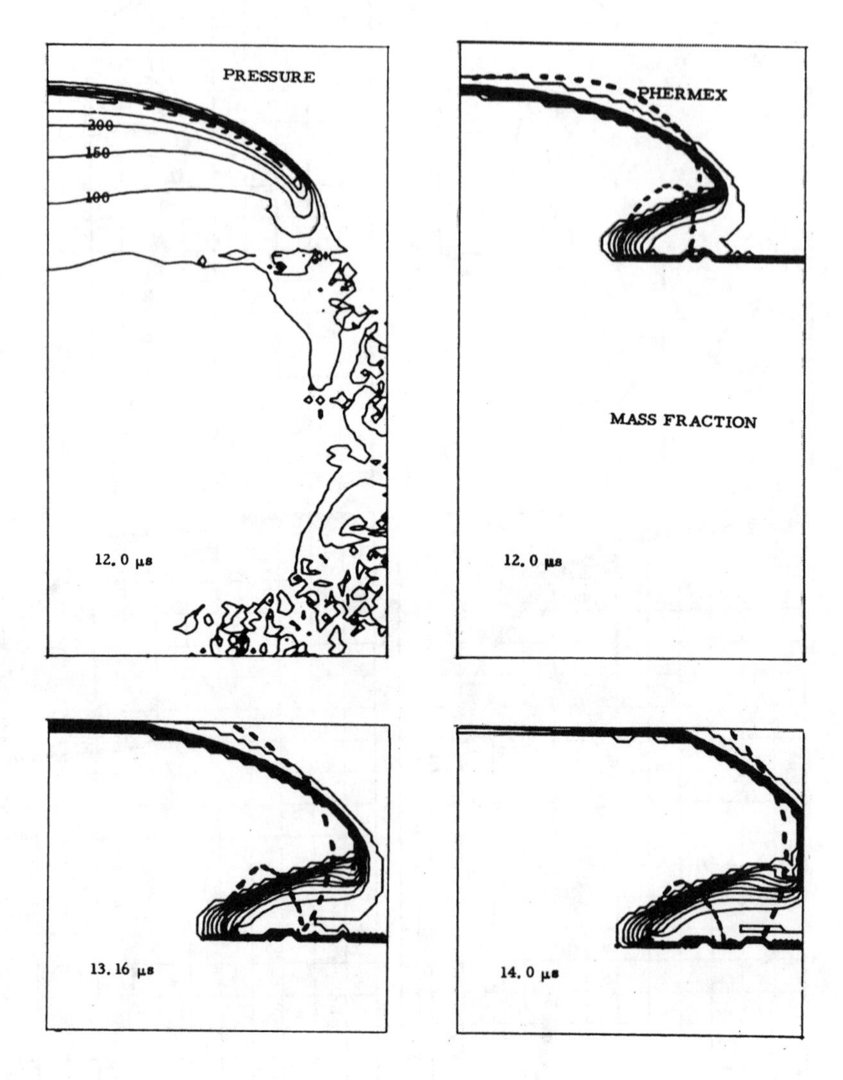

Figure 4.19 Calculated Eulerian 2DE code profiles of detonation propagating around an air-filled corner of X-0219.

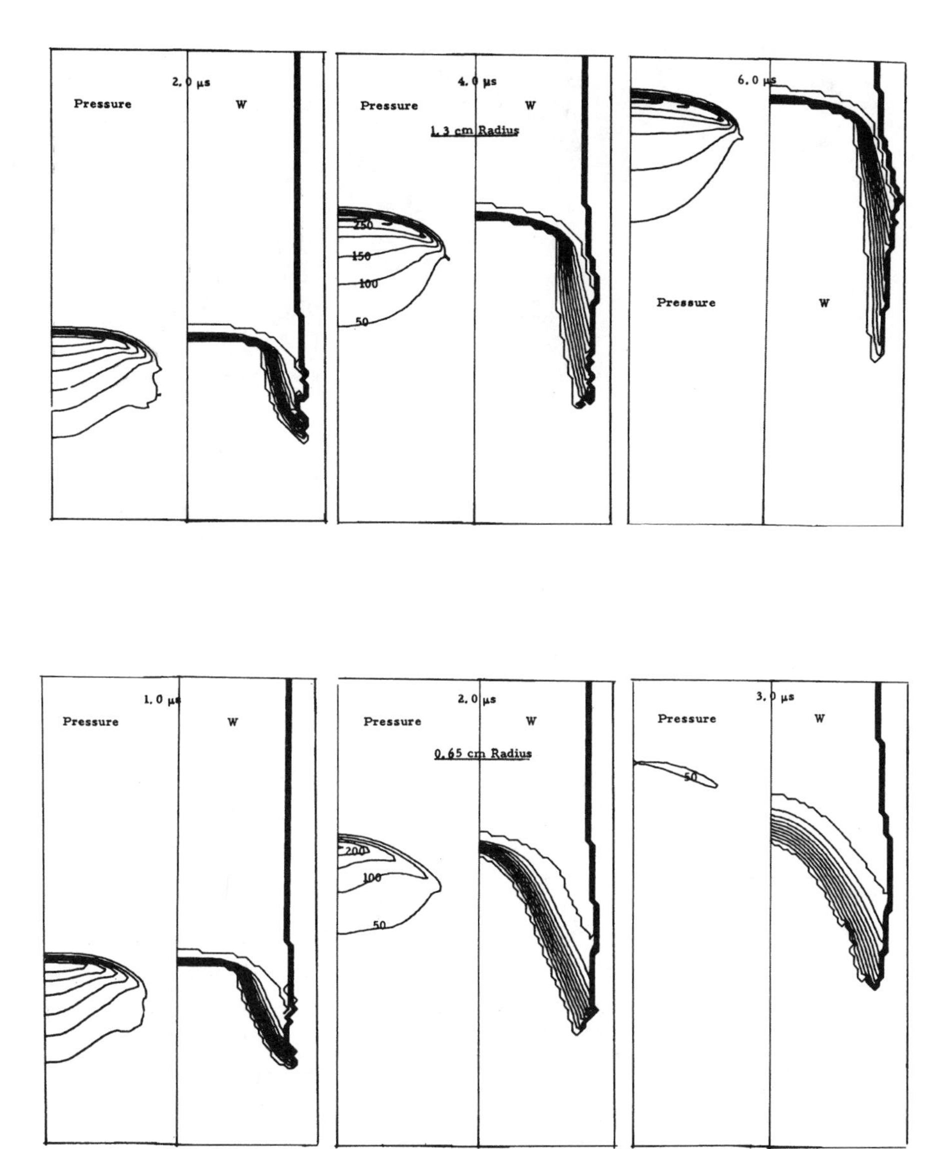

Figure 4.20 Pressure and mass fraction profiles of 0.65 and 1.3 cm radius cylinder of X-0219, calculated using 2DE code with Forest Fire burn model. The 1.3 cm radius cylinder is above and the 0.65 cm radius cylinder is below the experimental failure radius of 0.7 cm.

TABLE 4.2 Failure Radii

Explosive	Experimental Failure Radius (cm)	Calculated Results
PBX-9404	0.06 ± 0.01	0.1 cm Propagated 0.05 cm Failed
Comp B	0.214 ± 0.03	0.3 cm Propagated 0.2 cm Failed
PBX-9502	0.45 ± 0.05	0.5 cm Propagated 0.3 cm Failed
X-0219	0.75 ± 0.05	1.3 cm Propagated 0.7 cm Failed
X-0288	2.54 to 5.0	2.5 cm Propagated 0.7 cm Failed

TABLE 4.3 PBX-9502 Slab Failure Thickness

Confinement	Experimental Failure Thickness (cm)	Calculated Results
Air	1.75	2.5 cm Propagated 2.0 cm Failed
Plexiglas	1.75	2.5 cm Propagated 2.0 cm Failed
Aluminum	0.75	1.0 cm Propagated 0.75 cm Failed
Copper	0.55	0.75 cm Propagated 0.5 cm Failed
Tungsten		0.5 cm Propagated 0.3 cm Failed

Detonation initiation and propagation of heterogeneous explosives cannot be described adequately using Arrhenius kinetics. The Forest Fire model can describe the decomposition that occurs from hot spots formed by shock interactions with density discontinuities in heterogeneous explosives, and can also describe the passage of heterogeneous detonation waves around corners and along surfaces. Failure or propagation of a heterogeneous detonation wave depends upon the interrelated effects of the wave curvature and the shock sensitivity of the explosive. Increasing the density of the confining material results in increasing the support to the detonation wave so that the detonation will propagate in smaller geometries.

Failure or propagation as a function of diameter, pulse width of initiating shock, wave curvature, confinement, and propagation of detonations along surfaces are all dominated by the heterogeneous initiation mechanism. The Pop plot is the important experimental parameter that correlates with all of these explosive phenomena. The hot spot decomposition reaction dominates not only the build-up to detonation, but also the way in which the detonation propagates when side rarefactions are affecting the flow. The consequences are important for many practical problems of detonation chemistry and physics.

4.3 Desensitization by Preshocking

It has been observed that preshocking a heterogeneous explosive with a shock pressure too low to cause propagating detonation in the time of interest can cause a propagating detonation in unshocked explosive to fail to continue propagating when the detonation front arrives at the previously shocked explosive. The initiation of explosives by jets that first penetrate barriers of inert materials in contact with the explosive has been observed to result in decreased sensitivity of the explosive to jet impact. The bow shock wave that travels ahead of the jet through the barrier can desensitize the explosive sufficiently that it cannot be initiated by the higher pressure generated near the jet interface. The desensitization of the explosive by the weak bow shock results in the failure to establish an explosion in the Composition B.

To model desensitization by preshocking, the modification indicated by the three-dimensional study, described in Chapter 3, to the Forest Fire decomposition rate was to limit the rate by the initial shock pressure and to add the Arrhenius rate law to the limited Forest Fire rate. The Forest Fire rate for TATB is shown in Figure 4.21, along with the Arrhenius rate calculated using the temperatures from the HOM equation of state for the partially burned TATB associated with the pressure, as determined by Forest Fire. The multiple shock Forest Fire model (MSFF) uses a burn rate determined by Forest Fire, limited to the initial shock pressure, and the Arrhenius rate using local partially burned explosive temperature.

Richard Dick[18] performed a PHERMEX radiographic study of detonation waves in PBX-9502, proceeding up a 6.5 by 15. cm block of explosive that was preshocked by a 0.635 cm steel plate moving at 0.08 (Shot 1697) or 0.046 cm/μsec (Shot 1914). The static and dynamic radiographs for Shot 1697 are shown in Figure 4.22. The preshocked PBX-9502 explosive quenches the detonation wave as it propagates into the block of explosive.

The experimental geometries studied using PHERMEX were numerically modeled using the reactive hydrodynamic code 2DL, described in Appendix B. For explosives that had been previously shocked, Craig observed experimentally that the distance of run to detonation for several multiply shocked explosives was determined primarily by the distance after

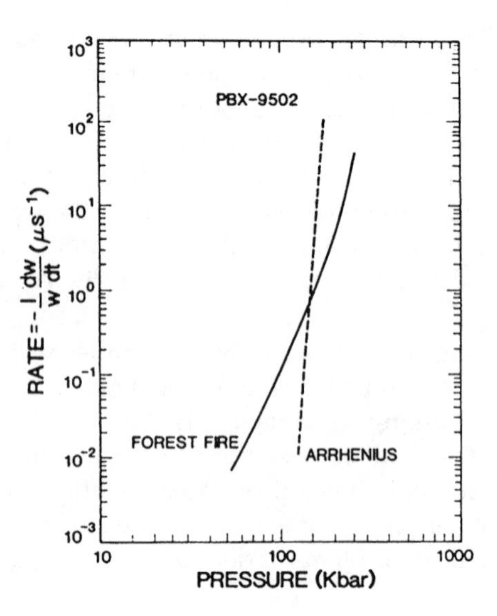

Figure 4.21 The burn rate as a function of pressure for the Forest Fire burn model and the Arrhenius rate law using the HOM temperatures associated with the Forest Fire pressures.

a second shock had overtaken the lower pressure shock wave (the preshock). To approximate this experimental observation and the results obtained from the three-dimensional numerical model study, the multiple shock Forest Fire model was used. The Forest Fire rate was determined by the first shock wave or the rates determined by any subsequent release waves that result in lower pressures and lower decomposition rates. As suggested by the three-dimensional study described in Chapter 3, the Arrhenius rate was added using the local partially burned explosive temperatures to the Forest Fire rate.

The calculated pressure and mass fraction contours for PHERMEX Shot 1697 are shown in Figure 4.23, along with the radiographic interfaces. The 2DL calculation had 50 by 33 cells to describe the PBX-9502 and 50 by 5 cells to describe the steel plate. The mesh size was 0.2 cm and the time step was 0.04 μsec.

The PHERMEX shot was numerically modeled using various velocity steel plates. The results are shown in Table 4.4. They agree with the experimental evidence that detonation wave failure occurs in preshocked TATB shocked by steel plates with velocities of 0.046 and 0.08 cm/μsec. The 9 kbar shock pressure is too low to result in an appreciable decrease in the second shock temperature, so it does not desensitize the explosive enough to prevent a propagating detonation from occurring in the preshocked explosive. The 90 kbar first shock pressure builds to propagating detonation after 1.5 cm of run, and the first shock temperature is high enough that the second shock temperature increase can result in propagating detonation in the multiply shocked explosive. The first shock temperature is even higher for the 130 and 180 kbar first shocks, and a detonation wave will propagate through the preshocked explosive if it has not already detonated. The multiple shock Forest Fire model thus models both the desensitization and failure to desensitize that results from either too low or too high preshock pressures.

Chick and Hatt[19] and Chick and MacIntyre[20] have studied the effects of desensitization by preshocking that occur when jets interact with barriers in contact with explosives. A

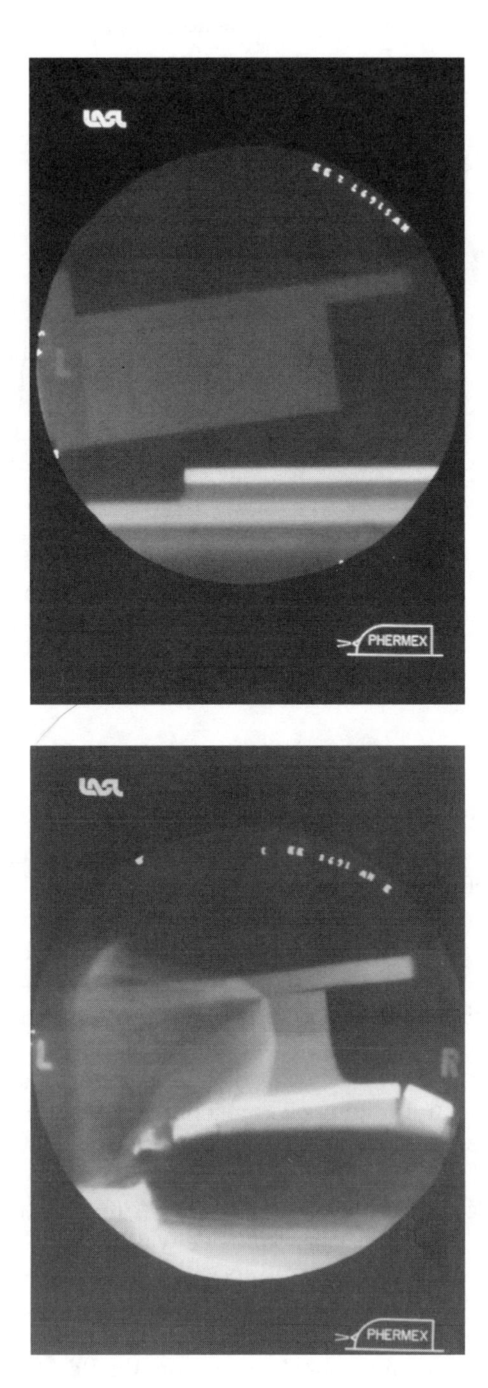

Figure 4.22 Static and dynamic radiograph 1697 of PBX-9502 shocked by a 0.635-cm thick steel plate going 0.08 cm/μsec and initiated by 2.54 cm of TNT and a P-40 lens.

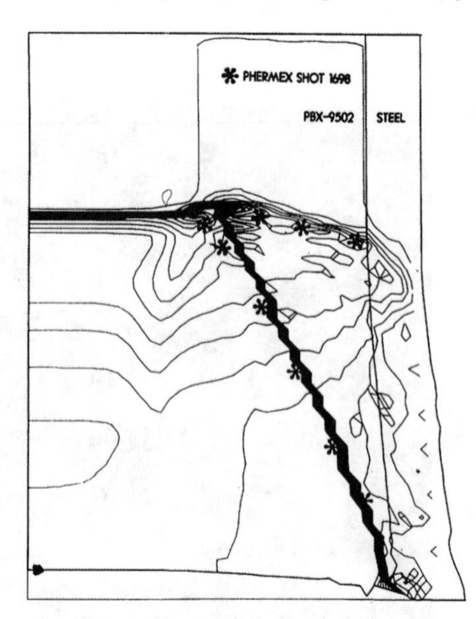

Figure 4.23 The pressure and mass fraction contours for a detonation wave in PBX-9502 interacting with explosive that had been previously shocked to 50 kbar. The PHERMEX radiographic interfaces are shown. The mass fraction contour interval is 0.1 and shown as a thick and almost solid line. The pressure contour interval is 40 kbar.

steel jet with an initial velocity of 0.75 cm/μsec and diameter of 0.15 cm initiated detonation in bare Composition B, but failed to initiate the Composition B if it first interacted with a 0.3 cm steel barrier on the surface of the Composition B.

The system studied by Chick and MacIntyre was modeled using the Eulerian reactive hydrodynamic code 2DE described in Appendix C with the multiple shock Forest Fire rate. Figure 4.24 shows the calculated initiation of detonation by the steel jet interacting with bare Composition B. Figure 4.25 shows the calculated initiation of detonation by the jet interacting with a 0.3 cm steel barrier and Composition B, if desensitization by preshocking does not occur. Prompt initiation of the Composition B is predicted to occur if the explosive is not desensitized. Figure 4.26 shows the calculated failure to initiate the Composition B when desensitization by the bow shock ahead of the jet is permitted to occur by the inclusion into the calculation of the multiple shock Forest Fire burn model. The effect of barriers that results in decreased sensitivity of explosives to projectile or to jet impact can be numerically modeled.

Travis and Campbell[21] performed a series of experiments studying the desensitization of PBX-9404 by shocks. The PBX-9404 explosive was 8 x 4 x 1/3 inch and cemented to a thick sheet of Plexiglas. It was immersed in water at various distances from a 15.24 cm diameter sphere of PBX-9205 initiated at its center, which served as the preshock generator. When the arrangement was fired, a detonation swept downward through the PBX-9404 and encountered an upward-spreading shock wave from the PBX-9205 generator. Events were photographed with a framing camera. They concluded that the detonation in PBX-9404 was not quenched by a preshock of 7.5 kbars. An 11 kbar initial shock pressure required a time lapse of about 6 μsec before the detonation wave failed and a 25 kbar preshock required less than 1 μsec.

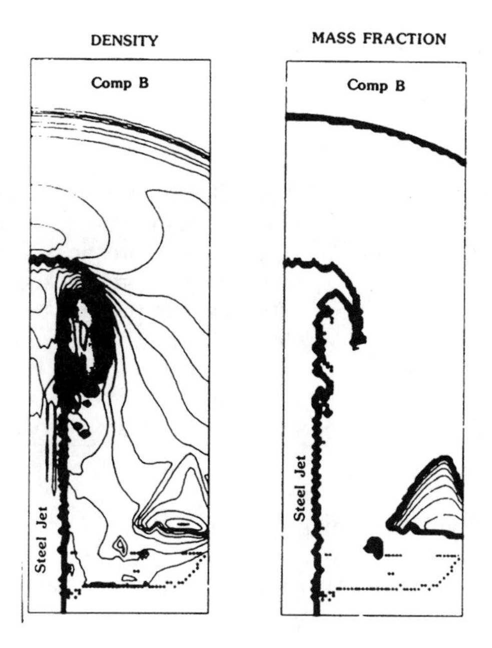

Figure 4.24 The density and mass fraction contours at 1.0 μsec for a 0.15-cm diameter steel rod initially moving at 0.75 cm/μsec penetrating Composition B and initiating detonation. The density contour interval is 0.2 g/cc. The mass fraction contour interval is 0.1.

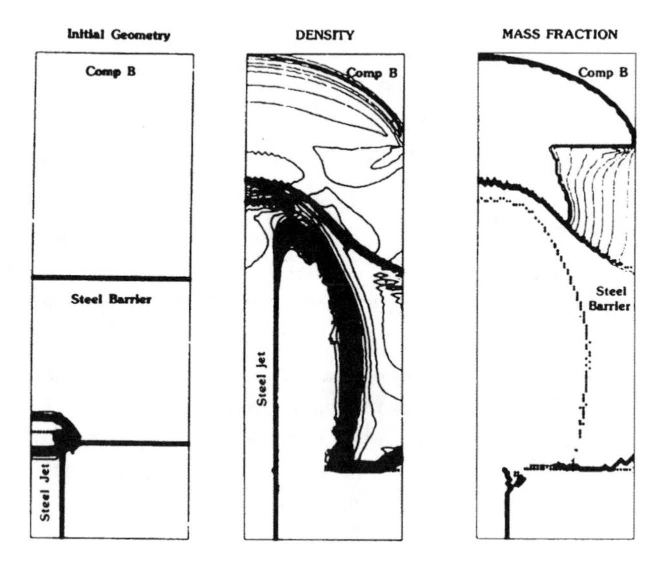

Figure 4.25 The density and mass fraction contours at 1.5 μsec for a 0.15-cm diameter steel rod initially moving at 0.75 cm/μsec penetrating a 0.3-cm thick steel plate in contact with Composition B. The density contour interval is 0.2 g/cc. The mass fraction contour interval is 0.1. The calculation assumes that desensitization does not occur. The jet initiated propagating detonation in the Composition B.

TABLE 4.4 PBX 9502 Desensitization Calculations

Steel Plate Velocity (cm/ μsec)	Preshock Pressure (kbar)	Result Upon Arrival of Detonation Wave
0.02	9.	Detonates preshocked HE
0.03	14.	Fails in preshocked HE
0.045	23.	Fails in preshocked HE
0.080	50.	Fails in preshocked HE
0.10	70.	Fails in preshocked HE
0.12	90.	Detonates preshocked HE and after 1.5 cm run
0.16	130.	Detonates preshocked HE
0.20	180.	Detonates preshocked HE

The Travis and Campbell experiments can be evaluated using calculated multiple shock temperatures and solid HMX Arrhenius constants determined from Craig's single crystal initiation data. His data and the Walsh and Christian shock temperatures for HMX are given below.

Pressure (kbars)	Induction Time (μsec)	Calculated Shock Temperature (^{o} K)
420	0.050	1809.6
358	0.272	1489.3
320	Failed	1340.0

Using the adiabatic explosion time equation from Chapter 3 with $C = 0.4$, and $Q = 800$, the calculated activation energy, E, is 34.8 kcal/mole, and the frequency factor Z is 3.0×10^4 μsec^{-1}. This is much slower than the liquid HMX constants given in Table 3.1 of an activation energy of 52.7 kcal/mole and a frequency factor of 5.0×10^{13} μsec^{-1}.

The solid kinetics of HMX is very complicated with an initial slow reaction succeeded by acceleratory and deceleratory periods. The initial slow reaction has been studied by Maksimov.[22] He reported an activation energy of 37.9 kcal/mole and a frequency factor of 1.58×10^5 μsec^{-1}. Rogers and Janney found an "initial" activation energy of 33.9 kcal/mole and a frequency factor of 4.8×10^5 μsec^{-1}.

Within the experimental errors and the uncertainty associated with the calculated Hugoniot temperatures, the initial solid Arrhenius constants agree with the constants derived from the shock initiation data. The liquid constants are clearly not appropriate for the single crystal HMX shock initiation experiment or for shock desensitization studies.

The calculated 9404 multiple shock results using solid Arrhenius constants obtained from Craig's single crystal shock initiation data are shown below.

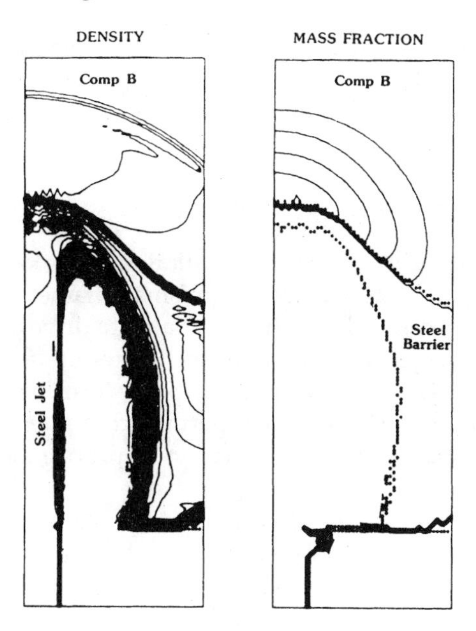

Figure 4.26 The density and mass fraction contours at 1.5 μsec for a 0.15-cm diameter steel rod initially moving at 0.75 cm/μsec penetrating a 0.3-cm thick steel plate in contact with Composition B. The density contour interval is 0.2 g/cc. The mass fraction contour interval is 0.1. The calculation includes desensitization by preshocking. The Composition B is not initiated.

First Shock (kbars)	Second Shock (kbars)	Temperature ^{o}K	Induction Time (μsec)
360	0	1669	0.051
7	360	1442	0.198
10	360	1368	0.345
15	360	1267	0.821
20	360	1185	1.87
25	360	1118	4.04
50	360	917	84.4

Induction times less than 0.2 μsec were calculated for the 7 kbar preshock that was observed to fail to quench a detonation wave. Induction times of 0.35 to 4 μsec were calculated for the preshock pressures observed to have a time lapse before the detonation wave failed. Larger induction times were calculated for a 50 kbar preshock pressure; however, the 50 kbar preshock would build to detonation in 0.4 cm or 0.75 μsec.

The experimental observations of Travis and Campbell are consistent with the desensitization being caused by the preshock making the HMX explosive more homogeneous and reducing the explosive temperature upon arrival of the detonation wave by the multiple shock process.

Desensitization of heterogeneous explosives by shocks too weak to initiate propagating detonation is an important feature that must be included in numerical modeling of energetic materials under projectile or to jet impact if the system being modeled includes layers of inert materials in contact with the energetic material. The multiple shock Forest Fire model will model both desensitization and failure to desensitize effects that occur when explosives or propellants are multiply shocked.

4.4 Projectile Initiation of Explosives

Most of the experimental studies of projectile initiation of explosives have been performed using shaped charges with a metal cone to form small diameter, high velocity projectiles. Held[23] investigated the process using a shaped charge to drive a 0.15 cm thick, 60° Copper cone. The resulting Copper jet was shot into steel plates of varying thicknesses to obtain different exit jet velocities. The jet sizes and velocities were measured using velocity screens and X-ray flash photography. The jets were permitted to interact with an explosive, and the critical velocity was determined for initiation of propagating detonation by various diameter jets.

Held observed that his data could be correlated by assuming that the critical jet value for explosive initiation was the jet velocity squared (V^2) times the jet diameter (d). If the velocity is in millimeters per microsecond and the jet diameter is in millimeters, he reported that the critical V^2d was about 5.8 for Copper jets initiating 60/40 RDX/TNT at 1.70 g/cc.

Similar experiments have been performed by A. W. Campbell and published in a report by Mader and Pimbley.[24] Campbell studied Copper jets initiating PBX-9404 (94/3/3 HMX/nitrocellulose/tris-β-chloroethyl phosphate at 1.844 g/cc) and PBX-9502 (95/5 TATB/Kel-F at 1.894 g/cc) and his data are summarized in Table 4.5. Critical V^2d values are 125 ± 5 mm^3/μs^2 for PBX-9502, 16 ± 2 mm^3/μs^2 for PBX-9404, 37 mm^3/μs^2 for 75/25 Cyclotol, and 29 mm^3/μs^2 for Composition B-3.

If the pressure-particle velocity is matched, as shown in Figure 4.27 for Copper at 0.5 cm/μsec shocking inert PBX-9502, the pressure match is 680 kbar; at 0.6 cm/μs, the pressure match is 920 kbar. The C-J pressure of PBX-9502 is 285 kbar, so the Copper jet must initiate a strongly overdriven detonation. However, the jet diameter ($\sim$ 0.4 cm) is less than half the failure diameter (0.9 cm) of unconfined PBX-9502.

Also shown in Figure 4.27, the pressure match is 340 kbar for PBX-9404 shocked initially by Copper at 0.3 cm/μsec and is 250 kbar at 0.25 cm/μsec. The run distance for PBX-9404 at 250 kbar is about 0.05 cm, so detonation would occur very quickly. The jet diameter ($\sim$ 0.2 cm) is larger than the unconfined PBX-9404 failure diameter of 0.12 cm.

To examine the dominant mechanisms of jet initiation of PBX-9404 and PBX-9502, the two-dimensional reactive Eulerian hydrodynamic code 2DE, described in Appendix C, was used to describe the reactive fluid dynamics. The Forest Fire description of heterogeneous shock initiation was used to describe the explosive burn. The HOM equation of state and Forest Fire rate constants for PBX-9502 and PBX-9404 were identical to those used earlier in this chapter. The Pop plots are shown in Figure 4.28, and the Forest Fire rates are shown in Figure 4.29.

The jet was described as a cylinder with a radius of 10 or 20 cells and 40 cells long and with the appropriate initial velocity. The cell size was .01, .02, or .005 cm, depending on the jet radius. The maximum burn pressure in the Forest Fire burn was set to 1.5 Mbar to permit overdriven detonations, and the rate was limited to e^{20}. The viscosity coefficient used for PBX-9502 was 0.25 and for PBX-9404 was 0.75. Sufficient viscosity to result in a resolved burn was necessary, just as in the detonation failure calculations described earlier in this chapter.

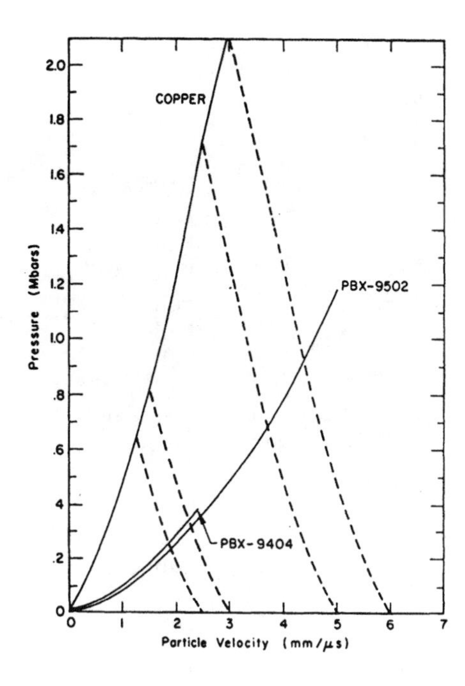

Figure 4.27 The pressure-particle velocity Hugoniots for Copper, PBX-9502, and PBX-9404 with the reflected shock states for Copper with initial free-surface velocities of 0.6, 0.5, 0.3, and 0.25 cm/μsec.

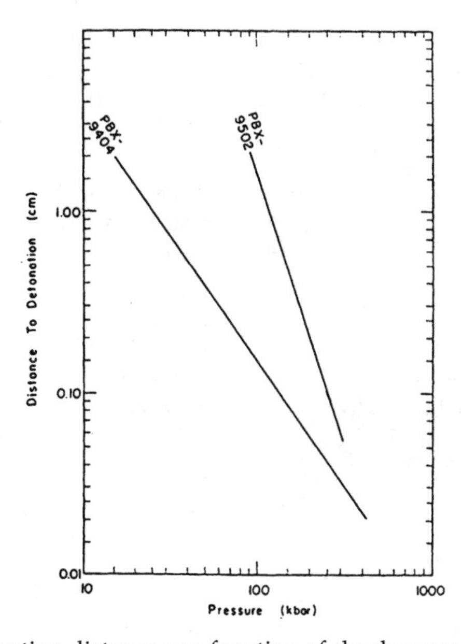

Figure 4.28 The run to detonation distance as a function of shock pressure.

TABLE 4.5 Campbell Experimental Data

Copper Jet Tip Velocity (cm/μs)	Jet Diameter (cm)	Result	V^2d (mm^3/μs^2)
PBX 9502		Critical	127± 5
.459	.588	Detonation	124
.447	.596	Fail	119
.465	.584	Detonation	126
.452	.592	Fail	121
.494	.415	Fail	101
.556	.395	Fail	122
.555	.395	Fail	122
.582	.380	Detonation	128
.593	.385	Detonation	135
PBX-9404		Critical	16± 2
.287	.204	Detonation	16.8
.251	.225	Fail	14.2
.276	.211	Fail	16.0
.307	.195	Detonation	18.4

An uncertainty in the calculations is the approximation of the jet as a cylinder of uniform velocity colliding end-on with the explosive. Although the jet is actually many small metal pieces, the calculations indicate that the critical conditions are determined by the first piece of about the same length as its radius. Side rarefactions dominate the flow after the reflected shock wave travels one diameter length back into the jet.

The proper initial conditions for the hot jet must be used. Because it was formed by being shocked and then rarefied, it will have a residual temperature greater than ambient and a density less than the initial density of the jet material.

Estimates of the Copper residual state can be made by assuming the material is at the same state as if it had achieved its final velocity by a single-shock and then had rarefied to one atmosphere. The residual temperature of the Copper initially shocked to 830 kbar and then rarefied to a free-surface velocity of 0.3 cm/μsec is 768 K; the residual density is 8.688 g/cc, comparable to the initial density of 8.903. Calculations were performed with Copper initial conditions of 8.903 g/cc and 300 K and of 8.688 g/cc and 768 K. Since the changed equation of state results in only slightly changed explosive shock pressure, the calculated results were insensitive to the Copper jet initial conditions.

Another uncertainty is how accurately the Forest Fire burn rate can be extrapolated to overdriven detonations. Since the overdrive decays rapidly and the side rarefactions quickly reduce the failure region pressure into the pressure range where the Pop plot was determined, the accuracy of the extrapolated burn rates is not important to the determination of the critical conditions for establishing propagating detonations.

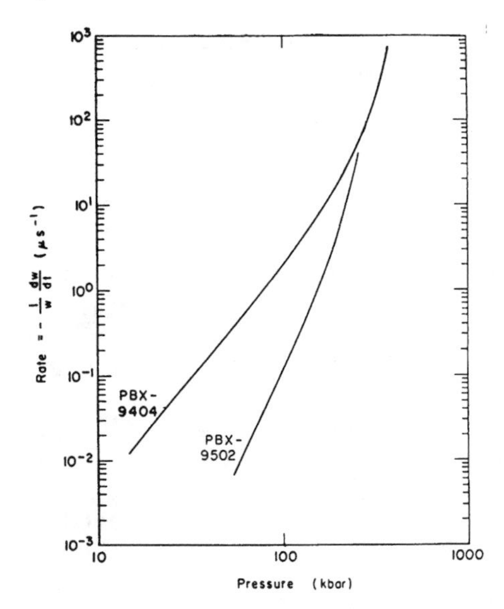

Figure 4.29 The Forest Fire decomposition rates as a function of shock pressure.

The calculated PBX-9502 results are presented in Tables 4.5 and 4.6. The effects of jet composition, diameter, and velocity were examined. The calculated profiles for a 0.4 cm diameter Copper jet with a 0.7 cm/μsec initial velocity are shown in Figure 4.30. The Copper jet initiates an overdriven detonation smaller than the critical diameter, which enlarges to greater than the critical diameter of self-confined PBX-9502 when shocked by a 0.7 cm/μs Copper jet. When the PBX-9502 is shocked by a 0.5 cm/μsec Copper jet, the detonation is decayed by side and rear rarefactions before it enlarges to the critical diameter, as shown in Figure 4.31.

As shown in Table 4.5, the numerical results bound the Campbell experimental critical V^2d of 127 for Copper jets. The shock pressure sent into the explosive depends on the jet material, so the Aluminum jet must have a greater velocity or diameter than the Copper jet to furnish an equivalent impulse to the PBX-9502. The critical V^2d would be expected to be larger for Aluminum than for Copper and larger still for a water jet. These conclusions are confirmed by the calculations with the predicted V^2d for Aluminum being 325$\pm$ 75 and for water about 800, as shown in Table 4.6.

The calculated PBX-9404 results are presented in Table 4.7. The effects of jet composition, diameter, and velocity were examined. The calculated profiles for a 0.2 cm diameter Copper jet with a 0.3 cm/μsec initial velocity are shown in Figure 4.32, and a velocity of 0.2 cm/μsec in Figure 4.33. The 0.3 cm/μsec Copper jet initiated detonation, which propagates only if it is maintained by the jet long enough to establish a stable, curved detonation front.

As shown in Table 4.7, the numerical results bound the Campbell experimental critical V^2d of 16 for Copper jets initiating propagating detonation in PBX-9404. The calculated results suggest a lower value for V^2d but one as accurate as the data and the approximations of the calculations. Again, the calculated critical V^2d is larger for Aluminum jets (70 $\pm$ 20) than for Copper jets and larger still for water jets (150 $\pm$ 50).

The V^2d criterion used by Held and Campbell relates to a critical value, above which a propagating detonation is produced in a given explosive when shocked by a projectile composed of a given material. Mader and Pimbley[24] proposed a more general critical

TABLE 4.6 PBX 9502 Calculations

Jet Material	Jet Diameter (cm)	Jet Velocity (cm/μs)	Result	V^2d
Copper	0.4	0.4	Failed	64.
Copper	0.4	0.5	Failed	100.
Copper	0.4	0.6	Marginal	144.
Copper	0.4	0.7	Propagated	196.
Copper	0.8	0.4	Marginal	128.
Copper	0.8	0.5	Propagated	200.
Aluminum	0.4	0.7	Failed	196.
Aluminum	0.4	0.8	Failed	256.
Aluminum	0.4	1.0	Propagated	400.
Aluminum	0.8	0.7	Propagated	392.
Water	0.4	0.6	Failed	144.
Water	0.4	1.0	Failed	400.
Water	0.4	1.4	Propagated	784.

condition of $\rho V^2 d$, where V is the velocity of the projectile, d is the diameter of the projectile, and ρ is the density of the projectile. A critical value, CV, of $\rho V^2 d$ (such that if $\rho V^2 d >$ CV, the explosive goes off, and does not go off otherwise) is intrinsic for the explosive and independent of the projectile material. The critical $\rho V^2 d$ value for PBX-9404 is 150 and for PBX-9502 is 800.

The initiations of propagating detonation in PBX-9502 and PBX-9404 by jets of Copper, Aluminum, and water have been numerically modeled using the two-dimensional Eulerian hydrodynamic code 2DE with the Forest Fire heterogenous shock initiation burn rate.

Shock initiation by jets near the critical $\rho V^2 d$ contrasts with other shock initiation studies. In the latter, if detonation occurred, it was because the initiating shock wave was of sufficient strength and duration to build up to detonation. The propagating detonation was assured by the large geometry. In near-critical jet initiation, however, a *prompt* detonation of the explosive always results, which builds to a propagating detonation only if the shock wave produced by the jet is of sufficient magnitude and duration. Therefore, jets may produce significant amounts of energy from an explosive even *below* the critical value of $\rho V^2 d$ required for propagating detonation.

The Forest Fire model has been used to solve many other scientific and engineering problems. The minimum priming charge test determines the least mass of booster explosive required to induce high-order detonation in a test charge. The hemispherical booster charge usually is small in size compared to the test sample, so the shock emanating from the booster is highly divergent and the shock pressure drops rapidly unless sufficient reaction occurs in the test charge. In Reference 25, the minimum priming charge test for PBX-9404 and Composition B were modeled in one dimensional spherical symmetry using the SIN

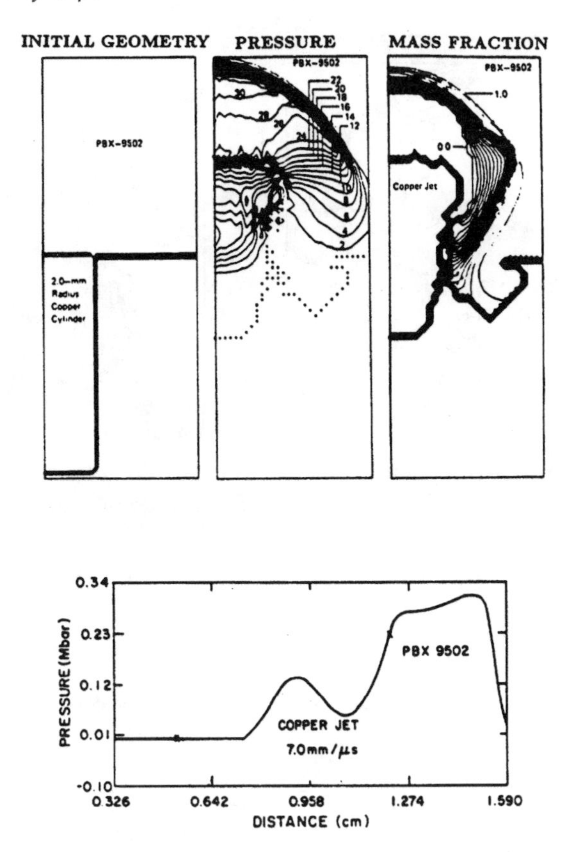

Figure 4.30 The initial profile and the pressure and mass fraction profiles for a 0.4-cm diameter Copper jet of 0.7 cm/μsec shocking PBX-9502 at 0.75 μsec. The jet initiates propagating detonation in the PBX-9502. The isobar contour interval is 20 kbar, and the mass fraction contour interval is 0.05. The pressure profile along the axis is also shown.

hydrodynamic code and the Forest Fire heterogeneous explosive decomposition rate. The minimum detonation initiation pressure was found to be greater than 60 kbars for PBX-9404 and greater than 70 kbar for Grade A Composition B, indicating that the minimum priming charge test is a high pressure prompt shock initiation experiment.

The relative shock sensitivities of explosive compositions are commonly assessed by means of gap tests. In these tests, the shock from a standard donor explosive is transmitted to the test explosive through an inert barrier (the so-called "gap"). The shock sensitivity of the test explosive is characterized by the gap thickness for which the probability of detonation is 50%. In Reference 26, the Los Alamos standard gap test and the Naval Ordnance Laboratory (NOL) large scale gap test were modeled using the 2DE code with Forest Fire burn rates. The Los Alamos gap test uses Dural for the inert barrier while the NOL gap test uses Plexiglas. The test explosive is unconfined in the Los Alamos gap test. In the NOL gap test the test explosive is confined with steel. The model showed good agreement between the calculated and experimental gap test values for PBX-9404, PBX-9502, Pentolite, Composition B, and an HMX based propellant, VTQ-2. The gap test is strongly two-dimensional as evidenced by the calculated distances of run to detonation in

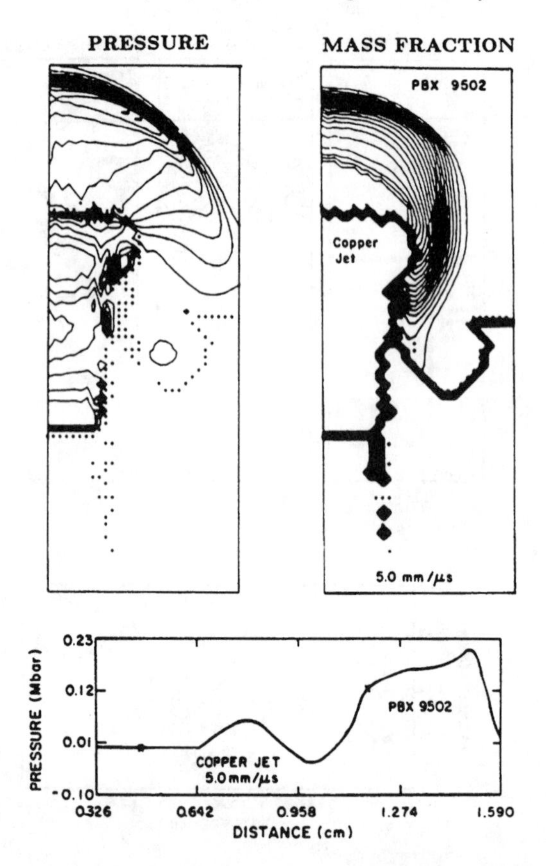

Figure 4.31 The pressure and mass fraction profiles for a 0.4 cm diameter Copper jet of 0.5 cm/μsec shocking PBX-9502 at 0.90 μsec. The jet fails to initiate propagating detonation in the PBX-9502. The isobar contour interval is 10 kbar, and the mass fraction contour interval is 0.05. The pressure profile along the axis is also shown.

the test samples which were significantly longer than those from the Pop plots at induced pressures near the critical gap thickness.

The shotgun test involves shooting a cylinder of explosive or rocket propellant from a shotgun at a target. The sympathetic detonation test involves two cubes of explosive or propellant, one of which is detonated. The extent of reaction in the acceptor cube is determined by its effect on a lead witness cylinder in contact with the acceptor cube. The critical separation distance at which no detonation is observed is determined. These tests were modeled using the 2DE code with Forest Fire burn in Reference 27. The gap test, shotgun test, and sympathetic detonation test were modeled. They are shock initiation experiments with a substained shock pulse. A detonation will occur when the shock wave is of sufficient strength and duration to build up to detonation.

The effect of the impact of steel cylinders and spheres on PBX-9502 was modeled by Fu and Cort in Reference 28. They used the 2DE code with the Forest Fire heterogeneous shock initiation burn. They found good agreement between the numerical model and

TABLE 4.7 PBX-9404 Calculations

Jet Material	Jet Diameter (cm)	Jet Velocity (cm/μs)	Result	V^2d
Copper	0.2	0.2	Failed	8.0
Copper	0.2	0.25	Marginal	12.5
Copper	0.2	0.3	Propagated	18.0
Copper	0.4	0.2	Propagated	16.0
Copper	0.4	0.25	Propagated	25.0
Copper	0.15	0.3	Marginal	13.5
Aluminum	0.2	0.3	Failed	18.0
Aluminum	0.2	0.6	Propagated	72.0
Aluminum	0.4	0.3	Failed	36.0
Aluminum	0.4	0.4	Marginal	64.0
Aluminum	0.4	0.6	Propagated	144.0
Water	0.4	0.3	Failed	36.0
Water	0.4	0.6	Marginal	144.0
Water	0.4	0.7	Marginal	196.0

experiments involving instrumented mock and high explosives, with projectiles of varying sizes, shapes, and velocities.

The Forest Fire burn model in reactive hydrodynamic codes, SIN, 2DL, 2DE, and 3DE has been used to model many initiation tests and vulnerability problems involving shock initiation that occur in times of order of microseconds. Accidental initiation of explosives often are caused initially by processes that take milliseconds or longer such as burning.

4.5 Burning to Detonation

An interesting problem of initiation of detonation is the burning of a confined piece of explosive or propellant. If the explosive or propellant is unconfined, it usually does not undergo transition to detonation; however, with strong confinement detonation occurs for most explosives and for many propellants. With Forest Fire to describe the transition once pressure excursions occur from the confined burning, numerical modeling of the process of burning requires only a model for burning.

The usual burn under confinement proceeds on the surface of the explosive with a rate that increases with pressure to some power. It is important to describe both the surface area available for burning and the pressure-dependent rate law of burning. This is called the bulk burn model.

Charles Forest [29–31] constructed the model for bulk burning assuming that the mass of propellant burns on a surface area S such that the burn proceeds normal to the surface

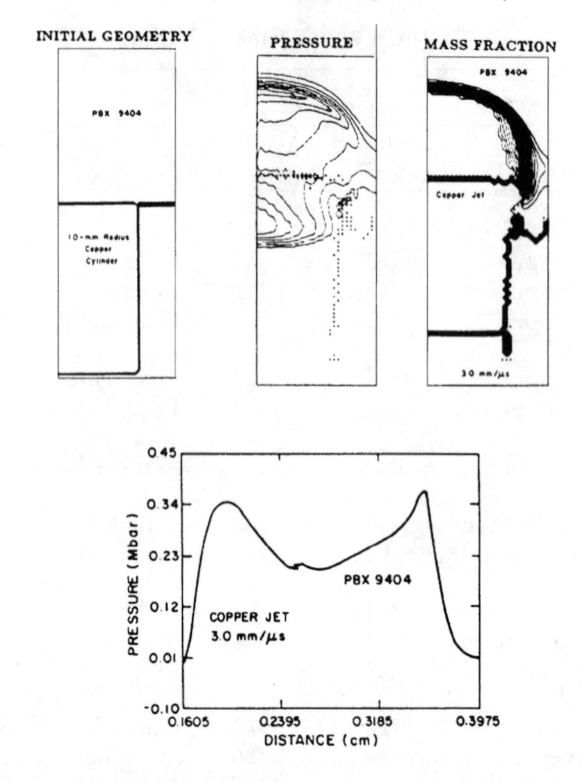

Figure 4.32 The initial profile and the pressure and mass fraction profiles for a 0.2-cm diameter Copper jet of 0.3 cm/μs shocking PBX-9404 at 0.20 μsec. The jet initiates propagating detonation in the PBX-9404. The isobar contour interval is 50 kbar, and the mass fraction contour interval is 0.05. The pressure profile along the axis is shown.

according to a linear burn rate. He also assumed that the density of the burning explosive is constant.

Equation Nomenclature

ρ	= Density of Unburned Explosive
S	= Surface Area Burning
V	= Total Volume of Unburned Explosive
M	= Total Mass of Explosive
R_b	= Linear Burn Rate $dX/dt = cP^n$
X	= Distance
t	= Time
P	= Pressure
$c,\ n$	= Burn Rate Experimental Constants
M_o, S_o, V_o	= Initial Values
S_o/V_o	= Initial Burn Surface-to-Volume Ratio
$W = M/M_o$	= Mass Fraction of Unburned Explosive

The time derivative of M is

$$\frac{dM}{dt} = -S\rho R_b\,.$$

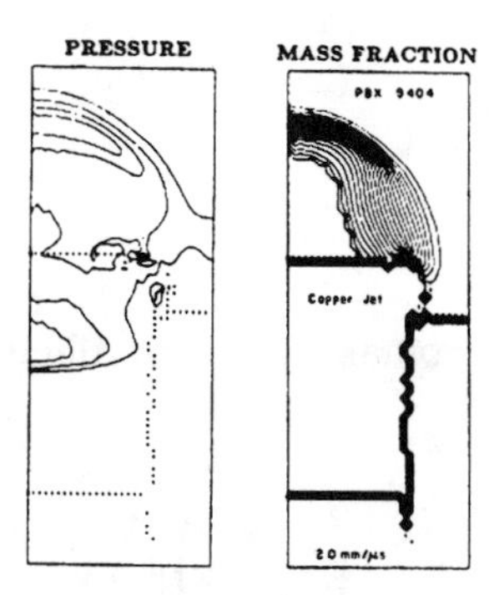

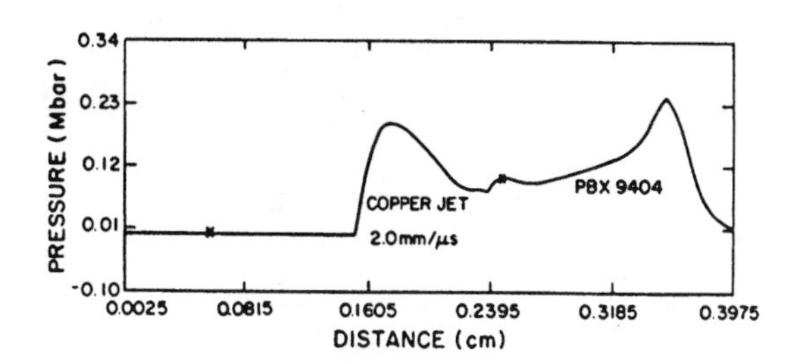

Figure 4.33 The pressure and mass fraction profiles for a 0.2 cm diameter Copper jet of 0.2 cm/μs shocking PBX-9404 at 0.25 μsec. The jet fails to initiate propagating detonation in the PBX-9404. The isobar contour interval is 50 kbar, and the mass fraction contour interval is 0.05. The pressure profile along the axis is shown.

Then

$$\frac{dM}{dt} = \left(\frac{S_o}{V_o}\right)(V_o\rho)\left(\frac{S}{S_o}\right)R_b$$

and

$$\frac{1}{M_o}\frac{dM}{dt} = -\left(\frac{S}{V}\right)_o\left(\frac{S}{S_o}\right)R_b\,.$$

An approximation for (S/S_o) is needed since there is no direct way to follow the burning surface area, except in idealized cases. For this purpose then, let

$$(S/S_o) = (M/M_o)^q.$$

The motivation for this approximation is found in the following cases:

a. For all polyhedral volumes containing an inscribed sphere of radius r, for some k

$$\left(\frac{S}{S_o}\right) = \left(\frac{3kr^2}{3kr_o^2}\right) = \left(\frac{r}{r_o}\right)^2 = \left(\frac{\rho kr^3}{\rho kr_o^3}\right)^{2/3} = \left(\frac{M}{M_o}\right)^{2/3}.$$

Also note that

$$(S/V_o)_o = \frac{3}{r_o}.$$

b. For all polyhedral volumes containing an inscribed cylinder (ignoring the ends), for some k

$$\left(\frac{S}{S_o}\right) = \left(\frac{2kLr}{2kLr_o}\right) = \left(\frac{r}{r_o}\right) = \left(\frac{\rho Lkr^2}{\rho Lkr_o^2}\right)^{1/2} = \left(\frac{M}{M_o}\right)^{1/2},$$

and

$$(S/V_o)_o = \frac{2}{r_o}.$$

c. For constant surface area volumes, for instance slabs,

$$\left(\frac{S}{S_o}\right) = 1 = \left(\frac{M}{M_o}\right)^o.$$

In summary then

$$\frac{dW}{dt} = -(S/V)_o W^q R_b ,$$

where

$(S/V)_o$	= Initial Burn Surface to Volume Ratio
R_b	= Linear Burn Rate
q	= Geometry Dependent Exponent
$W = M/M_o$	= Mass Fraction of Unburned Explosive

Some specific cases for q:

$q = 0$	Constant Surface Area Burn
$q = 1/2$	Cylindrical-like Particles Burn
$q = 2/3$	Sphere-like Particles Burn

A full hydrodynamic simulation of burning to detonation was performed for a burning segment of low density explosive adjacent to high density explosive, all confined by two metal plates. The explosive parameters approximated those of HMX.

Reflective Boundary	Aluminum Plate (2 cm)	Low-Density Explosive (2 cm) Porous Bed	High Density Explosive (10 cm) Forest Fire	Aluminum Plate (2 cm)

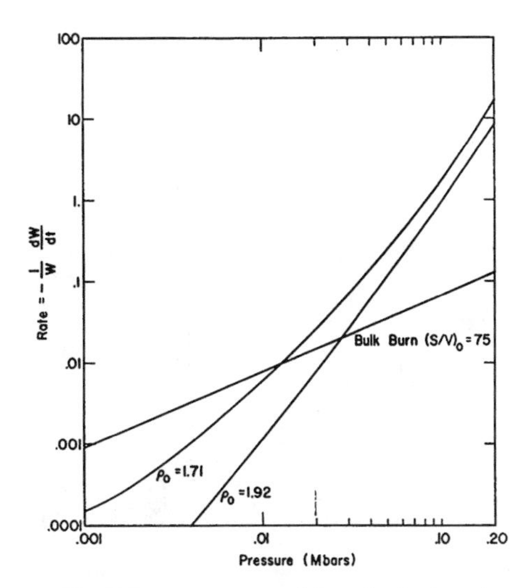

Figure 4.34 Forest Fire rates and bulk burn rates used in calculation shown in Figure 4.35.

The low-density explosive was assumed to be all burning according to the bulk burning model. The parameters used were

$$(S/V)_o = 75$$

$$q = 2/3$$

$$R_b = 0.0007728\ P^{0.942}$$

for the low-density explosive. In the low density region the initial pressure was 0.5 kbar. The initial density was 1.91 in the high density explosive and 1.72 in the low density explosive. Forest Fire shock initiation decomposition was turned on for the low density explosive at 10 kbar. Figure 4.34 shows the Forest Fire rate for the high and low density explosive. The higher turn-on pressure for Forest Fire in the low density explosive was used to estimate when shocks would be present during burning. The bulk burn rate is also shown in Figure 4.34.

The pressure and mass fraction of undecomposed explosive at various times are shown in Figure 4.35.

The Aluminum plate confines the burning low density explosive, permitting the burn rate to increase, which increases the pressure transmitted to the high density explosive.

The pressure gradient steepens in the high density explosive, and shock induced decomposition occurs according to the Forest Fire model, finally building to a detonation inside the high density explosive.

By changing the surface area of the low density explosive, one can change the timing and location of the transition from burning to detonation. An $(S/V)_o$ of 100 results in the transition occuring at the interface between the low and high density explosive. An $(S/V)_o$ of 200 results in initiation in the low density explosive. Low burning rates [$(S/V)_o$ less than 50] send pressure pulses reflecting back and forth between the Aluminum plates, which either slowly build up or fail to undergo transition to a detonation.

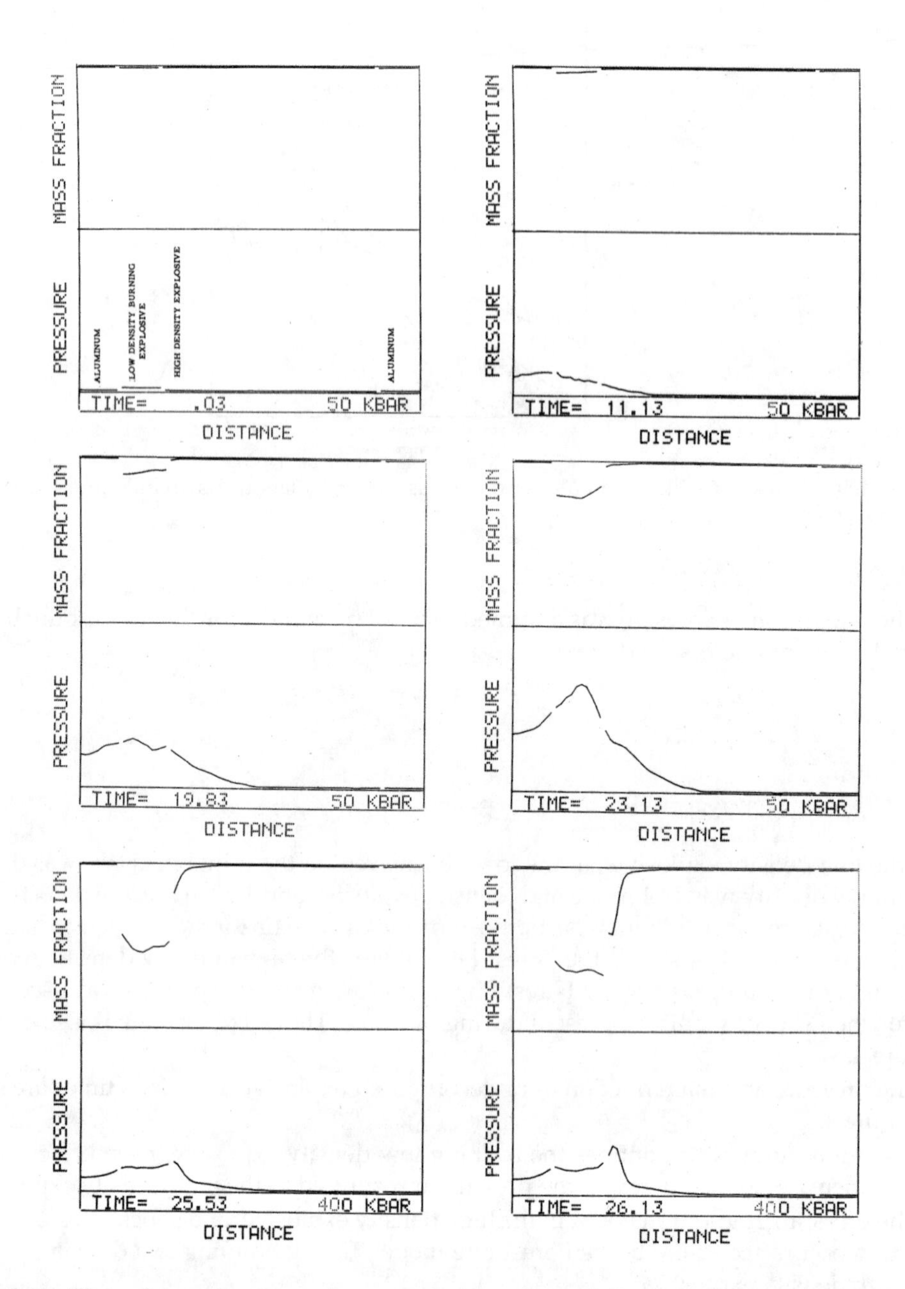

Figure 4.35 Plots of mass fraction and pressure as functions of distance. The total distance is 16 cm. The scale for the mass fraction of undecomposed explosive has a range of 0 to 1.0. The pressure scale is noted in the lower right corner of each frame, changing from 50 to 400 kbar. The time in microseconds is noted at the bottom left of each frame.

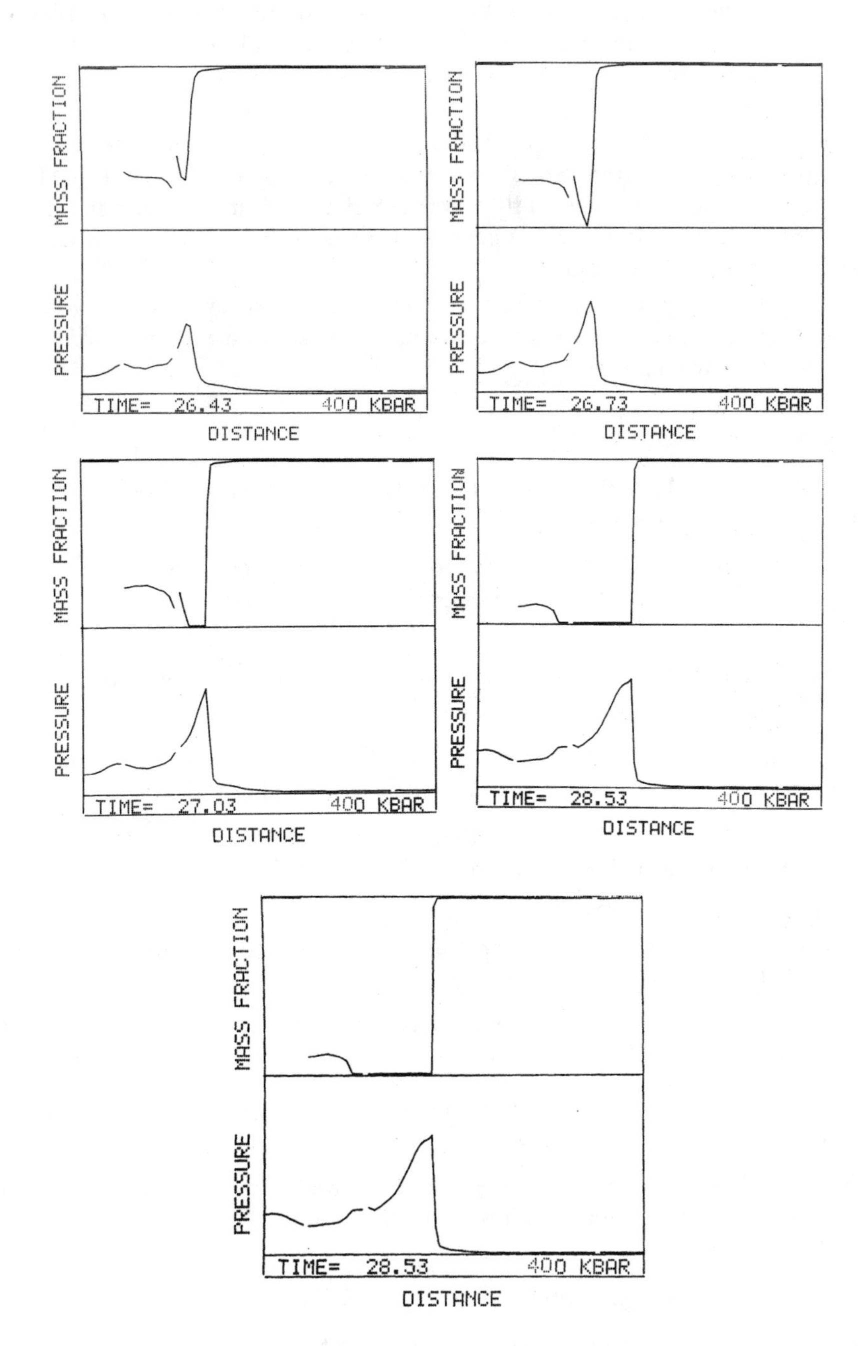

Figure 4.35 *(Continued)* Plots of mass fraction and pressure as functions of distance. The total distance is 16 cm. The scale for the mass fraction of undecomposed explosive has a range of 0 to 1.0. The pressure scale is noted in the lower right corner of each frame, changing from 50 to 400 kbar. The time in microseconds is noted at the bottom left of each frame.

Realistic numerical modeling of the burning to detonation process is possible. The bulk burn and Forest Fire model has been applied to plane and cylindrical one dimensional propellant burning to detonation problems in Reference 29.

In Reference 30, the burning and detonation in PETN hot-wire assemblies was modeled using the bulk burn and Forest Fire model. The surface-to-volume ratio of the burning particles and the ignition front velocity were varied in a parameter study. The transition to detonation was achieved more rapidly where a high mass burn rate occured in compacted explosive. Slow ignition delayed the gas production, and allowed greater expansion and lower density in the burning bed.

In Reference 31, the bulk burn and Forest Fire model was applied to two-dimensional rocket motor geometries with regions of damaged lower density propellant between the case and the main propellant bed.

References

1. A. W. Campbell, W. C. Davis, J. B. Ramsay, and J. R. Travis, "Shock Initiation of Solid Explosives," Physics of Fluids 4, 511 (1961).
2. B. G. Craig and E. F. Marshall, "Behavior of a Heterogeneous Explosive When Shocked but Not Detonated," Fifth International Symposium on Detonation, ACR-184, 321 (1970).
3. A. N. Dremin and S. A. Koldunov, "Initiation of Detonation by Shock Waves in Cast and Pressed TNT," Vzryvnoe Delo 63, 37 (1967).
4. J. E. Kennedy, "Quartz Gauge Study of Upstream Reaction in a Shocked Explosive," Fifth International Symposium on Detonation, ACR-184, 435 (1970).
5. Charles L. Mader, "An Empirical Model of Heterogeneous Shock Initiation of 9404," Los Alamos Scientific Laboratory report LA-4475 (1970).
6. Charles L. Mader and Charles A. Forest, "Two-Dimensional Homogeneous and Heterogeneous Detonation Wave Propagation," Los Alamos Scientific Laboratory report LA-6259 (1976).
7. J. B. Ramsay and A. Popolato, "Analysis of Shock Wave and Initiation Data for Solid Explosives," Fourth International Symposium on Detonation, ACR-126,233 (1965).
8. I. E. Lindstrom, "Plane Shock Initiation of an RDX Plastic Bonded Explosive," Journal of Applied Physics 37, 4873 (1966).
9. Charles L. Mader, "The Two-Dimensional Hydrodynamic Hot-Spot - Volume II," Los Alamos Scientific Laboratory report LA-3235 (1965).
10. E. F. Gittings, "Initiation of a Solid Explosive by a Short-Duration Shock," Fourth International Symposium on Detonation, ACR-126, 373 (1965).
11. B. D. Trott and R. G. Jung, "Effect of Pulse Duration of Solid Explosives," Fifth International Symposium on Detonation," ACR-184, 191 (1970).
12. J. E. Kennedy, "Pressure Field in a Shock Compressed High Explosive," Fourteenth International Symposium on Combustion, 1251 (1973).
13. D. Venable, "PHERMEX," Physics Today 17, 19-22 (1964).

14. Charles L. Mader "LASL PHERMEX Data, Volume II." University of California Press, Berkeley, California (1980).

15. Charles L. Mader, "Detonation Induced Two-Dimensional Flows," Acta Astronautica 1, 375 (1974).

16. P. Donguy and N. Legard, "Numerical Simulations of Non-Ideal Detonations of a Heterogeneous Explosive with the Two Dimensional Eulerian Code C.E.E.," Seventh Symposium (International) on Detonation, 695-702 (1981).

17. A. W. Campbell and Ray Engelke, "The Diameter Effect in High-Density Heterogeneous Explosives," Sixth International Symposium on Detonation, ACR-221, 642 (1976).

18. Charles L. Mader, "LASL PHERMEX Data, Volume III." University of California Press, Berkeley, California (1980).

19. M. C. Chick, and D. J. Hatt,"The Mechanism of Initiation of Composition B by a Metal Jet." Seventh Symposium (International) on Detonation, 352-361 (1981).

20. M. C. Chick and I. B. MacIntyre "The Jet Initiation of Solid Explosives," Eighth Symposium (International) on Detonation, 318-327 (1985).

21. James R. Travis and Arthur W. Campbell, "The Shock Desensitization of PBX-9404 and Composition B," Eighth Symposium (International) on Detonation 1057-1067 (1985).

22. I. I. Maksimov, Khimiko-Tekhnologicheskogo Instituta imeni D. I. Mendeleeva, TRUDY No. 53, 73-84 (1967).

23. M. Held, "Initiating of Explosives, A Multiple Problem of the Physics of Detonation," Explosivstoffe, 5, 98-113 (1968).

24. Charles L. Mader and George H. Pimbley, "Jet Initiation and Penetration of Explosives," Journal of Energetic Materials, 1, 3 (1983).

25. Charles Forest, "Numerical Modeling of the Minimum Priming Charge Test," Los Alamos Scientific Laboratory Report LA-8075 (1980).

26. Allen L. Bowman, James D. Kershner, and Charles L. Mader, "A Numerical Model of the Gap Test," Los Alamos Scientific Laboratory report LA-8408 (1980).

27. Allen L. Bowman, Charles A. Forest, James D. Kershner, Charles L. Mader, and George H. Pimbley, "Numerical Modeling of Shock Sensitivity Experiments," Seventh Symposium (International) on Detonation, 479-487 (1981).

28. J. H. M. Fu and G. E. Cort, "Numerical Calculation of Shock-to-Detonation from Projectile Impact," Los Alamos Scientific Laboratory report LA-8816-MS (1981).

29. Charles Forest, "Burning and Detonation," Seventh Symposium (International) on Detonation, 234-243 (1981).

30. Charles Forest, "A Numerical Model Study of Burning and Detonation in Small PETN-Loaded Assemblies," Los Alamos National Laboratory Report LA-8790 (1981).

31. Charles Forest, "A Model of Burning and Detonation in Rocket Motors," Los Alamos Scientific Laboratory report LA-8141-MS (1980).

chapter five

Interpretation of Experiments

The major use of numerical modeling of detonations is in predicting or interpreting the results of the interaction of a detonation with some other material. Another important use is to predict or interpret the interaction of some outside source of energy with an explosive. In this chapter, modeling of the interaction of a plane wave detonation with a slab of metal, the interaction of a spherically diverging detonation with water, the interaction of the end of detonating explosive with a metal plate in the plate dent test, jet penetration of inerts and explosives, regular reflection and mach stems formed by interacting detonation waves, the plane wave lens, and insensitive high explosive detonators will be described.

The numerical description of the explosive and material with which it interacts must be properly calibrated using experimental data designed to characterize the materials of interest. The Los Alamos dynamic material property data volumes are an extensive collection of data that have been used to calibrate the numerical models for the equation of state, the elastic-plastic properties, melting properties, decomposition rate properties both for homogeneous and heterogeneous explosives, and propellants and other materials that decompose from heating. The data volumes also furnish data that may be used to compare with the results of numerical models for many problems in shock wave physics and chemistry. The LASL Explosive Property Data volume[1] contains the physical, chemical, and explosive properties for many high performance explosives. The collection of data is expanded in the Los Alamos Explosive Performance Data[2] volume to include detailed plate acceleration data, aquarium data, and detonation velocity data for 55 explosives and explosive mixtures. The radiographic facility PHERMEX[3] has generated several thousand radiographs of shock physics experiments. The three volumes of PHERMEX data[4,5,6] furnish information needed for the calibration of spalling of materials under explosive shock loading and unloading and data about many problems of shock physics. Shock Hugoniots are calibrated using data from LASL Shock Hugoniot Data volume.[7] The elastic-plastic, melting, spalling, and viscous properties are calibrated using data from LASL Shock Wave Profile Data volume.[8]

The calibration of the dynamic material properties using these data bases is required for realistic numerical modeling of explosive or propellant driven systems. The experiments modeled in this book used descriptions of the materials that were previously calibrated using the dynamic materials data base. After all the materials have been characterized then *and only then* can other systems, for which data may or may not be available, be studied numerically.

5.1 Plane Wave Experiments

A simple experimental system to model is a slab of 5 cm of Composition B explosive at 1.713 g/cc initiated by a plane wave initiator at one end and in contact with a 1-cm-thick plate of Aluminum at the other end. The calculated pressure vs. distance profiles are shown in Figure 5.1. The Composition B explosive detonates with a velocity of 8000 m/sec and has a peak C-J pressure of 300 kbar, a C-J temperature of 2700° K, and a C-J density of 2.3 g/cc. When the detonation wave interacts with the Aluminum plate, a shock wave travels into the Aluminum with a pressure of 350 kbars at 7590 m/sec. Since a pressure gradient is present in the explosive, one also develops in the Aluminum. When the shock wave arrives at the surface of the Aluminum plate, the surface starts moving at about 3300 m/sec. This is faster than the bulk of the plate can move because of the gradient present, so the Aluminum plate goes into tension and eventually breaks or spalls. When the rarefaction travels back through the Aluminum plate and arrives at the explosive-Aluminum interface, a second small shock is sent into the Aluminum plate which, upon arriving at the surface, increases the surface velocity. The shocks and rarefactions pass through the plate with smaller and smaller amplitudes until the plate approaches a maximum velocity. An animation of the flow is on the CD-ROM.

Although this simple picture is often adequate, the material properties such as elastic-plastic flow, viscosity, and spalling, rather than the explosive properties, often dominate the details of the experimental observations. If these material properties are ignored and the calculated results are compared with experimental observations, it is likely that the calculation will predict different pull-back velocity and second shock arrival behavior. These differences have been misinterpreted as evidence that either the explosive or the metal equation of state is in gross error.

As an example of this problem, Hayes and Fritz[9] have studied the velocity of explosively driven Aluminum plates using their magnetic probe technique. The velocity as a function of time for 5 cm of Composition B shocking 0.6295 cm of Aluminum is shown in Figure 5.2. Also shown is the velocity-time history calculated with the Aluminum fluid model that does not include elastic-plastic or spall behavior. Not only is the velocity pull-back not observed experimentally, but at first glance one would conclude that a second shock is not received by the metal from the explosive products!

It is results such as these that make many experimentalists dubious about the results from any numerical model. However, if we include the gradient spall model calibrated for Aluminum[10] we find that the metal plate spalls at 0.19 cm, which significantly changes the reverberation time seen by the metal surface. The velocity profile is further damped by including the previously calibrated Aluminum shear modulus of 0.25, yield of 5.5 kbar[11], and real viscosity of 2000 poises.[12] The calculated and experimental free surface velocity profiles are shown in Figure 5.3. The spalling behavior is a dominant feature of the flow and since the experiments are only approximately one-dimensional, the calculated small reverberations are lost in the roughness of the experimental flow. The second shock arrives after the experimental measurements are completed because it takes considerable additional time for the second shock to close the gap closed by the fast moving surface spall layer.

Another example that Fritz studied did observe the second shock arrival. The experimental system was 5 cm of Composition B explosive and 0.3167 cm of Aluminum. The calculated and experimental profiles are shown in Figure 5.4 with the calculation being the fluid model, and in Figure 5.5 including the material properties. The observed smeared profile is reproduced by the calculation including the material properties.

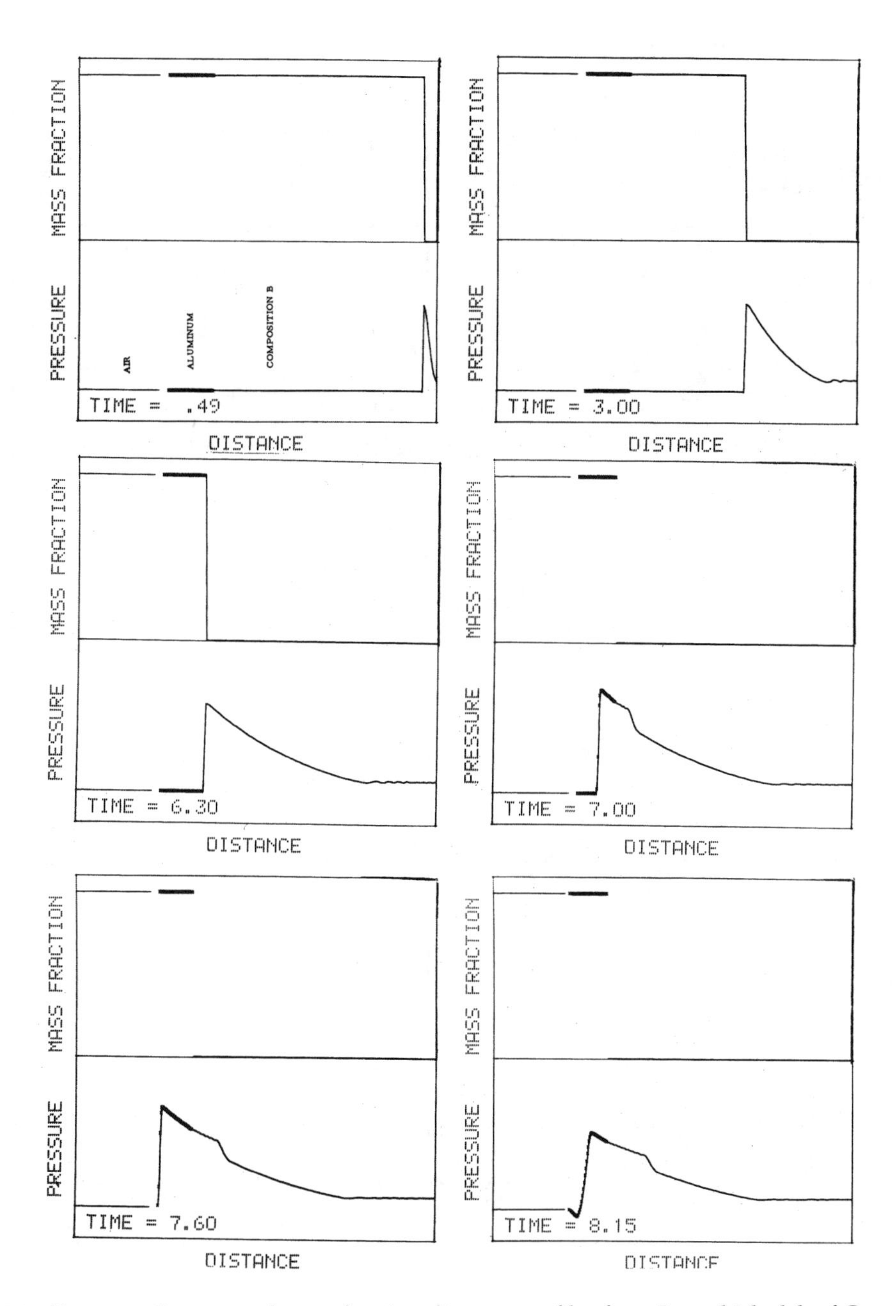

Figure 5.1 Pressure-distance and mass fraction-distance profiles for a 5-cm-thick slab of Composition B explosive detonating and interacting with a 1-cm-thick Aluminum plate. Time is in μsec. The initial distance scale is 0 to 8 cm, the pressure scale is -100 to + 500 kbar, and the mass fraction scale is from 0.0 to 1.1. An animation of the calculation is on the CD-ROM.

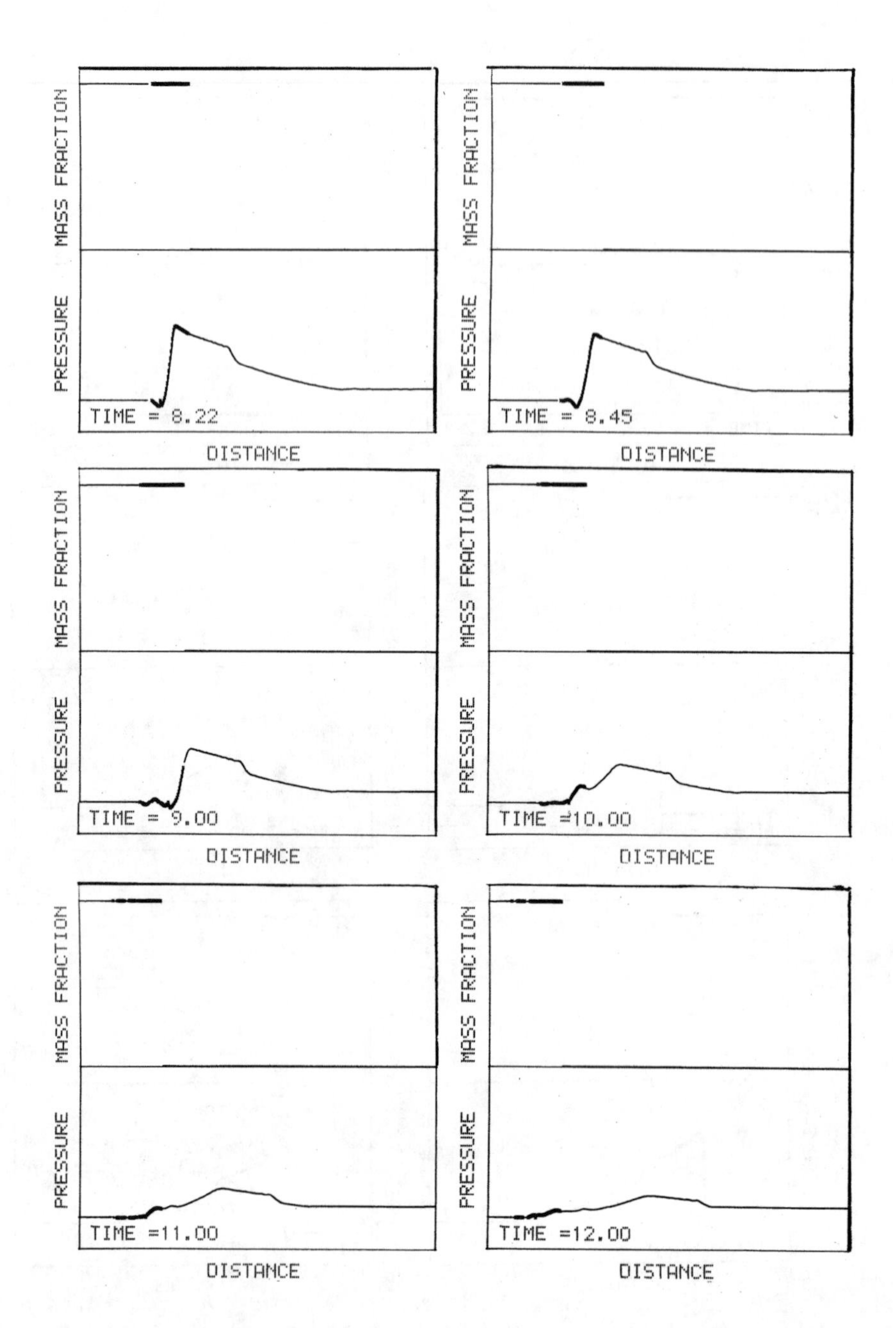

Figure 5.1 *(Continued)* Pressure-distance and mass fraction-distance profiles for a 5-cm-thick slab of Composition B explosive detonating and interacting with a 1-cm-thick Aluminum plate. Time is in μsec. The initial distance scale is 0 to 8 cm, the pressure scale is -100 to + 500 kbar, and the mass fraction scale is from 0.0 to 1.1. An animation of the calculation is on the CD-ROM.

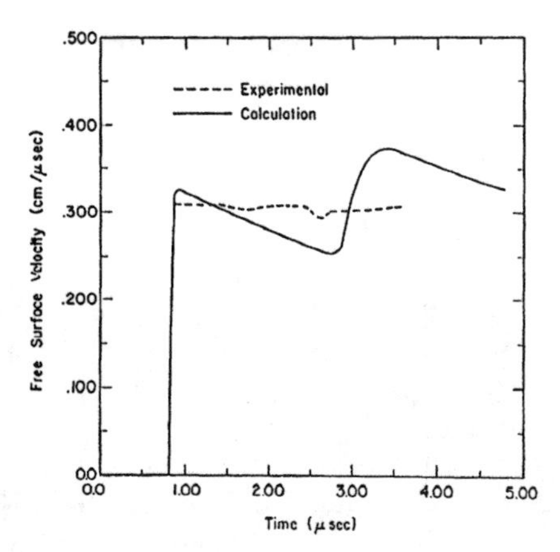

Figure 5.2 Experimental velocity-time profile of surface of 0.6295-cm-thick Aluminum plate in contact with detonated slab of 5.08-cm-thick Composition B. The calculated profile treating the metal as a fluid is also shown.

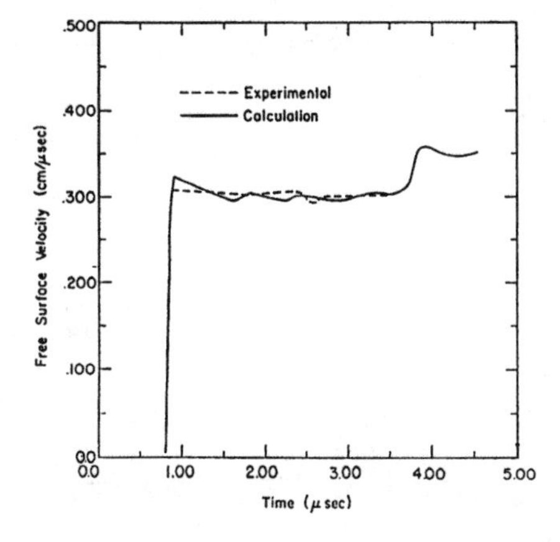

Figure 5.3 Same experimental profile as shown in Figure 5.2 and calculated profile treating the Aluminum as a viscous, elastic-plastic material with pressure gradient spalling properties.

If anything is to be inferred about an explosive from its interaction with a metal plate, the material properties of that plate must be known. Unfortunately, only Aluminum has been studied sufficiently. The material properties are often used as numerical knobs to obtain any desired result. Greater effort needs to be applied to determining the elastic-plastic, viscous, melting, and spalling properties of explosively loaded materials. One of the most difficult problems in the design of an explosive interaction experiment which will need numerical modeling is to find materials that are understood well enough to permit accurate descriptions of their equation of state and material properties.

5.2 Explosions in Water

One of the easiest and best methods of evaluating the low pressure end of the detonation product isentrope of an explosive is to study its effect under water. Although military laboratories throughout the world have been studying the effect of explosives under water

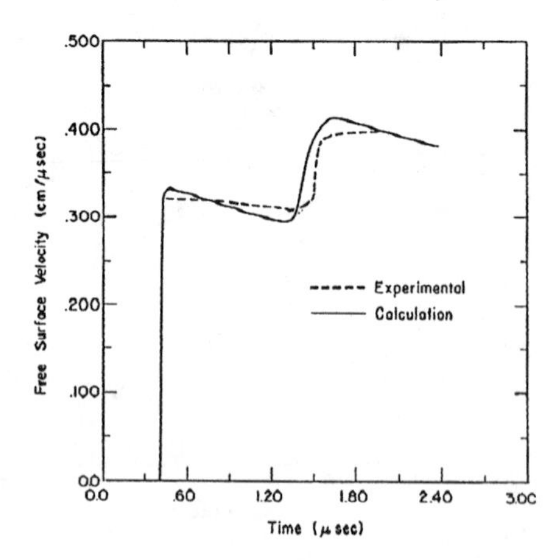

Figure 5.4 Experimental velocity-time profile of surface of 0.3167-cm-thick Aluminum plate in contact with detonated slab of 5.08-cm-thick Composition B. The calculated profile treating the metal as a fluid is also shown.

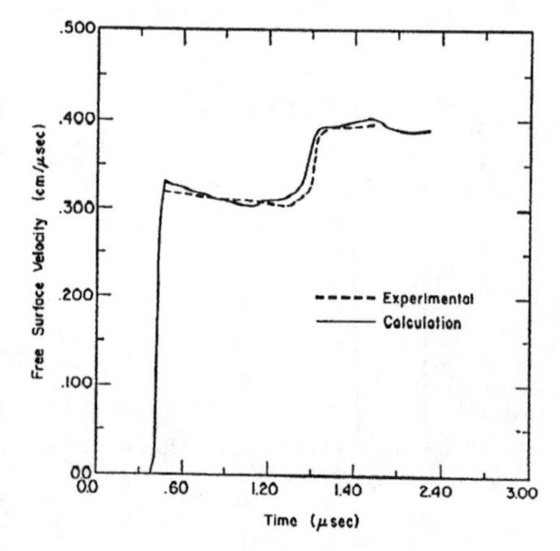

Figure 5.5 Same experimental profile as shown in Figure 5.4 and calculated profile treating the Aluminum as a viscous, elastic-plastic material with pressure gradient spalling properties.

for many decades, the problem of what type of an explosive is best for particular underwater applications is still being debated.

A good method of testing an equation of state for detonation products and the numerical method of describing the explosive burn is to compare the calculated behavior of an explosive bubble under water with that observed experimentally.

The BKW equation of state constants used for the explosives and the HOM equation of state are given in Table 5.1. The equation of state for water was $U_s = 0.1483 + 2.0U_p$, $\gamma_s = 1.0$, $C_v = 1.0$, and $\alpha = 0.0001$. The method of burning the explosive was the C-J burn technique found adequate for diverging detonations in Chapter 2.

The SIN code described in Appendix A was used to study the flow resulting from detonation of a 3.27-cm-radius sphere of Tetryl in water at various initial pressures and densities. The calculation model used 1000 cells with the explosive initially resolved to 0.1 cm and the first 10 cm of water resolved to 0.25 cm. For the low initial pressure calculations, the water cell width was then increased to allow sufficient distance to follow the water shock during the time of interest. This resolution was adequate to ensure that the important features of the flow were mesh and viscosity independent.

Numerical calculations of underwater detonations have been performed by many investigators; the work of Sternberg and Walker[13] is an example. Using the proper method for burning the explosive results in lower initial water shock pressures than reported in Reference 13, which used the Taylor self-similar solution to describe the explosive that was shown to be incorrect in Chapter 2.

The details of the equation of state of the explosive, the treatment of diverging detonations, and the calculation to bubble collapse for low hydrostatic pressures are the major differences between this study and most of the other studies.

TABLE 5.1 BKW Equation of State Parameters

	PETN	EXP		PBX-9404	EXP
A	−3.10639868833	+000	A	−2.88303447687	+000
B	−2.25218297095	+000	B	−2.25910150671	+000
C	+1.93865645401	−001	C	+2.09836811364	−001
D	+1.06761114309	−002	D	−1.62402872478	−002
E	−5.71317097698	−005	E	+4.14247701072	−004
K	−1.43880401718	+000	K	−1.27244575845	+000
L	+4.17630232758	−001	L	+4.27159472916	−001
M	+4.43146793248	−002	M	+4.61539702874	−002
N	+2.43302842995	−003	N	+2.54544398316	−003
O	+5.15057824089	−005	O	+5.31474988838	−005
Q	+8.10009012302	+000	Q	+8.24707528084	+000
R	−3.67433055630	−001	R	−4.89534325865	−001
S	+1.38196579791	−003	S	+6.12169699021	−002
T	+8.14028829459	−003	T	−3.22067926443	−003
U	−7.34294504930	−004	U	−5.13495324073	−006
C'_v	+5.0	−001	C'_v	+5.0	−001
Z	+1.0	−001	Z	+1.0	−001
ρ_o	+1.55 g/cc		ρ_o	+1.844	
P_{CJ}	+0.231 Mbar		P_{CJ}	+0.363	
D_{CJ}	+0.7378 cm/μsec		D_{CJ}	+0.8880	
T_{CJ}	+3369° K		T_{CJ}	+2460	
P_{Max}	+0.40		P_{Max}	+0.40	
P_{Min}	+1.00000000000	−008	P_{Min}	+1.00000000000	−008
Isentrope Pressure	+0.40		Isentrope Pressure	+0.40	

TABLE 5.1 BKW Equation of State Parameters (continued)

	Air (0.76 Bar)	EXP		Tetryl	EXP
A	−2.36733372864	+000	A	−3.63800897230	+000
B	−1.23356432554	+000	B	−2.45393338654	+000
C	+2.15170143603	−002	C	+3.10500324662	−001
D	−2.95528542190	−003	D	−3.05988910545	−002
E	+1.22549782445	−004	E	+8.47652818961	−004
K	−5.53376189904	−001	K	−1.61514480846	+000
L	+2.44880013455	−003	L	+4.45469837845	−001
M	−1.80516553555	−002	M	+5.81234373927	−002
N	−1.21968671688	−003	N	+3.69359509121	−003
O	−2.53726183472	−005	O	+8.90241616462	−005
Q	+9.88588851357	+000	Q	+7.55699503132	+000
R	−2.35014643148	−001	R	−4.24673177898	−001
S	+3.36987666054	−002	S	+8.78387835435	−002
T	−4.21156020156	−003	T	−9.19711030854	−003
U	+1.63045512702	−004	U	+8.51766433852	−005
C_v'	+5.0	−001	C_v'	+5.0	−001
Z	+1.0	−001	Z	+1.0	−001
ρ_o	+0.00107567		ρ_o	+1.70	
P_{CJ}			P_{CJ}	+0.2515	
D_{CJ}			D_{CJ}	+0.7629	
T_{CJ}			T_{CJ}	+2917	
P_{Max}	+0.001		P_{Max}	+0.40	
P_{Min}	+1.00000000000	−008	P_{Min}	+5.00000000000	−007
Isentrope Pressure	+0.0005		Isentrope Pressure	+0.2515	

The explosive system modeled was a 0.55-pound sphere of Tetryl at various depths. The system was studied experimentally in detail in the 1940s and is described by Cole.[14] The BKW C-J pressure for 1.70 g/cc Tetryl is 251 kbar, and the calculated peak detonation pressure of a 3.27-cm-radius Tetryl sphere is 200 kbar. The results of the study are summarized below.

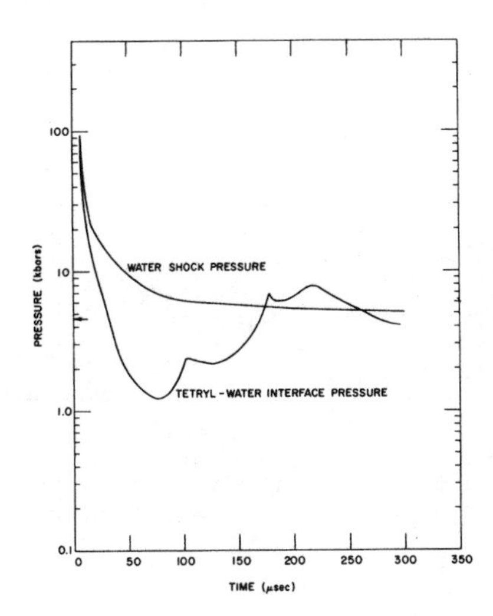

Figure 5.6 Water shock and Tetryl water interface pressure as a function of time for a 3.27-cm-radius Tetryl sphere in water at 4660 bars.

No.	Hydrostatic Pressure (kbar)	Depth (feet)	Period (μsec)	Maximum Radius (cm)
1	4660.	~156,000	200	6.3
2	462.	~15,500	1225	12.5
3	74.6	~ 2,500	5400	23.5
4	9.91	~ 300	25,500	46.5

The calculated pressure-time histories and bubble radius-time histories for 3.27-cm-radius Tetryl spheres in water with hydrostatic pressures of 4660, 462, 74.6, and 9.91 bars are shown in Figures 5.6-5.13. The experimental bubble radius-time profile as reported by Cole[14] is also shown in Figure 5.13. The calculated results reproduce the experimental observations. Figure 5.14 shows the calculated and experimental bubble periods as a function of water depth.

Figures 5.15 and 5.16 show early time profiles for several of the cases studied. Note that higher hydrostatic pressure systems have nearly the same pressure-time profiles as do the lower hydrostatic pressure systems until the Tetryl-water interface pressure drops to near the hydrostatic pressure. The pressure gradient behind the water shock becomes less steep for the higher hydrostatic pressure systems and determines when the bubble will collapse. At lower hydrostatic pressures, many reverberations occur during expansion and collapse of the bubble. The pressure-time and pressure-distance profiles become very complicated.

As Pritchett[15] and Cole[14] have described, the simple incompressible model predicts that the maximum bubble radius is inversely proportional to the hydrostatic pressure to the 1/3 power, and the period is inversely proportional to the hydrostatic pressure to the 5/6 power. The calculated results agree well with these predictions for the maximum bubble radius,

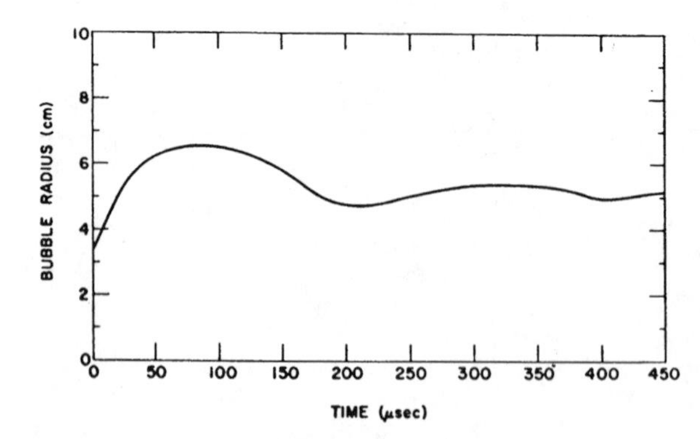

Figure 5.7 Bubble radius as function of time for a 3.27-cm-radius Tetryl sphere in water at 4660 bars.

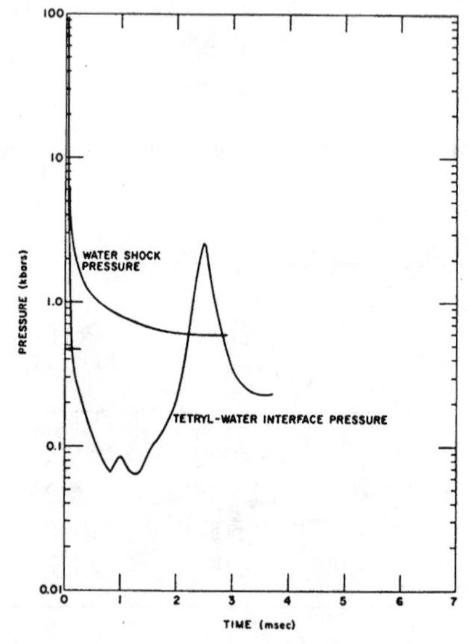

Figure 5.8 Water shock and Tetryl-water interface pressure as a function of time for a 3.27-cm-radius Tetryl sphere in water at 462 bars.

but less well with those for the periods, as shown in the following table. The calculated period is inversely proportional to the hydrostatic pressure to the 0.788 power.

Calculations	Ratio of Max Radii	Ratio ($P^{1/3}$)	Ratio of Periods	Ratio ($P^{5/6}$)
4/1	7.38	7.78	127.5	168.6
4/2	3.72	3.60	20.8	24.6
4/3	1.98	1.96	4.7	5.4
3/2	1.88	1.84	4.4	4.6
3/1	3.73	3.97	27.0	31.3
2/1	1.98	2.16	6.1	6.8

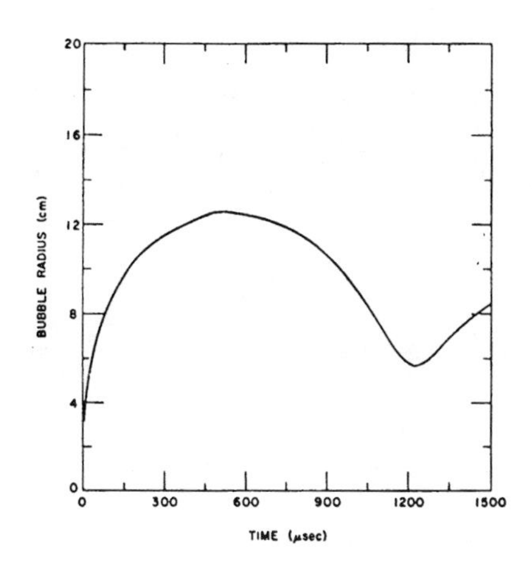

Figure 5.9 Bubble radius as a function of time for a 3.27-cm-radius Tetryl sphere in water at 462 bars.

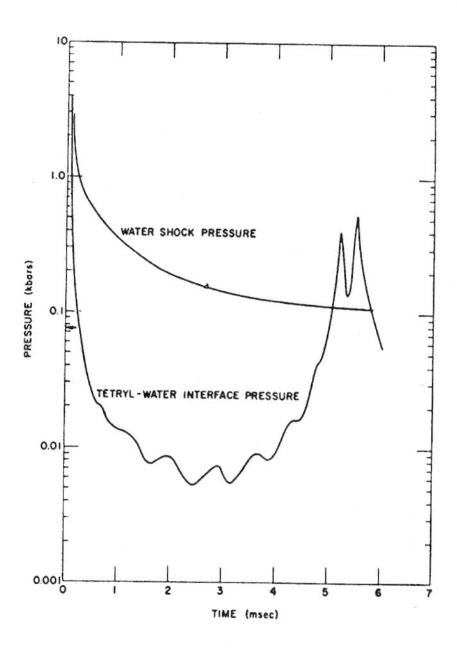

Figure 5.10 Water shock and Tetryl-water interface pressure as a function of time for a 3.27-cm-radius Tetryl sphere in water at 74.6 bars.

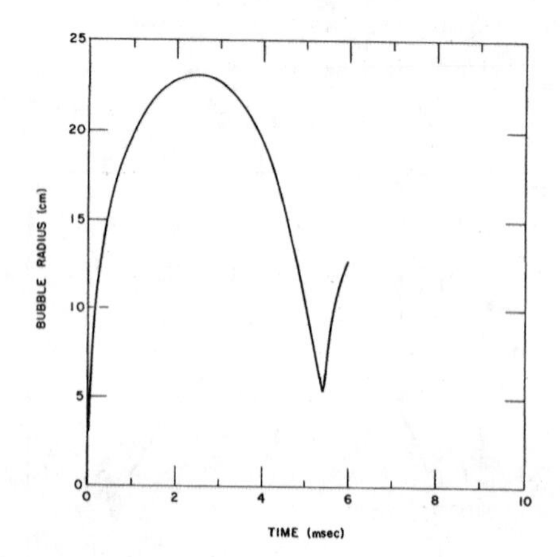

Figure 5.11 Bubble radius as a function of time for a 3.27-cm-radius Tetryl sphere in water at 74.6 bars.

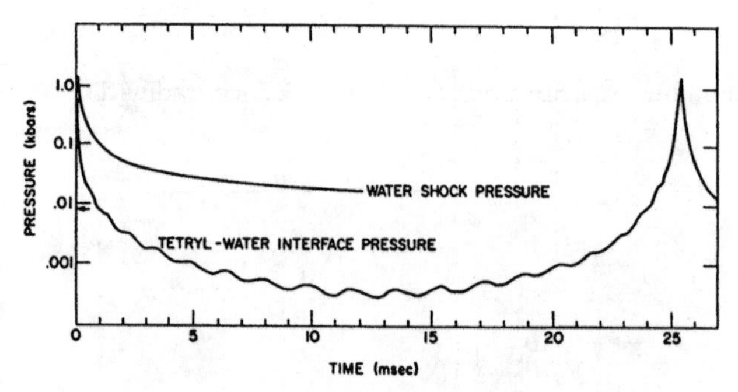

Figure 5.12 Water shock and Tetryl-water interface pressure as functions of time for a 3.27-cm-radius Tetryl sphere in water at 9.91 bars.

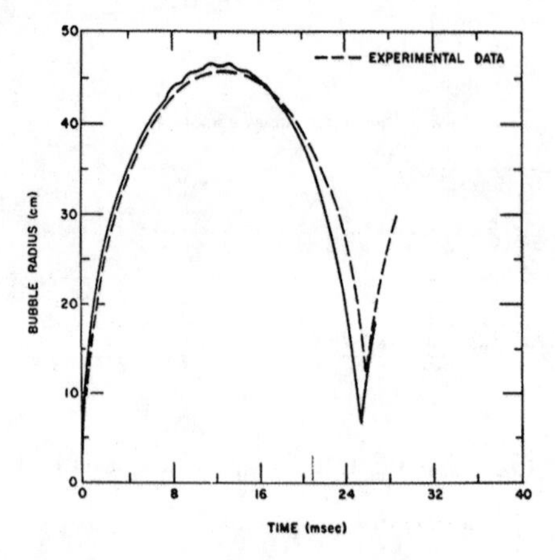

Figure 5.13 Bubble radius as a function of time for a 3.27-cm-radius tetryl sphere in water at 9.91 bars. The experimental data from Reference 14 are also shown.

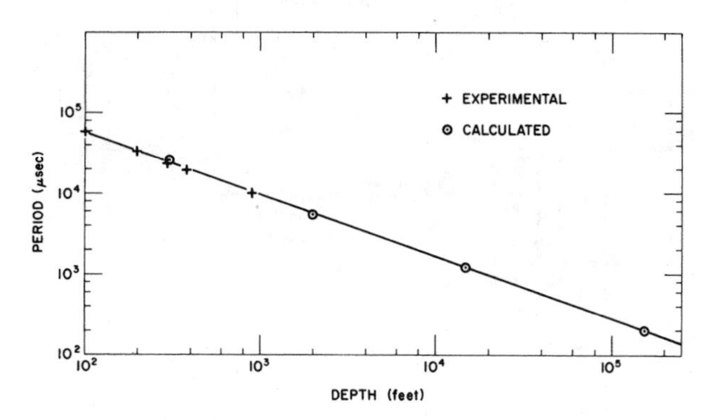

Figure 5.14 Calculated and experimental bubble periods as a function of water depth.

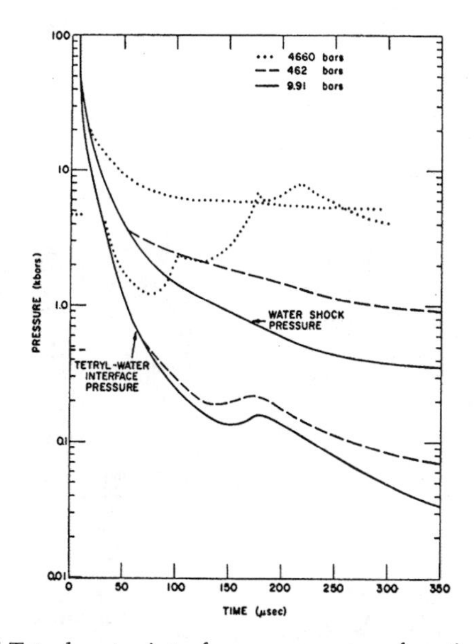

Figure 5.15 Water shock and Tetryl water interface pressure as a function of time.

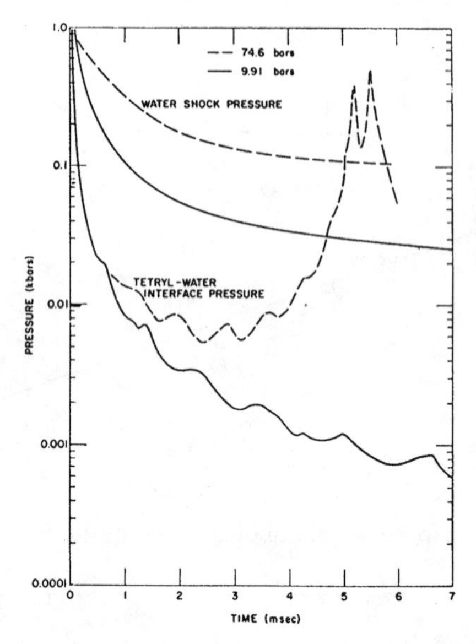

Figure 5.16 Water shock and Tetryl-water interface pressure as functions of time at early times.

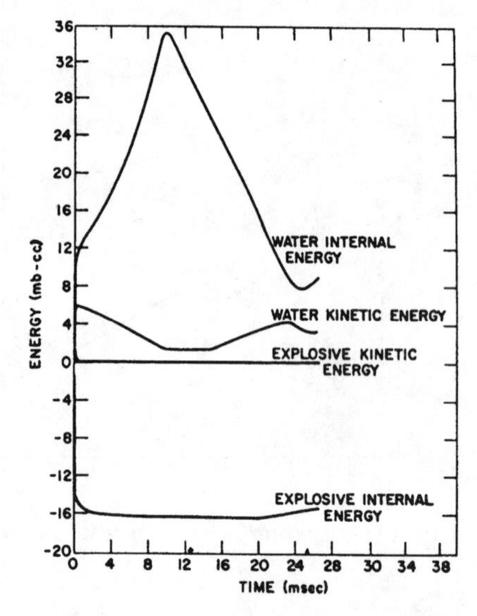

Figure 5.17 Internal and kinetic energy of a 3.27-cm-radius Tetryl sphere in water at 9.91 bars.

The calculated energy partition for the Tetryl experiment is shown in Figure 5.17. The total energy apparently increases from the hydrostatic pressure. The partition of energy in the detonation products is highly dependent upon the details of the detonation product equation of state, as Lambourn[16] has shown.

The observed agreement between the short and long time behavior of an underwater detonation and the detailed one-dimensional compressible hydrodynamic calculations indicates that the calculated energy partition between the detonation products and the water is accurate enough to be used in multidimensional studies of water wave generation mechanisms.

The modeling of water waves generated by explosions above the surface, on the surface, and under the surface have been described in the monograph "Numerical Modeling of Water Waves."[17] The two-dimensional Eulerian code, 2DE, has been used to model a 1.27-cm-radius PBX-9404 explosive sphere initiated at its center and immersed to a depth of 1.59 cm. The explosive is at the "upper critical depth" where the experimentally observed water wave maximum height occurs. The observed upper critical depth phenomenon is a result of a partition of energy near the water surface, which results in high amplitude, deep water waves (of high potential and low kinetic energy) and not the shallow water waves required for tsunamis. When the explosive is detonated under the water surface, more energy is imparted to the water resulting in waves that have smaller amplitude. More of the energy is present in the water as kinetic energy rather than as potential energy. Large explosions located under the surface of the ocean will be more likely to result in shallow water tsunami waves than cavities formed on the ocean surface.

5.3 The Plate Dent Experiment

The most useful and simplest experiment that can be performed to obtain a good estimate of the detonation C-J pressure is the plate dent test described by Smith[18] and performed by M. Urizer for almost 40 years at Los Alamos. Of the usual experiments used to study detonation performance, this experiment is also one of the most difficult to simulate numerically. Lagrangian codes, such as TDL, cannot describe the highly distorted flow around the surface of the dent. The Eulerian code, 2DE, described in Appendix C has realistic calibrated equations of state and elastic-plastic treatments for solid materials, so it has been used to model the plate dent test.

The plate dent test merely involves detonating a cylindrical charge of explosive in contact with a heavy steel plate and measuring the depth of the dent produced in the plate. The charges used are of a diameter and length sufficient to ensure establishment of a steady detonation wave of almost infinite-diameter velocity. The witness plates are massive and strong enough to limit the damage to the dent area so that the depth of the dent does not depend upon any gross distortions of the entire plate. Several test plates are stacked up on the ground, and the upper surface of the top plate is lightly greased. The explosive test charge is centered on the plate with a large enough booster to initiate the test charge and a detonator. Figure 5.18 is a diagram of the assembly. After the shot has been fired, the test plate is recovered and the depth of the dent is measured by placing a ball bearing in the

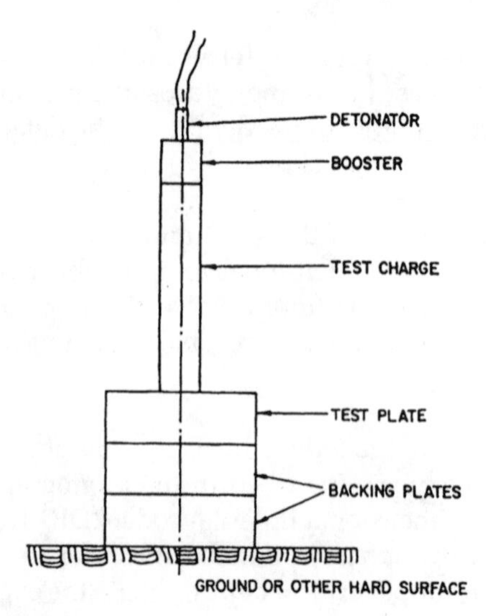

Figure 5.18 The Plate Dent Experiment.

dent (and measuring from the plate surface to the top of the ball bearing) to eliminate the effects of irregularities in the dent.

A dented plate is shown in Figure 5.19. The plate dent was cut in half to show the cross-section of the dent. The radial spalls below the dent are explained by the large tensions observed in that region in the numerical model.

The plate dent experiment would be just another integral experiment except that the depth of the dent has been observed to correlate with experimentally determined C-J pressures of large charges of explosives as discussed in Chapter 2 and shown in Figure 2.21.

The material properties of the metal plate are important in determining when the dent stops expanding. If the metal plate does not have strength and is treated only hydrodynamically, the plate dent will continue to expand indefinitely. As described in the previous section, Aluminum (Dural 2024 T4) is about the only metal whose material properties are adequately known. Urizar performed his plate dent test replacing the steel plates with thick blocks of Dural and reported a 0.90-cm-deep, 3.7-cm-wide dent at the plate surface from a 2.54 cm diameter charge of TNT. His observations can be reproduced numerically using the 2DE code described in Appendix C with a BKW equation of state for TNT, and the TNT burned using the C-J volume burn. The Aluminum equation of state was described using the experimental Hugoniot in Reference 7, and the previously determined[8,11] 5.5 kbar yield and 0.25-Mbar shear modulus described in the section on plane wave experiments.

The calculated isopycnic contours are shown in Figure 5.20 for a 1.27-cm-radius, 2.54-cm-long charge of TNT interacting with an Aluminum plate. The calculated results were similar for 1.27 and 3.8-cm-long charges, and the results scaled with charge radius. The calculation was performed with 70 cells in the radial direction and 90 in the Z direction. The cells were 0.1 cm square. The top, right, and bottom boundaries were continuum boundaries, and the left boundary was an axis. The explosive was initiated by initially detonating the bottom 0.3 cm of TNT. The profile of the dent in a calculation without elastic-plastic flow is shown in Figure 5.21. The calculated axial velocity of the explosive-steel interface is shown as a function of time in Figure 5.22, both with and without treatment of the Aluminum as the elastic-plastic material.

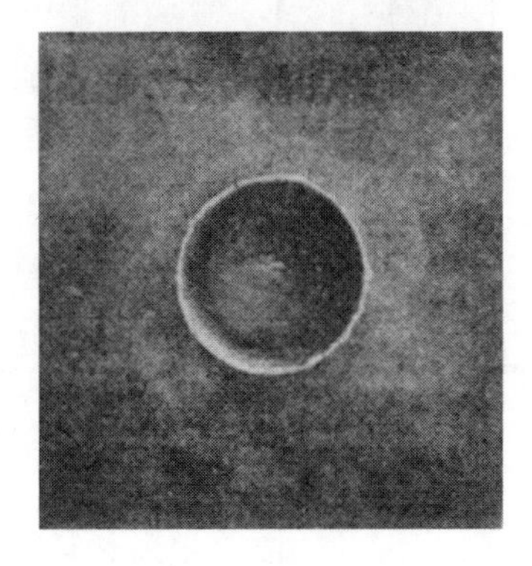

Figure 5.19 A dented steel plate from a plate dent test. The plate dent was cut in half to show the cross-section of the dent and the radial spalls below dent.

Similar calculations were performed with the Aluminum plate replaced by steel. The elastic-plastic properties used for steel (1018 CRS) were a yield of 7.5 kbar and a shear modulus of 0.987 Mbar taken from Reference 8.

Diameter (cm)	Explosive	Metal	Experimental Plate Dent (cm)	Calculated Plate Dent (cm)
2.54	TNT	Dural	0.90	0.95
2.54	9404	Dural	1.66	1.70
4.13	TNT	Steel	0.67	0.75
4.13	9404	Steel	1.12	1.20

The plate dent experiment has been numerically modeled for conventional explosives. Since these explosives have similar isentrope slopes in the pressure range of interest, the major difference among them is a function of the peak detonation pressure. The observed correlation of the plate dent depth with C-J pressure (Figure 2.21) is simply a consequence of similar expansion of the detonation products of most explosives down to about 10 kbar.

If the plate dent depth fails to correlate with the C-J pressure, as for the inert metal loaded explosives discussed in Chapter 2, the explosive is exhibiting unique isentropic expansion properties. Such an explosive needs to be examined more closely to determine if its unique properties may be useful for certain applications. For example, the metal loaded explosives permit the tailoring of time history of the energy delivered as discussed in Chapter 2. The

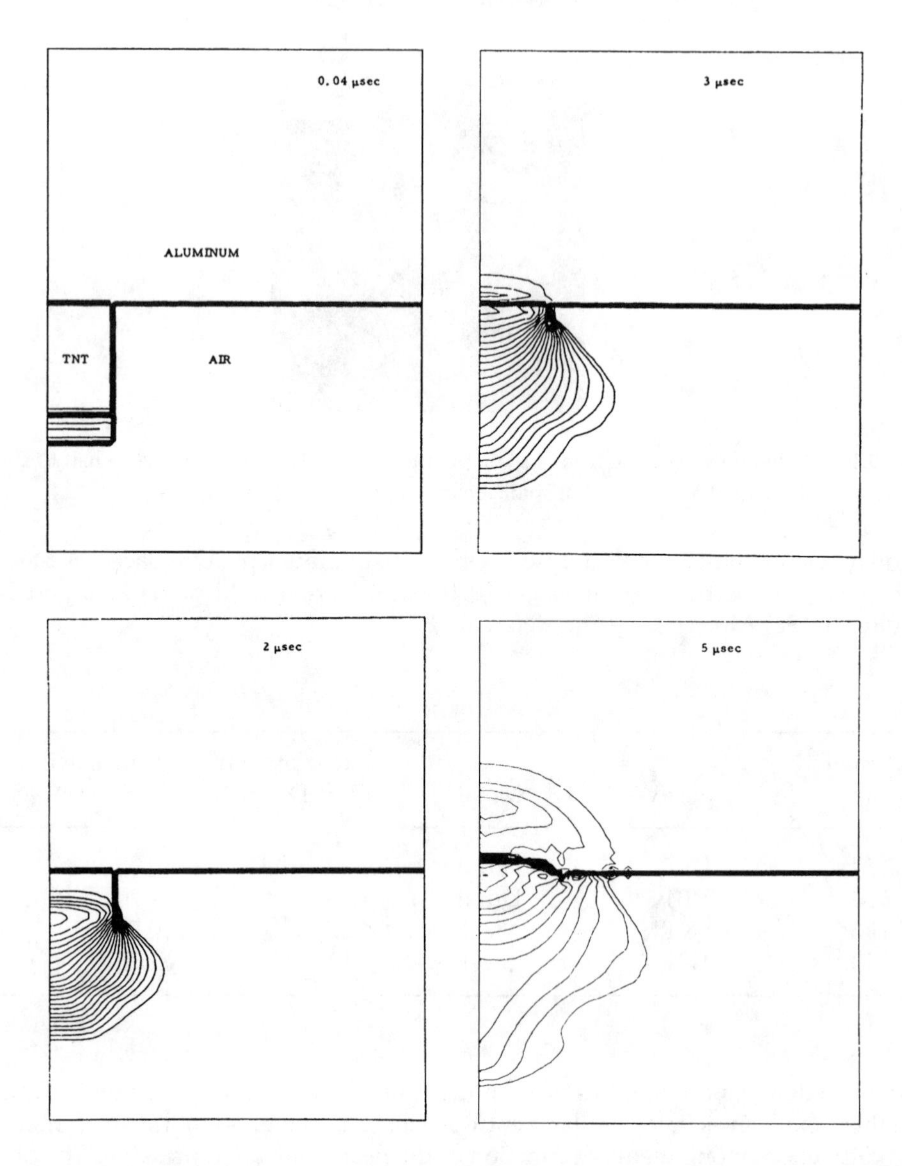

Figure 5.20 Isopycnic contours of 0.1 g/cc, for a 2.37-cm-radius, 2.54-cm-long cylinder of TNT interacting with an Aluminum plate. The bottom boundary is 7 cm long and the height is 9 cm.

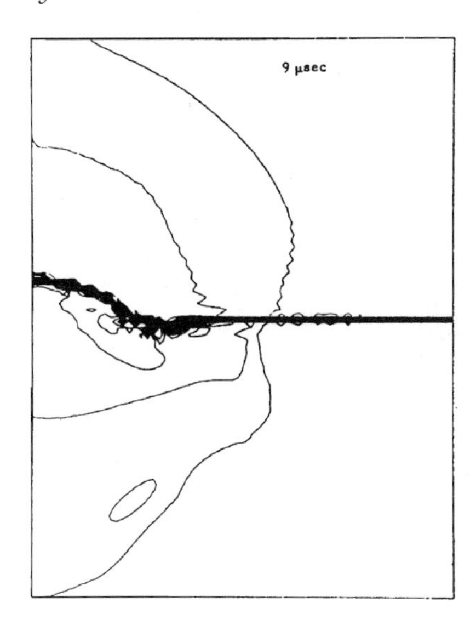

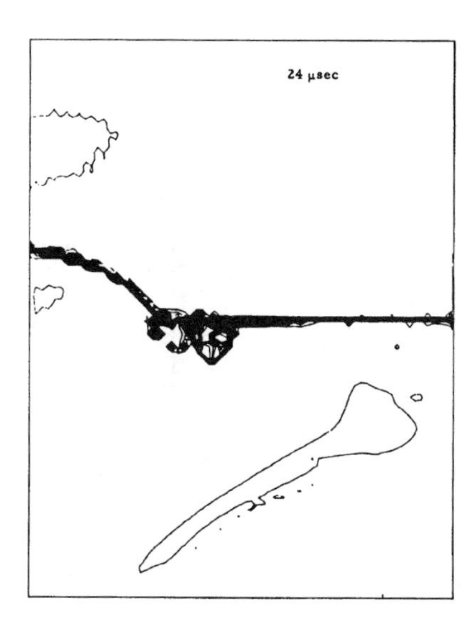

Figure 5.20 *(Continued)* Isopycnic contours of 0.1 g/cc, for a 2.37-cm-radius, 2.54-cm-long cylinder of TNT interacting with an Aluminum plate. The bottom boundary is 7 cm long and the height is 9 cm.

failure of the Nitroguanidine plate dent to correlate with its C-J pressure is a result of its low energy and the resulting steep isentrope compared to most other explosives.

The plate dent is also used to study the properties of a material by the measurement of the dent depth generated by well characterized explosives such as TNT or PBX-9404. This is called the "Inverse Plate Dent" experiment and described in Reference 19. The inverse problem has to be computational, because the yield strength is not an available experimental independent variable to correlate with dent depth observations. The plate dent has been used to determine the yield strength in many common materials and even unusual materials such as Beryllium.

5.4 *The Cylinder Test*

The cylinder test consists of detonating a cylinder of explosive confined by copper and measuring the velocity of the expanding copper wall until it fractures. The cylinder test is commonly used to evaluate the performance and calibrate the equation of state of explosives using the JWL fitting form. The cylinder test is usually interpreted using Lagrangian codes with programmed explosive burns. Fickett and Scherr[20] studied the accuracy of the calculation and found that it compared poorly with calculations of the steady flow near the front by the method of characteristics. Adjustments of the metal wall strength and the explosive equation of state could give late-time motions within 1% of the observations; however, comparison of the calculated initial wall velocity of a copper cylinder driven by 9404 with Campbell and Engelke's precise measurements showed 10% differences. Using one-dimensional cylindrical calculations with the explosive decomposing at constant volume will give late-time wall motions that agree as well with the experimental data as the two-dimensional calculations.

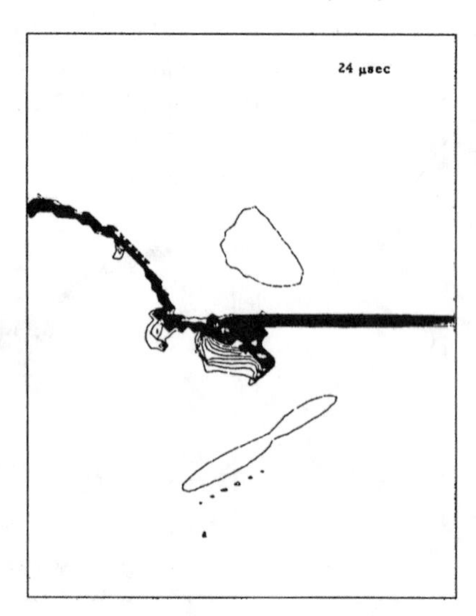

Figure 5.21 Isopycnic contours at 24 μsec for the same calculation as shown in Figure 5.20 but with the Aluminum treated as a fluid.

The numerical model commonly used to interpret cylinder wall expansion experiments must include a realistic description of the explosive burn and detonation wave curvature. A problem in numerical simulation of long cylinders of explosive confined by thin metal walls is to obtain sufficient numerical resolution to describe the explosive burn properly and also to follow the simulation of long cylinders.

The cylinder wall motion correlates with the C-J pressure just as the plate dent depth correlates with the C-J pressure. Using the cylinder wall "energy"[21] after a 0.26 cm thick copper wall surrounding a 2.54-cm-diameter explosive cylinder has traveled 6 to 19 mm, one gets an excellent correlation with the C-J pressures as shown in Figure 5.23. The error bars show the range of experimental C-J pressures measured in various geometries. The correlation is good for all the high performance explosives except PETN. Why PETN is different will make an interesting study because PETN is also the most oxygen-rich, has the fastest kinetics, smallest reaction zone, fastest Forest Fire rate, and the least build-up of any of the explosives. Our present understanding permits the reader to choose any, all, or none of the possible explanations.

The cylinder test furnishes information for only a limited range of the detonation product isentrope similar to the plate dent test. The cylinder test at 9 and 18 mm of copper wall expansion correlates as well as the plate dent experiment with the C-J pressure. The observed correlation of the cylinder wall velocity with C-J pressure is simply a consequence of similar expansion of the detonation products of most explosives down to about 1 to 10 kbar, after which the cylinder wall has fractured. Thus, the cylinder test is of no more use in calibrating an empirical equation of state than the plate dent test. As shown in Chapter 2, the aquarium test does furnish information that can be used to *evaluate* an equation of state over several orders of magnitude of pressure along the isentrope. Fitting an explosive equation of state to any single integral experiment is of use only for describing that experiment. William C. Davis has often stated, "JWL fit to a cylinder test may be useful for describing that particular cylinder test, but it will not be useful for describing anything else." As shown in Chapter 2, experimental data along the single shock Hugoniot and expansion isentrope over the pres-

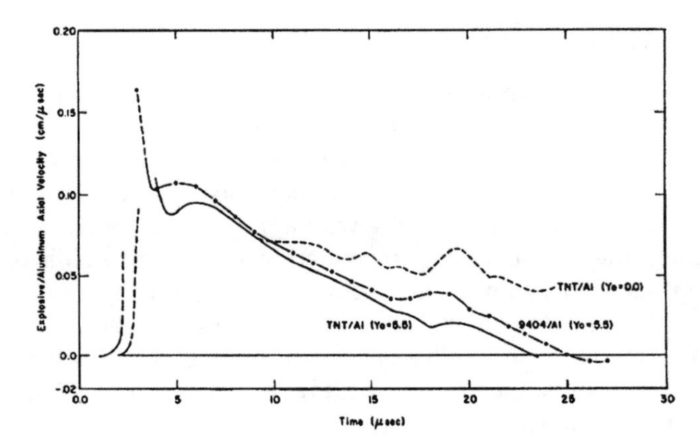

Figure 5.22 Velocity as a function of time of the axial explosive-Aluminum interface for TNT and 9404. Also shown is the TNT-Aluminum interface velocity for fluid Aluminum.

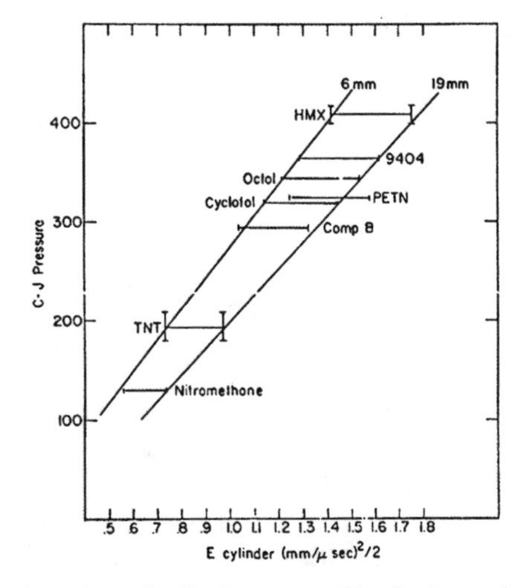

Figure 5.23 C-J pressure as a function of cylinder test wall velocity at 6 and 19 mm wall expansion.

sure range of interest is the minimum data needed to describe the equation of state for an explosive. Experimental data off the single shock Hugoniot (multiple shock data) and off the expansion isentrope would also be needed to calibrate a detonation product equation of state.

The BKW equation of state was shown earlier in this chapter to be adequate to describe the expansion of an explosion under water from several hundred kilobars to a tenth of a bar. The BKW equation of state was also shown to be adequate to describe the plate dent test from over 500 kbars to less than 10 kbars. While it increases the confidence in using the BKW equation of state for describing such integral experiments, there is nothing unique about the BKW equation of state. The experimental data could also be described, within experimental error, by explosive equations of state that describe the pressure-volume-temperature-energy relationship for detonation products quite differently.

5.5 Jet Penetration of Inerts and Explosives

The penetration velocities of projectiles interacting with explosives initiated by the projectile have been found to be much lower than the penetration velocities of inerts of the same density.[22] Studies of projectile penetration dynamics in inert and reactive targets have been performed using the Eulerian reactive hydrodynamic code 2DE described in Appendix C.

Because jets formed by a shaped charge with a metal cone contain small diameter, high velocity projectiles, the jets can be approximated by cylinders or balls of uniform velocity with the appropriate dimensions. The classic paper on jet formation and penetration by Birkhoff, MacDougall, Pugh, and Taylor[23] described what is called the "ideal" penetration velocity. They assumed that Bernoulli's theorem applies (that is, that the jet pressure is large compared to the target or jet material strengths and that pressure is the same in the jet and target near the interface), then

$$0.5\rho_j(V_j - V_p)^2 = 0.5\rho_t V_p^2,$$

where ρ is density, V is velocity, j implies jet, t implies target, and p signifies penetration. The expression is rearranged to find the ratio of the jet velocity to jet penetration velocity,

$$\frac{V_j}{V_p} = 1 + \sqrt{\frac{\rho_t}{\rho_j}}.$$

Using this expression to determine the penetration velocity for a 1-cm/μs steel jet interacting with various targets gives the following results.

Target	Penetration Velocity
Steel	0.500
Aluminum	0.628
Comp B	0.682
Water	0.738

Bernoulli's theorem is for steady motion of an incompressible, uniform fluid. For a compressible fluid, the ideal penetration velocity expression is exact for only a constant density system; however, the experimental penetration data from many studies of inert jets penetrating inerts are adequately described using the ideal model.

The penetration of an aluminum target by an aluminum jet and a steel target by a steel jet were modeled using the particle-in-cell (PIC) technique by Harlow and Pracht.[24] They concluded that their calculated velocities approached the ideal values.

Copper jet penetration into an aluminum target was calculated by Johnson[25] using his two- and three-dimensional Eulerian hydrodynamic codes. His calculated penetration velocity agreed with the ideal values.

The penetration velocity of a 1.3-cm-diameter steel ball moving at various velocities and impacting a 2.5-cm-thick cylinder of PBX-9404 or Composition B was reported by Rice.[22] The data were generated at Ballistic Research Laboratory by R. Frey. Experimental data and Eulerian calculations by Rice indicated that the penetration velocity was markedly decreased, beyond the critical projectile velocity for initiation of detonation, and was significantly less (less than half) than predicted by the ideal model.

The reactive hydrodynamic code 2DE was used to examine the jet penetration of inerts and explosives by Mader, Pimbley, and Bowman.[26,27]

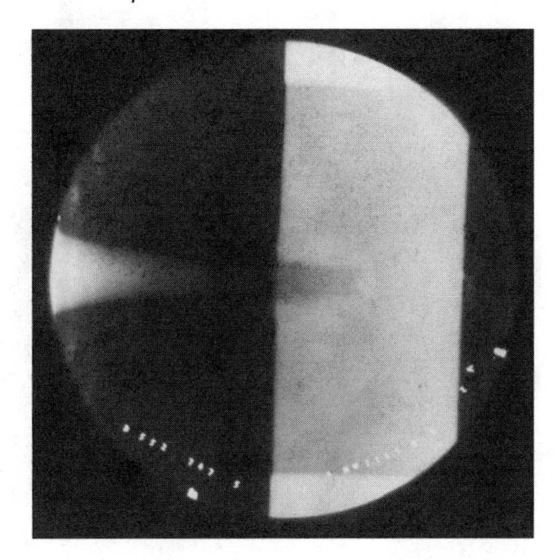

Figure 5.24 Shot 1185. A steel jet formed by a 0.4-cm-thick steel hemishell, which was driven by a 6.00-cm-thick PBX-9404 hemisphere, penetrated a 304 stainless steel block. The jet traveled for 35.57 μs. The steel block was 30.8 cm from the center of the steel hemishell. The radiographic time was 79.02 μs.

Two PHERMEX radiographs of a steel jet penetrating a steel block have been published in the Los Alamos PHERMEX data volumes.[6] The experiments were performed by Douglas Venable. The steel jet was formed by a 0.4-cm-thick steel hemishell driven by a 6.0-cm-thick PBX-9404 hemisphere. The jet interacted with a steel block after 30.8 cm of free run from the center of the steel hemisphere.

The jet tip velocity was 1.8 cm/μs and the jet velocity decreased approximately 0.13 cm/μs for each cm of jet length. Radiograph 1185 was taken after the jet had penetrated 4.5 cm of steel in 6 μs with an average penetration velocity of 0.75 cm/μs. Radiograph 1181 was taken after the jet had penetrated an additional 3.5 cm of steel in 7.0 μs with an average penetration velocity of 0.5 cm/μs. The jet velocity decays from 1.8 cm/μs to 1 cm/μs after a run of 24 feet in the air. The jet diameter is uncertain because of its diffuse boundaries in the radiograph. The jet diameter increases and the jet interface is more diffuse along the length of the jet. Both 0.8- and 1.2-cm-diameter jets were numerically modeled.

The radiographs are shown in Figures 5.24 and 5.25. Calculated density profiles, with the experimental shock front and interface, are shown in Figures 5.26 and 5.27. The 2DE code was used to model the flow. The steel shock Hugoniot was described by $U_s = 0.458 + 1.51U_p$, where U_s is shock velocity and U_p is particle velocity in cm/μs. The initial density was 7.917 g/cc and the Grüneisen gamma was 1.25. Elastic-plastic terms were negligible, and tension was permitted only in the target.

The calculated profiles are sensitive to the jet description. The jet initially has steel's standard density, but the velocity gradient results in the jet density decreasing with time.

The experimentally observed steel jet penetration into a steel plate appears to be adequately reproduced, considering the complicated nature of the jet. The hole observed between the jet and target walls also appears to be substantially reproduced. The jet material appears primarily along the side of the hole made in the target and in the splash wave. Birkhoff, MacDougall, and Pugh[23] state that "careful weighings have shown that a metal jet is captured by a metal target, which loses no weight except a very small amount at the front

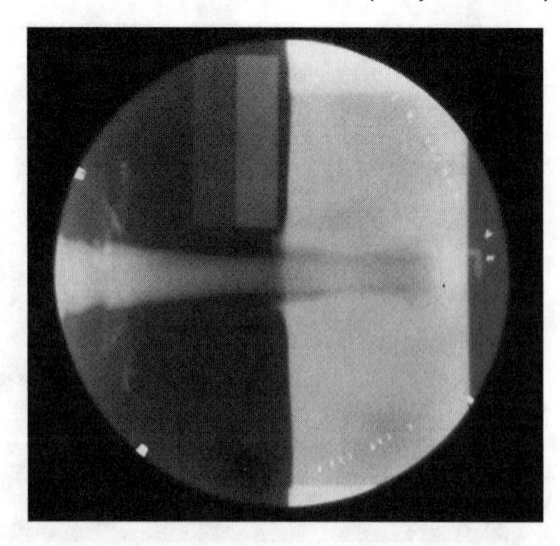

Figure 5.25 Shot 1181. A steel jet formed by a 0.4-cm-thick steel hemishell, which was driven by a 6.00-cm-thick PBX-9404 hemisphere, penetrated a 304 stainless steel block. The jet traveled for 42.53 μs. The steel block was 30.8 cm from the center of the steel hemishell. The radiographic time was 85.99 μs.

surface." Mautz has reported that high velocity steel jets penetrating steel partly remain on the target, and partly leave the target in a vapor or liquid form.

The observed agreement between the radiographs and the numerical model suggests that the important features of jet penetration are described by the fluid dynamics of the process. Since the steel target penetration by a steel jet of decreasing velocity could be modeled numerically, it was possible to examine the penetration physics of simpler systems.

A 1-cm-diameter steel rod with a 1.5 cm/μs velocity penetrating a steel block 11 cm long by 5 cm wide has been modeled. The mesh cells were 0.1 cm square and the time step was 4. x $10^{-3}\mu$s. The system was described by 50 cells in the radial direction and 100 cells along the z-axis.

Contours of pressure, density, energy, and velocity in the R and Z directions after 8.0 μs of penetration are shown in Figure 5.28. The steel rod sends an initial shock pressure of about 9.5 Mbar into the steel target with a 1.6 cm/μs shock velocity and a 0.75 cm/μs particle velocity. Radial motion of the steel target, and side rarefactions, decreased the interface pressure to about 2.5 Mbar, and the particle velocity remained at about 0.75 cm/μs. A diverging, approximately steady-state profile developed near the jet-target interface, which continued until the rod length was consumed. The shock wave was supported by the higher pressure at the rod-target interface.

The penetration quickly approached the "ideal" penetration velocity (half the initial rod velocity) for a steel rod penetrating a steel target. The initial high pressure shock wave formed at the rod-target interface was quickly degraded by side rarefactions and divergence to the "ideal" interface pressure, P, of $0.5\rho_t V_p^2$, whereas the particle velocity remained unchanged, as expected for jets and targets of the same material. The shock impedance relationship may be expressed as

$$\frac{V_j}{V_p} = 1 + \frac{\rho_t U_{s_t}}{\rho_j U_{s_j}},$$

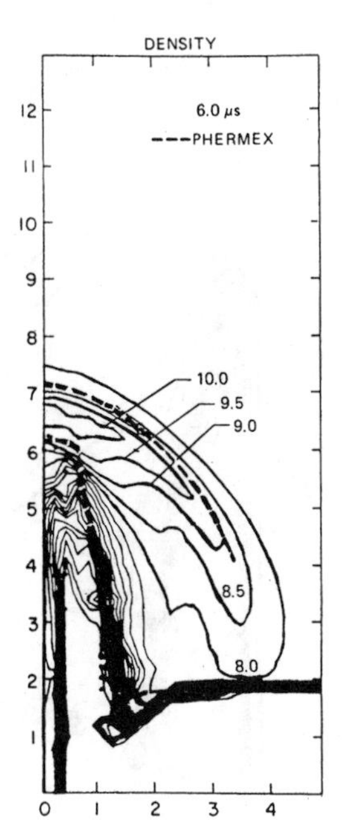

Figure 5.26 Calculated density profiles, after 6.0 μs of penetration, of a steel block 11.0 cm long and 5.0 cm wide, by an 0.8-cm-thick steel jet moving with a tip velocity that has decreased to 1.2 cm/μs from the initial velocity of 1.8 cm/μs. The experimental shock front profile and target interface are shown for shot 1185. The jet has penetrated 4.5 cm of steel. The density contour interval is 0.5 g/cc.

where U_s is the shock velocity and

$$P = \rho_t U_{s_t} V_p.$$

For jets and targets of the same material, $V_p = 0.5V_j$ for both the initial shock match across the interface and for the later penetration after steady-state is achieved, while the pressure decreases from $\rho_t U_{s_t} V_p$ to $0.5\rho_t V_p^2$. The high pressure near the rod-target interface supported a lower pressure shock wave that moved out into the target ahead of the rod-target interface with a shock speed similar to the penetration velocity of the rod.

Rice[22] reported Frey's data for the penetration velocity of a steel ball, 1.3 cm in diameter, moving at varying speeds and striking 2.5-cm-thick cylinders of either PBX-9404 or Composition B.

This system has been modeled numerically and the results have been compared with the experimental data of Frey. The mesh was 0.05616 cm by 0.05 cm and the time step was 0.005 μs. The computational problem was 78 cells in height and 20 cells in width. Figure 5.29 shows the experimental data and the calculated results of a steel ball interaction with PBX-9404 or Composition B. The ball velocity loss is defined as the initial ball velocity less the penetration velocity. The agreement demonstrates that the model describes the important process of the explosive penetration.

When the steel ball penetrates inert or nearly inert explosive, the penetration velocity could be described by the ideal model. When the ball velocity was just sufficient to cause propagating detonation, however, the observed and calculated penetration velocities were

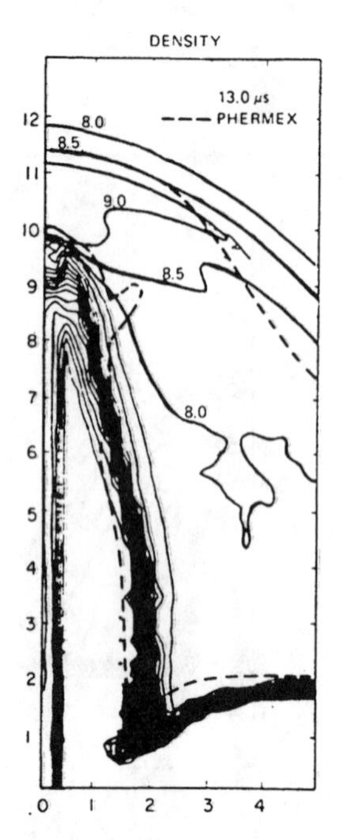

Figure 5.27 Calculated density profiles, after 13.0 μs of penetration, of a steel block 11.0 cm long and 5.0 cm wide, by an 0.8-cm-thick steel jet moving with a tip velocity that has decreased to 0.775 cm/μs from the initial velocity of 1.8 cm/μs. The experimental shock front profile and target interface are shown for shot 1181. The jet has penetrated 8.0 cm of steel. The density contour interval is 0.5 g/cc.

much less than predicted by the ideal model. As the ball velocity was increased, the difference between the actual and ideal velocities decreased. A summary of the ball penetration calculations is given in Table 5.2.

The 0.12-cm/μs ball results in prompt initiation of the PBX-9404. The interface pressure is much higher, and the velocity loss greater than for the 0.1-cm/μs ball, which does not cause prompt initiation.

As shown in Chapter 4, the Held[28] experimental critical condition for propagating detonation of V^2d adequately describes the jet initiation of explosives. The steel ball must present an "effective diameter" to the explosive. The critical ball velocity for PBX-9404 shown in Figure 5.29 is 0.114 cm/μs. The V^2d (in mm) for PBX-9404 is 16.0; therefore, the ball has an effective diameter of 1.23 cm. This is, within the calibration error, not significantly different from the actual ball diameter of 1.30 cm. The critical ball velocity for Composition B shown in Figure 5.28 is 0.18 cm/μs. The V^2d (in mm) for Composition B is 29; therefore, the ball has an effective diameter of 0.9 cm when it is shocking Composition B.

Because the steel ball exhibits a complicated flow, a simpler system was examined to determine the essential features of the flow. The ball was replaced by a steel rod of the same diameter, and rods with velocities of 0.2 and 0.6 cm/μs were calculated interacting with reactive and nonreactive Composition B. A summary of the rod calculations is given in Table 5.3

The lowered penetration velocity of a projectile moving into detonating (rather than nondetonating or inert) explosives is caused by the *higher pressure at the projectile-detonation product interface.* The ideal model assumes that the pressure at zero particle velocity is zero,

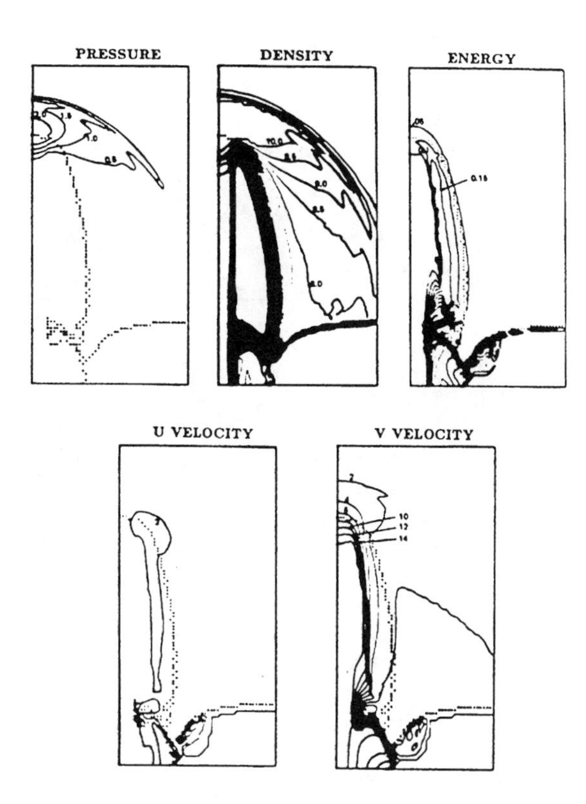

Figure 5.28 Pressure, density, energy, and U and V velocity contours at 8.0 μs for a steel rod with a 1.5 cm/μs initial velocity penetrating a steel target. The pressure contour interval is 500 kbar, the density contour interval is 0.5 g/cc, the energy contour interval is 0.05 Mbar-cm^3/g, and the velocity interval is 0.2 cm/μs. The true dimensions of the graph are 5.0 cm wide and 10.0 cm high.

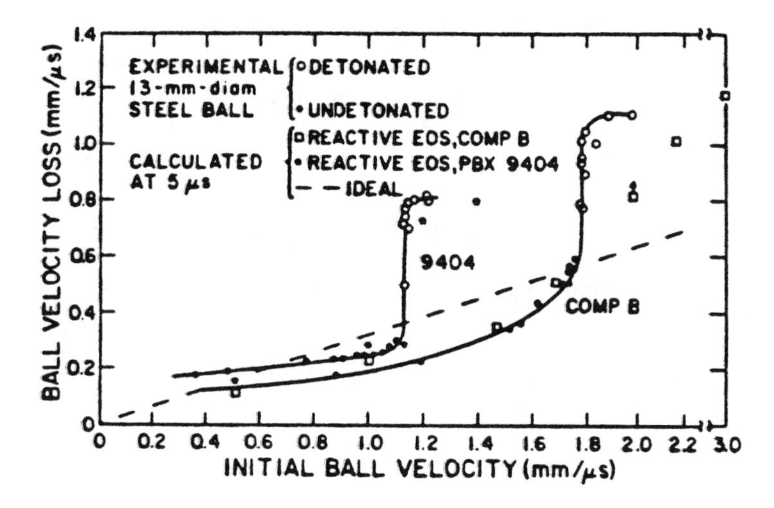

Figure 5.29 Initial ball velocity vs. the ball velocity loss for a 1.3-cm-diameter ball penetrating PBX-9404 or Composition B.

TABLE 5.2 Ball Penetration Calculations

	Velocity (cm/μs)	State	Velocity Loss (cm/μs)	V_p (cm/μs)	V_p Ideal[a]	V_{loss} Ideal (cm/μs)
Steel/Comp B	0.22	Reacted	0.100	0.120	0.1500	0.0700
	0.20	Late Reaction	>0.080	0.120	0.1365	0.0640
	0.17	No Reaction	0.050	0.120	0.1160	0.0540
	0.10	No Reaction	0.025	0.075	0.0680	0.0320
	0.05	No Reaction	0.010	0.040	0.0340	0.0160
	0.15	No Reaction	0.030	0.120	0.1020	0.0477
	0.30	Reacted	0.120	0.180	0.2047	0.0953
	0.20	No reaction permitted	0.050	0.150	0.1365	0.0635
	0.30	No reaction permitted	0.080	0.225	0.2047	0.0953
Steel/9404	0.14	Reacted	0.080	0.060	0.0944	0.0456
	0.12	Late reaction	>0.075	0.045	0.0810	0.0390
	0.10	No reaction	0.035	0.065	0.0675	0.0325
	0.05	No reaction	0.015	0.035	0.0337	0.0163

[a] V_p ideal for Composition B = 0.682 (ball velocity).
V_p ideal for PBX-9404 = 0.675 (ball velocity).

TABLE 5.3 Jet Penetration of Composition B

Condition	Initial Rod Velocity[a]	Penetration Velocity[b] Calc. Final	Penetration Velocity[b] Ideal	Interface Pressure[b] Calc. Final	Interface Pressure[b] Ideal
Reactive	0.6	0.40	0.4090	180.	144.
Reactive	0.2	0.10	0.1365	50.	16.
Inert	0.6	0.40	0.4090	150.	144.
Inert	0.2	0.14	0.1365	15.	16.

[a] Ideal penetration velocity = 0.682 (rod velocity) and ideal interface pressure = 857.5 (penetration velocity)^2.
[b] Velocities are in cm/μs and pressures in kbar.

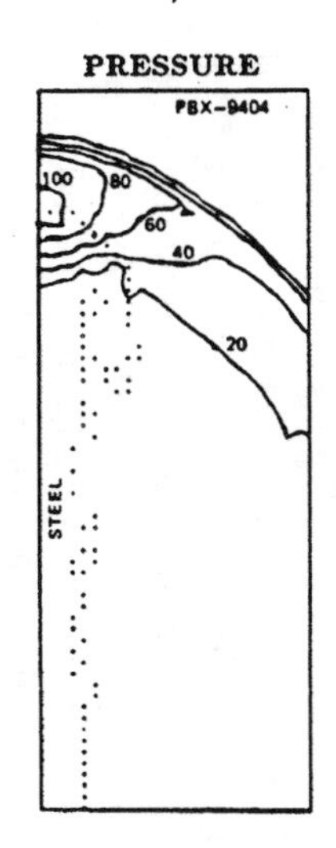

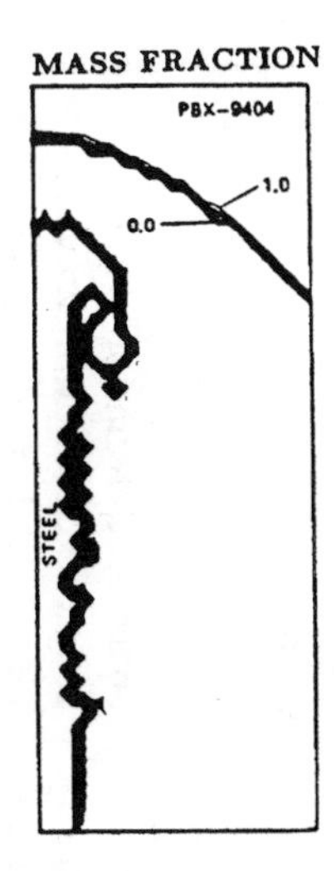

Figure 5.30 Pressure, and mass fraction contours at 8.8 μs, for a 1.6-cm-diameter steel rod initially moving at 1.5 cm/μs penetrating PBX-9404. The pressure contour interval is 200 kbar.

which is correct for inerts; however, for explosives the constant volume detonation pressure at zero particle velocity is approximately 100 kbar for a slab of Composition B. In diverging flow, the detonation product pressure at zero particle velocity is about 50 kbar. The effect is not included in the ideal model, so it fails to account for the velocity decrease.

If it is assumed that the ideal model is appropriate for the steel rod, and the calculated detonation product interface pressure, 50 kbar for the 0.2 cm/μs rod, is set equal to $0.5\rho_j(V_j - V_p)^2$, a 0.09-cm/μs penetration velocity is estimated. This is close to the 0.1 cm/μs calculated.

The relative difference between the ideal penetration velocity and the calculated penetration velocity decreases with increasing projectile velocity. The ideal model becomes better at higher projectile velocities, where the difference between the explosive reactive and nonreactive pressures (of about 40 kbar) becomes insignificant.

To examine the effect of projectile velocity in more detail, a range of velocities was studied. A 1.6-cm-diameter steel rod was modeled moving at velocities lower than those necessary for initiating propagating detonation (0.1 cm/μs) in PBX-9404 and at velocities great enough to penetrate the explosive faster than the detonation velocity (0.88 cm/μs). The study investigated the penetration velocity in explosives throughout the range of possible jet or projectile velocities and examined the steel rod and explosive interface pressure effect in more detail. The computational mesh was 0.2 cm by 0.2 cm square. The problem was described by 25 cells in the radial direction and 64 cells in the axial direction. The rod was 10 cells long and the steel target was 55 cells high. The time step for the calculation was 0.008 μs. A summary of the results of the calculations is given in Table 5.4

A steel rod with a 1.5-cm/μs initial velocity penetrating PBX-9404 is shown in Figure 5.30. The steel rod penetrates the PBX-9404 at 1.06 cm/μs, and it forms a steady, overdriven detonation wave moving at the same velocity as the steel rod, with an overdriven effective C-J pressure of 1 Mbar. The C-J pressure of PBX-9404 is 365 kbar and the C-J velocity is 0.88 cm/μs. The calculation demonstrates that high velocity jets *cannot* be used to destroy an explosive charge without initiating an overdriven propagating detonation.

TABLE 5.4 Steel Rod Penetrating PBX-9404

Initial Rod Velocity (cm/μs)	Final Calc. Penetration Velocity (cm/μs)	Ideal Penetration Velocity[a] (cm/μs)	Interface Pressure (kbar)	Ideal Interface Pressure[b] (kbar)	Diverging Effective C-J Pressure (kbar)	Detonation Velocity (cm/μs)	Comments
1.500	1.060	1.012	1000	944.	900	1.00	Steady overdriven detonation wave
1.300	0.920	0.877	750	709.	600	0.88	Steady overdriven detonation wave
1.100	0.780	0.742	550	507.	400	0.88	Decaying wave moving faster than rod
0.900	0.620	0.607	380	340.	360	0.87	Diverging det. moving faster than rod
0.800	0.550	0.540	310	268.	350	0.87	Diverging det. moving faster than rod
0.500	0.320	0.337	150	105.	350	0.87	Diverging det. moving faster than rod
0.300	0.180	0.202	80	38.	350	0.87	Diverging det. moving faster than rod
0.250	0.140	0.169	70	26.	350	0.87	Diverging det. moving faster than rod
0.200	0.110	0.135	60	17.	350	0.87	Diverging det. moving faster than rod
0.150	0.060	0.101	50	9.	350	0.87	Diverging det. moving faster than rod
0.100	0.025	0.067	50	4.	350	0.87	Diverging det. moving faster than rod
0.070	0.050	0.047	0	2.	–	—	No detonation, decaying shock wave
0.050	0.036	0.034	0	1.	–	—	No detonation, decaying shock wave
0.030	0.022	0.020	0	0.4	–	—	No detonation, decaying shock wave

[a] Ideal penetration velocity = 0.674 (rod velocity)
[b] Ideal interface pressure = 922. (penetration velocity)^2.

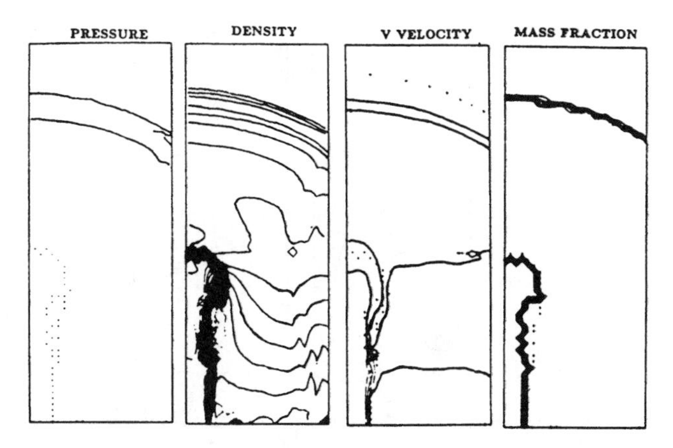

Figure 5.31 Pressure, density, V velocity, and mass fraction contours at 12 μs, for a 1.6-cm-diameter steel rod initially moving at 0.5 cm/μs penetrating PBX-9404.

When the rod penetration velocity becomes less than the C-J detonation velocity, the detonation wave moves away from the rod surface. To illustrate this, a steel rod with a 0.5-cm/μs initial velocity penetrating PBX-9404 was modeled. Contours are shown in Figure 5.31. The steel rod penetrates the PBX-9404 at 0.32 cm/μs. The detonation wave proceeds at 0.87 cm/μs with a diverging effective C-J pressure of 350 kbar. The interface detonation product pressure is 150 kbar, which is 45 kbar greater than the ideal interface pressure at the rod tip of 105 kbar. The ideal model assumes that the pressure at zero particle velocity is zero, which is incorrect for constant volume detonation. A slower penetration velocity results than would be expected from the ideal model.

The effect of the 40 kbar additional interface pressure is most apparent when the rod velocity just suffices to cause prompt propagating detonation. To illustrate this, a steel rod with a 0.15-cm/μs initial velocity penetrating PBX-9404 was modeled. Contour graphs are shown in Figure 5.32. The steel rod penetrates the PBX-9404 at 0.06 cm/μs, which is 0.04 cm/μs slower than it would penetrate the PBX-9404 if it were inert. The steel rod interface pressure is 50 kbar, which is 40 kbar higher than the 9 kbar ideal interface pressure. Illustrating the importance of this increased interface pressure on penetration velocity, the calculated penetration velocity is about the *same* for the 0.07-cm/μs steel rod, where the explosive does not decompose, as for the 0.15-cm/μs rod, where the explosive detonated.

For engineering purposes, the ideal jet penetration velocity into an inert can be estimated, using the shock matching relationship

$$\frac{V_j}{V_p} = 1 + \frac{\rho_t U_{s_t}}{\rho_j U_{s_j}},$$

where ρ is density, V is velocity, U_s is shock velocity, j implies jet, t implies target, and p signifies penetration. The initial shock pressure can be estimated using

$$P = \rho_t U_{s_t} V_p.$$

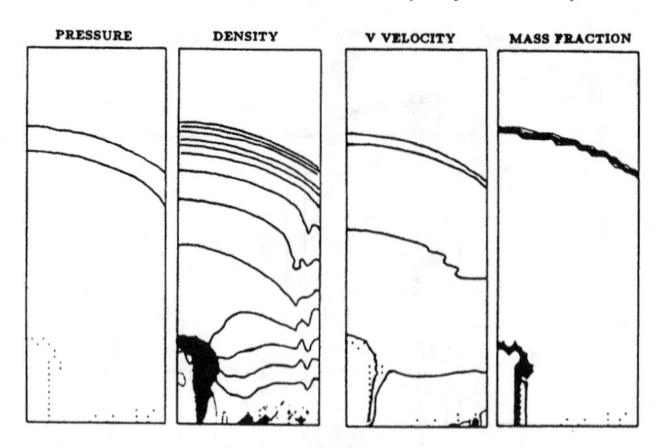

Figure 5.32 Pressure, density, V velocity, and mass fraction contours at 12 μs, for a 1.6-cm-diameter steel rod initially moving at 0.15 cm/μs penetrating PBX-9404.

Final penetration velocity can be estimated using Bernoulli's theorem, from

$$\frac{V_j}{V_p} = 1 + \sqrt{\frac{\rho_t}{\rho_j}},$$

and the interface pressure,

$$P = 0.5\rho_t V_p^2.$$

When an explosive or propellant is used as the target, the pressure expression needs an additional term P^*, and the Bernoulli equation for explosives becomes

$$0.5\rho_j(V_j - V_p)^2 = 0.5\rho_t V_p^2 + P^*.$$

P^* is approximately 40 kbar for the explosives studied.

5.6 Plane Wave Lens

Three types of plane wave generators for explosive systems used in generating dynamic material properties make use of the properties of Baratol and Composition B to form effective lens. Baratol's slow detonation velocity and Composition B's high detonation velocity, if used in the proper configuration, can convert a spherically diverging wave into an approximate plane wave. The P-40, P-081, and P-120 lens were used until the middle 1990s.

Of these, the P-040 has the simplest design, consisting of an inverted right circular Baratol cone encased in a coaxial Composition B cylinder as shown in Figure 5.33. The top face of the assembly consists of Baratol; the bottom surface is Composition B. The device is detonated axially symmetrically from beneath. The detonation is much smaller in diameter than the upper assembly and is often encased in Composition B, making a uniform cylinder.

The P-081 and P-120 generators each retain the inner Baratol cone, but they are larger and heavier than the P-040. A sketch of the lens is shown in Figure 5.34. The P-040 has a nominal 4-inch-diameter top surface, the P-081 an 8 inch diameter, and the P-120 a 12 inch diameter. The outer Composition B conical shell is truncated. The detonator is smaller in diameter than the truncated end and is encased in the Composition B.

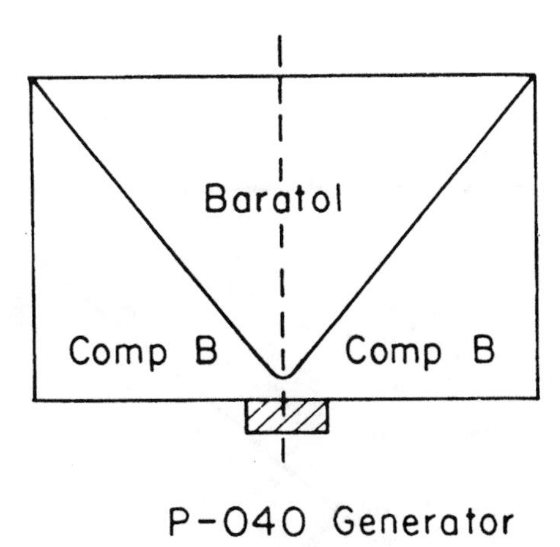

Figure 5.33 A sketch of the P-040 plane wave lens.

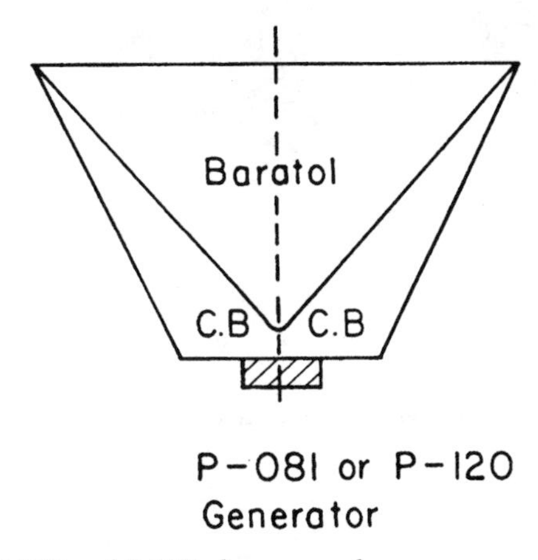

Figure 5.34 A sketch of the P-080 and P-120 plane wave lenses.

Each plane wave generator is designed to produce a plane shock wave at the upper Baratol surface of the device. The generators use the lens effect of a Baratol cone, and the lower transit velocity therein, to mold the somewhat spherical spreading detonation waves into planar patterns as they proceed.

Each of the plane wave lens was modeled[29] using the 2DE code. A radiographic study of the plane wave lens is presented in Reference 5.

The calculated density contour plots for a P-040 lens are shown in Figures 5.35 through 5.37 at 2.625, 4.725, and 6.825 microseconds. The radius is plotted horizontally and the axial Z is plotted vertically. The Baratol cone remains approximately static throughout the sequence. The circles on the figures mark the experiment points read from the radiographs and correlated in time with the numerical calculations. The radiographs were from Shots 630, 631, and 632 described in detail in Reference 5.

The calculated results for the P-080 and P-120 lenses are given in Reference 29. Although the detonation wave arrives almost planar across the top of the lens, a significant pressure gradient is present across the lens and behind the wave surface. The pressure gradients depend significantly upon the geometry and composition variables, which are difficult to

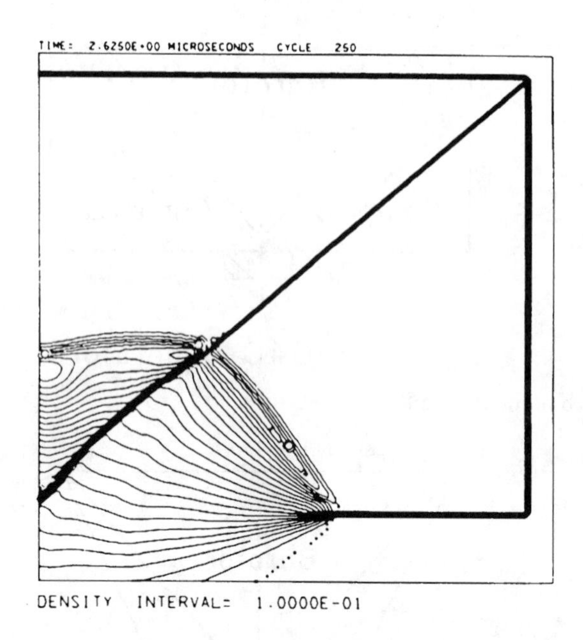

Figure 5.35 The calculated density contours for a P-040 plane wave lens after 2.625 μsec. The circles mark the experimental points read from the radiograph from Shot 630.

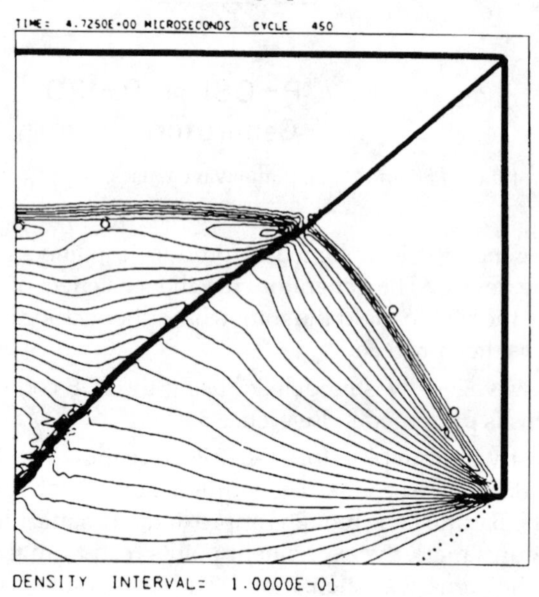

Figure 5.36 The calculated density contours for a P-040 plane wave lens after 4.725 μsec. The circles mark the experimental points read from the radiograph from Shot 631.

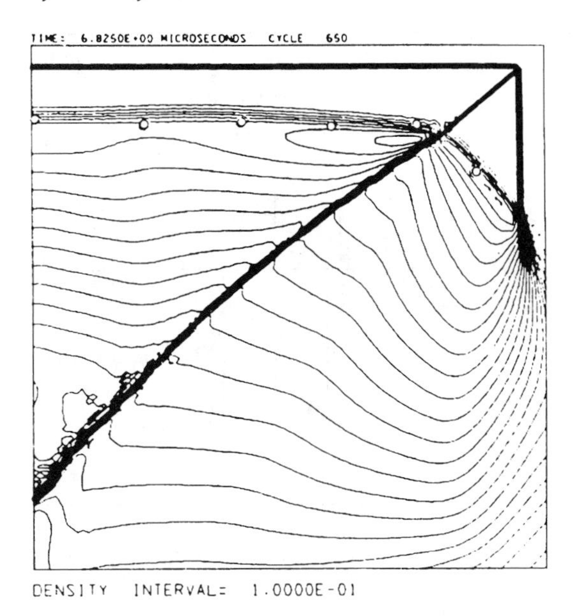

Figure 5.37 The calculated density contours for a P-040 plane wave lens after 6.825 μsec. The circles mark the experimental points read from the radiograph from Shot 632.

control during manufacture. Thus, the calculated pressure gradients across the lens are probably less steep than those actually present in most plane wave generators.

Experimental data indicate that the pressure profile behind the front is asymmetrically oriented with respect to the charge center. Pressure magnitude variations prevent obtaining high-quality shock initiation data, particularly for the more shock insensitive explosives.

The requirement to cease releasing Barium into the atmosphere has resulted in a redesigned plane wave lens. B. Olinger has developed an improved performing lens replacing Baratol with TNT and replacing Composition B with PBX-9501. Numerical modeling was used in the development of the new lens called the P-100. The new lens has significantly smaller pressure variations both across the surface of the lens and behind the detonation front than the P-040, P-080, or P-120 lens. An animation of the new lens generated during the design stage using the TDL code is included on the CD-ROM.

5.7 *Regular and Mach Reflection of Detonation Waves*

Radiographic studies of laterally colliding, diverging, cylindrical detonation waves in PBX-9404 have been studied using PHERMEX.[6] The PHERMEX shot numbers are 1019, 1037, 1038, 1130, and 1143. The flow has been modeled[30] using the 2DL code described in Appendix B.

The progress of two laterally colliding, diverging cylindrical detonation waves was studied using the experimental arrangement shown in Figure 5.38. Line generators served to initiate the detonations. The space-time histories of the detonation wave and of the interaction region that starts with regular reflection and subsequently develops into Mach reflection were measured. This interaction range covers angles of incidence, with respect to the plane of symmetry, from zero to about 85 degrees. The Mach wave started at about 50

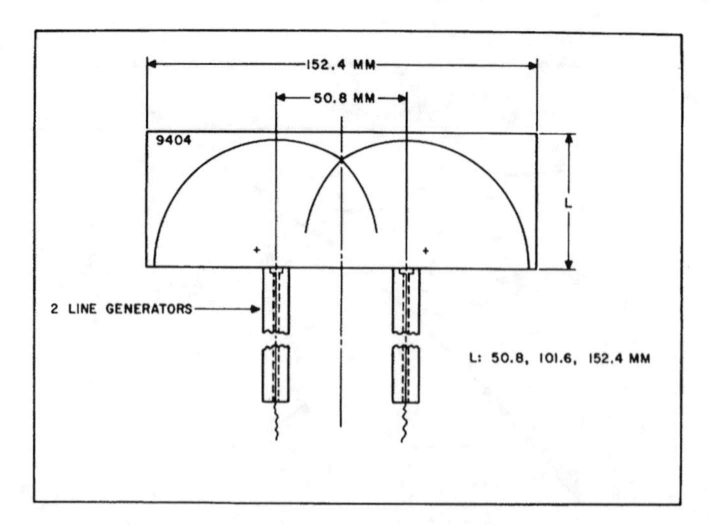

Figure 5.38 A sketch of the experimental geometry used to study colliding cylindrical detonation waves.

degrees. The radiographs are shown in Figures 5.39 through 5.43 at 1.06, 4.03, 5.61, 11.16, and 16.75 μsec after initiation at the PBX-9404 interface with the line generators.

Figure 5.44 shows the data on the radial motion, r(t), of the the detonation wave and a similar trajectory, y(t), of the Mach wave. The excursions were normalized with respect to the half distance between the line generators centers, called (a). Real time, t, was normalized with respect to a characteristic time, defined as $t\ D/a$, where D is the detonation velocity. Figure 5.44 also shows the computed trajectory of the interaction of the two waves in regular reflection mode with a dashed line. The detonation wave moved at 0.8745 ± 0.0016 cm/μsec which is 1% lower than the plane wave detonation velocity of 0.8880. The Mach wave velocity beginning at point of origin exceeds 1.2 cm/μsec and then slows approaching the normal detonation wave velocity. The velocities are shown in Figure 5.45.

The 2DL code was used to model the colliding detonation waves using the HOM PBX-9404 equation of state and Forest Fire burn. The same results would be obtained if the C-J volume burn had been used. The calculations were performed using 6400 cells each 0.2 cm square. The line initiation was simulated by starting the calculation with a four cell region of PBX-9404 initially exploded. Figures 5.46 through 5.50 show the calculated density contours at the same times as the radiographs in Figures 5.39 through 5.43. Also shown are the late time radiographic shock profiles.

The 3DE code described in Appendix D was used to model the interaction of three spherically diverging detonation waves in Reference 31. The formation of regular and Mach shock reflections in three-dimensional geometry was described. In Reference 32, the effect of multipoint initiation of an explosive on the motion of a thin metal plate was investigated in two- and three-dimensional geometry. The effect of two colliding diverging detonation waves on a 0.4-cm-thick copper plate driven by 6.2 cm of PBX-9404 initiated by two line generators located 5.08 cm apart is shown in Figure 5.51. A series of shocks and rarefactions travel back and forth in the metal plate, resulting in a very irregular free-surface profile. The location of the leading free surface changes from above the detonators to above the detonation wave interaction centers with increasing time. The relative position of the detonators and the run distance of the detonations after they interact and before they shock the

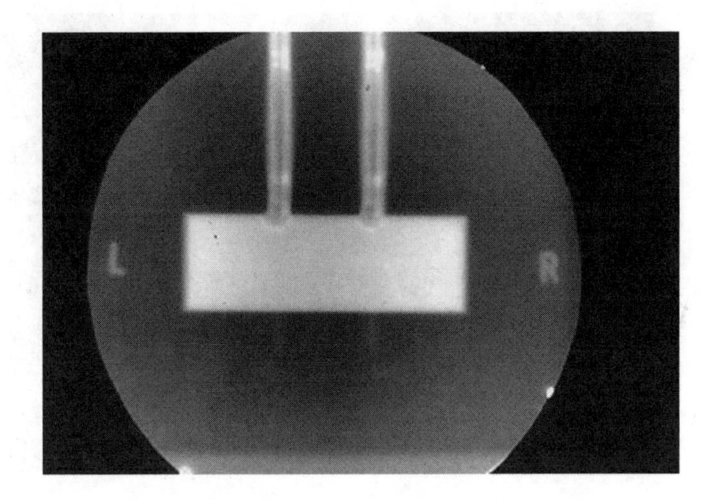

Figure 5.39 Radiograph 1130 at 1.06 μsec after arrival of the line generator shock wave which initiated two diverging cylindrical detonation waves.

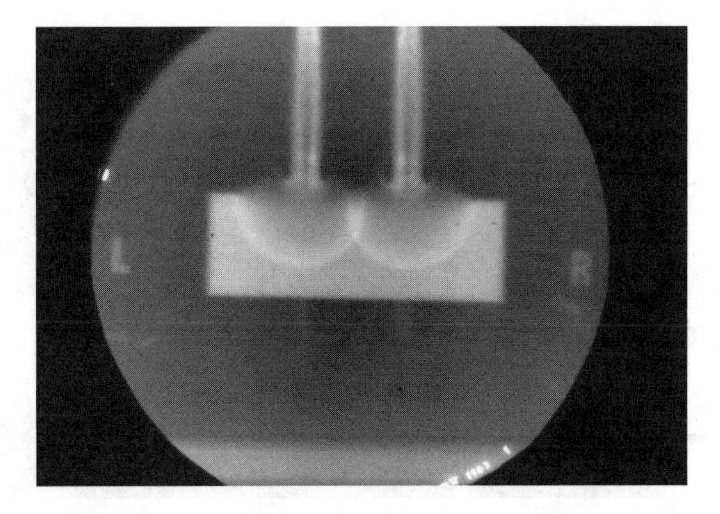

Figure 5.40 Radiograph 1143 at 4.03 μsec after arrival of the line generator shock wave which initiated two diverging cylindrical detonation waves. A regular shock reflection is shown.

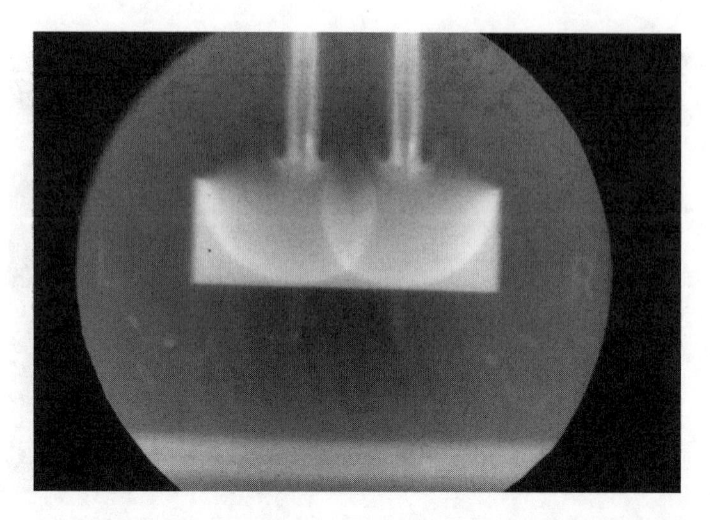

Figure 5.41 Radiograph 1019 at 5.61 μsec after arrival of the line generator shock wave which initiated two diverging cylindrical detonation waves. Transition from regular to Mach reflection is shown.

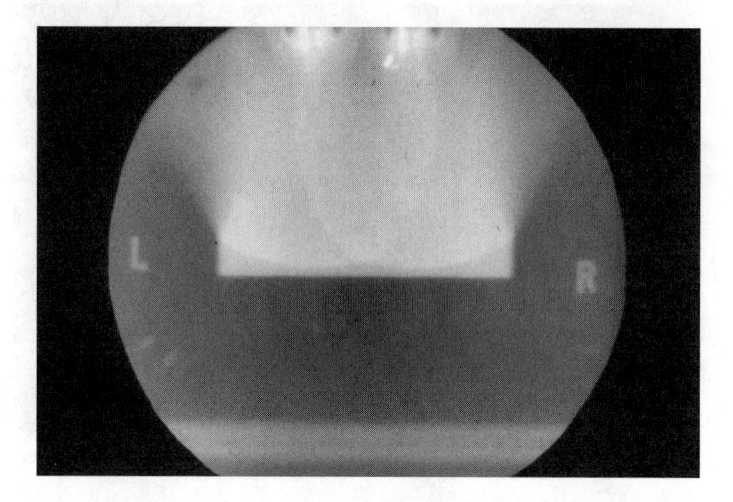

Figure 5.42 Radiograph 1037 at 11.16 μsec after arrival of the line generator shock wave which initiated two diverging cylindrical detonation waves. A Mach wave is shown.

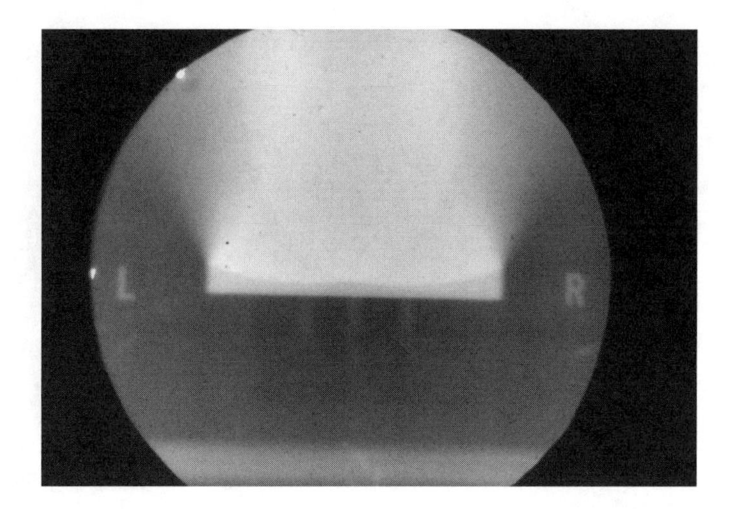

Figure 5.43 Radiograph 1038 at 16.75 μsec after arrival of the line generator shock wave which initiated two diverging cylindrical detonation waves. A Mach wave is shown.

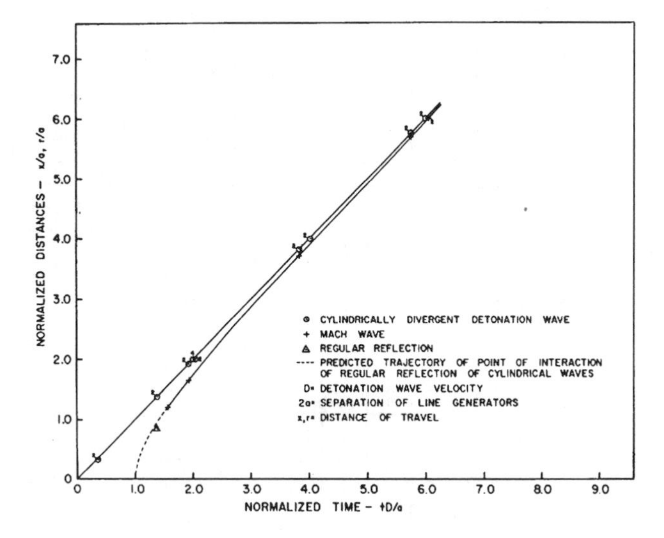

Figure 5.44 The progress of regular and Mach waves in the normalized distance and time planes. a is half the distance between the line generators.

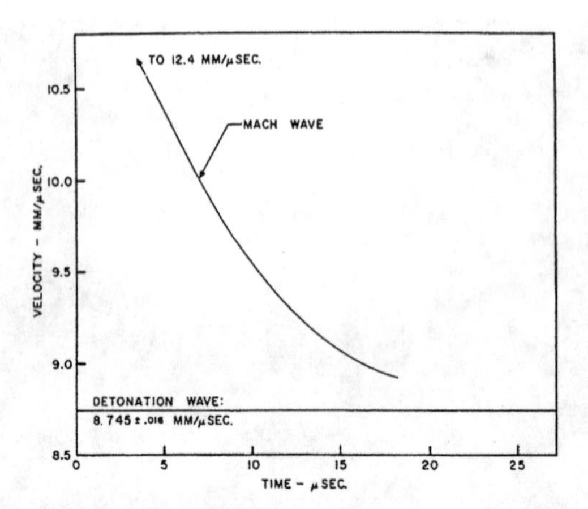

Figure 5.45 The detonation wave and Mach wave velocities as a function of distance.

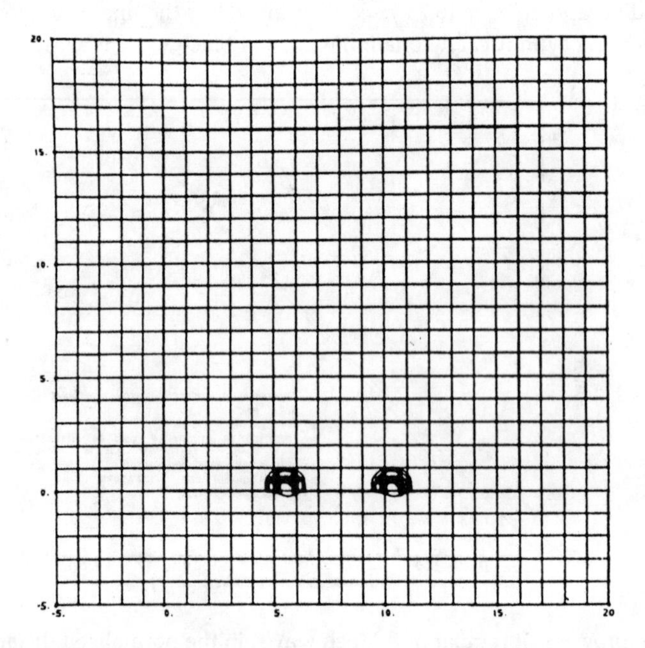

Figure 5.46 The 2DL calculated isopycnic contours at 1.0 μsec. The contour interval is 0.02 g/cc.

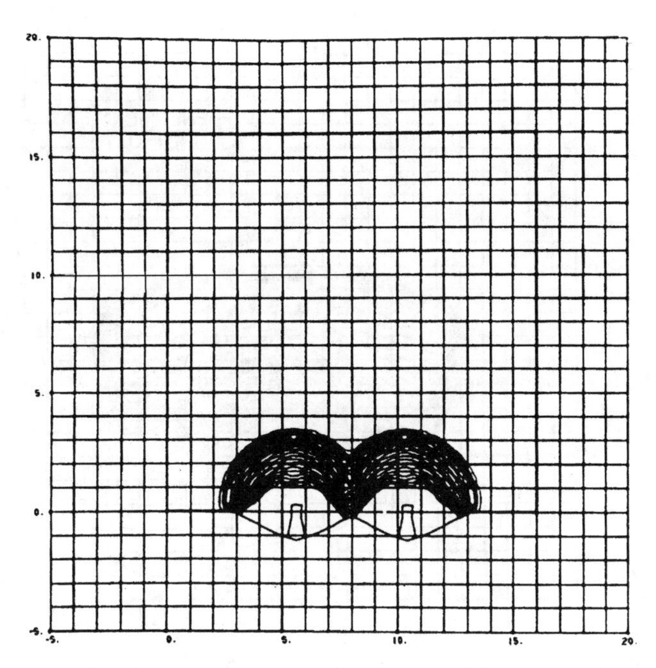

Figure 5.47 The 2DL calculated isopycnic contours at 4.0 μsec. The contour interval is 0.02 g/cc.

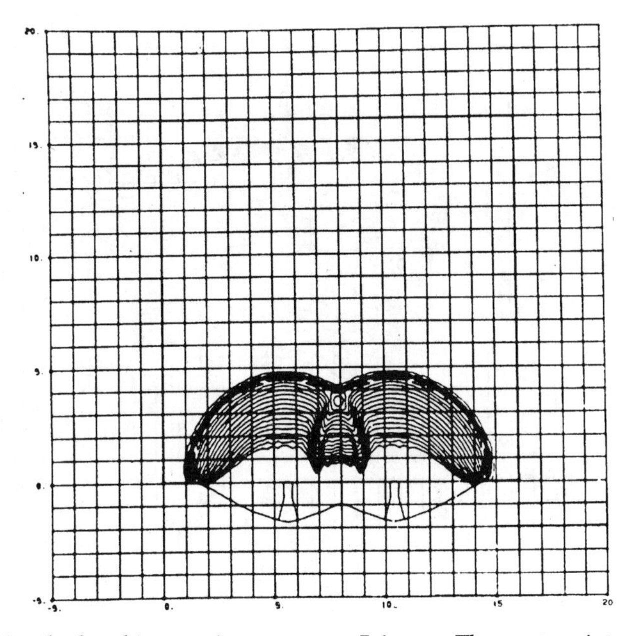

Figure 5.48 The 2DL calculated isopycnic contours at 5.6 μsec. The contour interval is 0.02 g/cc.

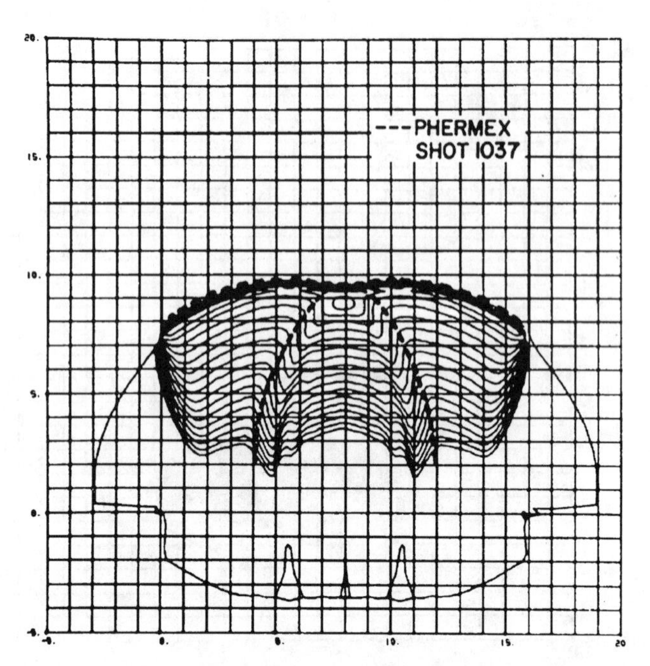

Figure 5.49 The 2DL calculated isopycnic contours at 11.2 μsec. The contour interval is 0.02 g/cc.

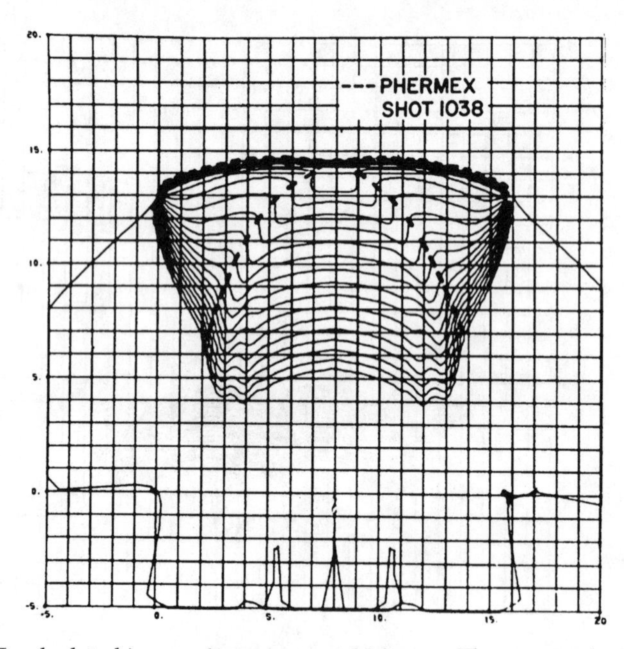

Figure 5.50 The 2DL calculated isopycnic contours at 16.8 μsec. The contour interval is 0.02 g/cc.

metal plate are important parameters in determining the magnitude of the plate free-surface perturbations.

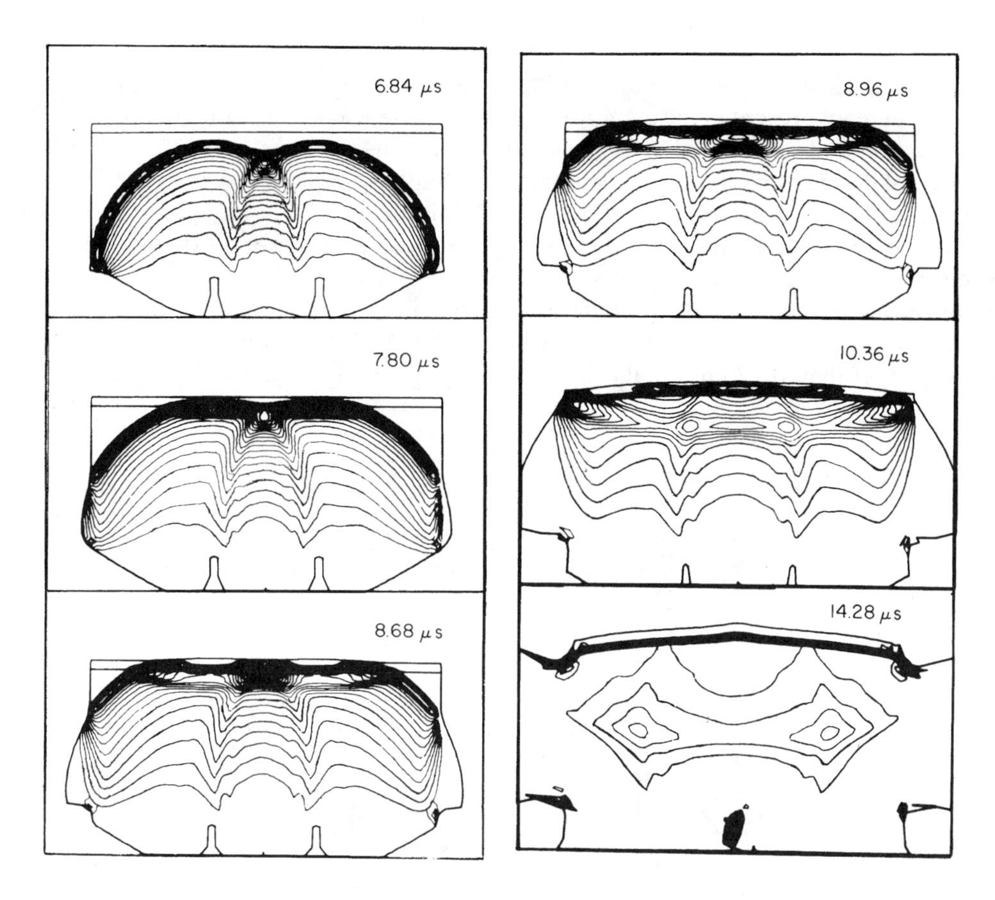

Figure 5.51 Calculated isobar contours with a 20 kbar contour interval for two cylindrical diverging detonations initially 5.0 cm apart, interacting with a 0.4 cm Copper plate after 6.2 cm of run.

5.8 *Insensitive High Explosive Initiators*

The initiation of propagating, diverging detonation is usually accomplished by using small conventional initiators; however, as the explosives to be initiated became more shock insensitive, the initiators must have larger diameters (> 2.54 cm) to be effective.

Travis[33] has used the image intensifier camera to examine the nature of the diverging detonation waves formed in PBX-9404, PBX-9502 (X0290) , and X0219 by hemispherical initiators. The geometries of the initiators were (1) a 0.635-cm-radius hemisphere of PBX-9407 at 1.61 g/cc surrounded by a 0.635-cm-radius hemisphere of PBX-9407, (2) a 0.635-cm-radius hemisphere of 1.7 g/cc TATB surrounded by a 1.905-cm-thick hemisphere of 1.8 gm/cc TATB, or (3) a 1.6-cm-radius hemisphere of X0351 (15/5/80 HMX/Kel-F/TATB) at 1.89 g/cc.

The 2DL code described in Appendix B was used to describe the reactive fluid dynamics. The Forest Fire description of heterogeneous shock initiation burn described in Chapter 4 was used to describe the explosive burn. The Pop plots are shown in Figure 4.2 and the Forest Fire rates in Figure 4.6. The BKW detonation product HOM equation of state constants for X0351, and for 1.7 and 1.8 g/cc TATB are given in Reference 33.

The calculations were done in cylindrical geometry with Lucite confinement rather than the air confinement present in the experimental study. The Lucite confinement prevents the mesh distortion that can be fatal to Lagrangian calculations.

The central 0.635 cm region of the detonator was initially exploded, which initiated the remaining explosive in the detonator using a C-J volume burn. For any given mesh size and time step, the viscosity must be adjusted to give a peak pressure at the detonation front near the effective C-J pressure.

The burn can become unstable when it turns a corner in Lagrangian hydrodynamic codes. The instability is eliminated by using an average of the nearby cell pressures for the Forest Fire burn, rather than the individual cell pressure.

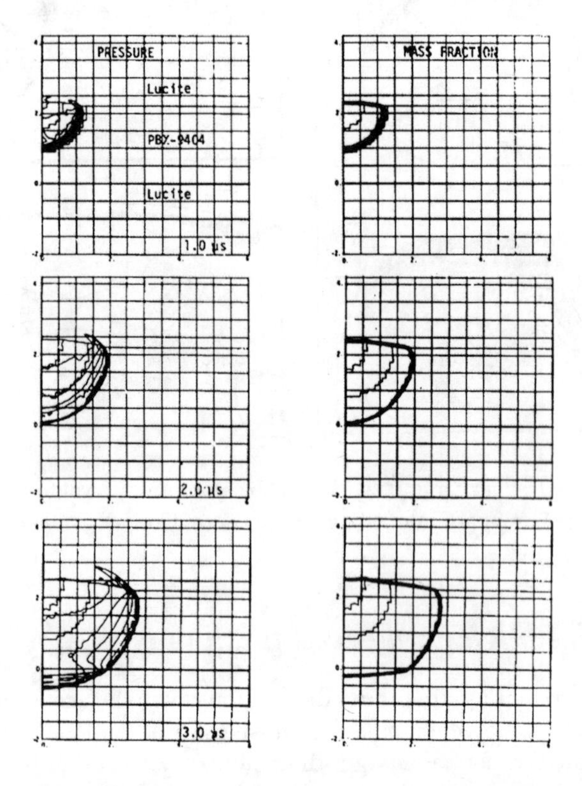

Figure 5.52 The pressure and mass fraction contours at various times for a hemispherical initiator of 0.635-cm-radius PBX-9407 surrounded by 0.635 cm of PBX-9404 initiating PBX-9404. The pressure contour interval is 50 kbar and the mass fraction contour interval is 0.1.

The pressure and mass fraction contours are shown for a PBX-9404 hemisphere initiating PBX-9404 in Figure 5.52, PBX-9502 (X0290) in Figure 5.53, and X0219 in Figure 5.54. The experimental and calculated position of the leading wave as a function of distance from the origin is shown in Figure 5.55. The PBX-9404 initiator initiates prompt propagating detonation in PBX-9404. The PBX-9404 initiator initiates propagating detonation in PBX-9502 leaving a large region of undecomposed explosive. The PBX-9404 hemispherical initiator fails to initiate propagating detonation in X0219.

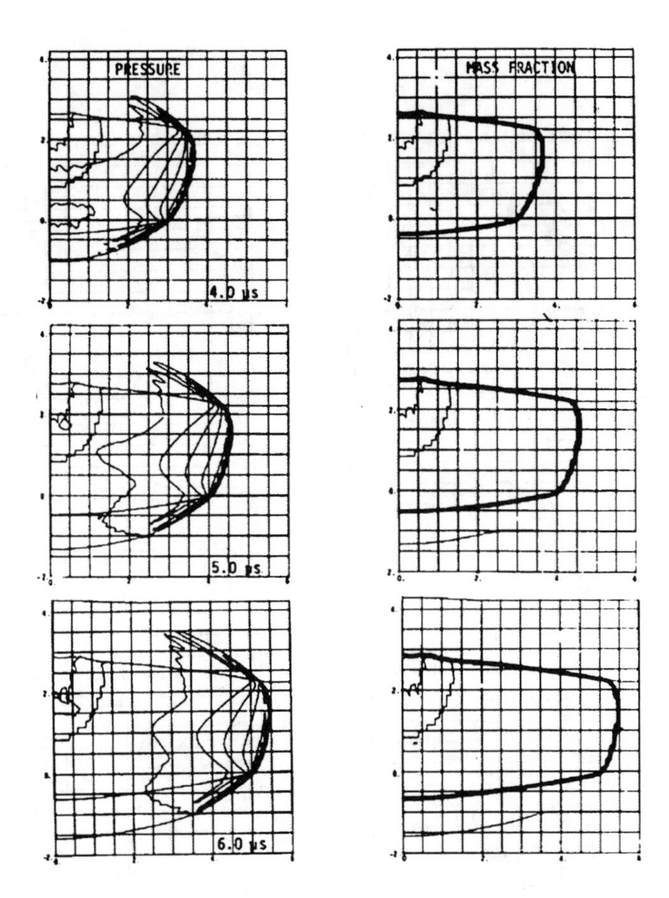

Figure 5.52 *(Continued)* The pressure and mass fraction contours at various times for a hemispherical initiator of 0.635-cm-radius PBX-9407 surrounded by 0.635 cm of PBX-9404 initiating PBX-9404. The pressure contour interval is 50 kbar and the mass fraction contour interval is 0.1.

The pressure and mass fraction contours are shown in Figure 5.56 for a large 1.8 g/cc TATB hemisphere initiating PBX-9502. Very little undecomposed explosive was observed experimentally, in agreement with the calculated results. The contours are shown in Figure 5.57 for an embedded X0351 hemispherical initiator initiating PBX-9502. The experimental and calculated regions of partially undecomposed PBX-9502 explosive are shown in Figure 5.58.

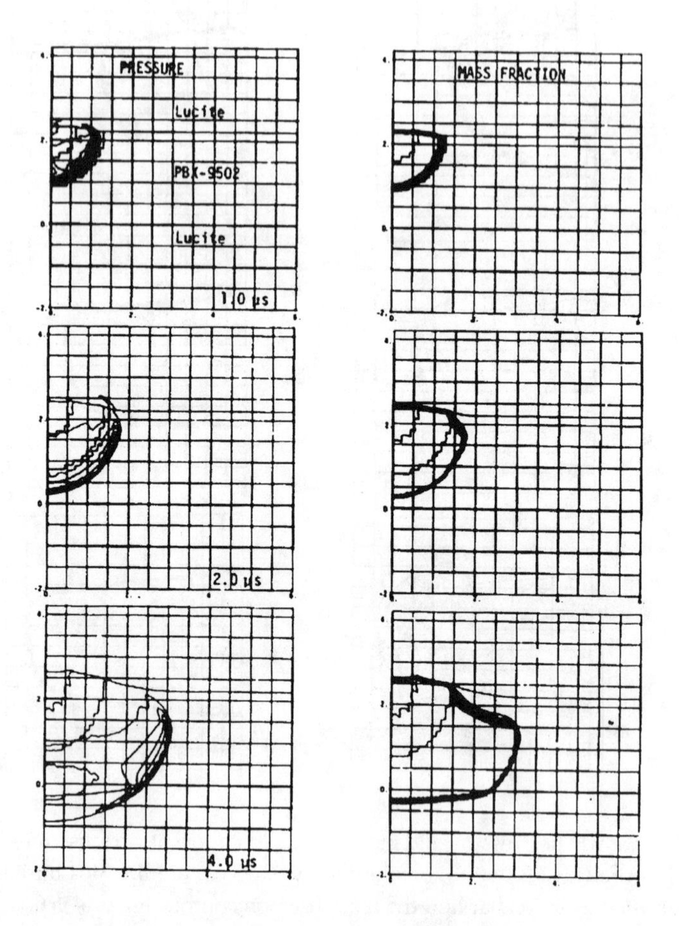

Figure 5.53 The pressure and mass fraction contours at various times for a hemispherical initiator of 0.635-cm-radius PBX-9407 surrounded by 0.635 cm of PBX-9404 initiating PBX-9502 (X0290). The pressure contour interval is 50 kbar and the mass fraction contour interval is 0.1.

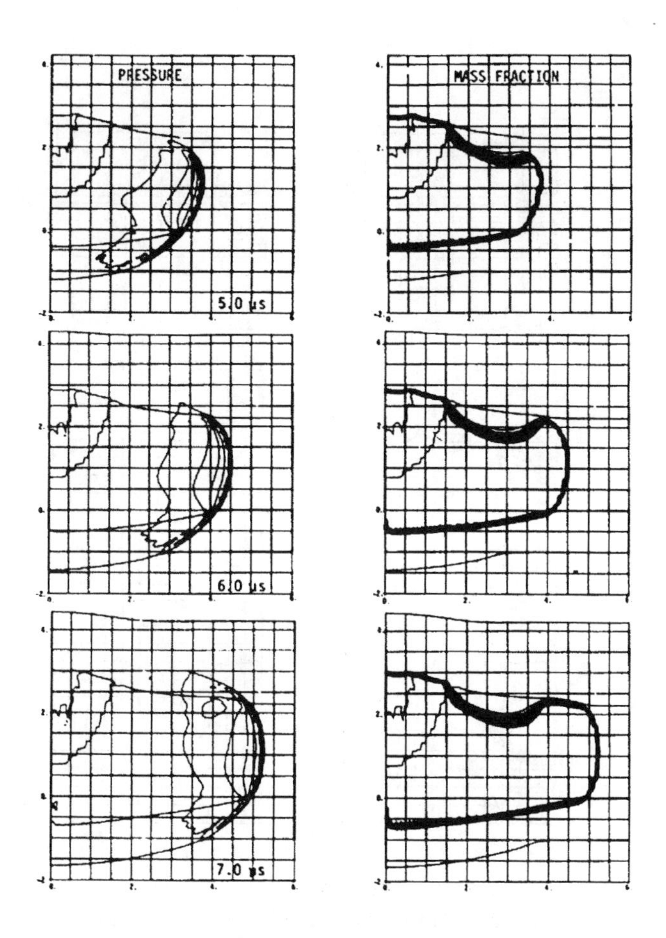

Figure 5.53 *(Continued)* The pressure and mass fraction contours at various times for a hemispherical initiator of 0.635-cm-radius PBX-9407 surrounded by 0.635 cm of PBX-9404 initiating PBX-9502 (X0290). The pressure contour interval is 50 kbar and the mass fraction contour interval is 0.1.

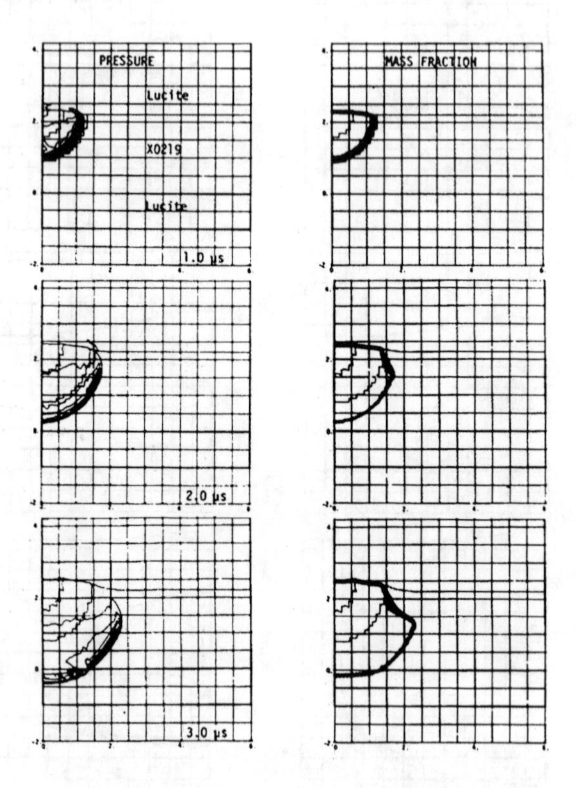

Figure 5.54 The pressure and mass fraction contours at various times for a hemispherical initiator of 0.635-cm-radius PBX-9407 surrounded by 0.635 cm of PBX-9404 failing to initiate X0219. The pressure contour interval is 50 kbar and the mass fraction contour interval is 0.1.

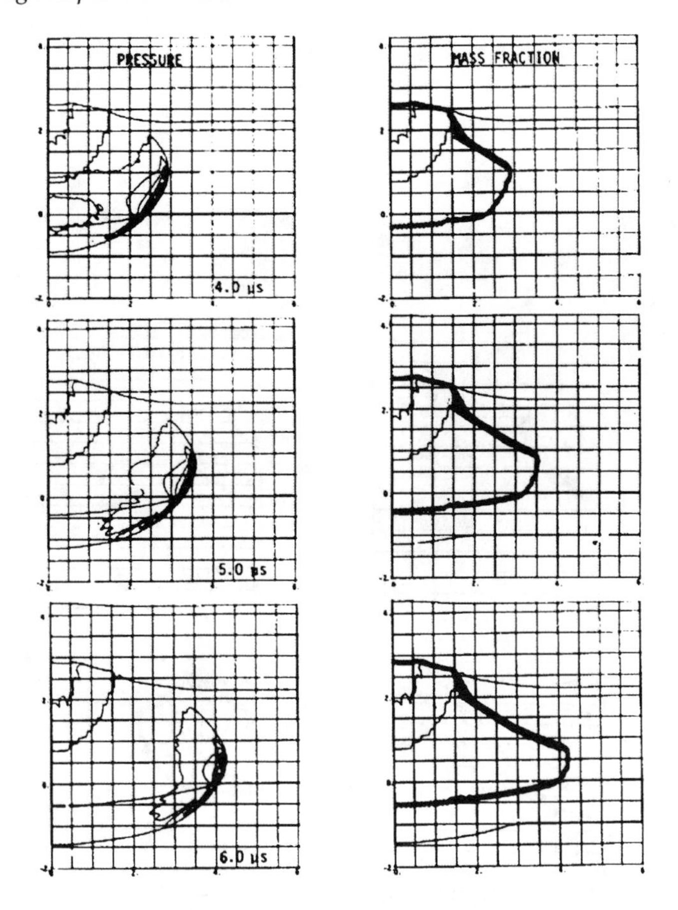

Figure 5.54 *(Continued)* The pressure and mass fraction contours at various times for a hemispherical initiator of 0.635-cm-radius PBX-9407 surrounded by 0.635 cm of PBX-9404 failing to initiate X0219. The pressure contour interval is 50 kbar and the mass fraction contour interval is 0.1.

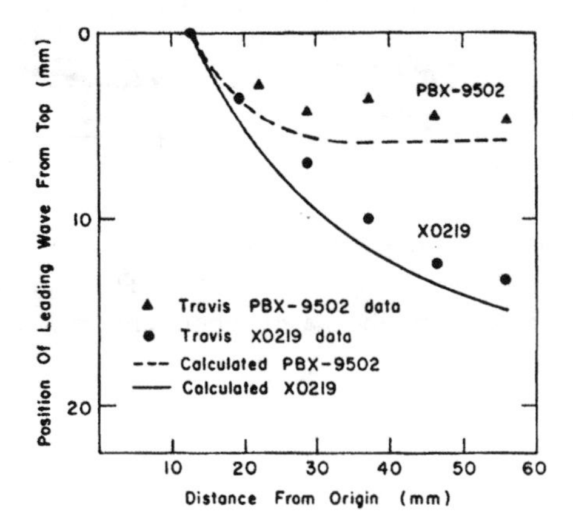

Figure 5.55 The experimental and calculated position of the leading wave from the top of the explosive block as a function of the distance of the leading front of the wave from the origin for the systems shown in Figures 5.53 and 5.54. The PBX-9502 is initiated by the detonator while the X0219 fails to be initiated by the detonator.

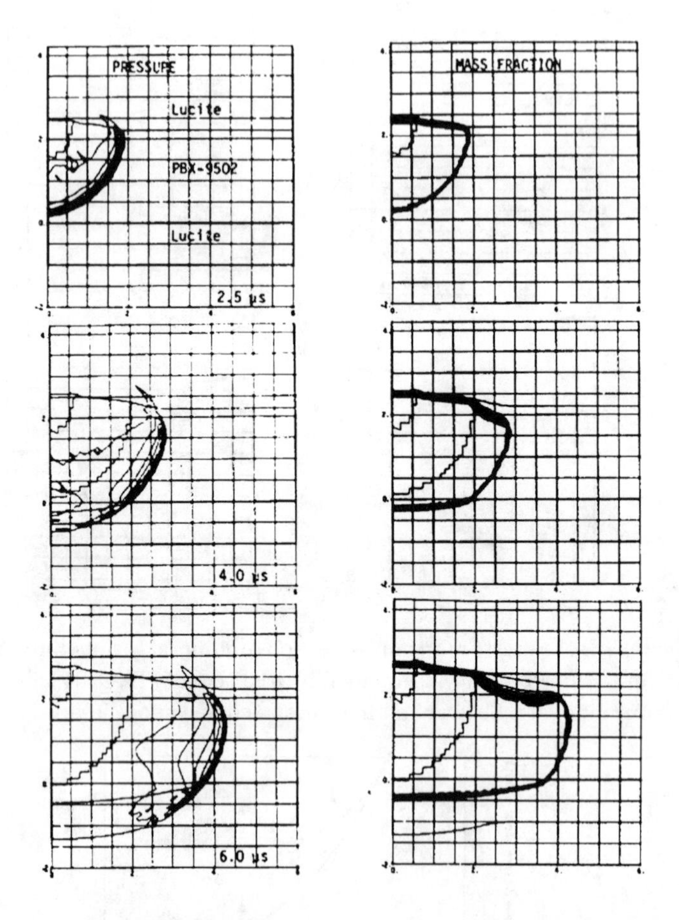

Figure 5.56 The pressure and mass fraction contours at various times for a hemispherical initiator of 0.635 cm TATB at 1.7 g/cc surrounded by 1.905 cm of TATB at 1.8 g/cc initiating PBX-9502. The pressure contour interval is 50 kbar and the mass fraction contour is 0.1.

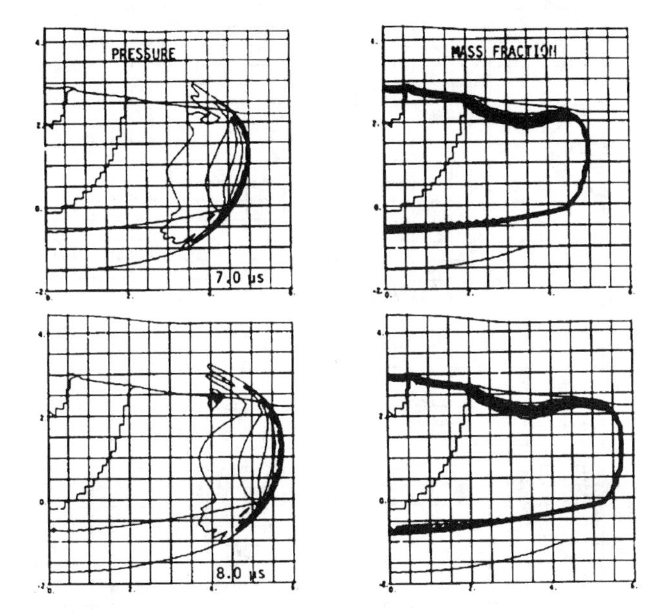

Figure 5.56 *(Continued)* The pressure and mass fraction contours at various times for a hemispherical initiator of 0.635 cm TATB at 1.7 g/cc surrounded by 1.905 cm of TATB at 1.8 g/cc initiating PBX-9502. The pressure contour interval is 50 kbar and the mass fraction contour is 0.1.

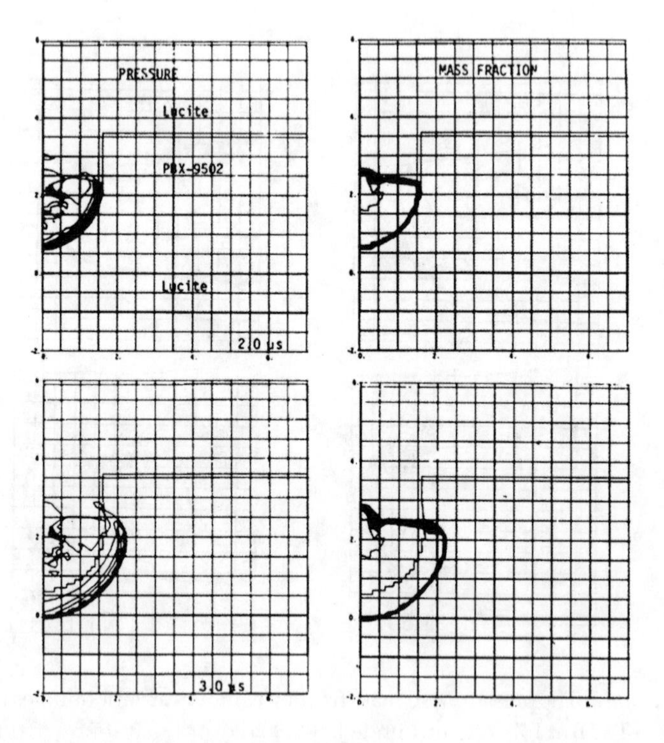

Figure 5.57 The pressure and mass fraction contours at various times for an embedded hemispherical initiator of 1.6-cm-radius X0351 initiating PBX-9502. The pressure contour interval is 50 kbar and the mass fraction contour is 0.1.

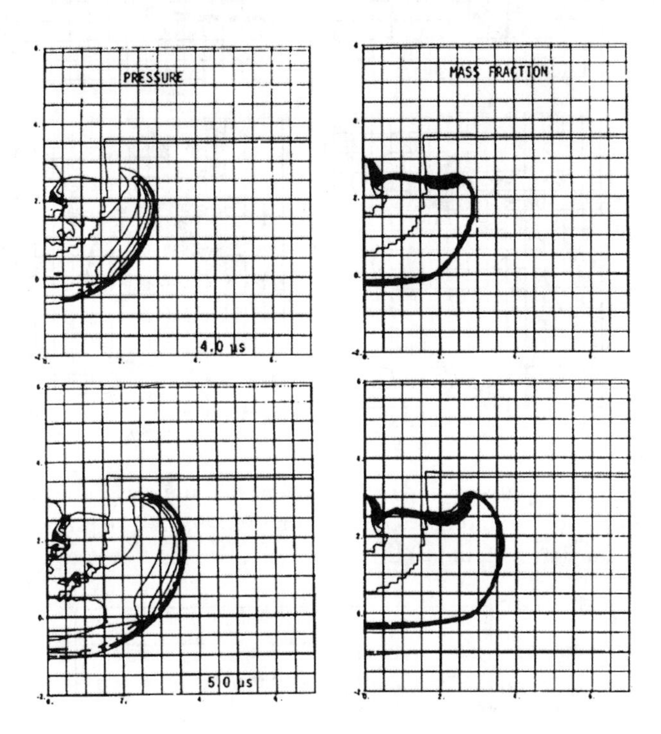

Figure 5.57 *(Continued)* The pressure and mass fraction contours at various times for an embedded hemispherical initiator of 1.6-cm-radius X0351 initiating PBX-9502. The pressure contour interval is 50 kbar and the mass fraction contour is 0.1.

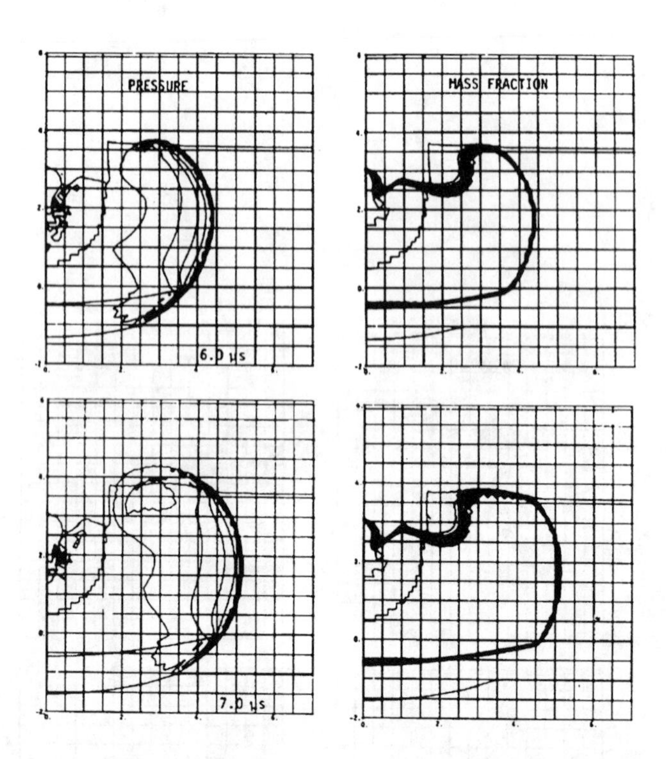

Figure 5.57 *(Continued)* The pressure and mass fraction contours at various times for an embedded hemispherical initiator of 1.6-cm-radius X0351 initiating PBX-9502. The pressure contour interval is 50 kbar and the mass fraction contour is 0.1.

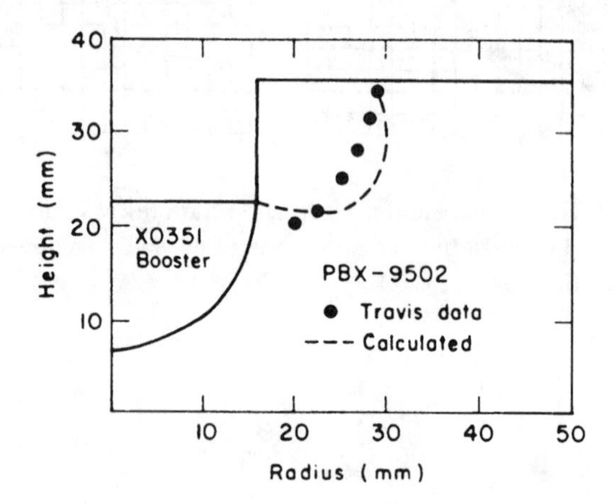

Figure 5.58 The calculated and experimental region of partially decomposed PBX-9502 when initiated by the X0351 initiator shown in Figure 5.57.

The initiation of propagating detonation in both sensitive and insensitive explosive by hemispherical initiators has been described numerically using the two-dimensional Lagrangian code 2DL and the Forest Fire rate. Large regions of partially decomposed explosive occur even when large initiators are used.

Since the initiators for insensitive high explosives must be very large to initiate propagating detonation, other methods have been studied to achieve the required high pressures of adequate duration. High pressures are achieved if two or more shock waves interact to form regular or Mach shock interactions.

The initiation of propagating detonation in PBX-9502 by double and triple wave interaction of shock waves formed by initiators that were too weak to initiate propagating detonation individually has been modeled.[34]

The use of multiple shock wave interactions to initiate propagating detonation has been studied experimentally by Goforth.[35] He observed that while double shock wave interactions were sometimes inadequate to initiate propagating detonation, triple shock wave interactions could be generated that initiated propagating detonation in insensitive high explosives.

The three-dimensional Eulerian hydrodynamic code, 3DE, described in Appendix D was used to model the interaction of shock waves in PBX-9502 formed by initiators that were too small to initiate propagating detonation individually. The Forest Fire rate used for PBX-9502 (X0290) is shown in Figure 4.6.

The geometry was two or three initiator cubes of 7 by 7 by 7 cells placed symmetrically in a PBX-9502 cube with continuum boundaries on its sides. The initiator cube centers were 1.6 cm apart and 1.09 cm from the cube bottoms. The PBX-9502 cube was described by 31 by 29 by 25 cells. The initiator cubes were initially decomposed PBX-9502 with a 2.5 g/cc initial density, which had an initial pressure of 245 kbar. This sent a diverging $\sim$ 100 kbar shock into the surrounding PBX-9502. The computational cell size was 0.114 cm, and the time step was 0.022 μsec.

The expected wave interactions are sketched in Figure 5.59. The sketch shows the waves,

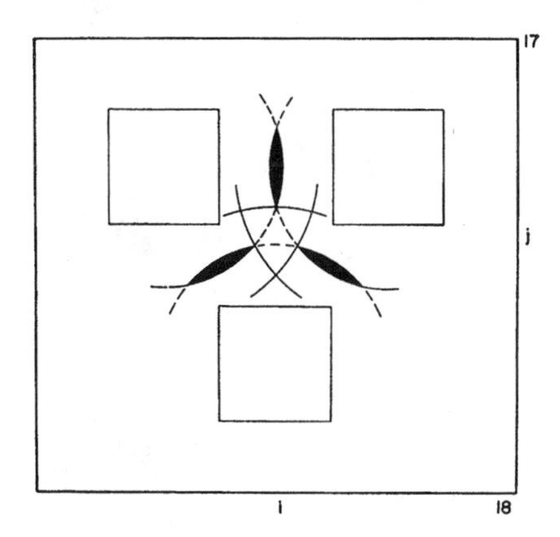

Figure 5.59 Sketch of expected wave interactions from three shock waves. The dashed lines show the shocks just after double interaction has occurred, and the dark regions show the double interaction locations. The solid lines and dotted region show the waves after triple wave interaction.

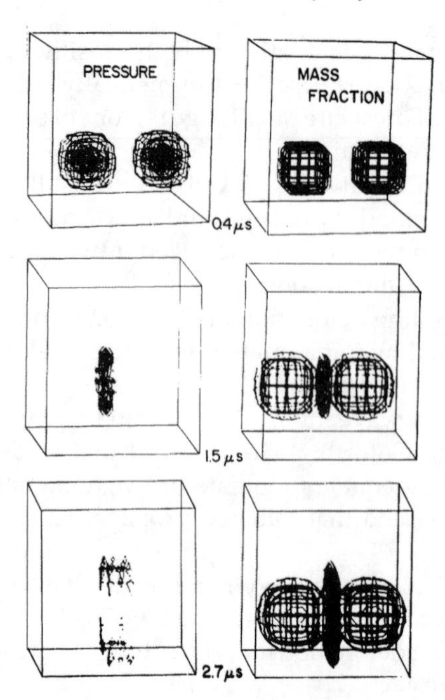

Figure 5.60 The calculated three-dimensional pressure and mass fraction contours for two initiators in PBX-9502. The pressure contours are shown for 200, 150, and 100 kbar at 0.4, 1.5, and 2.7 μsec. The mass fraction contours are 0.8 and 0.5. The two initiators fail to initiate propagating detonation in the PBX-9502.

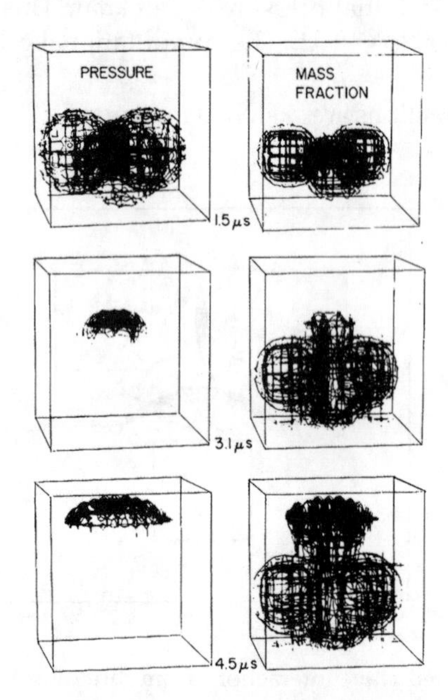

Figure 5.61 The calculated three-dimensional pressure and mass fraction contours for three initiators in PBX-9502. The pressure contours are shown for 200, 150, and 100 kbar at 1.5, 3.1, and 4.5 μsec. The mass fraction contours are 0.8 and 0.5. The triple shock interaction initiates propagating detonation in the PBX-9502.

just after double wave interaction as dashed lines, and the dark region shows the double wave interactions. The solid lines and dotted regions show the waves after triple wave interaction. The pressures from the double wave interaction in inert PBX-9502 are about 200 kbar, and those from the triple wave interaction are about 300 kbar.

The calculated three-dimensional pressure and mass fraction contours for two initiators are shown in Figure 5.60 and for three initiators in Figure 5.61. Although two initiators cause double wave interaction that results in considerable decomposition over a small area, propagating detonation does not result.

Three initiators fail to initiate propagating detonation at the double wave interaction points but do initiate at the triple interaction region. The higher triple wave interaction pressure results in a shorter run to detonation. The detonation can be maintained long enough to become a propagating, diverging detonation.

References

1. Terry R. Gibbs and Alphonse Popolato, "LASL Explosive Property Data," University of California Press, Berkeley, California (1980).
2. Charles L. Mader, James N. Johnson, and Sharon L. Crane, "Los Alamos Explosive Performance Data," University of California Press, Berkeley, California (1980).
3. D. Venable, "PHERMEX," Physics Today 17, 19-22 (1964).
4. Charles L. Mader, Timothy R. Neal, and Richard D. Dick, "LASL PHERMEX Data, Volume I", University of California Press, Berkeley, California (1980).
5. Charles L. Mader, "LASL PHERMEX Data, Volume II", University of California Press, Berkeley, California (1980).
6. Charles L. Mader, "LASL PHERMEX Data, Volume III", University of California Press, Berkeley, California (1980).
7. Stanley P. Marsh,"LASL Shock Hugoniot Data", University of California Press, Berkeley, California (1980).
8. Charles E. Morris, "Los Alamos Shock Wave Profile Data", University of California Press, Berkeley, California (1981).
9. B. Hayes and J. N. Fritz, "Measurement of Mass Motion in Detonation Products by an Axially Symmetric Electromagnetic Technique," Fifth International Symposium on Detonation, ACR-184, 447 (1970).
10. B. R. Breed, Charles L. Mader, and Douglas Venable, "A Technique for the Determination of Dynamic Tensile Strength Characteristics," Journal of Applied Physics 38, 3271 (1967).
11. Charles L. Mader, "One-Dimensional Elastic-Plastic Calculations for Aluminum," Los Alamos Scientific Laboratory report LA-3678 (1967).
12. Richard E. Swanson and Charles L. Mader, "One-Dimensional Elastic-Plastic Calculations Involving Strain-Hardening and Strain-Rate Effects for Aluminum," Los Alamos Scientific Laboratory report LA-5831 (1975).

13. H. M. Sternberg and W. A. Walker, "Artificial Viscosity Method Calculation of an Underwater Detonation," Fifth International Symposium on Detonation, ACR-184, 597 (1970).

14. Robert H. Cole, "Underwater Explosions," University Press, Princeton, N. J. (1948).

15. John W. Pritchett, "Incompressible Calculations of Underwater Explosion Phenomena," Second International Conference on Numerical Methods in Fluid Dynamics (1970).

16. B. D. Lambourn, "The Effect of Explosive Properties on the Shock Wave Parameters of Underwater Explosions," Sixth International Symposium on Detonation, ACR-221, 561 (1976).

17. Charles L. Mader, "Numerical Modeling of Water Waves," University of California Press (1988).

18. Louis C. Smith, "On Brisance, and a Plate-Denting Test for the Estimation of Detonation Pressure," Explosivstoffe No 5, 6 (1967).

19. George H. Pimbley, Allen L. Bowman, Wayne P. Fox, James D. Kershner, Charles L. Mader, and Manuel J. Urizar, "Investigating Explosive and Material Properties by Use of the Plate Dent Test," Los Alamos Scientific Laboratory report LA-8591-MS (1980).

20. W. Fickett and L. M. Scherr, "Numerical Calculation of the Cylinder Test," Los Alamos Scientific Laboratory report LA-5906 (1975).

21. Brigitta M. Dobratz, "Properties of Chemical Explosives and Explosive Simulants," Lawrence Livermore Laboratory report UCID-16557 (1974).

22. M. H. Rice,"Penetration of High Explosives by Inert Projectiles," Systems, Science and Software report SS-R-78-3512 (1977).

23. Garrett Birkhoff, Duncan P. MacDougall, Emerson M. Pugh, and Sir Geoffrey Taylor, "Explosives with Lined Cavities," Journal of Applied Physics, 19, 563-582 (1948).

24. Francis H. Harlow and William E. Pracht, "Formation and Penetration of High-Speed Collapse Jets," Physics of Fluids, 9, 1951-1959 (1966).

25. Wallace E. Johnson,"Three-dimensional Computations in Penetrator-Target Interactions," Ballistic Research Laboratory report BRL-CR-338 (1977).

26. Charles L. Mader, George H. Pimbley, and Allen L. Bowman, "Jet Penetration of Inerts and Explosives," Los Alamos National Laboratory report LA-9527-MS (1982).

27. Charles L. Mader and George H. Pimbley, "Jet Initiation and Penetration of Explosives," Journal of Energetic Materials, 1, 3 (1983).

28. M. Held, "Initiating of Explosives, a Multiple Problem of the Physics of Detonation," Explosivstoffe, 5, 98-113 (1968).

29. George H. Pimbley, Charles L. Mader, and Allen L. Bowman, "Plane Wave Generator Calculation," Los Alamos National Laboratory report LA-9119 (1982).

30. Charles L. Mader and Douglas Venable, "Mach Stems Formed by Colliding Cylindrical Detonation Waves," Los Alamos Scientific Laboratory report LA-7869 (1979).

31. Charles L. Mader and James D. Kershner, "Three-Dimensional Eulerian Calculations of Triple-Initiated PBX-9404," Los Alamos Scientific Laboratory report LA-8206 (1980).

32. Charles L. Mader and James D. Kershner, "Two- and Three-Dimensional Detonation Wave Interactions with a Copper Plate," Los Alamos National Laboratory report LA-8989 (1981).

33. Charles L. Mader, "Numerical Modeling of Insensitive High-Explosive Initiators," Los Alamos Scientific Laboratory report LA-8437-MS (1980).

34. Charles L. Mader and James D. Kershner, "Three-Dimensional Modeling of Triple-Wave Initiation of Insensitive Explosives," Los Alamos Scientific Laboratory report LA-8655 (1981).

35. James H. Goforth, "Safe-Stationary Detonation Train for Army Ordnance," Los Alamos National Laboratory report LA-7123-MS (1978).

appendix A

Numerical Solution of One-Dimensional Reactive Flow

The numerical technique and computer code called SIN have been used as a research and engineering tool for over forty years. The name was suggested by early users who found that the detailed treatments of equation of state and the amount of numerical resolution required to obtain resolved reactive flow required large amounts of computer time and money. They suggested that it was a violation of divine law (a SIN) to attempt to solve problems by such an expensive technique. Thus the name SIN was accepted by early users who could not imagine that the problems that took hours to solve in the 1950s would require only minutes in the 1970s or run on inexpensive personal computers in the 1990s. Also during the 1950s the finite difference technique of solving the fluid dynamic equations was very suspect to those accustomed to analytical or characteristic methods of solution. Such wickedness as using finite differences was certainly "a sin," and the detractors were at least in agreement as to the suitability of the name.

The first version of SIN was written during the late 1950s in machine language for the IBM 7090.[1] The second was written during the early 1960s in machine language for the IBM 7030 and was called Stretch SIN.[2] By the middle 1960s, a FORTRAN version was available for the CDC 6600 and called FORTRAN SIN.[3] Versions of the FORTRAN SIN code were used by many scientists and engineers on various computers throughout the world. As the science of detonation physics advanced, additions to FORTRAN SIN included the sharp-shock burning of the explosive, the heterogeneous shock initiation burn called Forest Fire described in Chapter 4, the build-up of detonation model described in Chapter 2, thermal conduction for multicomponent systems with different thermal properties, and the Kennedy melting law to determine when to turn off the elastic-plastic flow. Special versions of the SIN code also included the SESAME[4] equation of state tables and the Barnes[5] equation of state for very high pressures (> 10 Mbar) and temperatures (> 10,000^{o}K).

The availability of inexpensive but powerful personal computers in the 1980s and FORTRAN compilers for personal computers, permitted the SIN code to be run on personal computers. A user-friendly interface called USERSIN was developed which reduced the skill and time required to prepare the input file necessary to describe a problem for the SIN code. HOM equation of state data files and Forest Fire burn rate data files were created and are used by the USERSIN code.

The MCGRAPH package of graphics programs was developed for personal computers with CGA, EGA, and VGA graphics by Mader Consulting Co., and was included in the SIN code. The MCGRAPH graphics package also includes the capability of generating PCL graphics files for printing on black and white laser printers or on HP color printers. The executable personal computer code SIN with MCGRAPH graphics is included on the CD-ROM. The executable personal computer code USERSIN along with the data files EOSDATA and FFDATA are also on the CD-ROM.

A.1 The Flow Equations

The Nomenclature

D_{CJ}	The C-J detonation velocity
E	Total Energy (Mbar-cc/gram)
E^*	Activation Energy
I	Internal Energy (Mbar-cc/gram)
j	Net point of Lagrangian mesh
K	Viscosity Constant
M	Mass per Unit Length or Unit of Solid Angle
n	Time Cycle
P	Pressure (Mbar)
q	Viscosity
R	Eulerian Radius
r	Lagrangian Radius
Sx	Elastic Stress Deviator in x or r Direction
Sz	Elastic Stress Deviator in z Direction
T	Temperature ^{o}K
U	Particle Velocity (cm/μsec)
U_{CJ}	The C-J particle velocity
V	Specific Volume (cc/gram)
V_{CJ}	The C-J Volume of Detonation Products
W	Mass Fraction of Undecomposed Explosive
Y_o	Yield Strength (Mbar)
Z	Frequency Factor
α	1 for Slabs, 2 for Cylinders, 3 for Spheres
γ	γ-law gas constant
Δt	Time Increment in μ seconds
$(\lambda)_i$	Thermal Conductivity Coefficient for ith Component (Mbar-cc/deg/cm/μsec)
μ	Shear Modulus
ρ	Density $= 1/V$
ρ_o	Initial Density $= 1/V_o$
$(\rho_o)_i$	Initial Density of ith Component (gram/cc)

The Lagrangian conservation equations in one dimension for slabs, cylinders, and spheres are:

Conservation of momentum,

$$\frac{\partial U}{\partial t} = -R^{\alpha-1}\frac{\partial \sigma}{\partial M} - (\alpha - 1)\frac{V\phi}{R} \quad ,$$

Conservation of mass,

$$V = R^{\alpha-1}\frac{\partial R}{\partial M} \quad ,$$

and conservation of energy with heat conduction term,

$$\frac{\partial E}{\partial t} = -\frac{\partial \sigma U R^{\alpha-1}}{\partial M} + \lambda\frac{\partial}{\partial M}\left(R^{\alpha-1}\frac{\partial T}{\partial R}\right) \quad ,$$

where

$$E = I + 0.5U^2 \quad .$$

$$\frac{\partial R}{\partial t} = U \quad ,$$

and

$$dM = \rho_o r^{\alpha-1}dr = \rho R^{\alpha-1}dR \quad ,$$

where dM is element of mass per unit of solid angle ($\alpha = 2$, or 3) or of surface ($\alpha = 1$). σ is the sum of the viscous pressure, the equation of state pressure (hydrostatic pressure), and the stress deviator, Sx. $\phi = 2Sx + Sz$, which for spherical geometry is $3/2Sx$ since $Sz = -\frac{1}{2}Sx$.

$$\frac{\partial Sx}{\partial t} = 2\mu\left(-\frac{\partial U}{\partial R} + \frac{1}{3V}\frac{\partial V}{\partial t}\right) \quad .$$

$$\frac{\partial S_z}{\partial t} = \frac{2}{3}\left(\frac{\mu}{V}\frac{\partial V}{\partial t}\right) \quad .$$

The convention used is that the stress deviators have the same sign as pressure; that is, positive in compression and negative in tension. This convention is the reverse of that used by many workers in the field.

A.2 The Difference Equations

The pressure, temperature, energy, specific volume, and mass fraction of the cells are considered to be located at the centers of mass of the elements, and the particle velocity is considered to be located at the boundaries between the cells.

A. The initial Conditions

$$R_{j+\frac{1}{2}} = \sum_{i=1}^{j} (\Delta R)_i \quad ,$$

where $j = 1, 2, \ldots\ldots$, and the ith component is located between j and $j + k$ where k is the number of cells for each component.

$$M_j = (\rho_o)_i \left[(R_{j-\frac{1}{2}} + R_{j+\frac{1}{2}}) 0.5 \right]^{\alpha-1} (\Delta R)_i \quad .$$

B. The Conservation Equations for Time Advancement

1. *Velocity*

$$U^{n+\frac{1}{2}}_{j+\frac{1}{2}} = U^{n-\frac{1}{2}}_{j+\frac{1}{2}} + \frac{(\Delta t)\left(R^n_{j+\frac{1}{2}}\right)^{\alpha-1}}{0.5(M_j + M_{j+1})} \left[\left(P^n_j - P^n_{j+1}\right) + \left(q^n_j - q^n_{j+1}\right) \right]$$

$$- \frac{(\alpha - 1)}{2} \frac{(\phi)\left(V^n_j + V^n_{j+1}\right)(\Delta t)}{R^n_{j+\frac{1}{2}}} \quad ,$$

Note: $P = \sigma = P_{hydrostat} + Sx$

where

$$\phi = \frac{3(Sx^n_j + Sx^n_{j+1})}{4} \quad for\ \alpha = 3 \quad ,$$

and

$$\phi = \frac{1}{2}\left[2(Sx^n_j + Sx^n_{j+1}) + (Sz^n_j + Sz^n_{j+1}) \right] \; for\ \alpha = 2 \quad .$$

2. *Radius*

$$R^{n+1}_{j+\frac{1}{2}} = R^n_{j+\frac{1}{2}} + U^{n+\frac{1}{2}}_{j+\frac{1}{2}} (\Delta t) \quad . \tag{A.1}$$

3. *Volume*

$$V_j^{n+1} = \left(\frac{R_{j-\frac{1}{2}}^{n+1} + R_{j+\frac{1}{2}}^{n+1}}{2} \right)^{\alpha-1} \frac{(R_{j+\frac{1}{2}}^{n+1} - R_{j-\frac{1}{2}}^{n+1})}{M_j} \quad . \tag{A.2}$$

4. *Energy*

$$I_j^{n+1} = I_j^n + \frac{(\Delta t)}{M_j} \left[\frac{M_j P_{j-1}^n + M_{j-1} P_j^n}{M_j + M_{j-1}} + 0.5 \left(q_j^n + q_{j-1}^n \right) \right] U_{j-\frac{1}{2}}^{n+\frac{1}{2}} \left(R_{j-\frac{1}{2}}^{n+1} \right)^{\alpha-1}$$

$$- \frac{(\Delta t)}{M_j} \left[\frac{M_{j+1} P_j^n + M_j P_{j+1}^n}{M_j + M_{j+1}} + 0.5 \left(q_j^n + q_{j+1}^n \right) \right] U_{j+\frac{1}{2}}^{n+\frac{1}{2}} \left(R_{j+\frac{1}{2}}^{n+1} \right)^{\alpha-1}$$

$$+ \frac{1}{8} \left[\left(U_{j+\frac{1}{2}}^{n-\frac{1}{2}} + U_{j-\frac{1}{2}}^{n-\frac{1}{2}} \right)^2 - \left(U_{j+\frac{1}{2}}^{n+\frac{1}{2}} + U_{j-\frac{1}{2}}^{n+\frac{1}{2}} \right)^2 \right] \quad . \tag{A.3}$$

5. *Heat Conduction*

Add to I_j^{n+1}

$$\frac{\Delta t}{M_j} \left[\frac{\left(R_{j+\frac{1}{2}}^{n+1} \right)^{\alpha-1} \left(T_{j+1}^n - T_j^n \right) \left(\lambda_{j+1} + \lambda_j \right)}{\left(R_{j+1\frac{1}{2}}^{n+1} - R_{j-\frac{1}{2}}^{n+1} \right)} \right]$$

$$- \frac{\Delta t}{M_j} \left[\frac{\left(\lambda_j + \lambda_{j-1} \right) \left(R_{j-\frac{1}{2}}^{n+1} \right)^{\alpha-1} \left(T_j^n - T_{j-1}^n \right)}{\left(R_{j+\frac{1}{2}}^{n+1} - R_{j-1\frac{1}{2}}^{n+1} \right)} \right] \quad .$$

C. *Viscosity*

1. *PIC Viscosity*

$$q_j^{n+1} = \frac{K}{V_j^{n+1}} (0.5) \left(U_{j-\frac{1}{2}}^{n+\frac{1}{2}} + U_{j+\frac{1}{2}}^{n+\frac{1}{2}} \right) \left(U_{j-\frac{1}{2}}^{n+\frac{1}{2}} - U_{j+\frac{1}{2}}^{n+\frac{1}{2}} \right) \quad ,$$

if $\left(U_{j-\frac{1}{2}}^{n+\frac{1}{2}} - U_{j+\frac{1}{2}}^{n+\frac{1}{2}} \right)$ is positive, otherwise $q_j^{n+1} = 0$. The absolute value of q is used.

2. *Landshoff Viscosity*

$$q_j^{n+1} = \frac{K}{V_j^{n+1}} \left(U_{j-\frac{1}{2}}^{n+\frac{1}{2}} - U_{j+\frac{1}{2}}^{n+\frac{1}{2}} \right) \quad .$$

Restrictive conditions are the same as for the PIC form.

3. *Real Viscosity*

$$q_j^{n+1} = 1.333 \frac{K}{V_j^{n+1}} \frac{\left(U_{j-\frac{1}{2}}^{n+\frac{1}{2}} - U_{j+\frac{1}{2}}^{n+\frac{1}{2}} \right)}{M_j} \quad ,$$

where K is the "coefficient of viscosity". This form is appropriate only for slab geometry.

A.3 Burn Techniques

A. *Arrhenius Burn*

$$W_j^{n+1} = W_j^n \; - \; \Delta t \; Z \; W_j^n \; e^{-E^*/R_g T_j^n} \quad ,$$

where $1 \geq W \geq 0$.

B. *C-J Volume Burn*

The C-J volume burn requires sufficient viscosity to smear the burn over several cells. It assumes that W varies linearly with V from V_o to V_{CJ}. The C-J volume burn technique described below is suitable for diverging, converging, or plane geometries. Since it requires an equation of state only for the detonation products, it is often used to burn explosives in codes that do not have a HOM-like equation of state option.

$$W_j^{n+1} = 1 \; - \; \frac{V_o \; - \; V_j^{n+1}}{V_o \; - \; V_{CJ}} \quad ,$$

where $1 \geq W_j^{n+1} \geq 0$, and

$$W_j^n \; \geq \; W_j^{n+1} \, .$$

$$W_j^{n+1} \; = 0 \text{ if } W_j^n \; < \; W_j^{n+1} \text{ and } W_j^{n+1} \; < \; 0.9.$$

$$P_j^{n+1} \; = \; \left(1 \; - \; W_j^{n+1} \right) \, P',$$

where P' is pressure of detonation products at V, I, and $W = 0$, if $W_j^{n+1} < 0.99$; otherwise $P_j^{n+1} = P_o$. The burn is best started by an applied piston initially moving at C-J particle velocity and then decreased to escape velocity. The burn may also be started by a small region of initially burned explosive.

C. Gamma Law Taylor Wave Burn

The explosive is burned before the first time interval by assuming it to be a gamma-law explosive that has been detonated with a rear boundary of constant velocity.

Knowing ρ_o, γ, and D_{CJ},

$$U_{CJ} = \frac{D_{CJ}}{\gamma + 1} ,$$

$$P_{CJ} = \frac{\rho_o D_{CJ}^2}{\gamma + 1} ,$$

$$V_{CJ} = \left(\frac{\gamma}{\gamma + 1}\right)\left(\frac{1}{\rho_o}\right) ,$$

$$C_{CJ} = D_{CJ} - U_{CJ} ,$$

$$k = \frac{\gamma - 1}{\gamma + 1} ,$$

$$L = U_{CJ} - \frac{2}{\gamma - 1} C_{CJ} ,$$

$$t = \frac{R_n - R_o}{D_{CJ}} ,$$

where $R_n - R_o$ is the thickness of the explosive.

For each cell with a radius of $R_{j-\frac{1}{2}}$, the following are computed:

$$Y = \frac{\left(R_n - R_{j-\frac{1}{2}}\right)}{t} ,$$

$$U_{j+\frac{1}{2}} = \frac{2}{\gamma + 1} Y + kL ,$$

$$C_j = \frac{\gamma - 1}{2}\left(U_{j+\frac{1}{2}} - L\right) ,$$

$$P_j = P_{CJ}(C_j/C_{CJ})^{2\gamma/(\gamma-1)} ,$$

$$V_j = \left[\frac{P_{CJ}(V_{CJ})^\gamma}{P_j}\right]^{1/\gamma} ,$$

$$I_j = \frac{P_j V_j}{\gamma - 1} - \frac{P_{CJ} V_{CJ}}{\gamma - 1} + \frac{P_{CJ}}{2}(V_o - V_{CJ}) ,$$

$$R_{j+\frac{1}{2}} = R_{j-\frac{1}{2}} + M_j V_j .$$

With the new $R_{j+\frac{1}{2}}$ set equal to $R_{j-\frac{1}{2}}$, the calculation is repeated to compute Y, etc. When $U_{j+\frac{1}{2}}$ is equal to the applied piston velocity, the rest of the explosive cells are made identical to the last cell calculated.

D. The Sharp Shock Burn

The sharp shock burn option is suitable for plane and converging geometry only. It cannot be used for diverging detonations.

A sharp shock burn completely burns a cell of explosive after compressing it by moving the right cell boundary at C-J particle velocity for an interval of time determined by the detonation velocity.

The cell energy is set equal to the Hugoniot energy at the compressed density. In slab geometry, the process continues through the explosive at the input detonation velocity. For converging geometry, the increased density resulting from convergence also results in an increased pressure from which a new detonation and particle velocity are calculated. These new velocities are used to compress the next cell.

The new detonation velocity, D_{CJ}, particle velocity, U_{CJ}, and time step, Δt, are calculated from

$$D_{CJ} = V_o[P_{CJ}/(V_o - V_{CJ})]^{0.5} ,$$

$$U_{CJ} = [P_{CJ}(V_o - V_{CJ})]^{0.5} ,$$

$$\Delta t = 0.25(\Delta X/D_{CJ}) ,$$

where P_{CJ} = Hugoniot P, V_{CJ} = V of cell just burned.

The time steps used in the calculation are one-fourth of the time required to compress the cell to C-J density; four time steps occur before the cell is burned. When the shock arrives at a new explosive, the new input C-J detonation and particle velocities are used to compress the new explosive.

The advantage of a sharp shock burn is that one can obtain excellent Taylor waves using a small number of cells to describe multiple layers of explosives in plane or converging geometry. The disadvantages are that it requires more information and more artificial constraints than do the C-J volume or Arrhenius burn techniques. The sharp shock burn will give incorrect results for systems in diverging geometry and for systems that are overdriven or significantly underdriven.

E. The Forest Fire Heterogeneous Explosive Burn

As described in Chapter 4 the Forest Fire rate describes the rate of burning of shock initiated heterogeneous explosives.

$$W_j^{n+1} = W_j^n \left(1 - \Delta t \, e^{A+BP+CP^2....XP^n}\right) ,$$

where

$$-(1/W)(dW/dt) = e^{A+BP+CP^2....XP^n} ,$$

from $P_{minimum}$ to P_{CJ}

$$(1/W)(dW/dt) = 0 \text{ if } P < P_{minimum},$$

$$W_j^{n+1} = 0 \text{ if } P > P_{CJ} \text{ or } W_j^n < 0.05.$$

If Forest Fire is used to calculate the detonation wave, the viscosity must be adjusted to smear the burn over several cells. The peak detonation pressure should not exceed the C-J pressure.

A.4 Elastic Plastic Model

A. The Elastic Stress Deviators

$$Sx_j^{n+1} = Sx_j^n + 2\mu \left[-\left(\frac{U_{j+\frac{1}{2}}^{n+\frac{1}{2}} - U_{j-\frac{1}{2}}^{n+\frac{1}{2}}}{R_{j+\frac{1}{2}}^{n+1} - R_{j-\frac{1}{2}}^{n+1}} \right) (\Delta t) + \frac{2}{3} \left(\frac{V_j^{n+1} - V_j^n}{V_j^{n+1} + V_j^n} \right) \right] .$$

$$Sz_j^{n+1} = Sz_j^n + \frac{4}{3} \mu \left(\frac{V_j^{n+1} - V_j^n}{V_j^{n+1} + V_j^n} \right) .$$

B. The Equation of State and Yield Calculation

From V_j^{n+1}, I_j^{n+1}, W_j^{n+1}, the P_j^{n+1} is calculated using the HOM equation of state. Since the experimentally observed single shock Hugoniot is a measure of the elastic-plastic components, the hydrostat (pure plastic component) must be calculated from the experimental Hugoniot. To describe the low-pressure hydrostat the following model is used:

$$P_{hydrostat} = P_{Hugoniot} - (\tfrac{2}{3}\,Y_o)(P_{Hugoniot}\ /\ PLAP\ ).$$

$$(P_{Hugoniot}\ /\ PLAP\ ) \leq 1.0\ .$$

PLAP is the pressure at which the Hugoniot stress minus the hydrostat pressure is equal to $\frac{2}{3}\,Y_o$. For Aluminum[6] the value used was 0.050 Mbar.

If $P_j^{n+1} > P_o$ (initial pressure),

$$(P_{hyd})_j^{n+1} = (P_{hydrostat})_j^{n+1} = P_j^{n+1} + (\tfrac{2}{3}\,Y_o)(P_j^{n+1}/0.05),$$

where

$$(P_j^{n+1}\ /\ 0.05) \leq 1.0\ .$$

The 0.050 value for PLAP is input, and varies with the material. If $P_j^{n+1} < P_o$, then

$$(P_{hyd})_j^{n+1} = P_j^{n+1}\ .$$

The Hooke's law-Von Mises yield model is used as follows. For $\alpha = 1$ or 3 if

$$\left|Sx_j^{n+1}\right| \geq \left|\tfrac{2}{3}\,Y_o\right|,\ \text{then}\ P_j^{n+1} = (P_{hyd})_j^{n+1} \pm \tfrac{2}{3}\,Y_o,$$

where the sign of $(\frac{2}{3}\,Y_o)$ is identical to the sign of Sx_j^{n+1}. Or if

$$\left|Sx_j^{n+1}\right| < \left|\tfrac{2}{3}\,Y_o\right|,$$

then

$$P_j^{n+1} = (P_{hyd})_j^{n+1} + Sx_j^{n+1}\ .$$

For $\alpha = 2$

$$f = 2\left[\left(Sx_j^{n+1}\right)^2 + Sx_j^{n+1}Sz_j^{n+1} + \left(Sz_j^{n+1}\right)^2\right]\ .$$

Then if

$$f < \tfrac{2}{3}(Y_o)^2\ ,$$

$$P_j^{n+1} = (P_{hyd})_j^{n+1} + Sx_j^{n+1}$$

and Sz_j^{n+1} remains unchanged. Or if

$$f > \tfrac{2}{3}(Y_o)^2\ ,$$

$$P_j^{n+1} = (P_{hyd})_j^{n+1} + \left(\sqrt{\frac{\frac{2}{3}\,(Y_o)^2}{f}}\right)(Sx_j^{n+1})\ ,$$

$$Sz_j^{n+1} = Sz_j^{n+1}\left(\sqrt{\frac{\frac{2}{3}(Y_o)^2}{f}}\right) .$$

If melting may occur, the yield is set to zero if

$$T_j > T_m + mc\big[(V_o - V_j)/V_o\big] ,$$

where T_m is the normal melting temperature and mc is the Kennedy melt constant ($\sim$ 5000 for many metals, including Aluminum, Copper, and Iron).

A.5 Total Energy of Components

$$Kinetic\ Energy = \sum_{1}^{j} \frac{\omega M_j\left(U_{j+\frac{1}{2}}^{n+\frac{1}{2}}\right)^2}{2} ,$$

$$Internal\ Energy = \sum_{1}^{j} \omega M_j\left(I_j^{n+1}\right) ,$$

where $\omega = 4\pi$ for spheres, 2π for cylinders, and 1 for slabs. The units of energy are Mbar-cc.

A.6 The Boundary Conditions

A. An Applied Piston on Right Boundary

$U_{j+\frac{1}{2}}^n = U_{piston}$ where the outside boundary is at $j+\frac{1}{2}$, if $W_{j-3}^n > 0.5$; otherwise, $U_{j+\frac{1}{2}}^n =$ final U_{piston} or $U_{piston} = A + (B)(\text{Time})$. For both,

$$P_{j+1}^n = P_j^n \text{ and } q_{j+1}^n = q_j^n .$$

B. An Applied Piston on Left Boundary

$U_{\frac{1}{2}}^n = U_{+\frac{1}{2}}^{n+1} = U_{piston}$ where the boundary is at $j = \frac{1}{2}$, if $W_{j+3}^n > 0.5$; otherwise, $U_{\frac{1}{2}}^n =$ final U_{piston} or $U_{piston} = A + (B)(\text{Time})$. For both,

$$P_{j=0}^n = P_1^n \text{ and } q_{j=0}^n = q_1^n .$$

$R_{\frac{1}{2}}$ is computed using Equation (A.1).

C. A Steady-State Reaction Zone Piston on Left Boundary

The steady-state reaction zone piston is computed by iteration for a given detonation velocity by using the amount of reaction that occurred in a cell near the piston to determine the proper particle velocity of the piston.

For a W, V is iterated on using linear feedback (described in Appendix E) by calculating

$$I = I_o + \tfrac{1}{2}(\rho_o)^2(D_{CJ})^2(V_o - V)^2$$

$$P_r = P_o + (\rho_o)^2(D_{CJ})^2(V_o - V) \quad ,$$

where P_r is the Rayleigh line pressure.

Using the HOM equation of state, P is calculated for V, I, and W. The iteration continues until $P_r - P \leq 1 \times 10^{-5}$. The final U_{piston} is calculated from

$$U = \sqrt{(P - P_o)(V_o - V)}$$

D. A Right Free-Surface Boundary

$P^n_{j+1} = 0$ where the outside boundary is at $j + \frac{1}{2}$.

$$U^n_{j+1\frac{1}{2}} = U^n_{j+\frac{1}{2}} \quad .$$

E. A Left Free-Surface Boundary

$$P^n_1 = 0.$$

$P^n_{j=0} = -P^n_i$ where the boundary is at $j = \frac{1}{2}$.

$$U^n_{\frac{1}{2}} = -U^n_{1\frac{1}{2}} \quad .$$

$$q^n_{j=0} = q^n_1 \quad .$$

F. A Right Continuum Boundary

$P^n_{j+1} = P^n_j$ where the outside boundary is at $j + \frac{1}{2}$. $U^n_{j+1\frac{1}{2}} = U^n_{j+\frac{1}{2}}$.

G. A Left Continuum Boundary

$P^n_{j=0} = P^n_1$ where the outside boundary is at $j = \frac{1}{2}$. $q^n_{j=0} = q^n_1$. $U^n_{\frac{1}{2}} = U^n_{1\frac{1}{2}}$ unless

$\alpha = 2$ or 3, when $U^n_{\frac{1}{2}} = -U^n_{1\frac{1}{2}}$.

$$R^n_{\frac{1}{2}} = 2R^n_{1\frac{1}{2}} - R^n_{2\frac{1}{2}} \quad .$$

A.7 The HOM Equation of State

HOM calculates the pressure and temperature, given the internal energy, specific volume, and mass fraction of the solid for solids, gases, and mixtures.

The Nomenclature

$C,\ S$	Coefficients to the linear fit $U_S = C + S\,U_P$
$C1,\ S1$	Second set of Coefficients $U_S = C1 + S1\,U_P$ for Volumes less than MINV or VSW
C_v	Heat Capacity of Condensed Component (cal/g/deg)
C'_v	Heat Capacity of Gaseous Component (cal/g/deg)
I	Total Internal Energy (Mbar-cc/g)
P	Pressure (Mbar)
SPA	Spalling Constant
$SPALLP$	Interface Spalling Pressure
$SPMIN$	Minimum Pressure for Spalling
T	Temperature (^{o}K)
USP	Ultimate Spall Pressure
U_P	Particle Velocity (cm/μsec)
U_S	Shock Velocity (cm/μsec)
V	Total Specific Volume (cc/gram)
V_o	Initial Specific Volume of Condensed Explosive
W	Mass Fraction of Undecomposed Explosive
α	Linear Coefficient of Thermal Expansion

Subscripts

g	Gaseous Component
H	Hugoniot
i	Isentrope
s	Condensed Component

A.8 HOM for Condensed Components

The mass fraction, W, is 1; the internal energy, I, is I_s; and the specific volume, V, is V_s. For volumes less than V_o, the experimental shock Hugoniot data are expressed as linear fits of the shock and particle velocities. The shock Hugoniot temperatures are computed using the Walsh and Christian[7] technique and are fit to a fourth-degree polynomial in log of the volume as described in Appendix E.

$$U_S = C + S\,U_P \ . \tag{A.4}$$

$$P_H = [C^2(V_o - V_s)] \,/\, [V_o - S(V_o - V_s)]^2 \ .$$

$$ln\ T_H = F_s + G_s\, ln\ V_s + H_s\,(ln\ V_s)^2 + I_s\,(ln V_s)^3 + J_s\,(ln\ V_s)^4 \ .$$

$$I_H = (\frac{1}{2})\ P_H\,(V_o - V_s) \ .$$

$$P_s = (\gamma_s/V_s)\,(I_s - I_H) + P_H, \text{ where } \gamma_s = V(\partial P/\partial E)_V \ . \tag{A.5}$$

$$T_s = T_H + [(I_s - I_H)(23,890)/C_v] \quad . \qquad (A.6)$$

Two sets of C and S coefficients are sometimes necessary to fit the experimental Hugoniot data. For $V_s < MINV$, the fit $U_S = C1 + S1\,U_P$ is used. This may be used to describe phase changes such as exhibited by iron.[8] For volumes greater than V_o, the Grüneisen equation of state was used along with the $P = 0$ line as the standard curve.

$$P_s = \left[I_s - \frac{C_v}{(3)(23,890)(\alpha)} \left(\frac{V_s}{V_o} - 1 \right) \right] \frac{\gamma_s}{V_s} \quad .$$

$$T_s = \frac{(I_s)(23,890)}{C_v} + T_o \quad .$$

Gradient spalling does not occur if $SPA < 0.0001$. If $P_s \leq -SPA\sqrt{\Delta P/\Delta X}$ where $\Delta P/\Delta X$ is the tension rate, and $P_s \leq -SPMIN$ (5×10^{-3}), the P_s is set equal to $SPALLP$ and the spall indicator is set. Spalling does not occur if neither of the above conditions is satisfied. The gradient spall model is described in Reference 9. The spalling option is suitable only for problems with approximately constant tension gradients and many mesh points.

A.9 HOM for Gas Components

The mass fraction, W, is 0; the internal energy, I, is I_g; and the specific volume, V, is V_g. The pressure, volume, temperature, and energy values of the detonation products are computed using the BKW code described in Appendix E. They are fitted by the method of least squares to Equations (A.7)-(A.9). A gamma-law gas also may be fitted to these equations using the GLAW code included on the CD-ROM.

$$ln\ P_i = A + B\ ln\ V_g + C(ln\ V_g)^2 + D(ln\ V_g)^3 + E(ln\ V_g)^4 \quad . \qquad (A.7)$$

$$ln\ I_i = K + L\ ln\ P_i + M(ln\ P_i)^2 + N(ln\ P_i)^3 + O(ln\ P_i)^4 \quad . \qquad (A.8)$$

$I_i = I_i - Z$ (where Z is a constant used to change the standard state to be consistent with the solid explosive standard state. If the states are the same, Z is used to keep I positive when making a fit.)

$$ln\ T_i = Q + R\ ln\ V_g + S(ln\ V_g)^2 + T(ln\ V_g)^3 + U(ln\ V_g)^4 \quad . \qquad (A.9)$$

$$-(1/\beta) = R + 2\,S(ln\ V_g) + 3T(ln\ V_g)^2 + 4U(ln\ V_g)^3 \quad .$$

$$P_g = (1/\beta V_g)(I_g - I_i) + P_i \quad . \qquad (A.10)$$

$$T_g = T_i + [(I_g - I_i)(23,890)/C_v'] \ . \tag{A.11}$$

Mixture of Condensed and Gaseous Components

$$(0 < W < 1) \ .$$

$$V = WV_s + (1 - W)V_g \ .$$

$$I = WI_s + (1 - W)I_g \ .$$

$$P = P_s = P_g \ .$$

$$T = T_s = T_g \ .$$

Equation (A.6) is multiplied by (W/C_v) and Equation (A.11) by $(1 - W)C_v'$ and the resulting equations are added. T is substituted for T_s and T_g and I for $WI_s + (1 - W)I_g$ to get

$$T = \frac{23,890}{C_v W + C_v'(1 - W)}[I - (W\ I_H + I_i(1 - W))]$$

$$+ \frac{1}{C_v W + C_v'(1 - W)}\left[(T_H\ C_v\ W + T_i C_v'(1 - W))\right] . \tag{A.12}$$

Equation (A.5) is equated to (A.10) and substituted using (A.12) to get

$$\left(\frac{\gamma_s C_v}{V_s} - \frac{C_v'}{\beta V_g}\right)\left(\frac{1}{C_v W + C_v'(1 - W)}\right)[I - (W\ I_H + I_i(1 - W))]$$

$$+\left(\frac{\gamma_s C_v}{V_s} - \frac{C_v'}{\beta V_g}\right)\left(\frac{1}{C_v W + C_v'(1 - W)}\right)\frac{1}{23,890}[(T_H C_v W + T_i C_v'(1 - W))]$$

$$-\frac{1}{23,890}\left(\frac{\gamma_s\ C_v T_H}{V_s} - \frac{C_v' T_i}{\beta V_g}\right) + P_H - P_i = 0.0 \ . \tag{A.13}$$

Knowing V, I, and W, linear feedback (described in Appendix E) is used to iterate on either V_g or V_s until equation (A.13) is satisfied.

For $V < V_o$, iterate on V_s with an initial guess of $V_s = V_o$ and a ratio to get the second guess of 0.999. For $V \geq V_o$, iterate on V_g with an initial guess of $V_g = (V - 0.9V_o\ W)(1 - W)$ and a ratio to get the second guess of 1.002.

If the iteration goes out of the physical region ($V_g \leq 0$ or $V_s < 0$), that point is replaced by $V_s = V_g = V$. Then knowing V_s and V_g, P and T are calculated.

A.10 Build-Up of Detonation Equation of State

As described in Chapter 2, the detonation wave for some explosives is observed to have changing detonation front state values with distance of run even though the detonation velocity remains constant. This build-up may be described by assigning to each explosive cell a gamma (γ_g) for the detonation products that is a function of distance from the initiation point X.

$$\gamma_s = A + B/X \text{ and } \gamma_s \leq \gamma_{MAX}$$

The constants for PBX-9404 are $A = 2.68$, $B = 1.39$, $\gamma_{MAX} = 3.70$.

The equation of state is calculated for each cell using the γ_g for that cell, the gamma law equations to describe the isentrope, and the constant beta equation to calculate state points off the isentrope. The same equations are used in the GLAW code on the CD-ROM.

$$P_{CJ} = \frac{\rho_o D^2}{\gamma_g + 1} \; .$$

$$V_{CJ} = \frac{\gamma_g V_o}{\gamma_g + 1} \; .$$

$$C = P_{CJ} V_{CJ}^{\gamma_g} \; .$$

$$\beta = \frac{1}{\gamma_g} + \left(\frac{1}{\gamma_g}\right)\left[1 \Big/ \left(\frac{\gamma_g + 1}{1 + (\partial \ln D / \partial \ln \rho)} - 2.0\right)\right] \; .$$

$$K = \frac{-P_{CJ} V_{CJ}}{\gamma_g - 1} + \frac{P_{CJ}}{2}(V_o - V_{CJ}) \; .$$

$$P_i = e^{(\ln C - \gamma_g \ln V_g)} \; .$$

$$P = \frac{1}{\beta V_g}\left(1 - K - \frac{P_i V_g}{\gamma_g - 1}\right) + P_i \; .$$

If the C-J volume burn is used, the V_{CJ} calculated above is used for each burning cell.

The build-up of detonation model is particularly useful for explosives that exhibit build-up in converging geometries as discussed in Chapter 2.

A.11 Discussion of Difference Scheme

The difference scheme used for the conservation equation was first proposed by Fromm.[10] It is important to investigate any difference equation to determine how well it approximates the differential equation it is supposed to represent.

In the Fromm type of difference equations, the pressure, temperature, energy, and specific volume of the cells are considered to be located at the centers of mass of the elements, and the particle velocity is considered to be located at the boundary between the cells.

The energy is conserved globally in the Fromm type of difference equations. This may be shown by summing the total energy in the difference equation for conservation of energy, Equation (A.3), over all the mesh points, resulting in

$$\sum_{j=1}^{j=N} \left(E_j^{n+1} - E_j^n\right)$$

where $E = (I + 0.5U^2)M$. All the terms of the sum cancel except at the end points. Thus, the difference of the total energy of the space at two consecutive times is just the flux at the two boundaries.

The Fromm type of difference equations exhibits internal consistency for slabs. For example, the difference form of the conservation equation $dV/dT = dU/dM$ may be derived from equations (A.1) and (A.2).

By approximating the parameters as Taylor expansions it can be shown that the Fromm difference equations formally approach the differential equations as the time increment and cell mass decrease.

Substituting the Fromm difference equations into the thermodynamic equation $TdS = dI + PdV$ indicates that entropy is not conserved in the absence of viscosity.

Information about the sign and magnitude of the entropy change compared to the changes due to the "viscous" effects of q can be obtained by expanding P and U about $(j\Delta r,\ n\Delta t)$ in a Taylor series in one and two variables, respectively, and combining with the conservation of momentum equation. The difference effects are of order less than or equal to the order of the viscous effects and are positive in normal regions. Thus the entropy error will have the effect of adding more viscosity.

Fromm has shown that the difference equations are conditionally stable to infinitesimal fluctuations, using the von Neumann method, in which the perturbation is expanded in a Fourier series and a given Fourier component is examined.

Thus, the difference equations are sufficiently accurate and stable to infinitesimal fluctuations, and energy is conserved globally. It has been shown numerically that the difference equations with the PIC viscosity are conditionally stable.

Lagrangian hydrodynamics with smeared shocks give satisfactory results for systems with small density discontinuities. If an interface between two materials has an appreciable density gradient, the numerical solutions may result in incorrect densities and energies in the cells next to the interface. The errors may be reduced by proper selection of the cell size on the different sides of the interface. If the density interface gradient is not too great (for example, water and Aluminum), the cell sizes should be chosen so that a shock or rarefaction travels about the same number of cells in both materials in a given time. The objective is to have identical pressure and particle velocity across the boundary, and similar energy and density of cells next to the boundary and nearby cells of the same material. If the density interface gradient is large (for example, Aluminum and air), the energy of the air interface cell is usually too high and that of the Aluminum interface cell is too low. For most practical purposes, the calculation cannot be zoned adequately to make the energy errors across the interface small enough, so it becomes necessary to reset the energy of the interface cells to the energy of the closest cell of the same material.

Enough artificial or real viscosity must be used to smear the shock waves over at least three cells. The PIC form with a constant of 2.0 is useful for many problems because of its scaling as a function of particle velocity. The time step is often determined by the Courant condition; however, for most problems the time step can be estimated in microseconds as 0.2 of the cell width in cm.

References

1. Charles L. Mader, "The Hydrodynamic Hot Spot and Shock Initiation of Homogeneous Explosives," Los Alamos Scientific Laboratory report LA-2703 (1962).
2. Charles L. Mader, "STRETCH SIN–A Code for Computing One-Dimensional Reactive Hydrodynamic Problems," Los Alamos Scientific report LADC-5795 (1963).
3. Charles L. Mader and William R. Gage, "FORTRAN SIN - A One Dimensional Hydrodynamic Code for Problems which Include Chemical Reaction, Elastic-Plastic Flow, Spalling, and Phase Transitions," Los Alamos Scientific Laboratory report LA-3720 (1967).
4. B. I. Bennett, J. D. Johnson, G. I. Kerley, and G. T. Rood, "Recent Developments in the Sesame Equation of State Library," Los Alamos Scientific Laboratory report LA-7130 (1978).
5. John F. Barnes, "Statistical Atom Theory and the Equation of State of Solids," Physical Review 153, 269 (1967).
6. Charles L. Mader, "One-Dimensional Elastic Plastic Calculations for Aluminum," Los Alamos Scientific Laboratory report LA-3678 (1967).
7. John M. Walsh and Russell H. Christian, "Equation of State of Metals from Shock Wave Measurements," Physical Review 97, 1544 (1955).
8. Charles L. Mader, "An Equation of State for Iron Assuming an Instantaneous Phase Change," Los Alamos Scientific Laboratory Report LA-3599 (1966).
9. B. R. Breed, Charles L. Mader, and Douglas Venable, "A Technique for the Determination of Dynamic Tensile Strength Characteristics," Journal of Applied Physics 38, 3271 (1967).
10. Jacob E. Fromm, "Lagrangian Difference Approximation for Fluid Dynamics," Los Alamos Scientific Laboratory report LA-2535 (1961).

appendix B

Numerical Solution of Two-Dimensional Lagrangian Reactive Flow

Many finite difference analogs of the Lagrangian equations of motion of a compressible fluid in two dimensions have been proposed. The "Magee" method was developed and used during the last 45 years at the Los Alamos National Laboratory by Kolsky,[1] Orr,[2] Browne and Hoyt,[3] and others.

The first version of the 2DL code was written in machine language on the IBM 7030 and is described in Reference 4. A FORTRAN version was written by Dennis Simmonds in 1968. It included elastic-plastic flow, real viscosity, heat conduction, and gravity. Arrhenius, sharp-shock, C-J volume, and Forest Fire methods of burning an explosive have been used to model detonation phenomena.

Some versions of the code include techniques for allowing slip between interfaces,[5] techniques for rezoning regions undergoing distortion, and gradient spall models. These techniques are problem dependent and although they have been successfully used for a few types of problems, they must be used with caution. The Eulerian technique for solving the numerical fluid dynamics described in Appendices C and D is more useful for solving problems with large amounts of distortion, but gives less accurate results. A useful multi-dimensional model for spalling is still to be developed, so it is necessary to treat spalling on a case-by-case basis.

The availability of inexpensive but powerful personal computers in the 1980s and FORTRAN compilers for personal computers, permitted the 2DL code to be run on personal computers. The personal computer code is called TDL. A user-friendly interface called USERTDL was developed which reduced the skill and time required to prepare the input file necessary to describe a problem for the TDL code. HOM equation of state data files and Forest Fire burn rate data files were created and are used by the USERTDL code.

The MCGRAPH package of graphics programs was developed for personal computers with CGA, EGA, and VGA graphics by Mader Consulting Co., and was included in the TDL code. The MCGRAPH graphics package also includes the capability of generating PCL graphics files for printing on black and white laser printers or on HP color printers. The executable personal computer code TDL with MCGRAPH graphics is included on the CD-ROM. The executable personal computer code USERTDL along with the data files EOSDATA and FFDATA are also on the CD-ROM.

The Nomenclature

a_X	Acceleration in X direction
a_Z	Acceleration in Z direction
g_X, g_Z	Acceleration due to Gravity
I	Internal Energy (Mbar-cc/gram)
K	Artificial Viscosity Constant
M	Cell Mass$/(2\pi)^{\alpha-1} = \rho_o \Delta X \Delta Z$ for slabs; $= (X + (\Delta X/2))^{\alpha-1} \rho_o \Delta X \Delta Z$ for cylinders (if cells are rectangular).
P	Pressure (Mbar)
q_{XX}	XX Viscosity Deviator
q_{XZ}	XZ Viscosity Deviator
q_{ZZ}	ZZ Viscosity Deviator
R_g	Gas Constant
S_{XX}	XX Elastic Stress Deviator
S_{XZ}	XZ Elastic Stress Deviator
S_{ZZ}	ZZ Elastic Stress Deviator
T	Temperature ^{o}K
U_X	Particle Velocity in X or R direction (i)
U_Z	Particle Velocity in Z direction (j)
V	Specific Volume (cc/gram) $1/\rho$
W	Mass Fraction of Undecomposed Explosive
W'	Spin Tensor
X	Spatial (Eulerian) Coordinate in X Direction
X'	Material (Lagrangian) Coordinate in X Direction
Y_o	Yield Strength
Z	Spatial (Eulerian) Coordinate in Z Direction
Z'	Material (Lagrangian) Coordinate in Z Direction
α	1 for Slabs, 2 for Cylinders, 3 for Spheres
λ	Thermal Conductivity Coefficient
λ_V, μ_V	Real Viscosity Constants
σ_X, σ_Z	Sum of Pressure and Stress Deviator
μ	Shear Modulus
Δt	Time Increment in Microseconds
ρ	Density (g/cc)
Superscript	
n	The Cycle Number
Subscript	
o	Initial Condition

The Lagrangian conservation equations in two dimensions for slabs, cylinders, and spheres are given below. The sign convention of the stress deviators are reversed from those of Appendix A to be negative in compression and positive in tension so $\sigma_X = P - S_{XX}$ and $\sigma_Z = P - S_{ZZ}$. The conservation of mass equation is automatically satisfied. The conservation of momentum equation is

$$a_X = \frac{\partial U_X}{\partial t} = V\left[\frac{\partial(S_{XX} - P)}{\partial X} + \frac{\partial S_{XZ}}{\partial Z} + \frac{(\alpha - 1)}{X}(2S_{XX} + S_{ZZ})\right],$$

$$a_Z = \frac{\partial U_Z}{\partial t} = V\left[\frac{\partial (S_{ZZ} - P)}{\partial Z} + \frac{\partial S_{XZ}}{\partial X} + \frac{(\alpha - 1)}{X} S_{XZ}\right] .$$

In one-dimensional slab geometry

$$\frac{\partial U}{\partial t} = V \frac{\partial (-\sigma_X)}{\partial X} .$$

$$\frac{\partial I}{\partial t} = V\left(-\frac{P}{V}\frac{\partial V}{\partial t} + S_{XX}\left[\frac{\partial U_X}{\partial X} - (\alpha - 1)\frac{U_X}{X}\right]\right)$$

$$+ V\left(S_{ZZ}\left[\frac{\partial U_Z}{\partial Z} - (\alpha - 1)\frac{U_X}{X}\right]\right) + V\left(S_{XZ}\left[\frac{\partial U_X}{\partial Z} + \frac{\partial U_Z}{\partial X}\right]\right) .$$

In one-dimensional slab geometry

$$\frac{\partial I}{\partial t} = -\sigma_X \frac{\partial V}{\partial t} .$$

In addition, the following is added to the energy for heat conduction:

$$\frac{\partial I}{\partial t} = V\lambda\left[\frac{\partial^2 T}{\partial X^2} + \frac{\partial^2 T}{\partial Z^2} + \frac{(\alpha - 1)}{X}\frac{\partial T}{\partial X}\right] .$$

For gravity, g_X is added to a_X and g_Z to a_Z. The elastic-plastic equations are

$$W' = 0.5\left(\frac{\partial U_X}{\partial Z} - \frac{\partial U_Z}{\partial X}\right) ,$$

$$\frac{\partial S_{XX}}{\partial t} = 2\mu\left(\frac{\partial U_X}{\partial X} - \frac{1}{3V}\frac{\partial V}{\partial t}\right) + 2W' S_{XZ} ,$$

$$\frac{\partial S_{ZZ}}{\partial t} = 2\mu\left(\frac{\partial U_Z}{\partial Z} - \frac{1}{3V}\frac{\partial V}{\partial t}\right) - 2W' S_{XZ} ,$$

$$\frac{\partial S_{XZ}}{\partial t} = \mu\left(\frac{\partial U_X}{\partial Z} + \frac{\partial U_Z}{\partial X}\right) - W'(S_{XX} - S_{ZZ}) ,$$

$$f = 2\left[(S_{XX})^2 + (S_{XZ})^2 + (S_{ZZ})^2 + S_{XX}S_{ZZ}\right] .$$

If $f > \frac{2}{3}(Y_o)^2$, the material is plastic, so

$$S^n_{XX} = (\sqrt{((2/3)Y_o^2) / f})\, S_{XX} ,$$

$$S^n_{ZZ} = (\sqrt{((2/3)Y_o^2) / f})\, S_{ZZ} \ ,$$

$$S^n_{XZ} = (\sqrt{((2/3)Y_o^2) / f})\, S_{XZ} \ .$$

In one dimension, $S_{ZZ} = -1/2 S_{XX}$ and the equations for the deviators become as in Appendix A by substituting $(\partial U/\partial X) = (1/V)(\partial V/\partial t)$.

$$\frac{\partial S_{XX}}{\partial t} = 2\mu\left(\frac{2}{3V}\frac{\partial V}{\partial t}\right) .$$

$$\frac{\partial S_{ZZ}}{\partial t} = 2\mu\left(-\frac{1}{3V}\frac{\partial V}{\partial t}\right) .$$

For Artificial Viscosity

$$q = \frac{K}{V}\left|\frac{\partial V}{\partial t}\right|, \text{ and } P^n = P + q \ .$$

For real viscosity

$$q = -\left(\lambda_V + \frac{2}{3}\mu_V\right)\left(\frac{1}{V}\frac{\partial V}{\partial t}\right) ,$$

$$q_{XX} = 2\mu_V\frac{\partial U_X}{\partial X} - \frac{2}{3}\mu_V\left(\frac{1}{V}\frac{\partial V}{\partial t}\right) ,$$

$$q_{ZZ} = 2\mu_V\frac{\partial U_Z}{\partial Z} - \frac{2}{3}\mu_V\left(\frac{1}{V}\frac{\partial V}{\partial t}\right) ,$$

$$q_{XZ} = \mu_V\left(\frac{\partial U_X}{\partial Z} + \frac{\partial U_Z}{\partial X}\right) ,$$

$$S^n_{XX} = S_{XX} + q_{XX} \ ,$$

$$S^n_{ZZ} = S_{ZZ} + q_{ZZ} \ ,$$

$$S^n_{XZ} = S_{XZ} + q_{XZ} \ ,$$

$$P^n = P + q \ .$$

If the Stokes assumption is used, $\lambda_V + (2/3)\mu_V = 0$ and $q = 0$, so in one dimension only q_{XX} is left, and $\sigma = P - S_{XX} - q_{XX}$.

B.1 The Difference Equations

The problem is divided into Lagrangian cells. At the intersections of the lines forming the Lagrangian cells are located the X and Z components of velocity $(U_X,\ U_Z)$ and the coordinates of the intersections $(X,\ Z)$. It is necessary to have $N+1$ sets of velocity components and coordinates for each N cell. The rest of the cell quantities $(P,\ T,\ W,\ M,\ I,\ V)$ are cell centered, and it is necessary to have only N of each except that the cell mass must have an extra boundary quantity.

The calculations are performed in three phases for each time step.

PHASE I

A. Volumes

The quadrilateral cell is divided into two triangles.

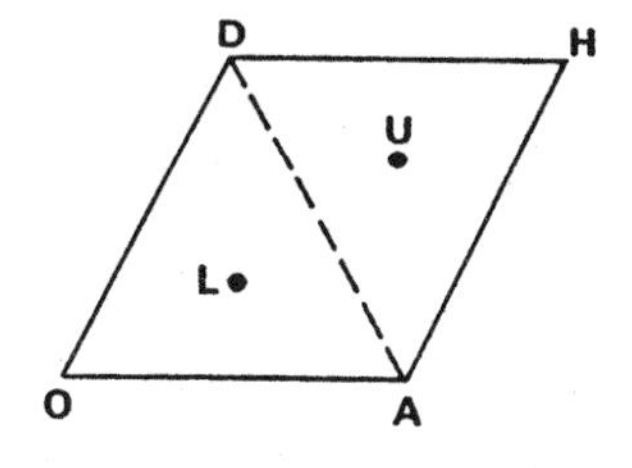

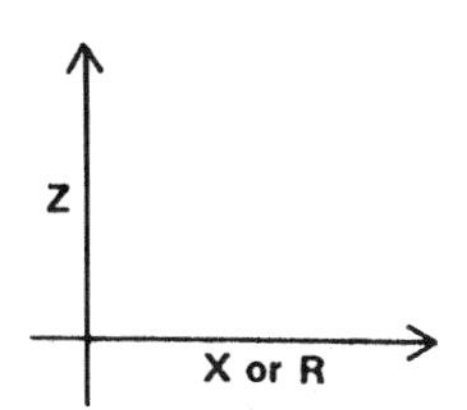

The areas of the triangles are

$$A_U = 0.5[(Z_A - Z_H)(X_D - X_H) - (X_A - X_H)(Z_D - Z_H)] ,$$

$$A_L = 0.5[(Z_D - Z_O)(X_A - X_O) - (X_D - X_O)(Z_A - Z_O)] ,$$

$$A = A_U + A_L .$$

The radii of the centroids are

$$X_U = \frac{1}{3}(X_H + X_D + X_A) ,$$

$$X_L = \frac{1}{3}(X_D + X_O + X_A) .$$

The specific volume of the cell is

$$V^{n+1} = \frac{A_L(X_L)^{\alpha-1} + A_U(X_U)^{\alpha-1}}{M} .$$

The constant 2π is not included.

B. Viscosity

For artificial viscosity

$$q^{n+1} = \frac{K}{V^n}|(V^{n+1} - V^n)| \ .$$

If $(V^{n+1} - V^n) > 0$ then $q^{n+1} = 0.$

$$q_{XX} = q_{ZZ} = q_{XZ} = 0 \ .$$

The partial derivatives were formed using the Green transformation.

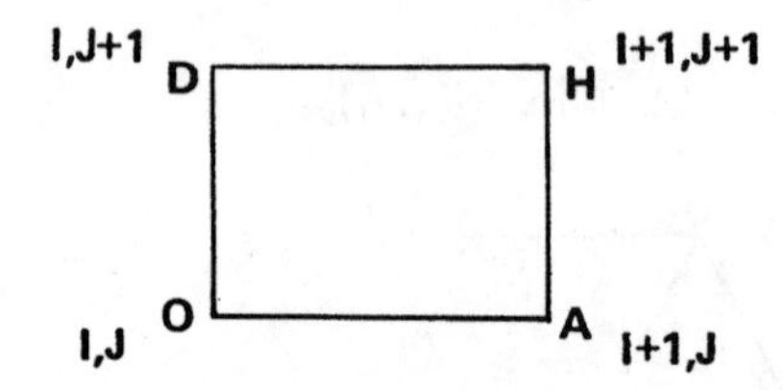

$$PUXPX = \frac{1}{2A}\left[(U_X^A - U_X^D)(Z_H - Z_O) + (U_X^H - U_X^O)(Z_D - Z_A)\right] = \frac{\partial U_X}{\partial X} \ .$$

$$PUZPZ = -\frac{1}{2A}\left[(U_Z^A - U_Z^D)(X_H - X_O) + (U_Z^H - U_Z^O)(X_D - X_A)\right] = \frac{\partial U_Z}{\partial Z} \ .$$

$$PUXPZ = -\frac{1}{2A}\left[(U_X^A - U_X^D)(X_H - X_O) + (U_X^H - U_X^O)(X_D - X_A)\right] = \frac{\partial U_X}{\partial Z} \ .$$

$$PUZPX = \frac{1}{2A}\left[(U_Z^A - U_Z^D)(Z_H - Z_O) + (U_Z^H - U_Z^O)(Z_D - Z_A)\right] = \frac{\partial U_Z}{\partial X} \ .$$

$$PVPT = \frac{2}{\Delta t}\left(\frac{V^{n+1} - V^n}{V^{n+1} + V^n}\right) = \frac{1}{V}\frac{\partial V}{\partial t} \ .$$

For real viscosity

$$q^{n+1} = -\left(\lambda_V + \frac{2}{3}\mu_V\right)(PVPT) \ ,$$

$$q_{XX}^{n+1} = 2\mu_V\left(PUXPX - \frac{1}{3}PVPT\right) ,$$

$$q_{ZZ}^{n+1} = 2\mu_V\left(PUZPZ - \frac{1}{3}PVPT\right) ,$$

$$q_{XZ}^{n+1} = \mu_V(PUXPZ + PUZPX) .$$

C. Elastic-Plastic

For elastic-plastic flow

$$W' = 0.5(PUXPZ - PUZPX) ,$$

$$S_{XX}^{n+1} = S_{XX}^n + 2\Delta t\left[\mu(PUXPX - \frac{1}{3}PVPT) + W'S_{XZ}^n\right] ,$$

$$S_{ZZ}^{n+1} = S_{ZZ}^n + 2\Delta t\left[\mu(PUZPZ - \frac{1}{3}PVPT) - W'S_{XZ}^n\right] ,$$

$$S_{XZ}^{n+1} = S_{XZ}^n + \Delta t\left[\mu(PUXPZ + PUZPX) - W'(S_{XX}^n - S_{ZZ}^n)\right] ,$$

$$f = 2\left[\left(S_{XX}^{n+1}\right)^2 + \left(S_{XZ}^{n+1}\right)^2 + \left(S_{ZZ}^{n+1}\right)^2 + S_{XX}^{n+1}\,S_{ZZ}^{n+1}\right] .$$

If $f \le \frac{2}{3}(Y_o)^2$, the material is elastic and described by the above S_{XX}, S_{ZZ}, S_{XZ}. If $f > \frac{2}{3}(Y_o)^2$,

$$F = \sqrt{2/3(Y_o)^2 \,/\, f} ,$$

$$S_{XX}^{n+1} = (F)(S_{XX}^{n+1}) + q_{XX} ,$$

$$S_{ZZ}^{n+1} = (F)(S_{ZZ}^{n+1}) + q_{ZZ} ,$$

$$S_{XZ}^{n+1} = (F)(S_{XZ}^{n+1}) + q_{XZ} .$$

D. Energy

$$I^{n+1} = I^n - (V^{n+1} - V^n)(P^n + q^n)$$

$$+\left(\frac{V^{n+1} + V^n}{2}\right)\Delta t\, S_{XX}^{n+1}\left[PUXPX - (\alpha - 1)\left(\frac{U_X^A + U_X^O}{X_A + X_O}\right)\right]$$

$$+\left(\frac{V^{n+1} + V^n}{2}\right)\Delta t\, S_{ZZ}^{n+1}\left[PUZPZ - (\alpha - 1)\left(\frac{U_X^A + U_X^O}{X_A + X_O}\right)\right]$$

$$+\left(\frac{V^{n+1} + V^n}{2}\right)\Delta t\, S_{XZ}^{n+1}(PUXPZ + PUZPX) .$$

For heat conduction,

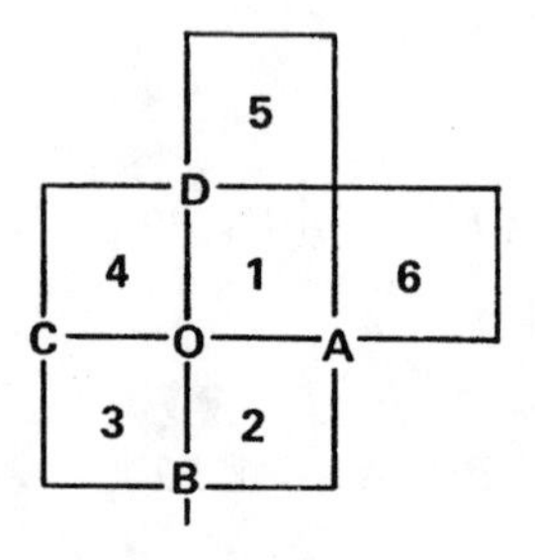

$$I^{n+1} = I^n + \lambda(\Delta t)\left(\frac{V^{n+1} + V^n}{2}\right)\left[\frac{T_4 - 2T_1 + T_6}{(X_A - X_O)^2}\right]$$

$$+ \lambda(\Delta t)\left(\frac{V^{n+1} + V^n}{2}\right)\left[\frac{T_5 - 2T_1 + T_2}{(Z_D - Z_O)^2} + \frac{\alpha - 1}{X_O}\left(\frac{T_4 - T_6}{X_C - X_A}\right)\right] .$$

E. Equation of State

Knowing I^{n+1}, V^{n+1}, W^{n+1}, the HOM equation of state described in Appendix A is used to find P_{HOM}^{n+1}, T^{n+1}. For spalling, $\Delta P/\Delta X = a_X/V$ and $\Delta P/\Delta Z = a_Z/V$ are calculated and the largest (most negative) value of $\Delta P/\Delta X$ or $\Delta P/\Delta Z$ is taken as the spalling gradient if the cell is undergoing increasing tension (ΔV is positive). The one-dimensional gradient model is used to determine if spalling should occur. The cell is flagged if spalled, and the cell pressure is set equal to zero and kept at zero as long as $V > V_o$. For elastic-plastic flow

$$P^{n+1} = P_{HOM}^{n+1} - \frac{2}{3}Y_o\left(\frac{P_{HOM}^{n+1}}{PLAP}\right) ,$$

where

$$\frac{P_{HOM}^{n+1}}{PLAP} \leq 1.0 \ ,$$

as discussed in Appendix A.

F. Explosive Burn

The Arrhenius, C-J volume, sharp shock in axial direction only, and Forest Fire burns are identical to those described in Appendix A.

PHASE II

G. Acceleratons

The accelerations are calculated using the force gradient method derived in Reference 5.

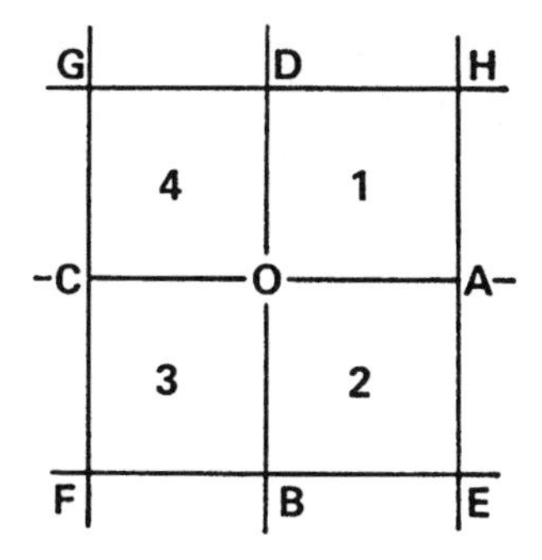

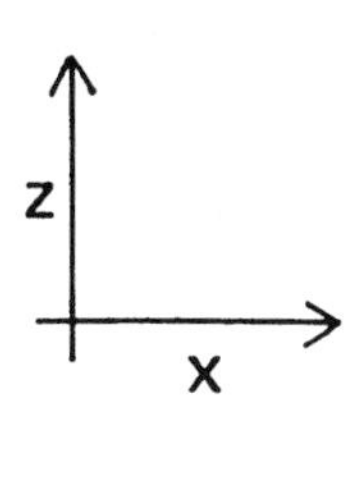

$$P_{1X} = P_1 + q_1 - S_{XX}^1 \ ,$$

$$P_{2X} = P_2 + q_2 - S_{XX}^2 \ ,$$

$$P_{3X} = P_3 + q_3 - S_{XX}^3 \ ,$$

$$P_{4X} = P_4 + q_4 - S_{XX}^4$$

$$a_X = -\frac{(P_{1X} - P_{4X})(Z_D - Z_O)}{(M_1 + M_4)} \left(\left| \frac{X_D + X_O}{2} \right| \right)^{\alpha-1}$$

$$-\frac{(P_{2X} - P_{3X})(Z_O - Z_B)}{(M_2 + M_3)} \left(\left| \frac{X_O + X_B}{2} \right| \right)^{\alpha-1}$$

$$+\frac{(P_{1X} - P_{2X})(Z_A - Z_O)}{(M_1 + M_2)}\left(\left|\frac{X_A + X_O}{2}\right|\right)^{\alpha-1}$$

$$+\frac{(P_{4X} - P_{3X})(Z_O - Z_C)}{(M_4 + M_3)}\left(\left|\frac{X_O + X_C}{2}\right|\right)^{\alpha-1} .$$

If elastic-plastic or real viscosity exists, add

$$-\frac{(S^1_{XZ} - S^4_{XZ})(X_D - X_O)}{(M_1 + M_4)}\left(\left|\frac{X_D + X_O}{2}\right|\right)^{\alpha-1}$$

$$-\frac{(S^2_{XZ} - S^3_{XZ})(X_O - X_B)}{(M_2 + M_3)}\left(\left|\frac{X_O + X_B}{2}\right|\right)^{\alpha-1}$$

$$+\frac{(S^1_{XZ} - S^2_{XZ})(X_A - X_O)}{(M_1 + M_2)}\left(\left|\frac{X_A + X_O}{2}\right|\right)^{\alpha-1}$$

$$+\frac{(S^4_{XZ} - S^3_{XZ})(X_O - |X_C|)}{(M_4 + M_3)}\left(\left|\frac{X_O + X_C}{2}\right|\right)^{\alpha-1}$$

$$+\frac{(V^{n+1})(\alpha - 1)}{((X_O + X_A)/2)}(2S^1_{XX} + S^1_{ZZ}) + g_X .$$

$$P_{1Z} = P_1 + q_1 - S^1_{ZZ} ,$$

$$P_{2Z} = P_2 + q_2 - S^2_{ZZ} ,$$

$$P_{3Z} = P_3 + q_3 - S^3_{ZZ} ,$$

$$P_{4Z} = P_4 + q_4 - S^4_{ZZ} .$$

$$a_Z = +\frac{(P_{1Z} - P_{4Z})(X_D - X_O)}{(M_1 + M_4)}\left(\left|\frac{X_D + X_O}{2}\right|\right)^{\alpha-1}$$

$$+\frac{(P_{2Z} - P_{3Z})(X_O - X_B)}{(M_2 + M_3)}\left(\left|\frac{X_O + X_B}{2}\right|\right)^{\alpha-1}$$

$$- \frac{(P_{1Z} - P_{2Z})(X_A - X_O)}{(M_1 + M_2)} \left(\left| \frac{X_A + X_O}{2} \right| \right)^{\alpha - 1}$$

$$- \frac{(P_{4Z} - P_{3Z})(X_O - |X_C|)}{(M_4 + M_3)} \left(\left| \frac{X_O + X_C}{2} \right| \right)^{\alpha - 1} ,$$

If elastic-plastic or real viscosity exists, add

$$+ \frac{(S_{XZ}^1 - S_{XZ}^4)(Z_D - Z_O)}{(M_1 + M_4)} \left(\left| \frac{X_D + X_O}{2} \right| \right)^{\alpha - 1}$$

$$+ \frac{(S_{XZ}^2 - S_{XZ}^3)(Z_O - Z_B)}{(M_2 + M_3)} \left(\left| \frac{X_O + X_B}{2} \right| \right)^{\alpha - 1}$$

$$- \frac{(S_{XZ}^1 - S_{XZ}^2)(Z_A - Z_O)}{(M_1 + M_2)} \left(\left| \frac{X_A + X_O}{2} \right| \right)^{\alpha - 1}$$

$$- \frac{(S_{XZ}^4 - S_{XZ}^3)(Z_O - Z_C)}{(M_4 + M_3)} \left(\left| \frac{X_O + X_C}{2} \right| \right)^{\alpha - 1}$$

$$+ \frac{(V^{n+1})(\alpha - 1)}{((X_O + X_A)/2)} (S_{XZ}^1) + g_Z .$$

H. Boundaries

The boundaries may be free-surface, piston, continuative, or axial. Corners require special treatments depending upon the particular combination of boundaries at the corner.

For free surfaces, the pressure and stress deviators in the cell adjacent to the boundary are set to the negative value of the state values in the boundary cell. For piston boundaries, the pressure in the adjacent cell is set equal to the piston pressure. For continuum boundaries, the pressure and stress deviators in the adjacent cell are set equal to the values in the boundary cell. For axis boundaries, the pressures and stress deviators in the adjacent cell across the axis are set equal to the values in axis cells adjacent to the axis. The coordinates of the cells adjacent to the boundary cells are calculated by linear extrapolation.

I. Cell Velocities

The cell particle velocities are calculated from

$$U_X^{n+1} = U_X^n + a_X \Delta t ,$$

$$U_Z^{n+1} = U_Z^n + a_Z \Delta t \ .$$

PHASE III

J. Positions

The cell boundary locations are calculated from

$$X^{n+1} = X^n + U_X^{n+1} \Delta t \ ,$$

$$Z^{n+1} = Z^n + U_Z^{n+1} \Delta t \ .$$

If the volume of the new cell boundaries is negative, the old boundaries are not changed and the U_X and U_Z velocities are set to zero. This is called a "poor man's rezone."

K. Time

The total time is incremented by Δt, and the calculation starts again with Phase I.

References

1. H. G. Kolsky, "A Method for Numerical Solution of Transient Hydrodynamic Shock Problems in Two Space Dimensions," Los Alamos Scientific Laboratory report LA-1867 (1955).
2. S. R. Orr, "F. Magee," Los Alamos Scientific Laboratory unpublished note (1961).
3. Phillip L. Browne and Martha S. Hoyt, "HASTI – A Numerical Calculation of Two-Dimensional Lagrangian Hydrodynamics Utilizing the Concept of Space-Dependent Time Steps," Los Alamos Scientific Laboratory report LA-3324-MS (1965).
4. Charles L. Mader, "The Two Dimensional Hydrodynamic Hot Spot, Volume III," Los Alamos Scientific Laboratory report LA-3450 (1966).
5. Charles L. Mader, "Numerical Modeling of Detonations," University of California Press (1979).

appendix C

Numerical Solution of Two-Dimensional Eulerian Reactive Flow

The Eulerian equations of motion are more useful for numerical solution of highly distorted fluid flow than are Lagrangian equations of motion. Multicomponent Eulerian calculations require equations of state for mixed cells and methods for moving mass and its associated state values into and out of mixed cells. These complications are avoided by Lagrangian calculations. Harlow's particle-in-cell (PIC) method[1] uses particles for the mass movement. The first reactive Eulerian hydrodynamic code EIC (Explosive-in-cell) used the PIC method, and it is described in Reference 2. The discrete nature of the mass movement introduced pressure and temperature variations from cycle to cycle of the calculation that were unacceptable for many reactive fluid dynamic problems. A one-component continuous mass transport Eulerian code[3] developed in 1966 proved useful for solving many one-component problems of interest in reactive fluid dynamics. The need for a multicomponent Eulerian code resulted in a second 2DE code, described in Reference 4. Elastic-plastic flow and real viscosity were added in 1976. The technique was extended to three-dimensions in the 1970s and the resulting 3DE code is described in Appendix D.

The 2DE code used techniques for solving the equations of motion investigated by Rich;[5] Gentry, Martin, and Daly;[6] and Johnson and Walsh.[7] For multicomponent systems the donor-acceptor method developed by Johnson[8] was used to determine the mass flux. The treatment of elastic-plastic flow used was similar to that of Hageman and Walsh.[9] The unique feature of the technique used in 2DE was the mixed equation of state treatments and the use of the associated state values in mixture cells in the mass movement across cell boundaries. This permitted solution of interface flow within the cell resolution and with an accuracy of state values almost as good as one could obtain with Lagrangian treatments. The complication of keeping track of the mixed cell properties and the logic associated with the mass movement made the 2DE code longer and slower than the 2DL code by about an order of magnitude.

The Nomenclature

g_X, g_Z	Acceleration due to Gravity
I	Cell Internal Energy (Mbar-cc/gram)
K	Artificial Viscosity Constant
P	Pressure (Mbar)
q	Bulk or Artificial Viscosity
q_{XX}	XX Viscosity Deviator
q_{XZ}	XZ Viscosity Deviator
R_g	Gas Constant
q_{ZZ}	ZZ Viscosity Deviator
S_{XX}	XX Elastic Stress Deviator
S_{XZ}	XZ Elastic Stress Deviator
S_{ZZ}	ZZ Elastic Stress Deviator
T	Temperature ^{o}K
U_X	Particle Velocity in X or R Direction (i)
U_Z	Particle Velocity in Z Direction (j)
V	Specific Volume (cc/gram) $1/\rho$
W	Mass Fraction of Undecomposed Explosive
W'	Spin Tensor
X or R	Spatial (Eulerian) Coordinate in X Direction
Z or Y	Spatial (Eulerian) Coordinate in Z Direction
Y_o	Yield Strength
ΔM	Density Increment for Mass Movement
ΔW	Mass Fraction Increment
ΔPUX	Momentum Increment in the X or R Direction
ΔPUZ	Momentum Increment in Z Direction
α	1 for Slab, 2 for Cylindrical Geometry
Δt	Time Increment in Microseconds
ρ	Density (g/cc)
λ	Thermal Conductivity Coefficient
λ_V, μ_V	Real Viscosity Constants
μ	Shear Modulus
Superscript	
n	The Cycle Number
Subscript	
o	Initial Condition

The Eulerian conservation equations in two dimensions for slabs and cylinders are as follows.

The conservation of mass equation is

$$\frac{\partial \rho}{\partial t} + U_X \frac{\partial \rho}{\partial X} + U_Z \frac{\partial \rho}{\partial Z} = -\rho \left[\frac{\partial U_X}{\partial X} + \frac{\partial U_Z}{\partial Z} + (\alpha - 1)\frac{U_X}{X} \right].$$

The conservation of momentum equations are

$$\rho\left(\frac{\delta U_X}{\delta t} + U_X\frac{\partial U_X}{\delta X} + U_Z\frac{\delta U_X}{\delta Z}\right) = \frac{\delta(S_{XX} - P)}{\delta X} + \frac{\partial S_{XZ}}{\partial Z}$$

$$+\frac{(\alpha - 1)}{X}(2S_{XX} + S_{ZZ}) + \rho g_X .$$

$$\rho\left(\frac{\delta U_Z}{\delta t} + U_X\frac{\partial U_Z}{\delta X} + U_Z\frac{\delta U_Z}{\delta Z}\right) = \frac{\delta(S_{ZZ} - P)}{\delta Z} + \frac{\partial S_{XZ}}{\partial X}$$

$$+\frac{(\alpha - 1)}{X}(S_{XZ}) + \rho g_Z .$$

The conservation of energy equation is

$$\rho\left(\frac{\delta I}{\delta t} + U_X\frac{\partial I}{\delta X} + U_Z\frac{\delta I}{\delta Z}\right) = -P\frac{\partial U_X}{\partial X} - P\frac{\partial U_Z}{\partial Z} - P(\alpha - 1)\frac{U_X}{X}$$

$$+S_{XX}\left[\frac{\partial U_X}{\partial X} - (\alpha - 1)\frac{U_X}{X}\right] + S_{ZZ}\left[\frac{\partial U_Z}{\partial Z} - (\alpha - 1)\frac{U_X}{X}\right]$$

$$+S_{XZ}\left(\frac{\partial U_X}{\partial Z} + \frac{\partial U_Z}{\partial X}\right) + \lambda\left(\frac{\partial^2 T}{\partial X^2} + \frac{\partial^2 T}{\partial Z^2} + \frac{(\alpha - 1)}{X}\frac{\partial T}{\partial X}\right).$$

$$W' = 0.5\left(\frac{\partial U_X}{\partial Z} - \frac{\partial U_Z}{\partial X}\right),$$

$$\frac{\partial S_{XX}}{\partial t} = 2\mu\left(\frac{\partial U_X}{\partial X} - \frac{1}{3V}\frac{\partial V}{\partial t}\right) + 2W'S_{XZ} ,$$

$$\frac{\partial S_{ZZ}}{\partial t} = 2\mu\left(\frac{\partial U_Z}{\partial Z} - \frac{1}{3V}\frac{\partial V}{\partial t}\right) - 2W'S_{XZ} ,$$

$$\frac{\partial S_{XZ}}{\partial t} = \mu\left(\frac{\partial U_X}{\partial Z} + \frac{\partial U_Z}{\partial X}\right) - W'(S_{XX} - S_{ZZ}) ,$$

$$f = 2\left[(S_{XX})^2 + (S_{XZ})^2 + (S_{ZZ})^2 + S_{XX}S_{ZZ}\right].$$

If $f > \frac{2}{3}(Y_o)^2$, the material is plastic, so

$$S_{XX}^n = (\sqrt{((2/3)Y_o^2) / f})\, S_{XX} ,$$

$$S_{ZZ}^n = (\sqrt{((2/3)Y_o^2) / f})\, S_{ZZ} ,$$

$$S_{XZ}^n = (\sqrt{((2/3)Y_o^2) / f})\, S_{XZ} .$$

For artificial viscosity

$$q = K\rho \frac{dU_X}{dX} \text{ or } K\rho \frac{dU_Z}{\Delta Z} .$$

$$P^n = P + q .$$

For real viscosity

$$q = -\left(\lambda_V + \frac{2}{3}\mu_V\right)\left(\frac{1}{V}\frac{\partial V}{\partial t}\right) ,$$

$$q_{XX} = 2\mu_V \frac{\partial U_X}{\partial X} - \frac{2}{3}\mu_V\left(\frac{1}{V}\frac{\partial V}{\partial t}\right) ,$$

$$q_{ZZ} = 2\mu_V \frac{\partial U_Z}{\partial Z} - \frac{2}{3}\mu_V\left(\frac{1}{V}\frac{\partial V}{\partial t}\right) ,$$

$$q_{XZ} = \mu_V\left(\frac{\partial U_X}{\partial Z} + \frac{\partial U_Z}{\partial X}\right) ,$$

$$S_{XX}^n = S_{XX} + q_{XX} ,$$

$$S_{ZZ}^n = S_{ZZ} + q_{ZZ} ,$$

$$S_{XZ}^n = S_{XZ} + q_{XZ} .$$

The equations, written in finite-difference form appropriate to a fixed (Eulerian) mesh, are used to determine the dynamics of the fluid. The fluid is moved by a continuous mass transport method.

The first of the above equations, that of mass conservation, is satisfied automatically. The momentum and energy equations are treated as follows. In the first step, contributions to the time derivatives which arise from the terms involving pressure are calculated. Mass is not moved in this step; thus, the transport terms are dropped. Tenative new values of velocity and internal energy are calculated for each cell.

In the second step, the mass is moved according to the cell velocity. The mass that crosses cell boundaries carries with it into the new cells appropriate fractions of the mass, momentum, and energy of the cells from which it came. This second step accomplishes the transport that was neglected in the first step.

In the third step, the amount of chemical reaction is determined, and the new cell pressure is computed using the HOM equations of state.

The equations to be differenced are of the form

$$\rho\frac{\delta U_X}{\delta t} = -\frac{\delta(P+q-S_{XX})}{\delta X} + \frac{\partial S_{XZ}}{\partial Z} + \frac{(\alpha-1)}{X}(2S_{XX}+S_{ZZ}) + \rho g_X \,.$$

$$\rho\frac{\delta U_Z}{\delta t} = -\frac{\delta(P+q-S_{ZZ})}{\delta Z} + \frac{\partial S_{XZ}}{\partial X} + \frac{(\alpha-1)}{X}(S_{XZ}) + \rho g_Z \,.$$

and

$$\rho\frac{\delta I}{\delta t} = -\frac{P}{X^{\alpha-1}}\frac{\partial(U_X\,X^{\alpha-1})}{\partial X} - \frac{1}{X^{\alpha-1}}\frac{\partial(q\,U_X\,X^{\alpha-1})}{\partial X}$$

$$+U_X\frac{\partial q}{\partial X} - P\frac{\partial U_Z}{\partial Z} - \frac{\partial q U_Z}{\partial Z} + U_Z\frac{\partial q}{\partial Z}$$

$$+S_{XX}\left[\frac{\partial U_X}{\partial X} - (\alpha-1)\frac{U_X}{X}\right] + S_{ZZ}\left[\frac{\partial U_Z}{\partial Z} - (\alpha-1)\frac{U_X}{X}\right]$$

$$+S_{XZ}\left(\frac{\partial U_X}{\partial Z} + \frac{\partial U_Z}{\partial X}\right) + \lambda\left(\frac{\partial^2 T}{\partial X^2} + \frac{\partial^2 T}{\partial Z^2} + \frac{(\alpha-1)}{X}\frac{\partial T}{\partial X}\right).$$

C.1 The Numerical Technique

The problem is divided into Eulerian cells. The state values are cell centered, the velocities and positions are centered at cell boundaries.

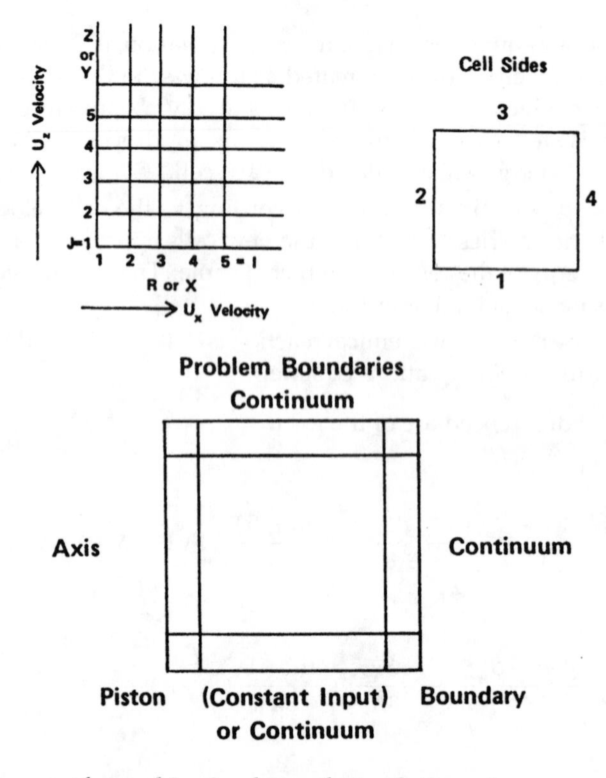

The calculations are performed in six phases for each time step.

PHASE I

Phases I and II are skipped if $\rho^n_{i,j} \leq$ MINGRHO $+ (\rho_o -$ MINGRHO$)(W^n_{i,j})$. MINGRHO has a value of 0.5 and is a minimum density at which the calculation will proceed. This is for handling free surfaces and to eliminate false diffusion. For gases at free surfaces, $W^n_{i,j}$ is replaced by one.

A. The Equation of State

The pressure and temperature are calculated from the density, internal energy, and cell mass fractions using the subroutines HOM, HOM2S, HOMSG, HOM2G, or HOM2SG described in Section C.3. Mixed cells carry the individual component densities and energies as calculated from the mixture equation of state.

For single material elastic-plastic flow

$$P^{n+1} = P^{n+1}_{HOM} - \frac{2}{3} Y_o \left(\frac{P^{n+1}_{HOM}}{PLAP} \right) \text{ where } \left(\frac{P^{n+1}_{HOM}}{PLAP} \right) \leq 1.0 \ .$$

For spalling, the gradient is calculated by SIN and used by HOM to determine whether spalling has occurred using the tension gradient model described in Appendix A.

$$\frac{\Delta P}{\Delta X} = -\sqrt{\left(\frac{P_{i+1,j} - P_{i-1,j}}{2\Delta X} \right)^2 + \left(\frac{P_{i,j+1} - P_{i,j-1}}{2\Delta Z} \right)^2} \ .$$

The cell is flagged if it is spalled, and the pressure is set to zero if $P \leq 0$ thereafter.

B. Explosive Burn

The Arrhenius, C-J volume, and Forest Fire burns are identical to those described in Appendix A.

C. Real Viscosity Deviators

The finite difference velocity and volume derivatives calculated from Phase VI on the previous cycle are called $DUDX$, $DVDZ$, $DUDZ$, $DVDT$, etc. and the initial space increments are called ΔX and ΔZ.

$$QXX = 2\mu_V(DUDX - 1/3\ DVDT)\,.$$

$$QZZ = 2\mu_V(DVDZ - 1/3\ DVDT)\,.$$

$$QXZ = \mu_V(DUDZ + DVDX)\,.$$

$$q_{i,j} = -\left(\lambda_V + \frac{2}{3}\mu_V\right)(DVDT_{i,j})\,.$$

$$q_{i,j+1} = -\left(\lambda_V + \frac{2}{3}\mu_V\right)(DVDT_{i,j+1})\,.$$

$$q_{i+1,j} = -\left(\lambda_V + \frac{2}{3}\mu_V\right)(DVDT_{i+1,j})\,.$$

D. Stress Deviators

To take approximate convective derivatives, $DSXX$ and $DSXZ$ calculated in the previous cycle are weighted by volume.

$$DX = +U_{X\,i,j}\,\Delta t\,,$$

$$DY = +U_{Z\,i,j}\,\Delta t\,.$$

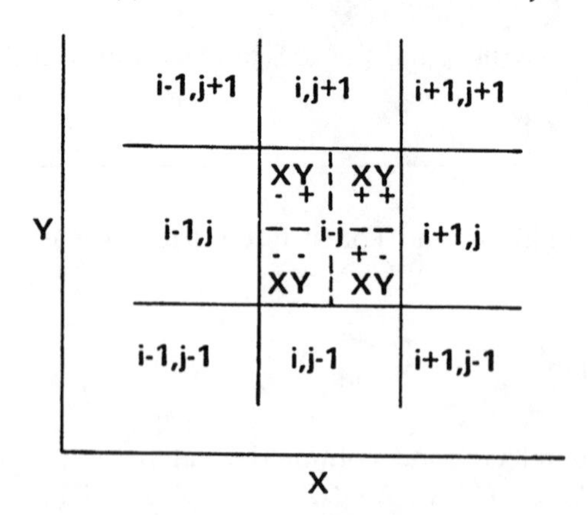

If DX and DY are positive:

$$S_{XX}^{n+1} = S_{XX}^{n} + \frac{1}{\Delta X \Delta Z}[\ DSXX_{i,j}(\Delta X - DX)(\Delta Z - DY)$$

$$+DSXX_{i+1,j}(DX)(\Delta Z - DY) + DSXX_{i,j+1}(DY)(\Delta X - DX)$$

$$+DSXX_{i+1,j+1}(DX)(DY)\ ]\ ,$$

$$S_{ZZ}^{n+1} = S_{ZZ}^{n} + \frac{1}{\Delta X \Delta Z}[\ DSZZ_{i,j}(\Delta X - DX)(\Delta Z - DY)$$

$$+DSZZ_{i+1,j}(DX)(\Delta Z - DY) + DSZZ_{i,j+1}(DY)(\Delta X - DX)$$

$$+DSZZ_{i+1,j+1}(DX)(DY)\ ]\ ,$$

$$S_{XZ}^{n+1} = S_{XZ}^{n} + \frac{1}{\Delta X \Delta Z}[\ DSXZ_{i,j}(\Delta X - DX)(\Delta Z - DY)$$

$$+DSXZ_{i+1,j}(DX)(\Delta Z - DY) + DSXZ_{i,j+1}(DY)(\Delta X - DX)$$

$$+DSXZ_{i+1,j+1}(DX)(DY)\ ]\ .$$

On the boundaries in the above equations where KK is XX, XZ, or ZZ:

On boundary 3 , $DSKK_{i,j+1} = DSKK_{i,j}$ and $DSKK_{i+1,j+1} = DSKK_{i+1,j}$;

on boundary 4 , $DSKK_{i+1,j+1} = DSKK_{i,j+1}$ and $DSKK_{i+1,j} = DSKK_{i,j}$;

on corner 3/4 , $S_{KK}^{n+1} = S_{KK}^{n} + DSKK_{i,j}$.

If DX and DY are both negative:

$$S_{XX}^{n+1} = S_{XX}^{n} + \frac{1}{\Delta X \Delta Z}[\, DSXX_{i,j}(\Delta X - |DX|)(\Delta Z - |DY|)$$

$$+DSXX_{i-1,j}(|DX|)(\Delta Z - |DY|) + DSXX_{i,j-1}(|DY|)(\Delta X - |DX|)$$

$$+DSXX_{i-1,j-1}(|DX|)(|DY|)\,] \ ,$$

$$S_{ZZ}^{n+1} = S_{ZZ}^{n} + \frac{1}{\Delta X \Delta Z}[\, DSZZ_{i,j}(\Delta X - |DX|)(\Delta Z - |DY|)$$

$$+DSZZ_{i-1,j}(|DX|)(\Delta Z - |DY|) + DSZZ_{i,j-1}(|DY|)(\Delta X - |DX|)$$

$$+DSZZ_{i-1,j-1}(|DX|)(|DY|)\,] \ ,$$

$$S_{XZ}^{n+1} = S_{XZ}^{n} + \frac{1}{\Delta X \Delta Z}[\, DSXZ_{i,j}(\Delta X - |DX|)(\Delta Z - |DY|)$$

$$+DSXZ_{i-1,j}(|DX|)(\Delta Z - |DY|) + DSXZ_{i,j-1}(|DY|)(\Delta X - |DX|)$$

$$+DSXZ_{i-1,j-1}(|DX|)(|DY|)\,] \ .$$

On the boundaries in the above equations where KK is XX, XZ, or ZZ:

On boundary 2 , $DSKK_{i-1,j} = DSKK_{i,j}$ and $DSKK_{i-1,j-1} = DSKK_{i,j-1}$;

on boundary 1 , $DSKK_{i,j-1} = DSKK_{i,j}$ and $DSKK_{i-1,j-1} = DSKK_{i-1,j}$;

on corner 1/2 , $S_{KK}^{n+1} = S_{KK}^{n} + DSKK_{i,j}$.

If DX is negative and DY is positive:

$$S_{XX}^{n+1} = S_{XX}^{n} + \frac{1}{\Delta X \Delta Z}[\, DSXX_{i,j}(\Delta X - |DX|)(\Delta Z - |DY|)$$

$$+DSXX_{i-1,j}(|DX|)(\Delta Z - |DY|) + DSXX_{i,j+1}(|DY|)(\Delta X - |DX|)$$

$$+DSXX_{i-1,j+1}(|DX|)(|DY|)\,] \; ,$$

$$S_{ZZ}^{n+1} = S_{ZZ}^{n} + \frac{1}{\Delta X \Delta Z}[\, DSZZ_{i,j}(\Delta X - |DX|)(\Delta Z - |DY|)$$

$$+DSZZ_{i-1,j}(|DX|)(\Delta Z - |DY|) + DSZZ_{i,j+1}(|DY|)(\Delta X - |DX|)$$

$$+DSZZ_{i-1,j+1}(|DX|)(|DY|)\,] \; ,$$

$$S_{XZ}^{n+1} = S_{XZ}^{n} + \frac{1}{\Delta X \Delta Z}[\, DSXZ_{i,j}(\Delta X - |DX|)(\Delta Z - |DY|)$$

$$+DSXZ_{i-1,j}(|DX|)(\Delta Z - |DY|) + DSXZ_{i,j+1}(|DY|)(\Delta X - |DX|)$$

$$+DSXZ_{i-1,j+1}(|DX|)(|DY|)\,] \; .$$

On the boundaries in the above equations where KK is XX, XZ, or ZZ:

On boundary 2 , $DSKK_{i-1,j} = DSKK_{i,j}$ and $DSKK_{i-1,j+1} = DSKK_{i,j+1}$;

on boundary 3 , $DSKK_{i,j+1} = DSKK_{i,j}$ and $DSKK_{i-1,j+1} = DSKK_{i-1,j}$;

on corner 2/3 , $S_{KK}^{n+1} = S_{KK}^{n} + DSKK_{i,j}$.

If DX is positive and DY is negative:

$$S_{XX}^{n+1} = S_{XX}^{n} + \frac{1}{\Delta X \Delta Z}[\, DSXX_{i,j}(\Delta X - |DX|)(\Delta Z - |DY|)$$

$$+DSXX_{i+1,j}(|DX|)(\Delta Z - |DY|) + DSXX_{i,j-1}(|DY|)(\Delta X - |DX|)$$

$$+DSXX_{i+1,j-1}(|DX|)(|DY|)\ ] \ ,$$

$$S_{ZZ}^{n+1} = S_{ZZ}^{n} + \frac{1}{\Delta X \Delta Z}[\ DSZZ_{i,j}(\Delta X - |DX|)(\Delta Z - |DY|)$$

$$+DSZZ_{i+1,j}(|DX|)(\Delta Z - |DY|) + DSZZ_{i,j-1}(|DY|)(\Delta X - |DX|)$$

$$+DSZZ_{i+1,j-1}(|DX|)(|DY|)\ ] \ ,$$

$$S_{XZ}^{n+1} = S_{XZ}^{n} + \frac{1}{\Delta X \Delta Z}[\ DSXZ_{i,j}(\Delta X - |DX|)(\Delta Z - |DY|)$$

$$+DSXZ_{i+1,j}(|DX|)(\Delta Z - |DY|) + DSXZ_{i,j-1}(|DY|)(\Delta X - |DX|)$$

$$+DSXZ_{i+1,j-1}(|DX|)(|DY|)\ ] \ .$$

On the boundaries in the above equations where KK is XX, XZ, or ZZ:

On boundary 1 , $DSKK_{i,j-1} = DSKK_{i,j}$ and $DSKK_{i+1,j} = DSKK_{i+1,j-1}$;

on boundary 4 , $DSKK_{i+1,j-1} = DSKK_{i,j-1}$ and $DSKK_{i+1,j} = DSKK_{i,j}$;

on corner 1/4 , $S_{KK}^{n+1} = S_{KK}^{n} + DSKK_{i,j}$.

$$f = 2\left[\left(S_{XX}^{n+1}\right)^2 + \left(S_{XZ}^{n+1}\right)^2 + \left(S_{ZZ}^{n+1}\right)^2 + S_{XX}^{n+1}\, S_{ZZ}^{n+1}\right] .$$

If $f \leq \frac{2}{3}(Y_o)^2$, the material is elastic and described by the above S_{XX}, S_{ZZ}, S_{XZ}. If $f \geq \frac{2}{3}(Y_o)^2$,

$$F = \sqrt{((2/3)Y_o)^2 \ / \ f} \ ,$$

$$S_{XX}^{n+1} = (F)(S_{XX}^{n+1}) + QXX \ ,$$

$$S_{ZZ}^{n+1} = (F)(S_{ZZ}^{n+1}) + QZZ\,,$$

$$S_{XZ}^{n+1} = (F)(S_{XZ}^{n+1}) + QXZ\,.$$

PHASE II

E. Artificial Viscosity

$$Q1_{i,j} = q^n_{i,j-\frac{1}{2}} = Q3_{i,j-1}\,,$$

3
2 i,j 4
1

except on boundary 1, when $Q1_{i,1} = 0.0$, or on piston boundary, when

$$Q1_{i,1} = K(\rho^n_{i,j} + MAPP)(VAPP - U^n_{Z\,i,1}) \text{ if } VAPP > U^n_{Z\,i,1}$$

$$Q1_{i,1} = 0 \quad \text{if } VAPP \leq U^n_{Z\,i,1}$$

$IAPP$, $VAPP$, $PAPP$, $UAPP$, and $MAPP$ are applied piston values.

$$Q2_{i,j} = q^n_{i-\frac{1}{2},j} = Q4_{i-1,j}\,,$$

except on axis boundary 2, when

$$Q2_{1,j} = 2Q4_{1,j} = -Q4_{2,j}\,.$$

$$Q3_{i,j} = q^n_{i,j+\frac{1}{2}} = K(\rho^n_{i,j} + \rho^n_{i,j+1})(U^n_{Z\,i,j} - U^n_{Z\,i,j+1}) \text{ if } U^n_{Z\,i,j} \geq U_{Z\,i,j+1}$$

$$Q3_{i,j} = 0 \quad \text{if } U^n_{Z\,i,j} < U^n_{Z\,i,j+1}\,,$$

except for $Q3_{i,JMAX} = 0$ on continuative boundary 3.

$$Q4_{i,j} = q^n_{i+\frac{1}{2},j} = K(\rho^n_{i,j} + \rho^n_{i+1,j})(U^n_{X\,i,j} - U^n_{X\,i+1,j}) \text{ if } U^n_{X\,i,j} \geq U^n_{X\,i+1,j}$$

$$Q4_{i,j} = 0 \quad \text{if } U^n_{X\,i,j} < U^n_{X\,i+1,j}\,,$$

except on continuative boundary 4 when $Q4_{IMAX,j} = 0$.

F. Real Viscosity

$$Q1_{i,j} = Q3_{i,j-1} ,$$

except on boundary 1, $Q1_{i,1} = q_{i,2}$, or piston boundary, $Q1_{i,1} = q_{i,1}$.

$$Q2_{i,j} = Q4_{i-1,j} ,$$

except on axis boundary 2, $Q2_{i,j} = q_{1,j}$.

$$Q3_{i,j} = (q_{i,j} + q_{i,j+1})/2 ,$$

except on continuative boundary 3, $Q3_{i,JMAX} = q_{i,JMAX-1}$.

$$Q4_{i,j} = (q_{i,j} + q_{i+1,j})/2 ,$$

except on continuative boundary 4, $Q4_{IMAX,j} = q_{IMAX-1,j}$.

G. Velocity

$$P1 = P^n_{i,j-1} ,$$

except on piston boundary 1, $P1 = PAPP$ or on continutative boundary 1, $P1 = P^n_{i,1}$.

$$P2 = P^n_{i-1,j} ,$$

except on axis boundary 2, $P2 = P^n_{i,j}$.

$$P3 = P^n_{i,j+1} ,$$

except on continuative boundary 3, $P3 = P^n_{i,JMAX}$.

$$P4 = P^n_{i+1,j} ,$$

except on continuative boundary 4, $P4 = P^n_{1,IMAX}$.

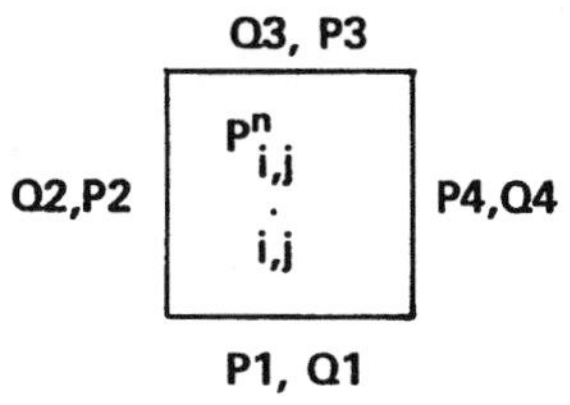

Q3, P3

Q2,P2 $P^n_{i,j}$ i,j P4,Q4

P1, Q1

where

$$i = 1, 2, \ldots\ldots\ IMAX ,$$

$j = 1, 2, \ldots\ldots\; JMAX$.

$$\tilde{U}^n_{X\,i,j} = U^n_{X\,i,j} - \frac{\Delta t}{\rho^n_{i,j}\Delta X}\left(\left[\frac{(i-1)(P4 - P2) + P4 - P^n_{i,j}}{2i-1}\right] + Q4_{i,j} - Q2_{i,j}\right).$$

$$\tilde{U}^n_{Z\,i,j} = U^n_{Z\,i,j} - \frac{\Delta t}{\rho^n_{i,j}\,\Delta Z}\left(\frac{P3 - P1}{2} + Q3_{i,j} - Q1_{i,j}\right).$$

In slab geometry, the quantity in [ ] = $(P4 - P2)/2$ for $\tilde{U}^n_{X\,i,j}$.

H. Elastic Plastic or Real Viscosity

$$P1 = P^n_{i,j-1} - S_{ZZ\,i,j-1}\,.$$

On piston or continuative boundary, $S_{ZZ\,i,j-1} = S_{ZZ\,i,j}$.

$$P2 = P^n_{i-1,j} - S_{XX\,i-1,j}\,.$$

On axis boundary, $S_{XX\,i-1,j} = S_{XX\,i,j}$.

$$P3 = P^n_{i,j+1} - S_{ZZ\,i,j+1}\,.$$

On continuative boundary 3, $S_{ZZ\,i,j+1} = S_{ZZ\,i,JMAX}$.

$$P4 = P^n_{i+1,j} - S_{XX\,i+1,j}\,.$$

On continuative boundary 4, $S_{XX\,i+1,j} = S_{XX\,IMAX,j}$.

$$P^n_{i,j} = P^n_{i,j} - S_{XX\,i,j} \text{ for } \tilde{U}_{X\,i,j} \text{ calculation only}.$$

Add to $\tilde{U}^n_{X\,i,j}$

$$\frac{\Delta t}{\rho^n_{i,j}}\left[\frac{1}{2\Delta Z}(S_{XZ\,i,j+1} - S_{XZ\,i,j-1}) + \frac{(\alpha-1)}{i\Delta X}(2S_{XX\,i,j} + S_{ZZ\,i,j})\right] + g_X\Delta t\,.$$

Add to $\tilde{U}^n_{Z\,i,j}$

$$\frac{\Delta t}{\rho^n_{i,j}}\left(\frac{1}{\Delta X}\left[\frac{(i-1)(S_{XZ\,i+1,j} - S_{XZ\,i-1,j}) + S_{XZ\,i+1,j} - S_{XZ\,i,j}}{2i-1}\right]\right)$$

$$+\frac{\Delta t}{\rho^n_{i,j}}\left(\frac{(\alpha-1)}{i\Delta X}(S_{XZ\,i,j})\right) + g_Z\Delta t\,.$$

In slab geometry $[\ ] = S_{XZ\ i+1,j} - S_{XZ\ i-1,j}$.
On axis boundary $[\ ] = 2(S_{XZ\ i+1,j} - S_{XZ\ i,j})$.

PHASE III

I. $\hat{\rho}$ Calculation

$$\hat{\rho}^n_{i,j} = \rho^n_{i,j}\left[1.0 - \frac{\Delta t}{2\Delta X}(U^n_{X\ i+1,j} - U^n_{X\ i-1,j}) - \frac{\Delta t}{2DZ}(U^n_{Z\ i,j+1} - U^n_{Z\ i,j-1})\right].$$

Exceptions:
Piston boundary, $U_{Z\ i,j-1} = VAPP$.
Axis boundary, $U_{X\ i-1,j} = -U_{X\ i+1,j} + 2U_{X\ i,j}$.
Boundary 3, $U_{Z\ i,j+1} = U_{Z\ i,j}$.
Boundary 4, $U_{X\ i+1,j} = U_{X\ i,j}$.

J. Piston Energy Constraints

For the first VCNT cycles,

$$\tilde{I}^n_{i,j} = 1/2(P^n_{i,j} + Q3_{i,j})(V_o - 1/\rho^n_{i,j}) + KE_{ij} \quad \text{if } V_o \geq 1/\rho^n_{i,j}\ ,$$

$$\tilde{I}^n_{i,j} = I^n_{i,j} \quad \text{if } V_o < 1/\rho^n_{i,j}\ ,$$

where

$$KE_{i,j} = 1/2[(\tilde{U}^n_{X\ i,j})^2 + (\tilde{U}^n_{Z\ i,j})^2]\ .$$

K. Internal Energy Calculations

For $\rho^n_{i,j} \leq MINGRHO + (\rho_o - MINGRHO)(W^n_{i,j})$, $E^n_{i,j} = I^n_{i,j}$, and the rest of Phase III is skipped.

$$U1 = (U^n_{X\ i-1,j} + \tilde{U}^n_{X\ i-1,j})\ .$$

$$U2 = (U^n_{X\ i+1,j} + \tilde{U}^n_{X\ i+1,j})\ .$$

$$V1 = (U^n_{Z\ i,j-1} + \tilde{U}^n_{Z\ i,j-1})\ .$$

$$V2 = (U^n_{Z\ i,j+1} + \tilde{U}^n_{Z\ i,j+1})\ .$$

$$T3 = (U^n_{X\,i,j} + \tilde{U}^n_{X\,i,j})\,.$$

$$T1 = (U^n_{Z\,i,j} + \tilde{U}^n_{Z\,i,j})\,.$$

Exceptions:

Axis Boundary, $U1 = 2(U^n_{X\,1,j} + \tilde{U}^n_{X\,1,j}) - U2$.

Continuative boundary 4, $U2 = (U^n_{X\,IMAX,j} + \tilde{U}^n_{X\,IMAX,j})$.

Piston boundary, $V1 = 2VAPP$.

Continuative boundary 1, $V1 = (U^n_{Z\,i,1} + \tilde{U}^n_{Z\,i,1})$.

Continuative boundary 3, $V2 = (U^n_{Z\,i,JMAX} + \tilde{U}^n_{Z\,i,JMAX})$.

$$\tilde{I}^n_{i,j} = I^n_{i,j} - \frac{\Delta t}{4\rho_{i,j}}\frac{P^n_{i,j}}{\Delta X}\left[U2 - U1 + \left(\frac{U2 + 2T3 + U1}{2i-1}\right)\right]$$

$$-\frac{\Delta t}{4\rho_{i,j}}\frac{Q4_{i,j}}{\Delta X}\left[U2 - T3 + \left(\frac{2U2}{2i-1}\right)\right] + \frac{\Delta t}{4\rho_{i,j}}\frac{Q2_{i,j}}{\Delta X}\left[U1 - T3 + \left(\frac{2U1}{2i-1}\right)\right]$$

$$-\frac{\Delta t}{4\rho_{i,j}}\frac{1}{\Delta Z}\left[P^n_{i,j}(V2 - V1) + Q3_{i,j}(V2 - T1) + Q1_{i,j}(T1 - V1)\right].$$

In slab geometry the quantities in ( ) are set equal to zero.

L. Elastic-Plastic

If the calculation is elastic-plastic, add to $\tilde{I}_{i,j}$ the term

$$\frac{\Delta t}{\rho_{i,j}}\left(S_{XX\,i,j}\left[DUDX_{i,j} - \frac{(\alpha-1)U_{X\,i,j}}{i\Delta X}\right] + S_{ZZ\,i,j}\left[DVDZ_{i,j} - \frac{(\alpha-1)U_{X\,i,j}}{i\Delta X}\right]\right)$$

$$+\frac{\Delta t}{\rho_{i,j}}S_{XZ\,i,j}(DUDZ_{i,j} + DVDX_{i,j})\,.$$

If heat conduction occurs, add to $\tilde{I}^n_{i,j}$

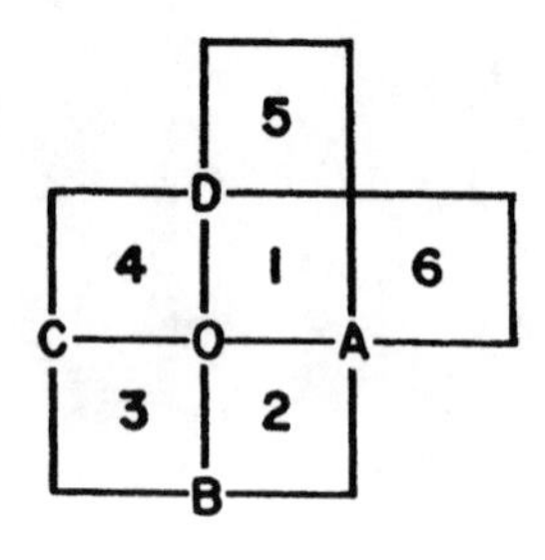

$$\frac{\lambda \Delta t}{\rho_{i,j}}\left[\frac{T_4 - 2T_1 + T_6}{(\Delta X)^2} + \frac{T_5 - 2T_1 + T_2}{(\Delta Z)^2} + \frac{(\alpha - 1)}{i\ \Delta X}\left(\frac{T4 - T6}{2\Delta X}\right)\right].$$

On boundaries, set T equal to nearest T.

Boundary 4, $T_6 = T_1$

Boundary 1, $T_3 = T_4$, $T_2 = T_1$

M. Total Energy Calculation

$$E^n_{i,j} = \tilde{I}^n_{i,j} + 1/2\left[(\tilde{U}^n_{Z\ i,j})^2 + (\tilde{U}^n_{X\ i,j})^2\right].$$

PHASE IV

N. Mass Movement

Mass is not moved if the pressure of the cell from which it moves is less than FREPR (0.0005) or if the cell is in tension. FREPR is the pressure below which mass is not moved.

Nomenclature for Phase IV and V:

DE Change in energy

DW Change in mass fraction

DPU Change in U_X momentum

DPV Change in U_Z momentum

DMASS Change in density

GEO Geometry factor

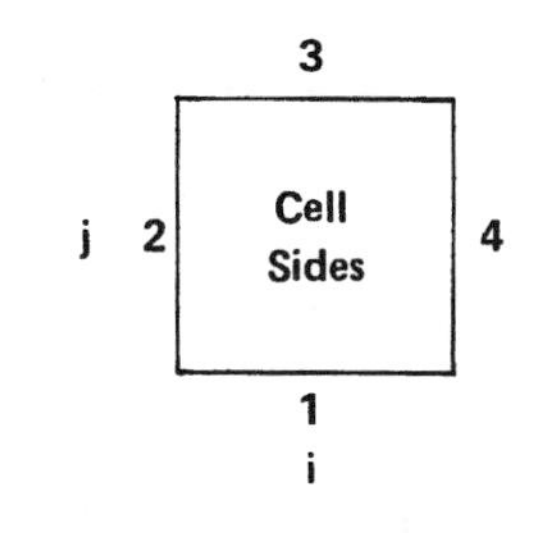

1. Mass Movement Across Side 2

For axis boundary $DMASS = 0$, otherwise

$$\Delta = \frac{1/2(\tilde{U}^n_{X\ i,j} + \tilde{U}^n_{X\ i-1,j})(\Delta t/\Delta X)}{1 + (\tilde{U}^n_{X\ i,j} - \tilde{U}^n_{X\ i-1,j})(\Delta t/\Delta X)}.$$

If $\Delta \geq 0$ the mass moves from donor cell $i-1, j$ to cell i, j.

$$\text{Cylindrical } DMASS = \hat{\rho}^n_{i-1,j}\left(\frac{2i - 2 - \Delta}{2i - 3}\right)(\Delta),\quad GEO = \frac{2i-3}{2i-1}.$$

$$\text{Slab } DMASS = \hat{\rho}^n_{i-1,j}(\Delta),\quad GEO = 1.0\,.$$

$$DE_{i,j} = E^n_{i-1,j}(DMASS)(GEO),\quad DE^n_{i-1,j} = DE_{i-1,j} - E^n_{i-1,j}(DMASS)\,.$$

Mixed-cell modification of $DMASS$ and DE occurs, if required, as described in the mixed-cell section.

$$DW_{i,j} = W^n_{i-1,j}(DMASS)(GEO), \quad DW^n_{i-1,j} = DW_{i-1,j} - W^n_{i-1,j}(DMASS)\,.$$

$$DPU_{i,j} = \tilde{U}^n_{X\ i-1,j}(DMASS)(GEO)\,.$$

$$DPU^n_{i-1,j} = DPU_{i-1,j} - \tilde{U}^n_{X\ i-1,j}(DMASS)\,.$$

$$DPV_{i,j} = \tilde{U}^n_{Z\ i-1,j}(DMASS)(GEO)\,.$$

$$DPV^n_{i-1,j} = DPV_{i-1,j} - \tilde{U}^n_{Z\ i-1,j}(DMASS)\,.$$

$$DM_{i,j} = DMASS(GEO), \quad DM^n_{i-1,j} = DM_{i-1,j} - DMASS\,.$$

If $\Delta < 0$, the mass moves from the donor cell i, j to the acceptor cell $i - 1, j$.

$$\text{Cylindrical } DMASS = \hat{\rho}^n_{i,j}\left(\frac{2i - 2 - \Delta}{2i - 1}\right)(\Delta), \quad GEO = \frac{2i - 1}{2i - 3}\,.$$

$$\text{Slab } DMASS = \hat{\rho}^n_{i,j}(\Delta), \quad GEO = 1.0\,.$$

$$DE_{i,j} = E^n_{i,j}(DMASS), \quad DE^n_{i-1,j} = DE_{i-1,j} - E^n_{i,j}(DMASS)(GEO)\,.$$

Mixed-cell modification of $DMASS$ and DE occurs, if required, as described in the mixed-cell section.

$$DW_{i,j} = W^n_{i,j}(DMASS), \quad DW^n_{i-1,j} = DW_{i-1,j} - W^n_{i,j}(DMASS)(GEO)\,.$$

$$DPU_{i,j} = \tilde{U}^n_{X\ i,j}(DMASS)\,.$$

$$DPU^n_{i-1,j} = DPU_{i-1,j} - \tilde{U}^n_{X\ i,j}(DMASS)(GEO)\,.$$

$$DPV_{i,j} = \tilde{U}^n_{Z\ i,j}(DMASS)\,.$$

$$DPV^n_{i-1,j} = DPV_{i-1,j} - \tilde{U}^n_{Z\ i,j}(DMASS)(GEO)\ .$$

$$DM_{i,j} = DMASS,\ \ DM^n_{i-1,j} = DM_{i-1,j} - DMASS(GEO)\ .$$

2. Mass Movement Across Side 1

For a piston boundary,

$$\Delta = \frac{1/2(VAPP + \tilde{U}^n_{Z\ i,j})(\Delta t/\Delta Z)}{1 + (\tilde{U}^n_{Z\ i,j} - VAPP)(\Delta t/\Delta Z)}\ .$$

For $\Delta < 0$, $DMASS = 0$. For $\Delta > 0$, $DMASS = (MAPP)(\Delta)$, and the mass moves from the piston to cell i, j.

$$DE^n_{i,j} = DE_{i,j} + (DMASS)(EAPP)\ .$$

$$DW^n_{i,j} = DW_{i,j} + (DMASS)(WAPP)\ .$$

$$DPV^n_{i,j} = DPV_{i,j} + (DMASS)(VAPP)\ .$$

$$DPU^n_{i,j} = DPU_{i,j} + (DMASS)(UAPP)\ .$$

$$DM^n_{i,j} = DM_{i,j} + (DMASS)\ .$$

For continuative boundary 1 cell,

$$\Delta = 1/2(3\tilde{U}^n_{Z\ i,j} - \tilde{U}^n_{Z\ i,j+1})(\Delta t/\Delta Z)\ .$$

$$DMASS = \hat{\rho}^n_{i,j}(\Delta)\ .$$

$$DE^n_{i,j} = DE_{i,j} + E^n_{i,j}(DMASS)\ .$$

Mixed-cell modifications of DMASS and DE occur, if required.

$$DW^n_{i,j} = DW_{i,j} + W^n_{i,j}(DMASS)\ .$$

$$DPV^n_{i,j} = DPV_{i,j} + \tilde{U}^n_{Z\ i,j}(DMASS)\ .$$

$$DPU^n_{i,j} = DPU_{i,j} + \tilde{U}^n_{X\ i,j}(DMASS)\ .$$

$$DM^n_{i,j} = DM_{i,j} + (DMASS)\ .$$

Otherwise,

$$\Delta = \frac{1/2(\tilde{U}^n_{Z\ i,j} + \tilde{U}^n_{Z\ i,j-1})(\Delta t/\Delta Z)}{1 + (\tilde{U}^n_{Z\ i,j} - \tilde{U}^n_{Z\ i,j-1})(\Delta t/\Delta Z)}\ .$$

If $\Delta \geq 0$, the mass moves from donor cell $i, j-1$ to cell i, j.

$$DMASS = \hat{\rho}^n_{i,j-1}(\Delta)\ .$$

$$DE^n_{i,j} = DE_{i,j} + E^n_{i,j-1}(DMASS)\ .$$

$$DE^n_{i,j-1} = DE_{i,j-1} - E^n_{i,j-1}(DMASS)\ .$$

Mixed-cell modification of $DMASS$ and DE occurs, if required, as described in the mixed-cell section.

$$DW^n_{i,j} = DW_{i,j} + W^n_{i,j-1}(DMASS)\ .$$

$$DW^n_{i,j-1} = DW_{i,j-1} - W^n_{i,j-1}(DMASS)\ .$$

$$DPV^n_{i,j} = DPV_{i,j} + \tilde{U}^n_{Z\ i,j-1}(DMASS)\ .$$

$$DPV^n_{i,j-1} = DPV_{i,j-1} - \tilde{U}^n_{Z\ i,j-1}(DMASS)\ .$$

$$DPU^n_{i,j} = DPU_{i,j} + \tilde{U}^n_{X\ i,j-1}(DMASS)\ .$$

$$DPU^n_{i,j-1} = DPU_{i,j-1} - \tilde{U}^n_{X\ i,j-1}(DMASS)\ .$$

$$DM^n_{i,j} = DM_{i,j} + DMASS\ .$$

$$DM^n_{i,j-1} = DM_{i,j-1} - DMASS\ .$$

If $\Delta < 0$, the mass moves from the donor cell i, j to the acceptor cell $i, j-1$.

$$DMASS = \hat{\rho}^n_{i,j}(\Delta) .$$

$$DE^n_{i,j} = DE_{i,j} + E^n_{i,j}(DMASS) .$$

$$DE^n_{i,j-1} = DE_{i,j-1} - E^n_{i,j}(DMASS) .$$

Mixed-cell modification of $DMASS$ and DE occurs, if required, as described in the mixed-cell section.

$$DW^n_{i,j} = DW_{i,j} + W^n_{i,j}(DMASS) .$$

$$DW^n_{i,j-1} = DW_{i,j-1} - W^n_{i,j}(DMASS) .$$

$$DPV^n_{i,j} = DPV_{i,j} + \tilde{U}^n_{Z\ i,j}(DMASS) .$$

$$DPV^n_{i,j-1} = DPV_{i,j-1} - \tilde{U}^n_{Z\ i,j}(DMASS) .$$

$$DPU^n_{i,j} = DPU_{i,j} + \tilde{U}^n_{X\ i,j}(DMASS) .$$

$$DPU^n_{i,j-1} = DPU_{i,j-1} - \tilde{U}^n_{X\ i,j}(DMASS) .$$

$$DM^n_{i,j} = DM_{i,j} + DMASS .$$

$$DM^n_{i,j-1} = DM_{i,j-1} - DMASS .$$

3. Mass Movement Across Side 3

Except on continuative boundary 3, this mass movement is taken care of by the mass movement across side 1 of the cell directly above.

On the boundary 3,

$$\Delta = 1/2(3\tilde{U}^n_{Z\ i,j} - \tilde{U}^n_{Z\ i,j-1})(\Delta t/\Delta Z) .$$

$$DMASS = \hat{\rho}^n_{i,j}(\Delta) .$$

$$DE^n_{i,j} = DE_{i,j} - E^n_{i,j}(DMASS)\ .$$

Mixed-cell modifications of DMASS and DE occur, if required.

$$DW^n_{i,j} = DW_{i,j} - W^n_{i,j}(DMASS)\ .$$

$$DPV^n_{i,j} = DPV_{i,j} - \tilde{U}^n_{Z\ i,j}(DMASS)\ .$$

$$DPU^n_{i,j} = DPU_{i,j} - \tilde{U}^n_{X\ i,j}(DMASS)\ .$$

$$DM^n_{i,j} = DM_{i,j} - (DMASS)\ .$$

4. Mass Movement Across Side 4

Except on the continuative boundary 4, this mass movement is taken care of by the mass movement across side 2 of the cell on its right.

For $i = IMAX$.

$$\Delta = 1/2(3\tilde{U}^n_{X\ i,j} - \tilde{U}^n_{X\ i-1,j})(\Delta t/\Delta X)\ .$$

if $\Delta \geq 0$ mass moves out of cell i, j.

$$\text{Cylindrical } DMASS = \hat{\rho}^n_{i,j}\frac{(2i - \Delta)(\Delta)}{(2i - 1)}\ .$$

if $\Delta \leq 0$ mass moves into cell i, j.

$$\text{Cylindrical } DMASS = \hat{\rho}^n_{i,j}\frac{(2i - \Delta)(\Delta)}{(2i + 1)}\ .$$

$$\text{Slab } DMASS = \hat{\rho}^n_{i,j}(\Delta)\ .$$

$$DE^n_{i,j} = DE_{i,j} - E^n_{i,j}(DMASS)\ .$$

Mixed-cell modifications of DMASS and DE occur, if required.

$$DW^n_{i,j} = DW_{i,j} - W^n_{i,j}(DMASS)\ .$$

$$DPV^n_{i,j} = DPV_{i,j} - \tilde{U}^n_{Z\ i,j}(DMASS)\ .$$

$$DPU^n_{i,j} = DPU_{i,j} - \tilde{U}^n_{X\ i,j}(DMASS)\ .$$

$$DM^n_{i,j} = DM_{i,j} - (DMASS) \ .$$

5. Mixed Cells

The composition of the mass to be moved from the donor to acceptor cell is determined as follows. Materials common to both the donor and acceptor cell are moved according to the mass fractions of common materials in the acceptor cell. If the donor and acceptor cell have no common materials, then mass is moved according to the mass fractions of the donor cell. The mass to be moved from the donor cell has the density and energy determined for that component or components by the mixture equation of state calculation in Phase I. Only one material is permitted to be depleted in one time step. Therefore, the $DMASS$ term is corrected by dividing by cell ρ used in Phase IV and replacing it with the ρ of the material being moved from the donor cell (ρ_k).

$$DMASS \ = \ DMASS\frac{\rho_k}{\hat{\rho}} \ .$$

If the mass of the material (k) in the donor cell is less than the total mass to be moved, the remainder of the mass moved has the remaining donor-cell composition, density, and energy, and a new $DMASS$.

The DE term is calculated using the internal energy of the component being moved from the donor cell as calculated from the mixture equation of state routines in Phase I.

$$DE^{n+1}_{DONOR} \ = \ DE^n_{DONOR} - (DMASS)(\text{Donor Component I} + \text{Donor K.E.}) \ .$$

$$DE^{n+1}_{ACCEPTOR} \ = \ DE^n_{ACCEPTOR} + (DMASS)(\text{Donor I} + \text{Donor K.E.}) \ .$$

The mass fraction of decomposing explosive between a mixed and an unmixed cell containing only inerts is adjusted by

$$W^n_{ACCEPTOR} \ = \ (W^n_{DONOR})(1 + DMASS_{ACCEPTOR})$$

to permit proper repartition in Phase V. By convention, W is set equal to 1.0 for an inert and 0.0 for a gas if the donor or acceptor does not contain an explosive.

The stress deviator for a mixed cell is chosen as the deviator of the largest component in a nearby cell of that component only.

PHASE V

O. Repartition

Add on mass-moved quantities.
For $\rho^n_{i,j} \ > \ MINGRHO \ + \ (\rho_o \ - \ MINGRHO) \ W^n_{i,j}$,

$$\rho^{n+1}_{i,j} \ = \ \rho^n_{i,j} \ + \ DM_{i,j} \ \ .$$

$$W^{n+1}_{i,j} = \frac{1}{\rho^{n+1}_{i,j}}(\rho^n_{i,j} \ W^n_{i,j} \ + \ DW_{i,j}) \ \ .$$

$$U_{Z\,i,j}^{n+1} = \frac{1}{\rho_{i,j}^{n+1}}(\tilde{U}_{Z\,i,j}^{n}\rho_{i,j}^{n} + DPV_{i,j}) \ .$$

$$U_{X\,i,j}^{n+1} = \frac{1}{\rho_{i,j}^{n+1}}(\tilde{U}_{X\,i,j}^{n}\rho_{i,j}^{n} + DPU_{i,j}) \ .$$

$$I_{i,j}^{n+1} = \frac{1}{\rho_{i,j}^{n+1}}(\rho_{i,j}^{n}\ E_{i,j}^{n} + DE_{i,j}) - \frac{1}{2}\left[(U_{Z\,i,j}^{n+1})^2 + (U_{X\,i,j}^{n+1})^2\right].$$

For $\rho_{i,j}^{n} < MINGRHO + (\rho_o - MINGRHO)(W_{i,j}^{n})$ and not mixed or in tension,

$$\rho_{i,j}^{n+1} = \rho_{i,j}^{n} + DM_{i,j} \ .$$

$$W_{i,j}^{n+1} = \frac{1}{\rho_{i,j}^{n+1}}(\rho_{i,j}^{n}\ W_{i,j}^{n} + DW_{i,j}) \ .$$

$$U_{Z\,i,j}^{n+1} = \frac{1}{\rho_{i,j}^{n+1}}(U_{Z\,i,j}^{n}\rho_{i,j}^{n} + DPV_{i,j}) \ .$$

$$U_{X\,i,j}^{n+1} = \frac{1}{\rho_{i,j}^{n+1}}(U_{X\,i,j}^{n}\rho_{i,j}^{n} + DPU_{i,j}) \ .$$

$$I_{i,j}^{n+1} = 0.0 \ .$$

P. Form Volume Derivative

$$DVDT_{i,j} = \frac{1}{V}\frac{\partial V}{\partial t} = \frac{V^{n+1} - V^{n}}{V^{n+1} + V^{n}}\left(\frac{2}{\Delta t}\right) \ .$$

For a mixed cell, the $DVDT_{i,j}$ is formed for largest mass fraction component in the cell.

PHASE VI

Q. Elastic-Plastic Flow

$$W' = 1/2(DUDZ - DVDX) \ .$$

$$DSXX = 2\Delta t\left[\mu\left(DUDX - \frac{1}{3}DVDT\right) + W'S_{XZ}\right] .$$

$$DSZZ = 2\Delta t\left[\mu\left(DVDZ - \frac{1}{3}DVDT\right) + W'S_{XZ}\right] .$$

$$DSXZ = \Delta t\left[\mu(DUDZ + DVDX) + W'(S_{XX} - S_{ZZ})\ \right] .$$

R. Form Derivatives

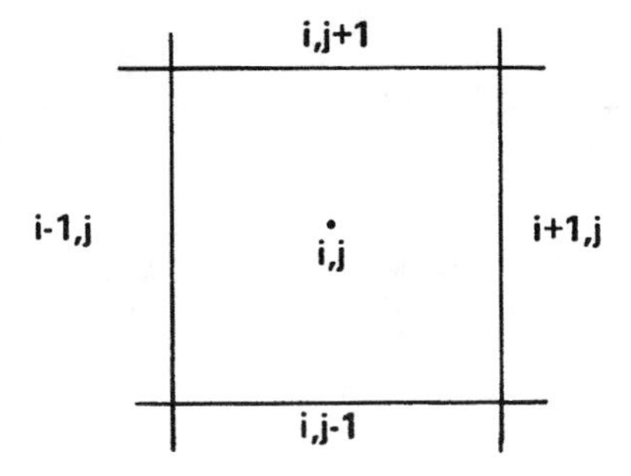

$$DUDX_{i,j} = -\left(\frac{U_{X\ i-1,j} - U_{X\ i+1,j}}{2\Delta X}\right) .$$

On axis boundary 2, $DUDX_{i,j} = \dfrac{U_{X\ i+1,j}}{2\Delta X}$.

On boundary 4, $DUDX_{i,j} = -\left(\dfrac{U_{X\ i-1,j} - U_{X\ i,j}}{\Delta X}\right)$.

$$DUDZ_{i,j} = -\left(\frac{U_{X\ i,j-1} - U_{X\ i,j+1}}{2\Delta Z}\right) .$$

On piston boundary 1, $DUDZ_{i,j} = -\left(\dfrac{UAPP - U_{X\ i,j+1}}{2\Delta Z}\right)$.

On continuative boundary 1, $DUDZ_{i,j} = -\left(\dfrac{U_{X\ i,j} - U_{X\ i,j+1}}{\Delta Z}\right)$.

On boundary 3, $DUDZ_{i,j} = -\left(\dfrac{U_{X\ i,j-1} - U_{X\ i,j}}{\Delta Z}\right)$.

$$DVDX_{i,j} = -\left(\frac{U_{Z\,i-1,j} - U_{Z\,i+1,j}}{2\Delta X}\right) .$$

On axis boundary 2, $DVDX_{i,j} = -\left(\frac{U_{Z\,i,j} - U_{Z\,i+1,j}}{\Delta X}\right)$.

On boundary 4, $DVDX_{i,j} = -\left(\frac{U_{Z\,i-1,j} - U_{Z\,i,j}}{\Delta X}\right)$.

$$DVDZ_{i,j} = -\left(\frac{U_{Z\,i,j-1} - U_{Z\,i,j+1}}{2\Delta Z}\right) .$$

On piston boundary 1, $DVDZ_{i,j} = -\left(\frac{VAPP - U_{Z\,i,j+1}}{2\Delta Z}\right)$.

On boundary 3, $DVDZ_{i,j} = -\left(\frac{U_{Z\,i,j-1} - U_{Z\,i,j}}{\Delta Z}\right)$.

C.2 Derivations of Momentum, Energy, and Mass Movement Equations

The derivations of the difference approximations used above to describe the momemtum and energy equations with transport terms dropped are presented in Reference 10. The derivation of the mass movement equations are presented in References 7, 8, and 10.

C.3 Equations of State

The HOM equation of state, described in Appendix A, was used for a cell containing a single component or any mixture of condensed explosive and detonation products. Temperature and pressure equilibrium was assumed for any mixture of detonation products and condensed explosive in a cell.

When there are two components present in a cell, separated by a boundary and not homogeneously mixed, it is reasonable to assume pressure equilibrium, but the temperatures may be quite different. For these systems it was assumed that the difference between the total Hugoniot energy and the total cell energy was distributed between the components according to the ratio of the Hugoniot energies of the components for two solids or liquids. For two gases it was assumed that the difference between the total isentrope energy and the total cell energy was distributed between the components according to the ratio of the isentrope energies of the components. For a solid and a gas it was assumed that the difference between the sum of the gas isentrope energy and the solid Hugoniot energy and the total cell energy was distributed between the components according to the ratio of the gas isentrope energy and the solid Hugoniot energy of the components.

When there are three components in a cell, an explosive, its detonation products, and a third solid or liquid nonreactive component, the equation of state was computed assuming temperature and pressure equilibrium for the detonation products and condensed explosive, and pressure equilibrium with the nonreactive component.

The equation of state subroutines require for input the specific volume, internal energy, and mass fractions of the components. The subroutines iterate to give as output the pressure, the individual densities, the temperatures, and the energies of the components.

This treatment of the equation of state of mixtures is appropriate because the state values of the components are usually close or identical to the state values of the standard curve (the Hugoniot for the solid or the isentrope for the gas). For gases an isentrope that passes through state values near those expected in the problem should be used for best results.

The Nomenclature

$C,\ S$	Coefficients to the linear fit $U_S = C + S\ U_P$
$C1,\ S1$	Second set of Coefficients $U_S = C1 + S1\ U_P$ for Volumes less than MINV or VSW
C_V	Heat Capacity of Condensed Component (cal/g/deg)
C'_V	Heat Capacity of Gaseous Component (cal/g/deg)
I	Total Internal Energy (Mbar-cc/g)
P	Pressure (Mbar)
SPA	Spalling Constant
$SPALL\ P$	Interface Spalling Pressure
$SPMIN$	Minimum Pressure for Spalling
T	Temperature (^{o}K)
USP	Ultimate Spall Pressure
U_P	Particle Velocity (cm/μsec)
U_S	Shock Velocity (cm/μsec)
V	Total Specific Volume (cc/gram)
V_o	Initial Specific Volume of Condensed Explosive
W	Mass Fraction of Undecomposed Explosive
X or χ	Mass Fraction of Solid or Gaseous Component
α	Linear Coefficient of Thermal Expansion
Subscripts	
g or G	Gaseous Component
H	Hugoniot
i	Isentrope
s or S	Condensed Component
A	Component A
B	Component B

A. HOM

HOM is used for a single solid or gas component and for mixtures of solid and gas components in pressure and temperature equilibrium. It is described in detail in Appendix A.

The Method

1. Condensed Components

The mass fraction, W, is 1; the internal energy, I, is I_s; and the specific volume, V, is V_s. For volumes less than V_o, the experimental shock Hugoniot data are expressed as linear fits of the shock and particle velocities. The shock Hugoniot temperatures are computed using the Walsh and Christian technique described in Appendix A, and are fit to a fourth-degree polynomial in log of the volume.

$$U_S = C + S\,U_P\,.$$

$$P_H = [C^2(V_o - V_s)]\,/\,[V_o - S(V_o - V_s)]^2\,.$$

$$\ln T_H = F_s + G_s \ln V_s + H_s (\ln V_s)^2 + I_s (\ln V_s)^3 + J_s (\ln V_s)^4\,. \tag{C.1}$$

$$I_H = (1/2)\,P_H\,(V_o - V_s)\,.$$

$$P_s = (\gamma_s/V_s)\,(I_s - I_H) + P_H,\ \text{where}\ \gamma_s = V(\partial P/\partial E)_V\,. \tag{C.2}$$

$$T_s = T_H + [(I_s - I_H)(23,890)/C_V]\,. \tag{C.3}$$

Two sets of C and S coefficients are sometimes necessary to fit the experimental Hugoniot data. For $V_s < MINV$, the fit $U_S = C1 + S1\,U_P$ is used. For volumes greater than V_o, the Grüneisen equation of state was used along with the $P = 0$ line as the standard curve.

$$P_s = \left[I_s - \frac{C_V}{(3)(23,890)(\alpha)}\left(\frac{V_s}{V_o} - 1\right)\right]\frac{\gamma_s}{V_s}\,.$$

$$T_s = \frac{(I_s)(23,890)}{C_V} + T_o\,.$$

Gradient spalling does not occur if $SPA < 0.0001$. If $P_s \leq -SPA\sqrt{\Delta P/\Delta X}$ where $\Delta P/\Delta X$ is the tension rate, and $P_s \leq -SPMIN$ (5×10^{-3}), the P_s is set equal to $SPALLP$ and the spall indicator is set. Spalling does not occur if neither of the above conditions is satisfied. The gradient spall model is described in Appendix A.

2. Gas Components

The mass fraction, W, is 0; the internal energy, I, is I_g; and the specific volume, V, is V_g. The pressure, volume, temperature, and energy values of the detonation products are computed using the BKW code described in Appendix E. They are fitted by the method of least squares to Equations (C.4)-(C.6).

$$ln\ P_i = A + B\ ln\ V_g + C(ln\ V_g)^2 + D(ln\ V_g)^3 + E(ln\ V_g)^4 . \quad (C.4)$$

$$ln\ I_i = K + L\ ln\ P_i + M(ln\ P_i)^2 + N(ln\ P_i)^3 + O(ln\ P_i)^4 . \quad (C.5)$$

$I_i = I_i - Z$ (where Z is a constant used to change the standard state to be consistent with the solid explosive standard state. If the states are the same, Z is used to keep I positive when making a fit.)

$$ln\ T_i = Q + R\ ln\ V_g + S(ln\ V_g)^2 + T(ln\ V_g)^3 + U(ln\ V_g)^4 . \quad (C.6)$$

$$-(1/\beta) = R + 2\,S(ln\ V_g) + 3T(ln\ V_g)^2 + 4U(ln\ V_g)^3 .$$

$$P_g = (1/\beta V_g)(I_g - I_i) + P_i . \quad (C.7)$$

$$T_g = T_i + [(I_g - I_i)(23,890)/C'_V] . \quad (C.8)$$

3. *Mixture of Condensed and Gaseous Components*

$$(0 < W < 1) .$$

$$V = WV_s + (1 - W)V_g .$$

$$I = WI_s + (1 - W)I_g .$$

$$P = P_s = P_g .$$

$$T = T_s = T_g .$$

Equation (C.3) is multiplied by (W/C_V) and Equation (C.8) by $(1 - W)C'_V$ and the resulting equations are added. T is substituted for T_s and T_g and I for $WI_s + (1 - W)I_g$ to get

$$T = \frac{23,890}{C_v W + C'_v(1 - W)}[I - (W\ I_H + I_i(1 - W))]$$

$$+\frac{1}{C_v W + C'_v(1-W)}\left[(T_H\, C_v\, W + T_i C'_v(1-W))\right]. \tag{C.9}$$

Equation (C.2) is equated to (C.7) and substituted using (C.9) to get

$$\left(\frac{\gamma_s C_v}{V_s} - \frac{C'_v}{\beta V_g}\right)\left(\frac{1}{C_v W + C'_v(1-W)}\right)[I - (W\, I_H + I_i(1-W))]$$

$$+\left(\frac{\gamma_s C_v}{V_s} - \frac{C'_v}{\beta V_g}\right)\left(\frac{1}{C_v W + C'_v(1-W)}\right)\frac{1}{23,890}\left[(T_H C_v W + T_i C'_v(1-W))\right]$$

$$-\frac{1}{23,890}\left(\frac{\gamma_s\, C_v T_H}{V_s} - \frac{C'_v T_i}{\beta V_g}\right) + P_H \;-\; P_i = 0.0\,. \tag{C.10}$$

Knowing $V, I,$ and W, linear feedback (described in Appendix E) is used to iterate on either V_g or V_s until equation (C.10) is satisfied.

For $V < V_o$, iterate on V_s with an initial guess of $V_s = V_o$ and a ratio to get the second guess of 0.999. For $V \geq V_o$, iterate on V_g with an initial guess of $V_g = (V - 0.9V_o\, W)(1-W)$ and a ratio to get the second guess of 1.002.

If the iteration goes out of the physical region ($V_g \leq 0$ or $V_s \leq 0$), that point is replaced by $V_s = V_g = V$. Then, knowing V_s and V_g, P and T are calculated.

B. HOM2S

HOM2S is used for two solids or liquids that are in pressure, but not temperature, equilibrium.

The Method

Knowing total energy I, total volume V, and the mass fraction X of component A present, iterate for the volume of A, V^A, if X is greater than 0.5, and for the volume of B, V^B, if X is less than 0.5.

To obtain the first guess it is assumed that the volumes of A and B are proportional to the initial specific volumes of the components.

$$V^A = V/\left(X + \frac{V_o^B}{V_o^A}(1-X)\right)$$

or

$$V^B = V/\left(W\frac{V_o^A}{V_o^B} + 1 - W\right)$$

Using the following relationships

$$V = X(V^A) + (1-X)(V^B)\,,$$

$$P = P^A = P^B \,. \tag{C.11}$$

Knowing V^A and V^B, I_H^A and I_H^B are calculated from Equation (C.1).

$$I_H = X(I_H^A) + (1 - X)(I_H^B) \,,$$

$$I^A = I_H^A + (I - I_H)\left[\frac{I_H^A}{I_H}\right] ,$$

and

$$I^B = I_H^B + (I - I_H)\left[\frac{I_H^B}{I_H}\right] .$$

From Equation (C.2) and Equation (C.11),

$$\frac{\gamma_S^A}{V^A}(I^A - I_H^A) + P_H^A - \frac{\gamma_S^B}{V^B}(I^B - I_H^B) - P_H^B = 0 \,. \tag{C.12}$$

Knowing V, I, and X, linear feedback is used to iterate on either V^A or V^B until Equation (C.12) is satisfied. The pressure and temperatures are calculated as in HOM.

C. HOMSG

HOMSG is used when there is one solid or liquid and one gas in pressure, but not temperature, equilibrium.

The Method

Knowing total energy I, total volume V, and the mass fraction X of the solid present, iterate on the solid volume V^S. To obtain the first guess, V^S is set equal to V if less than V_o^S, otherwise, V^S is set equal to $0.99V_o^S$. Using the following relationships

$$V = X(V^S) + (1 - X)(V^G) \,,$$

$$P = P^S = P^G \,. \tag{C.13}$$

Knowing V^S and V^G, I_H and I_i^G are calculated from Equations (C.2) and (C.5).

$$I' = X(I_H^S) + (1 - X)(I_i^G) \,,$$

$$I'' = X(I_H^S) + (1 - X)(|I_i^G|) \,,$$

$$I^S = I_H^S + (I - I')\left[\frac{I_H^S}{I''}\right] ,$$

and

$$I^G = I_i^G + (I - I')\left[\frac{|I_i^G|}{I''}\right].$$

From Equations (C.13), (C.2), and (C.7),

$$\frac{\gamma_S}{V^S}(I^S - I_H^S) + P_H^S - \frac{1}{\beta V^G}(I^G - I_i^G) - P_i^G = 0. \qquad (C.14)$$

Knowing V, I, and X, linear feedback is used to iterate on V^S until Equation (C.14) is satisfied. The pressure and temperatures are calculated as in HOM.

D. HOM2G

HOM2G is used for two gases that are in pressure, but not temperature, equilibrium.

The Method

Knowing total energy I, total volume V, and the mass fraction X of gas A present, iterate on V^A. To obtain the first guess, it is assumed that the volumes of A and B are proportional to the initial specific volumes of the components with the limitation that the ratio of the initial specific volumes is less than 10 or greater than 0.1.

$$V^A = V/\left(X + \frac{V_o^B}{V_o^A}(1 - X)\right)$$

Using the following relationships

$$V = X(V^A) + (1 - X)(V^B),$$

$$P = P^A = P^B. \qquad (C.15)$$

Knowing V^A and V^B, I_i^A and I_i^B are calculated from Equation (C.5).

$$I' = X(I_i^A) + (1 - X)(I_i^B),$$

$$I'' = X(|I_i^A|) + (1 - X)(|I_i^B|),$$

$$I^A = I_i^A + (I - I')\left[\frac{|I_i^A|}{I''}\right],$$

and

$$I^B = I_i^B + (I - I')\left[\frac{|I_i^B|}{I''}\right].$$

From Equation (C.15) and Equation (C.7),

$$\frac{1}{\beta^A V^A}(I^A - I_i^A) + P_i^A - \frac{1}{\beta^B V^B}(I^B - I_i^B) - P_i^B = 0 . \tag{C.16}$$

Knowing V, I, and X, linear feedback is used to iterate on V^A until Equation (C.16) is satisfied. The pressure and temperatures are calculated as in HOM.

E. HOM2SG

HOM2SG is used for mixtures of an undecomposed explosive (1), detonation products (2), and a nonreactive component (3) that is in pressure, but not thermal, equilibrium with the explosive and its products which are assumed to be in pressure and thermal equilibrium. W is the mass fraction of undecomposed explosive and χ is the mass fraction of the nonreactive component.

To obtain the initial guess for the volume of the undecomposed explosive $V^{(1)}$, the volume of the detonation products $V^{(2)}$ and the volume of the nonreactive component $V^{(3)}$, a volume for $V^{(2)}$ is assumed to be 0.9 of $V_o^{(1)}$. $V^{(1)}$ and $V^{(3)}$ are estimated assuming that the remaining volume is partitioned proportional to the initial specific volumes of components (1) and (3).

$$P = P^{(1)} = P^{(2)} = P^{(3)} , \tag{C.17}$$

$$T^{(1)} = T^{(2)} , \tag{C.18}$$

$$V^{(1+2)} = W(V^{(1)}) + (1 - W)(V^{(2)}) , \tag{C.19}$$

$$V = \chi(V^{(3)}) + (1 - \chi)(V^{(1+2)}) , \tag{C.20}$$

$$I^{(1+2)} = W(I^{(1)}) + (1 - W)(I^{(2)}) ,$$

$$I_H^{(1+2)} = W(I_H^{(1)}) + (1 - W)(I_i^{(2)}) ,$$

$$I_H = \chi(I_H^{(3)}) + (1 - \chi)(I_H^{(1+2)}) ,$$

$$I^{(1+2)'} = W(I_H^{(1)}) + (1 - W)(|I_i^{(2)}|) ,$$

$$I^{(3)} = (I_H^{(3)}) + (I - I_H)\left(\frac{I_H^{(3)}}{I'}\right) , \tag{C.21}$$

$$I' = \chi(I_H^{(3)}) + (1-\chi)(I^{(1+2)'}) ,$$

$$I^{(1+2)} = (I_H^{(1+2)}) + (I - I_H)\left(\frac{I^{(1+2)'}}{I'}\right) , \quad (C.22)$$

and

$$I = \chi(I^{(3)} - I_H^{(3)}) + (1-\chi)(I^{(1+2)} - I_H^{(1+2)}) + I_H .$$

From Equations (C.19) and (C.20)

$$V = \chi(V^{(3)}) + (1-\chi)[W(V^{(1)}) + (1-W)(V^{(2)})] .$$

From Equation (C.2)

$$P^{(3)} = P_H^{(3)} + \frac{\gamma^{(3)}}{V^{(3)}}(I^{(3)} - I_H^{(3)}) , \quad (C.23)$$

and from Equations (C.17) and (C.21)

$$f_1 = P = P_H^{(3)} + \frac{\gamma^{(3)}}{V^{(3)}}\left[(I - I_H)\frac{I_H^{(3)}}{I'}\right] \quad (C.24)$$

From Equation (C.7)

$$P^{(2)} = P_i^{(2)} + \frac{1}{\beta V^{(2)}}(I^{(2)} - I_i^{(2)}) , \quad (C.25)$$

and from Equations (C.23), (C.21) and (C.22)

$$f_2 = P = \left(I - I_H + \frac{\chi V^{(3)} P_H^{(3)}}{\gamma^{(3)}} + (1-\chi)\left[\frac{V^{(1)} P_H^{(1)} W}{\gamma^{(1)}} + (1-W)P_i^{(2)}\beta V^{(2)}\right]\right)$$

$$/\left(\frac{\chi V^{(3)}}{\gamma^{(3)}} + (1-\chi)\left[\frac{W(V^{(1)})}{\gamma^{(1)}} + (1-W)V^{(2)}\beta\right]\right) . \quad (C.26)$$

and from Equations (C.3), (C.18), (C.23), and (C.25)

$$f_3 = P = \left(\frac{T_i^{(2)} - T_H^{(1)}}{23,890} + \frac{P_H^{(1)} V^{(1)}}{C_V^{(1)}\gamma^{(1)}} - \frac{P_i^{(2)}\beta V^{(2)}}{C_V^{(2)}}\right)$$

$$/\left(\frac{V^{(1)}}{C_V^{(1)}\gamma^{(1)}} - \frac{\beta(V^{(2)})}{C_V^{(2)}}\right) . \quad (C.27)$$

Taking

$$F = f_1 - f_3 = 0$$

and

$$G = f_2 - f_3 = 0\,,$$

solve for $V^{(1)}$ and $V^{(2)}$ using the Newton-Raphson method. These equations were first derived by W. Gage.

	Eulerian	Lagrangian
Nitromethane Shock		
Pressure	0.0858	0.0857
Specific Volume	0.5455	0.5455
Energy	0.0146	0.0146
Temperature	1146.2	1181.9
Particle Velocity	0.171	0.171
Reflected Shock (Nitromethane)		
Pressure	0.178	0.1787
Specific Volume	0.4686	0.4858
Energy	0.02248	0.0225
Temperature	1365.5	1436.1
Particle Velocity	0.0967	0.0964
Aluminum Shock		
Pressure	0.1785	0.1786
Specific Volume	0.3071	0.3070
Energy	0.0046	0.0046
Temperature	511.0	518.3
Particle Velocity	0.0964	0.0964
Aluminum Rarefraction		
Pressure	0.008	0.0004
Specific Volume	0.3561	0.3603
Energy	0.0007	0.0005
Temperature	380.0	355.5
Particle Velocity	0.1873	0.1932

C.4 Comparison of SIN and 2DE

The 2DE code has been used to calculate the one-dimensional, multicomponent problem of an 85 kbar shock in 0.04 cm of nitromethane described by 100 cells in the Z direction,

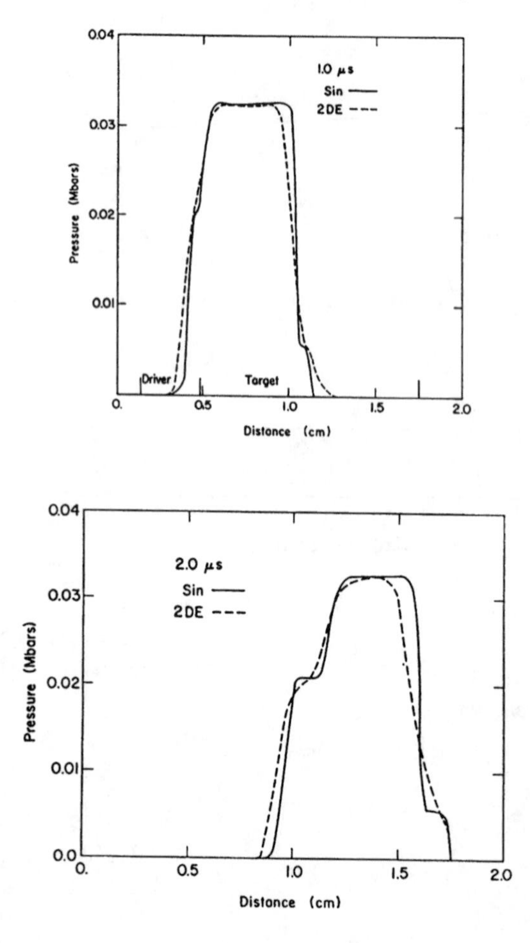

Figure C.1 Pressure distance profiles for a 32.5 kbar shock in Aluminum calculated using the 2DE and the SIN codes with elastic-plastic flow.

interacting with a 0.016 cm thick slab of Aluminum described by 40 cells, which had as its other interface 0.016 cm of air at 1-atmosphere initial pressure described by 40 cells. The results were compared with SIN one-dimensional Lagrangian calculations for the same problem.

The units are Mbar, cc/g, Mbar-cc/g, K, and cm/μsec, respectively, for pressure, specific volume, energy, temperature, and particle velocity.

The agreement between the Lagrangian and Eulerian calculations is adequate except at the nitromethane-Aluminum interface. In the Eulerian calculation, the internal energy is 11% too high in the nitromethane cell next to the interface and 10% too low in the Aluminum cell next to the interface. Such an error is in the expected direction because the reflected shock Hugoniot has less energy than the single shock Hugoniots used to partition the energy between the double shocked nitromethane and the singly shocked Aluminum in the mixture equation of state routines. This suggests that the mixture equation of state treatment could be improved by keeping track of whether the component had been shocked previously and by partitioning the energy accordingly.

Of course, it is not correct to assume that the Lagrangian calculation treated the boundary in an exact manner. Because of the large density difference, the Lagrangian calculation had nitromethane energies that were 6.6% too high at the interface and Aluminum energies that were 2.4% too low.

The elastic-plastic treatment in 2DE was compared to one-dimensional, Lagrangian SIN calculations[11] of a 0.3175 cm thick Aluminum plate initially traveling 0.0412 cm/μsec driving a 32.5 kbar shock into 1.27 cm of Aluminum. The yield was 2.5 kbar, the shear modulus was 0.25, and PLAP (see Appendix A) was 0.05 for both calculations. The shock profiles at 1 and 2 μsec are shown in Figure C.1 using the same number of cells (100) in the SIN and 2DE calculations. The Eulerian calculations are more smeared, but the magnitude of the elastic component of the rarefaction is reproduced. The elastic-plastic treatment in the Eulerian code gives realistic, if more smeared, results.

References

1. Francis H. Harlow, "The Particle-in-Cell Method for Numerical Solution of Problems in Fluid Dynamics," Proceedings of Symposia in Applied Mathematics 15 (1963).

2. Charles L. Mader, "The Two Dimensional Hydrodynamic Hot Spot", Los Alamos Scientific Laboratory report LA-3077 (1964).

3. William R. Gage and Charles L. Mader, "2DE - A Two-Dimensional Eulerian Hydrodynamic Code for Computing One Component Reactive Hydrodynamic Problems," Los Alamos Scientific Laboratory report LA-3629-MS (1966).

4. James D. Kershner and Charles L. Mader, "2DE - A Two-Dimensional Eulerian Hydrodynamic Code for Computing Multicomponent Reactive Hydrodynamic Problems," Los Alamos Scientific Laboratory report LA-4846 (1972).

5. M. Rich, "A Method for Eulerian Fluid Dynamics," Los Alamos Scientific Laboratory report LAMS-2826 (1962).

6. Richard A. Gentry, Robert E. Martin, and Bart J. Daly, "An Eulerian Differencing Method for Unsteady Compressible Flow Problems," Journal of Computational Physics 1, 87 (1966).

7. W. E. Johnson, "OIL: A Continuous Two-Dimensional Eulerian Hydrodynamic Code," General Atomic Division of General Dynamics Corporation report GAMD-5580 (1964).

8. W. E. Johnson, "Development and Application of Computer Programs Related to Hypervelocity Impact," Systems, Science and Software report 3SR-353 (1970).

9. Laura J. Hageman and J. M. Walsh, "HELP - A Multi-Material Eulerian Program for Compressible Fluid and Elastic-Plastic Flows in Two Space Dimensions and Time," Systems, Science and Software report 3SR-350 (1970).

10. Charles L. Mader, "Numerical Modeling of Detonations," University of California Press (1979).

11. Charles L. Mader, "One-Dimensional Elastic-Plastic Calculations for Aluminum," Los Alamos Scientific Laboratory report LA-3678 (1967).

appendix D

Numerical Solution of Three-Dimensional Eulerian Reactive Flow

D.1 The Flow Equations

The three-dimensional Eulerian code is called 3DE and is a three-dimensional version of the 2DE code described in Appendix C.

The Nomenclature

$E_{i,j,k}$	Total Energy (Mbar-cc/gram)
$I_{i,j,k}$	Internal Energy (Mbar-cc/gram)
$M_{i,j,k}$	Mass
$P_{i,j,k}$	Pressure
$P^{l}_{i,j,k}$	Pressure on Face l
$Q^{l}_{i,j,k}$	Viscous Pressure on Face l
$\Delta x,\ \Delta y,\ \Delta z$	Length of Cell Sides in x, y, z Directions
$T_{i,j,k}$	Temperature
$U_{x,i,j,k}$	x-Velocity
$U_{y,i,j,k}$	y-Velocity
$U_{z,i,j,k}$	z-Velocity
$W_{i,j,k}$	Mass Fraction of Undecomposed Explosive
$X_{x,i,j,k}$	Momentum in x-direction
$X_{y,i,j,k}$	Momentum in y-direction
$X_{z,i,j,k}$	Momentum in z-direction
K	Viscosity Coefficient
V_c	Cell Volume $= \Delta x \Delta y \Delta z$
Δt	Time Increment in Microseconds
ρ	Density (g/cc)
m	Mass Moved
$iMAX, jMAX, kMAX$	Maximum Number of Cells in i, j, k Directions
$PAPP, UAPP, MAPP$	Piston Values of Pressure, Velocity, Mass
$IAPP, WAPP$	Piston Values of Energy and Mass Fraction
Superscript n	The Cycle Number
Subscript o	Initial Condition

The three-dimensional partial differential equations for nonviscous, nonconducting, compressible fluid flow are

$$\frac{\partial \rho}{\partial t} + U_x\left(\frac{\partial \rho}{\partial x}\right) + U_y\left(\frac{\partial \rho}{\partial y}\right) + U_z\left(\frac{\partial \rho}{\partial z}\right) = -\rho\left(\frac{\partial U_x}{\partial x} + \frac{\partial U_y}{\partial y} + \frac{\partial U_z}{\partial z}\right) ,$$

$$\rho\left[\frac{\partial U_x}{\partial t} + U_x\left(\frac{\partial U_x}{\partial x}\right) + U_y\left(\frac{\partial U_x}{\partial y}\right) + U_z\left(\frac{\partial U_x}{\partial z}\right)\right] = -\frac{\partial P}{\partial x} ,$$

$$\rho\left[\frac{\partial U_y}{\partial t} + U_x\left(\frac{\partial U_y}{\partial x}\right) + U_y\left(\frac{\partial U_y}{\partial y}\right) + U_z\left(\frac{\partial U_y}{\partial z}\right)\right] = -\frac{\partial P}{\partial y} ,$$

$$\rho\left[\frac{\partial U_z}{\partial t} + U_x\left(\frac{\partial U_z}{\partial x}\right) + U_y\left(\frac{\partial U_z}{\partial y}\right) + U_z\left(\frac{\partial U_z}{\partial z}\right)\right] = -\frac{\partial P}{\partial z} ,$$

$$\rho\left[\frac{\partial I}{\partial t} + U_x\left(\frac{\partial I}{\partial x}\right) + U_y\left(\frac{\partial I}{\partial y}\right) + U_z\left(\frac{\partial I}{\partial z}\right)\right] = -P\left(\frac{\partial U_x}{\partial x} + \frac{\partial U_y}{\partial y} + \frac{\partial U_z}{\partial z}\right) .$$

In Phase II, the transport terms in the momentum and energy equations are dropped resulting in the following set of equations.

$$\rho\frac{\partial U_x}{\partial t} = -\frac{\partial P}{\partial x} ,$$

$$\rho\frac{\partial U_y}{\partial t} = -\frac{\partial P}{\partial y} ,$$

$$\rho\frac{\partial U_z}{\partial t} = -\frac{\partial P}{\partial z} ,$$

$$\rho\frac{\partial I}{\partial t} = -P\left(\frac{\partial U_x}{\partial x} + \frac{\partial U_y}{\partial y} + \frac{\partial U_z}{\partial z}\right) .$$

With artificial viscosity these equations become

$$\rho\frac{\partial U_x}{\partial t} = -\frac{\partial (P+Q)}{\partial x} ,$$

$$\rho\frac{\partial U_y}{\partial t} = -\frac{\partial (P+Q)}{\partial y} ,$$

$$\rho \frac{\partial U_z}{\partial t} = -\frac{\partial (P+Q)}{\partial z} \ ,$$

$$\rho \frac{\partial I}{\partial t} = -P\left(\frac{\partial U_x}{\partial x} + \frac{\partial U_y}{\partial y} + \frac{\partial U_z}{\partial z}\right)$$

$$-\frac{\partial (U_x Q)}{\partial x} + U_x \frac{\partial Q}{\partial x} - \frac{\partial (U_y Q)}{\partial y} + U_y \frac{\partial Q}{\partial y} - \frac{\partial (U_z Q)}{\partial z} + U_z \frac{\partial Q}{\partial z} \ .$$

Given some initial conditions, these equations are solved using finite-difference approximations. The fluid is moved by a continuous mass transport method identical to the one described in Appendix C.

D.2 The Numerical Technique

The problem is divided into Eulerian cells. The state values are cell-centered; the velocities and positions are at the cell corners.

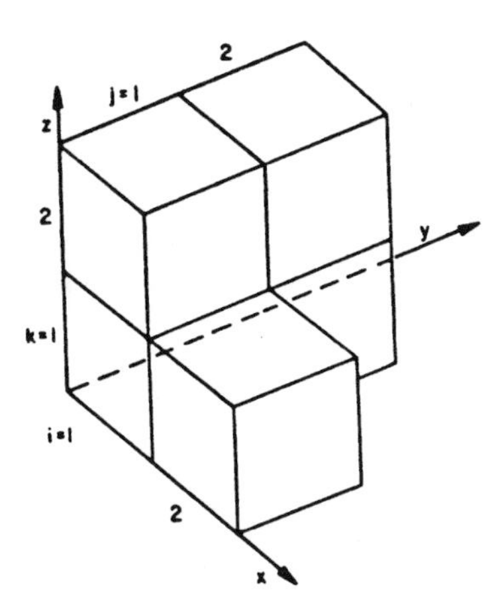

The problem is set up with a cubical lattice of $(iMAX * jMAX * kMAX)$ cells. Every cell has three indices $(i\ j\ k)$ associated with it. Each index has a range from 1 to $iMAX$, $jMAX$, or $kMAX$. The (x, y, z) coordinates of the corner nearest the origin of cell $(i\ j\ k)$ are then $(i-1)\Delta x$, $(j-1)\Delta y$, $(k-1)\Delta z$, where Δx, Δy and Δz are lengths of the side of a cell. The faces of each cell are numbered one through six.

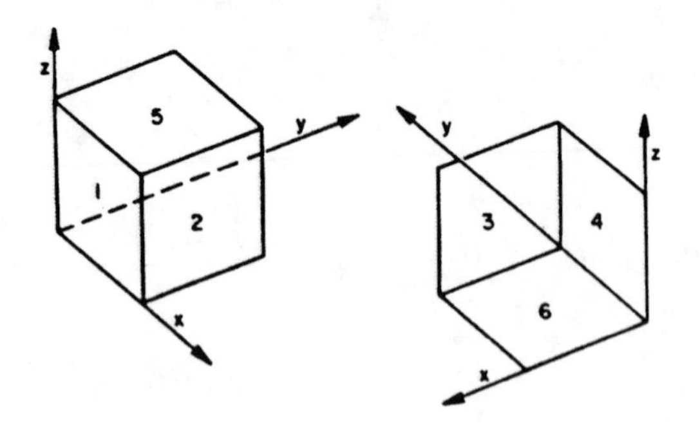

The faces of cell $(i\ j\ k)$ have indices

face 1 $\quad i,\ j-\frac{1}{2},\ k$

face 2 $\quad i+\frac{1}{2},\ j,\ k$

face 3 $\quad i,\ j+\frac{1}{2},\ k$

face 4 $\quad i-\frac{1}{2},\ j,\ k$

The center of the cell is at $(i+\frac{1}{2},\ j+\frac{1}{2},\ k+\frac{1}{2})$. Face boundaries 1, 4, and 6 are piston or continuum boundaries. Face boundaries 2, 3, and 5 are continuum boundaries only.

PHASE I

For the first cycle, skip to IA.

$$M^{n+1}_{i,j,k} = M^{n}_{i,j,k} + \Delta m_{i,j,k} \ ,$$

$$U_{x,i,j,k} = \frac{\bar{X}_{x,i,j,k}}{M_{i,j,k}} \ ,$$

$$U_{y,i,j,k} = \frac{\bar{X}_{y,i,j,k}}{M_{i,j,k}} \ ,$$

$$U_{z,i,j,k} = \frac{\bar{X}_{z,i,j,k}}{M_{i,j,k}} \ ,$$

$$\bar{E}_{i,j,k} = \Delta E_{i,j,k} + E_{i,j,k} \ ,$$

$$I_{i,j,k} = \frac{\bar{E}_{i,j,k}}{M_{i,j,k}} - \frac{1}{2}\left(U^2_{x,i,j,k} + U^2_{y,i,j,k} + U^2_{z,i,j,k}\right) .$$

IA. All Cycles

$$(time)^{n+1} = time^{n} + \Delta t \ ,$$

$$\Delta m_{i,j,k} = 0.0 \ ,$$

$$\Delta E_{i,j,k} = 0.0 \ .$$

The pressure and temperature are calculated from the density, internal energy, and cell mass fractions using the HOM, HOM2S, HOM2G, HOMSG, or HOM2SG described in Appendix C. The Arrhenius, C-J volume, and Forest Fire burns are identical to those described in Appendix A.

PHASE II

A. Calculate the pressures on the six faces of each cell.

If a cell face is on a boundary, its pressure is set equal to the cell pressure. Otherwise,

Face 1 of cell (i, j, k) $$P^1_{i,j,k} = \frac{1}{2}(P_{i,j,k} + P_{i,j-1,k}) \ ,$$

Face 2 of cell (i, j, k) $$P^2_{i,j,k} = \frac{1}{2}(P_{i,j,k} + P_{i+1,j,k}) \ ,$$

Face 3 of cell (i, j, k) $$P^3_{i,j,k} = \frac{1}{2}(P_{i,j,k} + P_{i,j+1,k}) \ ,$$

Face 4 of cell (i, j, k) $$P^4_{i,j,k} = \frac{1}{2}(P_{i,j,k} + P_{i-1,j,k}) \ ,$$

Face 5 of cell (i, j, k) $$P^5_{i,j,k} = \frac{1}{2}(P_{i,j,k} + P_{i,j,k+1}) \ ,$$

Face 6 of cell (i, j, k) $$P^6_{i,j,k} = \frac{1}{2}(P_{i,j,k} + P_{i,j,k-1}) \ .$$

B. Calculate the viscosities on the six faces of each cell.

If the viscosity is negative, it is set to zero. If a cell face is on a boundary, its viscosity is set to zero. K is the viscosity coefficient.

$$Q^1_{i,j,k} = \frac{K \ M_{i,j,k} M_{i,j-1,k} \ (U_{y,i,j-1,k} - U_{y,i,j,k})}{2(M_{i,j,k} + M_{i,j-1,k}) \ V_c} \ ,$$

$$Q^2_{i,j,k} = \frac{K\, M_{i,j,k} M_{i+1,j,k} \,(U_{x,i,j,k} - U_{x,i+1,j,k})}{2(M_{i,j,k} + M_{i+1,j,k})\, V_c} ,$$

$$Q^3_{i,j,k} = \frac{K\, M_{i,j,k} M_{i,j+1,k} \,(U_{y,i,j,k} - U_{y,i,j+1,k})}{2(M_{i,j,k} + M_{i,j+1,k})\, V_c} ,$$

$$Q^4_{i,j,k} = \frac{K\, M_{i,j,k} M_{i-1,j,k} \,(U_{x,i-1,j,k} - U_{x,i,j,k})}{2(M_{i,j,k} + M_{i-1,j,k})\, V_c} ,$$

$$Q^5_{i,j,k} = \frac{K\, M_{i,j,k} M_{i,j,k+1} \,(U_{z,i,j,k} - U_{z,i,j,k+1})}{2(M_{i,j,k} + M_{i,j,k+1})\, V_c} ,$$

$$Q^6_{i,j,k} = \frac{K\, M_{i,j,k} M_{i,j,k-1} \,(U_{z,i,j,k-1} - U_{z,i,j,k})}{2(M_{i,j,k} + M_{i,j,k-1})\, V_c} .$$

C. Calculate the tentative velocities according to momentum equations.

$$\bar{U}_{x,i,j,k} = U_{x,i,j,k} + \frac{\Delta t \Delta y \Delta z}{M_{i,j,k}} \left(P^4_{i,j,k} - P^2_{i,j,k} + Q^4_{i,j,k} - Q^2_{i,j,k} \right) ,$$

$$\bar{U}_{y,i,j,k} = U_{y,i,j,k} + \frac{\Delta t \Delta x \Delta z}{M_{i,j,k}} \left(P^1_{i,j,k} - P^3_{i,j,k} + Q^1_{i,j,k} - Q^3_{i,j,k} \right) ,$$

$$\bar{U}_{z,i,j,k} = U_{z,i,j,k} + \frac{\Delta t \Delta x \Delta y}{M_{i,j,k}} \left(P^6_{i,j,k} - P^5_{i,j,k} + Q^6_{i,j,k} - Q^5_{i,j,k} \right) .$$

D. Calculate tentative cell internal energies.

1. If a neighbor cell is on a boundary, the average velocity for that face is $\frac{1}{2}(\bar{U}_{i,j,k} + U_{i,j,k})$. If the boundary is a piston, the velocity for that face is UAPP.
2. For the first VCNT (25) cycles, the piston energy is constrained as follows:

$$\bar{I}_{i,j,k} = \frac{1}{2}(P_{i,j,k} + Q^5_{i,j,k}) \left(\frac{1}{\rho_o} - \frac{1}{\rho_{i,j,k}} \right) \quad \text{for } \frac{1}{\rho_o} > \frac{1}{\rho} ,$$

$$\bar{I}_{i,j,k} = I_{i,j,k} \quad \text{for } \frac{1}{\rho_o} < \frac{1}{\rho} .$$

3. Calculate the following quantities.

$$U^4_{x,i,j,k} = \frac{1}{4}(\bar{U}_{x,i,j,k} + U_{x,i,j,k} + \bar{U}_{x,i-1,j,k} + U_{x,i-1,j,k}) \ ,$$

$$U^2_{x,i,j,k} = \frac{1}{4}(\bar{U}_{x,i,j,k} + U_{x,i,j,k} + \bar{U}_{x,i+1,j,k} + U_{x,i+1,j,k}) \ ,$$

$$U^1_{y,i,j,k} = \frac{1}{4}(\bar{U}_{y,i,j,k} + U_{y,i,j,k} + \bar{U}_{y,i,j-1,k} + U_{y,i,j-1,k}) \ ,$$

$$U^3_{y,i,j,k} = \frac{1}{4}(\bar{U}_{y,i,j,k} + U_{y,i,j,k} + \bar{U}_{y,i,j+1,k} + U_{y,i,j+1,k}) \ ,$$

$$U^6_{z,i,j,k} = \frac{1}{4}(\bar{U}_{z,i,j,k} + U_{z,i,j,k} + \bar{U}_{z,i,j,k-1} + U_{z,i,j,k-1}) \ ,$$

$$U^5_{z,i,j,k} = \frac{1}{4}(\bar{U}_{z,i,j,k} + U_{z,i,j,k} + \bar{U}_{z,i,j,k+1} + U_{z,i,j,k+1}) \ .$$

4. The tentative cell energy is

$$\begin{aligned}
\bar{I}_{i,j,k} &= I_{i,j,k} + \frac{\Delta t}{M_{i,j,k}} P_{i,j,k}\Big[(U^4_{x,i,j,k} - U^2_{x,i,j,k})\Delta y \Delta z\Big] \\
&+ \frac{\Delta t}{M_{i,j,k}} P_{i,j,k}\Big[(U^1_{y,i,j,k} - U^3_{y,i,j,k})\Delta x \Delta z + (U^6_{z,i,j,k} - U^5_{z,i,j,k})\Delta x \Delta y\Big] \\
&+ \frac{\Delta t}{M_{i,j,k}}(Q^4_{i,j,k}U^4_{x,i,j,k} - Q^2_{i,j,k}U^2_{x,i,j,k})\Delta y \Delta z \\
&+ \frac{\Delta t}{M_{i,j,k}}(Q^1_{i,j,k}U^1_{y,i,j,k} - Q^3_{i,j,k}U^3_{y,i,j,k})\Delta x \Delta z \\
&+ \frac{\Delta t}{M_{i,j,k}}(Q^6_{i,j,k}U^6_{z,i,j,k} - Q^5_{i,j,k}U^5_{z,i,j,k})\Delta x \Delta y \\
&- \frac{\Delta t}{M_{i,j,k}}\left[\frac{\bar{U}_{x,i,j,k} + U_{x,i,j,k}}{2}\right](Q^4_{i,j,k} - Q^2_{i,j,k})\Delta y \Delta z
\end{aligned}$$

$$-\frac{\Delta t}{M_{i,j,k}}\left[\frac{\bar{U}_{y,i,j,k}+U_{y,i,j,k}}{2}\right](Q^1_{i,j,k} - Q^3_{i,j,k})\Delta x \Delta z$$

$$-\frac{\Delta t}{M_{i,j,k}}\left[\frac{\bar{U}_{z,i,j,k}+U_{z,i,j,k}}{2}\right](Q^6_{i,j,k} - Q^5_{i,j,k})\Delta x \Delta y \ .$$

5. Calculate total cell energy and momenta.

$$E_{i,j,k} = M_{i,j,k}[\bar{I}_{i,j,k} + \frac{1}{2}(\bar{U}^2_{x,i,j,k} + \bar{U}^2_{y,i,j,k} + \bar{U}^2_{z,i,j,k})] \ ,$$

$$X_{x,i,j,k} = M_{i,j,k}\bar{U}_{x,i,j,k} \ ,$$

$$X_{y,i,j,k} = M_{i,j,k}\bar{U}_{y,i,j,k} \ ,$$

$$X_{z,i,j,k} = M_{i,j,k}\bar{U}_{z,i,j,k} \ .$$

PHASE III

Mass movement occurs if the pressure of the cell from which it moves is greater than FREPR (1.1 x 10^{-6}) or if the cell is in tension.

Following the method described in Appendix C, the mass movement is calculated as in the case described for slab geometry. Mass is moved either into or out of a cell across each of its six cell faces. When mass moves into a cell it is designated an acceptor cell, and when mass moves out of a cell it is designated a donor cell. During a time cycle, Δm is the total change in a cell and is the sum of all mass movements across all faces of the cell. Since cells are contiguous, a mass movement across a cell face increases the Δm of the acceptor cell and decreases that of the donor cell.

A. Calculate $\hat{\rho}$ for donor cells.

$$\hat{\rho}^n_{i,j,k} = \rho^n_{i,j,k} - \frac{\rho^n_{i,j,k}\Delta t}{2\Delta x}(U^n_{x,i+1,j,k} - U^n_{x,i-1,j,k})$$

$$-\frac{\rho^n_{i,j,k}\Delta t}{2\Delta y}(U^n_{y,i,j+1,k} - U^n_{y,i,j-1,k}) - \frac{\rho^n_{i,j,k}\Delta t}{2\Delta z}(U^n_{z,i,j,k+1} - U^n_{z,i,j,k-1}) \ .$$

Piston boundary exceptions

face 1 $U_{y,i,j-1,k} = UAPP$
face 4 $U_{x,i-1,j,k} = UAPP$
face 6 $U_{z,i,j,k-1} = UAPP$

Continuum boundary exceptions

face 1 $U_{y,i,j-1,k} = U_{y,i,j,k}$
face 2 $U_{x,i+1,j,k} = U_{x,i,j,k}$
face 3 $U_{y,i,j+1,k} = U_{y,i,j,k}$
face 4 $U_{x,i-1,j,k} = U_{x,i,j,k}$
face 5 $U_{z,i,j,k+1} = U_{z,i,j,k}$
face 6 $U_{z,i,j,k-1} = U_{z,i,j,k}$

B. Calculate mass movement across face 1.

$$\Delta = \frac{\frac{1}{2}(\bar{U}_{y,i,j,k} + \bar{U}_{y,i,j-1,k})(\Delta t/\Delta y)}{1 + (\bar{U}_{y,i,j,k} - \bar{U}_{y,i,j-1,k})(\Delta t/\Delta y)} .$$

If $\Delta \geq 0$, the mass moves from donor cell $(i, j-1, k)$ to cell (i, j, k). Mass donated to cell (i, j, k) is subscripted (l, m, n).

$$m_{l,m,n} = (\hat{\rho}_{i,j-1,k})(\Delta)V_c .$$

If $\Delta \leq 0$, the mass moves from (i, j, k) to $(i, j-1, k)$

$$m_{i,j,k} = (\hat{\rho}_{i,j,k})(|\Delta|)V_c .$$

C. Calculate mass movement across face 4.

$$\Delta = \frac{\frac{1}{2}(\bar{U}_{x,i,j,k} + \bar{U}_{x,i-1,j,k})(\Delta t/\Delta x)}{1 + (\bar{U}_{x,i,j,k} - \bar{U}_{x,i-1,j,k})(\Delta t/\Delta x)} .$$

If $\Delta \geq 0$, the mass moves from donor cell $(i-1, j, k)$ to cell (i, j, k). Mass donated to cell (i, j, k) is subscripted (l, m, n).

$$m_{l,m,n} = (\hat{\rho}_{i-1,j,k})(\Delta)V_c .$$

If $\Delta \leq 0$, the mass moves from (i, j, k) to $(i-1, j, k)$

$$m_{i,j,k} = (\hat{\rho}_{i,j,k})(|\Delta|)V_c .$$

D. Calculate mass movement across face 6.

$$\Delta = \frac{\frac{1}{2}(\bar{U}_{z,i,j,k} + \bar{U}_{z,i,j,k-1})(\Delta t/\Delta z)}{1 + (\bar{U}_{z,i,j,k} - \bar{U}_{z,i,j,k-1})(\Delta t/\Delta z)} \ .$$

If $\Delta \geq 0$, the mass moves from donor cell $(i,\ j,\ k-1)$ to cell $(i,\ j,\ k)$. Mass donated to cell $(i,\ j,\ k)$ is subscripted $(l,\ m,\ n)$.

$$m_{l,m,n} = (\hat{\rho}_{i,j,k-1})(\Delta)V_c \ .$$

If $\Delta \leq 0$, the mass moves from $(i,\ j,\ k)$ to $(i,\ j,\ k-1)$

$$m_{i,j,k} = (\hat{\rho}_{i,j,k})(|\Delta|)V_c \ .$$

Mass movements across faces 2, 3, and 5 are described by faces 1, 4, 7 except at the boundaries.

E. Boundaries for B, C, and D are treated as follows.

Piston boundaries use applied mass ($MAPP$) and applied particle velocity ($UAPP$). The following are used to calculate Δ on continuum boundaries.

face 1 $\quad \Delta = \dfrac{1}{2}(3\bar{U}_{y,i,j,k} - \bar{U}_{y,i,j+1,k})\dfrac{\Delta t}{\Delta y}$,

face 4 $\quad \Delta = \dfrac{1}{2}(3\bar{U}_{x,i,j,k} - \bar{U}_{x,i+1,j,k})\dfrac{\Delta t}{\Delta x}$,

face 6 $\quad \Delta = \dfrac{1}{2}(3\bar{U}_{z,i,j,k} - \bar{U}_{z,i,j,k+1})\dfrac{\Delta t}{\Delta z}$,

face 2 $\quad \Delta = \dfrac{1}{2}(3\bar{U}_{x,i,j,k} - \bar{U}_{x,i-1,j,k})\dfrac{\Delta t}{\Delta x}$,

face 3 $\quad \Delta = \dfrac{1}{2}(3\bar{U}_{y,i,j,k} - \bar{U}_{y,i,j-1,k})\dfrac{\Delta t}{\Delta y}$,

face 5 $\quad \Delta = \dfrac{1}{2}(3\bar{U}_{z,i,j,k} - \bar{U}_{z,i,j,k-1})\dfrac{\Delta t}{\Delta z}$.

F. Calculate m for faces 1, 4, and 6

Add to cell quantities if the face is an acceptor cell and subtract if a face is a donor cell. Mass is donated by cell $(l,\ m,\ n)$ to cell $(i,\ j,\ k)$.

$$\Delta m_{i,j,k} = \sum m_{l,m,n} - \sum m_{i,j,k} \ .$$

For unmixed cells of the same material,

$$\Delta E_{i,j,k} = \sum m_{l,m,n} \frac{E_{l,m,n}}{M_{l,m,n}} - \sum m_{i,j,k} \frac{E_{i,j,k}}{M_{i,j,k}} \ .$$

For unmixed cells containing an explosive,

$$W^n_{i,j,k} = W_{i,j,k} + \sum \frac{m_{l,m,n}}{M_{l,m,n}} W_{l,m,n} - \sum \frac{m_{i,j,k}}{M_{i,j,k}} W_{i,j,k} \ .$$

$$\bar{X}^n_{x,i,j,k} = X_{x,i,j,k} + \sum \frac{m_{l,m,n}}{M_{l,m,n}} \bar{U}_{x,l,m,n} - \sum \frac{m_{i,j,k}}{M_{i,j,k}} \bar{U}_{x,i,j,k} \ ,$$

$$\bar{X}^n_{y,i,j,k} = X_{y,i,j,k} + \sum \frac{m_{l,m,n}}{M_{l,m,n}} \bar{U}_{y,l,m,n} - \sum \frac{m_{i,j,k}}{M_{i,j,k}} \bar{U}_{y,i,j,k} \ ,$$

$$\bar{X}^n_{z,i,j,k} = X_{z,i,j,k} + \sum \frac{m_{l,m,n}}{M_{l,m,n}} \bar{U}_{z,l,m,n} - \sum \frac{m_{i,j,k}}{M_{i,j,k}} \bar{U}_{z,i,j,k} \ .$$

G. Mixed cells

The mixed cells are treated as described in Appendix C. The composition of the mass to be moved from the donor to the acceptor cell is determined as follows. Materials common to both the donor and acceptor cells are moved according to the mass fractions of common materials in the acceptor cell. If the donor and acceptor cells have no common materials, then mass is moved according to the mass fractions of the donor cell. The mass to be moved from the donor cell has the density and energy determined for that component or components by the mixture equation of state calculation in Phase I.

appendix E

Numerical Solution of Equilibrium Detonation Properties Using the BKW Equation of State

The use of the BKW equation of state in calculating explosive and propellant performance is described in Chapter 2. The computer code used is called BKW. Its development began in the mid 1950s. Versions of the code in machine language and later in FORTRAN have been written for most of the computers that became available over the last 40 years.

The availability of inexpensive but powerful personal computers in the 1980s and FORTRAN compilers for personal computers, permitted the BKW code to be run on personal computers.

A user-friendly interface called USERBKW was developed which reduced the skill and time required to prepare the input file necessary to describe a problem for the BKW code. Detonation product data files include thermodynamic function coefficients, heat of formation, and co-volumes. A data file of explosives and additives composition, crystal density, and heat of formation is the source of the information required by USERBKW.

The MCGRAPH package of graphics programs was developed for personal computers with CGA, EGA, and VGA graphics by Mader Consulting Co., and was included in the BKW code. The MCGRAPH graphics package also includes the capability of generating PCL graphics files for printing on black and white laser printers or on HP color printers. The executable personal computer code BKW with MCGRAPH graphics are included on the CD-ROM. The executable personal computer code USERBKW along with the data files ZZZTHERC, ZZZSOLEQ, and ZZZCOMPS are also on the CD-ROM. A special version of the BKW code for calculating the performance of propellants is called ISPBKW. USERBKW can be used to generate the input file required to describe a propellant. The executable personal computer code ISPBKW is included on the CD-ROM.

E.1 Introduction

The historical background of the Becker-Kistiakowsky-Wilson (BKW) equation of state is of some value, since it gives an insight into its theoretical basis. Although also called the

Kistiakowsky-Wilson equation of state, it was attributed to Becker by Kistiakowsky and Wilson.[1]

Becker,[2] in 1921, proposed

$$\frac{PV}{RT} = (1 + xe^{x}) - \left(\frac{a}{V} + \frac{b}{V^{n+1}}\right)\left(\frac{1}{RT}\right) \quad , \quad \text{where } x = \frac{k}{V} \tag{E.1}$$

as an equation of state for Nitrogen at high densities. It was derived by assuming a virial equation of state and using a repulsive or "point centers of repulsion" potential to estimate the first term, xe^{x}. The second term, a/V, described the attractive forces, and Becker used the last term to obtain agreement at the critical point.

In 1922, Becker[3] used the equation

$$\frac{PV}{RT} = 1 + xe^{x} = F(x) \tag{E.2}$$

to compute the detonation velocities of nitroglycerine and mercury fulminate. He stated that the computed detonation velocities were determined "with an accuracy indicating the order, at least, of the magnitude observed."

Kistiakowsky and Wilson,[1,4] used for x the expression $K/VT^{1/3}$ and found that K, the covolume, could be approximated as an additive function of the covolumes of the constituent molecules of the product gases for a large number of explosives. R. S. Halford was an active contributor to this equation of state study.[4]

Kistiakowsky and Wilson[4] attribute to D. P. MacDougall and L. Epstein the addition of β to the repulsive term, resulting in an equation of state of the form

$$\frac{PV}{RT} = 1 + xe^{\beta x} \quad \text{with } x = \frac{K}{VT^{\alpha}} \ . \tag{E.3}$$

The values of α and β found satisfactory for reproducing the experimental detonation velocities for a number of explosives were 0.25 and 0.3, respectively.

Cowan and Fickett[5] added a Θ to T to prevent the pressure from tending to infinity as the temperature tends to zero and to keep $(\partial P/\partial T)_v$ positive over the range of volumes of interest. They found that the values of $\alpha = 0.5$ and $\beta = 0.09$ were satisfactory for reproducing the experimental detonation velocity-density curve and the C-J pressure of Composition B. The value of Θ they used was 400; K was defined as $\kappa \sum x_i\, k_i$, where κ was 11.85. The k_i were the individual detonation product geometrical covolumes, and x_i were the mole fraction.

With this historical background, it becomes apparent that the BKW equation of state is based upon a repulsive potential applied to the virial equation of state,

$$\frac{PV}{RT} = 1 + \frac{B}{V} + \frac{C}{V^2} + \ldots \ .$$

Let $x = B/V$, then, neglecting higher order terms,

$$\frac{PV}{RT} = 1 + x + \beta x^2 \ ,$$

or to a first approximation,

$$\frac{PV}{RT} = 1 + xe^{\beta x} .$$

Using a repulsive potential of the form $U = A/r^n$ where r is the separation distance, Jeans[6] shows that

$$B = \frac{K}{T^{3/n}} \quad \text{or} \quad = \frac{K}{T^{\alpha}}$$

if $\alpha = 3/n$ and $K \propto A^{3/n}$, thus, $x = K/VT^{\alpha}$.

In a parameter study of the BKW equation of state one may adjust the BKW parameters α, β, and κ, and the covolumes of the detonation products. Cowan and Fickett[5] showed that, for a given alpha and beta, one may adjust kappa to give the experimental velocity for a single explosive at a single density. One may change the slope of the detonation velocity-density curve by changing beta. With successive iterations on kappa and beta, one can reproduce the experimental detonation velocities at two densities for a single explosive.

For $CHNO$ explosives the covolumes of primary importance are those of Water, Carbon Dioxide, Carbon Monoxide, and Nitrogen. Since experimental Hugoniots were available for Water,[7] liquid Nitrogen,[8] and solid Carbon Dioxide,[9] in the pressure range of interest (100 to 500 kbar), the covolumes of these species were adjusted so as to best reproduce these Hugoniots for the alpha, beta, and kappa set found to reproduce the detonation velocities.[10] They are listed in Table E.1.

TABLE E.1 BKW Parameters

No.	Parameter Set	β	κ	α	Θ	Co volumes H_2O	CO_2	CO	N_2
1	Fitting RDX	0.181	14.15	0.54	400	250	600	390	380
2	Fitting TNT	0.09585	12.685	0.50	400	250	600	390	380
3	Best RDX Fit $(\partial P/\partial T)_v > 0$	0.16	10.91	0.50	400	250	600	390	380

E.2 Covolumes

The geometrical covolume is defined as the volume in A^3 occupied by a molecule rotating about its center of mass x 10.46. Fickett introduced the factor 10.46 to make the CO covolume the same as that originally used. There arises the question whether to calculate an actual volume of the nonrotating molecule, or a volume of the rotating molecule represented as a sphere whose radius is the maximum dimension of the molecule measured from its center of mass. The rotational motion may be inhibited at the high densities of interest. Since the high densities on the Hugoniot curve are accompanied by high temperatures which tend to maintain the rotation, the rotating-molecule model is usually chosen for estimating covolumes. This is consistent with the nature of the thermodynamic data which assume full thermal rotation.

The method used to calculate the geometrical covolumes in Table E.2 is demonstrated by several examples. Van der Waal radii used are from Pauling.[11]

H	1.20	B	1.70
C	1.60	Cl	1.80
N	1.50	Al	2.00
O	1.40	Br	1.54
F	1.35	S	1.60

Single bond covalent radii used are from Herzberg.[12]

P-Cl	2.03	H-F	0.917	C-C	1.54
P-O	1.45	C-F	1.317	C-H	1.093
N-F	1.37	C-Cl	1.988	S-H	1.334
C-N	1.17	N-H	1.014	S-O	1.46
Al-Cl	2.10	B-O	1.205	C-S	1.54
C≡O	1.1281	C=O	1.163	S-S	1.54
H-H	0.7416	B-F	1.29	Br-Br	1.887
O-O	1.2074	H-F	0.917	H-Br	1.413
N-N	1.094	H-Cl	1.275		

For Nitrogen (N_2):

Diameter = N-N bond radii + 2(N Van der Waal radii)
= 1.094 + 2 (1.50) = 4.094
Radius = 2.047
k_i = (10.46) (4.189) $(2.047)^3$ = 376
which is rounded to 380.

For Hydrogen Fluoride (HF):

The HF radius needed is the largest distance from the center of mass of the molecule. The $H-F$ bond distance is 0.9171 A. The center of mass from the H atom (X) is

$$(1.)(X) = (0.9171 - X)(19) .$$
$$X = 0.871 .$$
$$\text{Fluorine distance} = (0.9171 - 0.871) + 1.35 = 1.396 .$$
$$\text{Hydrogen distance} = (0.871) + 1.20 = 2.071 .$$

The largest distance from the center of mass is 2.071 which is used for the radius of the spherical volume occupied by HF.

$$k_i = (10.46)\,(4.189)\,(2.071)^3 = 389$$

Another method for estimating the covolume is to use the Fickett[13] technique for relating the covolume to the depth of the potential well $T_i^{1/2}$ and the separation radius r_i^* by

$$k_i \approx (r_i^*)^3$$
$$k_i \approx (r_i^*)^3\,(T_i)^{1/2} .$$

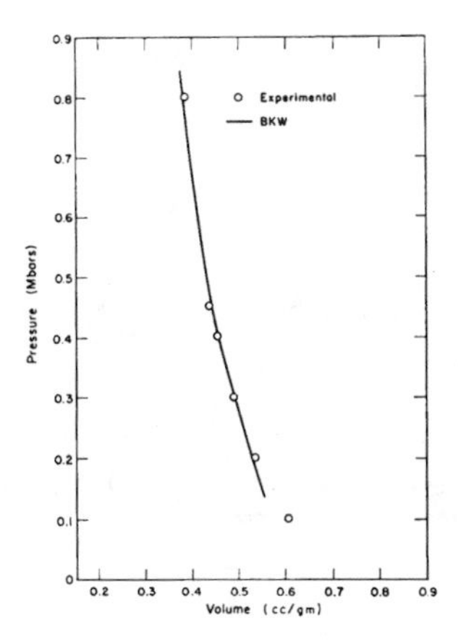

Figure E.1 The Water Hugoniot.

The constant of proportionality was chosen to give the geometric value of 380 for the Nitrogen covolume. The r_i^* and T_i used for Table E.2 are as follows.

Species	r_i^*	T_i
N_2	4.05	120
CO	4.05	120
H_2O	3.35	138
NO	3.56	131
H_2	3.34	37
CO_2	4.20	200
O_2	3.73	132
CH_4	4.30	154
He	3.23	6.48

The best method for calibrating or evaluating the covolume of a product molecule is by using the experimental shock Hugoniot. For the calibration described in Reference 10, the Hugoniots for liquid Water, liquid Nitrogen, and solid Carbon Dioxide are adequately reproduced for the gaseous region by the BKW calculations as shown in Figure E.1, E.2, and E.3.

As experimental Hugoniot data became available for other species; the covolume being used in BKW was evaluated. The initial states are given in Table E.3.

The inert gas Helium was compressed to 1.011 kbar at 22° C and shocked to 19 kbar by Wackerle.[14] He obtained a shock velocity of 0.54 and particle velocity of 0.305 cm/μsec. The BKW equation of state could reproduce the results using a covolume of 60 for Helium.

Wackerle, Seitz, and Jamieson[15] have performed experimental studies of the Hugoniot of liquid Oxygen. The geometrical covolume for Oxygen will reproduce the experimental Hugoniot as shown in Figure E.4.

Van Thiel and Wasley have made experimental studies[16] of the Hugoniot and reflected shock states of liquid Hydrogen, and Dick[17] has studied those of Hydrogen and Deuterium. The experimental states are listed in Table E.4. The experimental and calculated Hugoniots are shown in Figure E.5. The geometric covolume of 180 for Hydrogen is too hard. The

TABLE E.2 Covolumes

Species	Geometrical Covolumes	ΔH^o_{fo}	Fit To Hugoniot Data	LJD $\approx (r_i)^3$	LJD $\approx (r_i)^3(T_i)^{1/2}$
CH_4	528	−16.0	–	455	515
CO	390	−27.2	–	380	380
CO_2	735	−94.0	600	423	547
D_2	180	0.0	100	214	118
H_2	180	0.0	80	214	118
H_2O	420	−57.1	250	215	231
He	100	0.0	60	191	45
N_2	380	0.0	380	380	380
NO	386	21.5	–	258	270
O_2	350	0.0	–	296	311
Al	350	76.9	–	–	–
$AlCl_3$	2600	−138.0	–	–	–
AlO	800	20.8	–	–	–
Al_2O	1300	−33.0	–	–	–
Al_2O_2	1800	−93.1	–	–	–
Al_2O_3	1350	−240.0	–	–	–
B	215	137.0	–	–	–
B_2	674	205.0	–	–	–
BF	685	−41.0	–	–	–
BF_3	800	−265.0	–	–	–
BH	533	119.0	–	–	–
BHO_2	1270	−140.0	–	–	–
BN	619	158.0	–	–	–
BO	610	−13.5	–	–	–
B_2O_2	1740	−107.0	–	–	–
B_2O_3	730	−216.0	–	–	–
Br	160	26.4	–	–	–
Br_2	670	4.4	–	–	–
BrH	1112	−8.5	–	–	–
C	180	170.0	–	–	–
C_2	380	233.1	–	–	–
C_5	750	230.4	–	–	–
CCl_4	2000	−25.5	–	–	–
CF_2	1330	−23.0	–	–	–
CF_3	1330	−120.0	–	–	–

TABLE E.2 Covolumes (continued)

Species	Geometrical Covolumes	ΔH^o_{fo}	Fit To Hugoniot Data	LJD $\approx (r_i)^3$	LJD $\approx (r_i)^3(T_i)^{1/2}$
CF_4	1330	−218.0	–	–	–
CF_3H	1920	−163.0	–	–	–
CF_3H_2	1330	−105.0	–	–	–
CFH_3	1920	−59.0	–	–	–
CF_2O	1330	−150.0	–	–	–
CH	472	141.0	–	–	–
CH_2	520	96.0	–	–	–
CH_3	520	33.4	–	–	–
C_2H_2	1013	54.3	–	–	–
C_2H_4	1100	14.5	–	–	–
C_4H_{10}	6625	−23.7	–	–	–
CH_2O	815	−26.8	–	–	–
CN	486	107.0	–	–	–
CS_2	1354	27.5	–	–	–
Cl_2	956	0.0	–	–	–
ClF	610	−13.2	–	–	–
ClH	637	−22.0	–	–	–
F	108	18.4	–	–	–
F_2	387	0.0	–	–	–
HF	389	−65.1	–	–	–
HO	413	3.6	–	–	–
H_2S	680	−4.8	–	–	–
H_3N	476	−10.5	–	–	–
N	148	112.5	–	–	–
NO_2	650	8.7	–	–	–
N_2O	730	20.3	–	–	–
O	120	59.0	–	–	–
OS	584	19.0	–	–	–
O_2S	591	−70.9	–	–	–
S	180	53.2	–	–	–

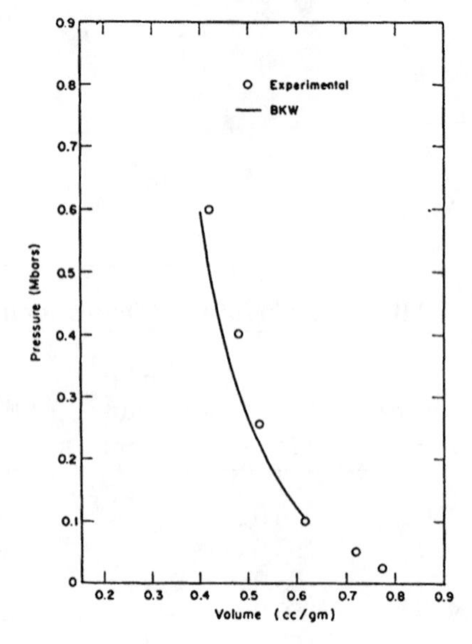

Figure E.2 The Liquid Nitrogen Hugoniot.

available data for Deuterium may be described with a covolume of 100 and those for Hydrogen with a covolume of 80. The experimentally observed reflected states are also described adequately as shown in Figure E.6.

TABLE E.3 Experimental Initial States

	T_o (^oK)	ρ_o (g/cc)	ΔH^o_{fo} $(kcal/mole)$	P_o $(kbar)$
N_2 (liquid)	77.4	0.808	-0.82	–
CO_2 (solid)	194.7	1.540	-98.50	–
H_2O (liquid)	293.0	1.000	-65.275	–
H_2 (liquid)	20.5	0.071	-0.0775	–
O_2 (liquid)	76.9	1.202	-1.413	–
NH_3 (liquid)	205.0	0.720	-13.324	–
He (gas)	295.0	0.1148	0.0	1.011
NO (liquid)	120.0	1.30	19.03	–
CO (liquid)	79.8	0.798	-28.09	–
D_2 (liquid)	23.5	0.165	0.0775	–

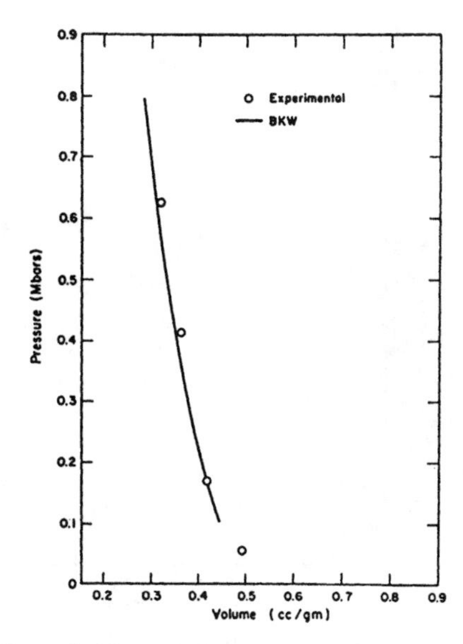

Figure E.3 The Solid Carbon Dioxide Hugoniot.

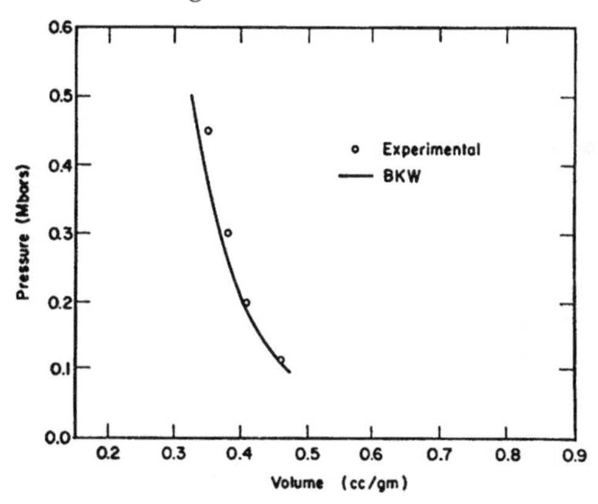

Figure E.4 The Liquid Oxygen Hugoniot.

TABLE E.4 Experimental D_2 and H_2 States

	Single-Shock Data				Magnesium States			Dural States			
	U_S (cm/μs)	U_P (cm/μs)	P (Mbar)	ρ (g/cc)	P (Mbar)	U_P (cm/μs)	ρ (g/cc)	P (Mbar)	U_P (cm/μs)	ρ (g/cc)	Ref.
D_2	1.094	0.727	0.130	0.491	–	–	–	–	–	–	17
D_2	0.954	0.596	0.094	0.439	0.281	0.215	0.67	0.338	0.160	0.68	17
D_2	0.831	0.499	0.069	0.412	0.221	0.180	0.58	0.236	0.120	0.65	17
D_2	0.794	0.499	0.058	0.379	0.163	0.142	0.56	0.205	0.107	0.53	17
H_2	1.154	0.827	0.067	0.250	–	–	–	–	–	–	16
H_2	0.940	0.592	0.040	0.192	–	–	–	–	–	–	17

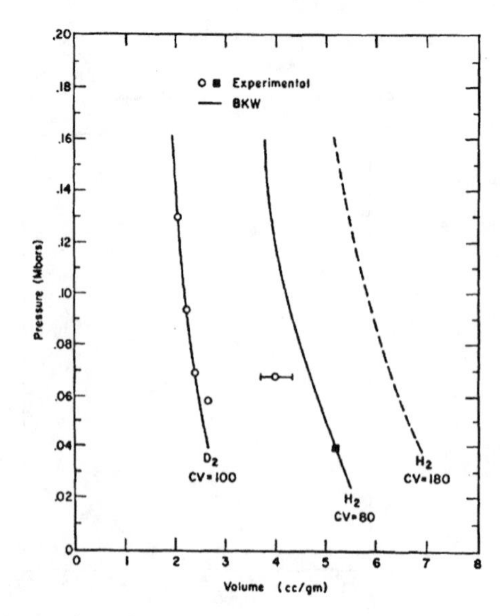

Figure E.5 Liquid Deuterium and Hydrogen Hugoniots.

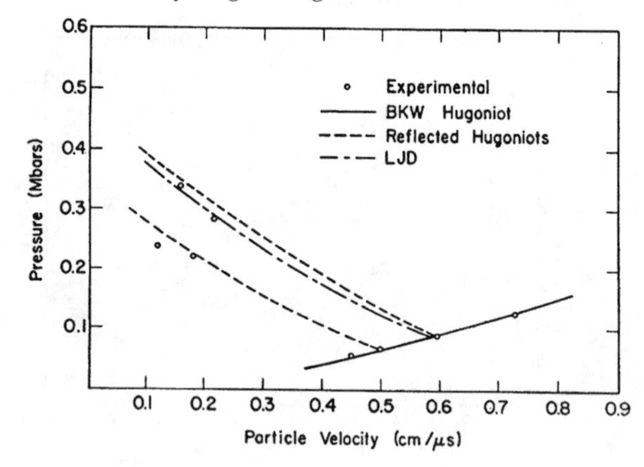

Figure E.6 Deuterium Shock States.

The liquid Ammonia Hugoniot studied by Dick[17] may be described by $U_S = 0.175 + 1.58U_P$. The experimental observations cannot be reproduced by BKW if it is assumed that only Ammonia gas is present. An equilibrium Hugoniot calculation indicates that Ammonia is decomposed above 150 kbars. The calculated equilibrium Hugoniot closely approximates the experimental one, as shown in Figure E.7. The covolume for Ammonia is uncalibrated; however, the Nitrogen and Hydrogen mixture that results from shocking Ammonia appears to be adequately described by the Nitrogen and Hydrogen covolumes.

Liquid nitric oxide has been studied by Ramsay and Chiles.[18] It was observed to detonate with a detonation pressure of 0.103 ± 0.015 Mbar and a velocity of 0.562 cm/μsec. The BKW calculated C-J pressure is 0.106, velocity is 0.5607, and temperature is 1850°K. The calculated detonation products are Nitrogen and Oxygen with 0.003 mole of NO gas and 10^{-5} moles of NO_2 gas. Although the experimental studies do not help us calibrate the covolume of Nitrogen Oxide, they do increase our confidence in the Nitrogen and Oxygen covolumes. As described in Chapter 2, Schott measured the shock Hugoniots of liquid

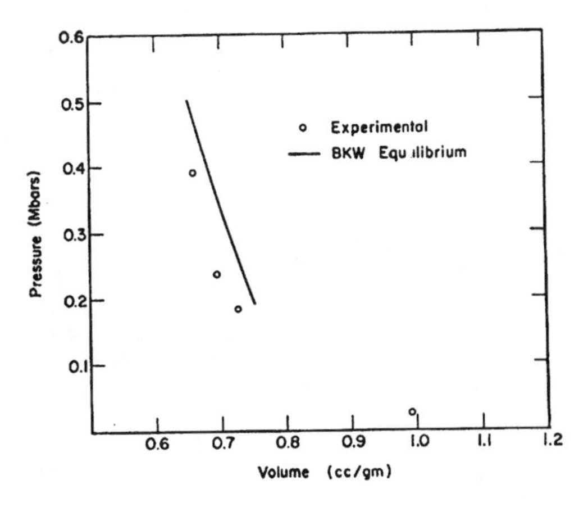

Figure E.7 Liquid Ammonia Hugoniot.

Nitrogen Oxide, Nitrogen, Oxygen, and synthetic equal molar mixture of Oxygen and Nitrogen. The experimental data for single and multiple shocked Nitrogen, Nitrogen plus Oxygen, and Nitrogen Oxide were found to be predicted by the BKW equation of state.

Also described in Chapter 2 are the results of recent experimental studies of shocked Carbon Dioxide. The BKW equation of state was found to give a good fit to the Hugoniot data for solid and liquid Carbon Dioxide.

Liquid Carbon Monoxide is calculated by BKW to decompose exothermally to Carbon Dioxide and solid Carbon. The calculated BKW C-J state is 0.0253 Mbar, 0.333 cm/μsec, and 2056°K. The calculated BKW Hugoniots for liquid Carbon Monoxide decomposing to Carbon Dioxide and Carbon, vaporizing to Carbon Monoxide gas only, and decomposing to the calculated equilibrium composition are shown in Figure E.8. The composition and temperatures are shown in Figure E.9. The Carbon Monoxide covolume cannot be calibrated from the experimental studies of the Hugoniot because the Carbon Monoxide is expected to decompose.

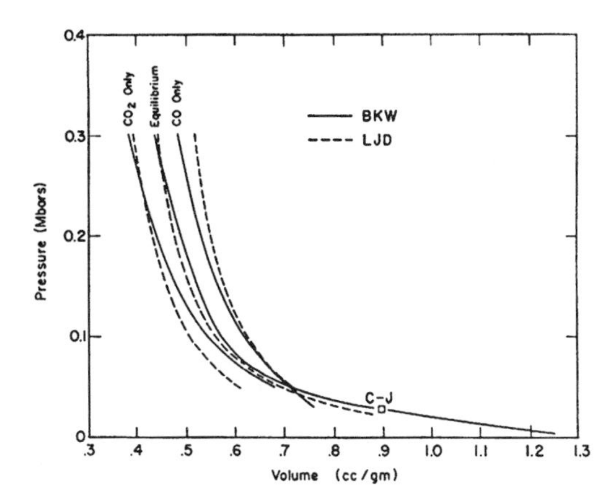

Figure E.8 Liquid Carbon Monoxide Hugoniots.

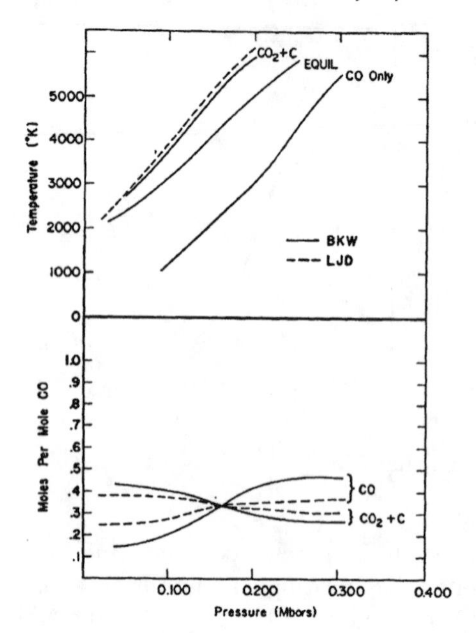

Figure E.9 Liquid Carbon Monoxide Hugoniot Composition and Temperature.

E.3 Solid Products

The detonation products of explosives include solid products such as graphite. Solid products are often assumed to be incompressible; however, the calculated detonation velocity as a function of density is sensitive to the equation of state of the detonation product graphite for explosives such as Composition B. The Cowan[5] solid equation of state is of the form

$$p \;=\; p_1(\eta) \;+\; a(\eta)T \;+\; b(\eta)T^2 \;, \qquad (E.4)$$

where $\eta = \rho/\rho_o$ is the compression of the solid graphite relative to its normal crystal density $\rho_o = 2.25g/cc$.

$$P_1(\eta) \;=\; -2.4673 \;+\; 6.7692\eta \;-\; 6.9555\eta^2 \;+\; 3.0405\eta^3 \;-\; 0.3869\eta^4 \;,$$

$$a(\eta) \;=\; -0.2267 \;+\; 0.2712\eta \;, \qquad (E.5)$$

$$b(\eta) \;=\; 0.08316 \;-\; 0.07804\eta^{-1} \;+\; 0.03068\eta^{-2} \;.$$

These coefficients give p in Mbar when T is in volts (i.e., in units of 11,605.6°K). The expression for $b(\eta)$ is an analytic fit to values of b obtained from the Thomas-Fermi-Dirac zero-temperature pressure. The electronic temperature dependence is calculated from the zero-temperature pressure by assuming the latter to be the pressure of a degenerate Fermi gas, and then determining the temperature dependence of pressure of a Fermi gas at any density of interest. This procedure is equivalent to obtaining b from temperature dependent Thomas-Fermi-Dirac calculations. The function $a(\eta)$ is of a form suggested by Reitz,[19] the second coefficient being $3R\rho_o/2$ and the first being chosen to make $a(1) = 3\alpha/\kappa$ where α is the linear coefficient of thermal expansion and κ is the isothermal compressibility at normal temperature and pressure (called standard temperature and pressure or STP). The range of applicability of these expressions is $0 < T < 2, \quad 0.95 < \eta < 2.5$.

E.4 Thermodynamic Theory

The BKW equation of state is again (where V_g is gas volume)

$$\frac{PV_g}{RT} = 1 + xe^{\beta x} = F(x) \ ,$$

where

$$x = \kappa k / V_g (T + \Theta)^{\alpha} \ .$$

In these expressions, V_g is the molar volume of the gaseous products (i.e,, the total volume of the reaction products less the volume occupied by any solid products, divided by the number of moles of gas, n_g). The quantity Θ is a constant which was introduced to prevent P from tending to infinity as T tends to zero. It was given the somewhat arbitrary value of 400°K.

The correction to the ideal gas equation of state is allowed to be a function of the temperature and also to depend on composition through the "constant" k:

$$k = \sum_{g} x_i \, k_i \ ,$$

where $x_i = n_i/n_g$ is the mole fraction of the gaseous compound i. The k_i is a constant covolume characteristic of that compound, and the summation is carried only over the gaseous components in the reaction products.

The calculations require a knowledge of the thermodynamic properties of the detonation products. The method for computing the thermodynamic functions of gaseous and solid detonation products is described in Appendix F.

In general, the detonation products are a gas-solid mixture for which specific quantities (per unit mass) are convenient. In deriving the properties for the separate phases, however, molar quantities are used. The relation between the two is

$$g = n_g \, G_g + n_s \, G_s \ , \tag{E.6}$$

where g and G represent any extensive property in specific and molar units, respectively. The subscripts g and s refer to gas and solid. n_g and n_s are the number of moles of gas and solid, respectively, per unit mass of mixture. Expressions will be derived from which G_g and G_s can be calculated for substances obeying the equations of state (E.2) and (E.4) in terms of tabulated data for the pure components in their standard reference states (ideal gas or real solid at unit pressure).

A. Internal Energy

The dependence of internal energy on volume is given by the thermodynamic relation[20]

$$\left(\frac{\partial E}{\partial V}\right)_T = T\left(\frac{\partial P}{\partial T}\right)_V - P \, .$$

Integration of this expression at constant temperature and fixed composition then gives the energy of a system at temperature and volume (T, V) in terms of its energy at temperature T and a reference volume V^*;

$$E(T, v) = E(T, V^*) + \int_{V^*}^{V} \left[T\left(\frac{\partial P}{\partial T}\right)_V - P\right] dV \, . \qquad (E.7)$$

Gaseous Components

The standard reference state of ideal gas at unit pressure is equivalent, in the case of internal energy, to ideal (or real) gas at zero pressure. Taking then the limit $V^* \to \infty$, the term $E(T, V^*)$ in Equation (E.7) may be written as a simple sum of the internal energies of the component substances:

$$E_g\,(T, \infty) = \sum_i x_i (E_T^o)_i = \sum_i x_i (E_T^o - H_o^o)_i + \sum_i x_i (H_o^o)_i \, , \qquad (E.8)$$

where $x_i = n_i/n_g$ is the mole fraction of component i and the sum is over all gaseous components of the mixture.

For equation of state (E.2),

$$\left(\frac{\partial P}{\partial T}\right)_{V_g} = \frac{P}{T} - \frac{\alpha\, R\, T\, x}{V_g(T+\Theta)} \frac{dF_{(x)}}{dx} = \frac{P}{T} + \frac{\alpha\, R\, T}{T+\Theta}\left(\frac{dF_{(x)}}{dV_g}\right)_T \, , \qquad (E.9)$$

since at constant temperature and composition

$$dx = -x\, dV_g \,/\, V_g \, . \qquad (E.10)$$

The integral in (E.7) can thus be written

$$\frac{\alpha\, R\, T^2}{T+\Theta}\int_{\infty}^{V_g}\left(\frac{dF}{dV_g}\right)_T dV_g = \frac{\alpha\, R\, T^2}{T+\Theta}\int_{1}^{F_{(x)}} dF_{(x)} = \frac{\alpha\, R\, T^2}{T+\Theta}(F_{(x)} - 1) \, . \qquad (E.11)$$

Substitution of (E.8) and (E.11) in (E.7) then gives

$$\frac{E_g(T, V)}{R\, T} = \sum_i x_i \frac{(E^o - H_o^o)_i}{R\, T} + \sum_i x_i \frac{(H_o^o)_i}{R\, T} + \frac{\alpha T}{T+\Theta}(F_{(x)} - 1) \, . \qquad (E.12)$$

Solid Component

For a solid, the standard reference state is at unit pressure. Using $V^* = V_s^o$ as the (molar) volume of the solid at temperature T and unit pressure P^o, equation (E.7) for a single solid component becomes

$$E_s(T, V) = (H_T^o - H_o^o)_s + (H_o^o)_s - P^o V_s^o + \int_{V_s^o}^{V_s} \left[T \left(\frac{\partial P}{\partial T} \right)_V - P \right] dV , \qquad (E.13)$$

or, for the solid equation of state (E.4), neglecting $P^o V_s^o$ and setting $V_s^o = V_s^o$ (25°C).

$$E_s(T, V) = (H_T^o - H_o^o)_s + (H_o^o)_s - P^o V_s^o + \int_{V_s^o}^{V_s} \left[b(V) T^2 - p_1(V) \right] dV . \qquad (E.14)$$

B. Entropy

The pressure dependence of entropy is given by the thermodynamic relation[20]

$$\left(\frac{\partial S}{\partial P} \right)_T = - \left(\frac{\partial V}{\partial T} \right)_P .$$

Integrating at constant temperature and fixed composition from the reference pressure $P_o = 1$ atmosphere gives

$$S(T, P) = S(T, P^o) - \int_{P^o}^{P} \left(\frac{\partial V}{\partial T} \right)_P dP . \qquad (E.15)$$

Gaseous Components

The entropy of a real gas at (T, P^o) may be expressed in terms of the entropy of the ideal gas at (T, P^o) through the relation

$$S_g(T, P^o) = S_I(T, P^o) + \int_o^{P^o} \left[\frac{R}{P} - \left(\frac{\partial V}{\partial T} \right)_P \right] dP . \qquad (E.16)$$

The entropy S_I of the mixture of ideal gases is, in turn, related to the entropies of the pure components, each at pressure P^o, by

$$S_I(T, P^o) = \sum_i x_i (S^o)_i - R \sum_i x_i \ln x_i , \qquad (E.17)$$

where the second sum represents the entropy of mixing. Combining (E.16) and (E.17) with (E.15) gives for the gaseous mixture,

$$\frac{S_g(T, P)}{R} = \sum_i x_i \frac{(S^o)_i}{R} - \sum_i x_i \ln x_i - \ln \frac{P}{P^o} + \int_o^P \left[\frac{1}{P} - \frac{1}{R} \left(\frac{\partial V_g}{\partial T} \right)_P \right] dP .$$

For the equation of state (E.2), it is convenient to transform the integral in (E.16) with the aid of the differential relation

$$\left(\frac{\partial V}{\partial T}\right)_P = -\left(\frac{\partial V}{\partial P}\right)_T \left(\frac{\partial P}{\partial T}\right)_V . \tag{E.18}$$

Using (E.9), (E.10), and the following relation for constant temperature and composition

$$\frac{dP}{P} = \frac{dF_{(x)}}{F_{(x)}} - \frac{dV_g}{V_g} , \tag{E.19}$$

the integral in (E.17) can be written

$$\int_1^F \frac{dF}{F} + \int_\infty^{V_g} \left[\frac{F_{(x)}-1}{V_g} + \frac{\alpha T}{T+\Theta}\left(\frac{dF_{(x)}}{dV_g}\right)_T\right] dV_g$$

$$= ln\ F - \int_o^x e^{\beta x}\,dx + \frac{\alpha T}{T+\Theta}\int_1^{F_{(x)}} dF_{(x)} .$$

Thus we obtain for the entropy of the gaseous components of the mixture:

$$\frac{S_g}{R} = \sum_i x_i \frac{(S^o)_i}{R} - \sum_i x_i\, ln\, x_i - ln\frac{P}{P^o}$$

$$+ ln\ F_{(x)} - \frac{e^{\beta x}-1}{\beta} + \frac{\alpha T}{T+\Theta}(F_{(x)} - 1) . \tag{E.20}$$

Solid Component

Converting the integral in (E.15) with the aid of the relation (E.18), the entropy of the solid component is

$$S(T, V_s) = (S^o)_s + \int_{V_s^o}^{V_s} \left(\frac{\partial P}{\partial T}\right)_{V_s} dV_s , \tag{E.21}$$

where V_s^o is, as before, the molar volume of the solid at temperature T and unit pressure P^o. For the solid equation of state (E.4), this becomes

$$\frac{S_S}{R} = \frac{(S^o)_s}{R} + \frac{1}{R}\int_{V_s^o}^{V_s} [a(V) + 2b(V)\,T]dV . \tag{E.22}$$

C. Free Energy

The equilibrium state of a system is characterized by the possibility of a number of chemical reactions among the components, each reaction being subject to the thermodynamic condition

$$\sum_i \nu_i \, \mu_i = 0 \; . \tag{E.23}$$

ν_i is the number of moles of component i involved in the reaction. ν_i is positive for products and negative for reactants, and

$$\mu_i = \left(\frac{\partial(nA)}{\partial n_i}\right)_{T,v,n_j} = \left(\frac{\partial(nF)}{\partial n_i}\right)_{T,P,n_j} \; . \tag{E.24}$$

μ is the chemical potential of component i. The subscripts v and n_j indicate differentiation with total volume and all n's other than n_i held constant. A and F are the molar Helmholtz and Gibbs free energies, respectively.

The conditions for the conservation of mass, together with all independent equations of the type (E.23) are just sufficient to determine the composition of the system, provided that the chemical potentials of the components are known.

Gaseous Components

For the gaseous components that obey the equation of state (E.2), μ_i is found most easily from the work function A, which, from (E.12) and (E.20), since $E^o/RT - S^o/R = F^o/RT - 1$, may be calculated as

$$\frac{n_g A_g}{RT} = \frac{n_g E_g}{RT} - \frac{n_g S_g}{R} \; .$$

From (E.24) the chemical potential of the ith gaseous component is

$$\frac{\mu_i}{RT} = \frac{(F^o - H_o^o)_i}{RT} + \frac{(H_o^o)_i}{RT} + ln\, n_i + ln \frac{P}{n_g P^o F} + \frac{e^{\beta x} - 1}{\beta}$$

$$+ n_g \, e^{\beta x} \left(\frac{\partial x}{\partial n_i}\right)_{T,v_g,n_j} \; .$$

$$\frac{\mu_i}{RT} = \frac{(F^o - H_o^o)_i}{RT} + \frac{(H_o^o)_i}{RT} + ln \frac{x_i \, P}{P^o} + \frac{e^{\beta x} - 1}{\beta}$$

$$- \, ln \, F_{(x)} + \frac{k_i}{k}(F_{(x)} - 1) \; . \tag{E.25}$$

Solid Component

For the solid component, the chemical potential is found more easily from the free energy F, which from (E.13) and (E.21) may be written

$$\frac{n_s F_s}{RT} = \frac{n_s E_s + n_s P V_s}{RT} - \frac{n_s S_s}{R}$$

$$\frac{n_s F_s}{RT} = \frac{n_s (F_s - H_o^o)_s}{RT} + \frac{n_s (H_o^o)_s}{RT} + \frac{n_s (P V_s - P^o V_s^o)}{RT} - \frac{n_s}{RT}\int_{V_s^o}^{V_s} P dV \qquad (E.26)$$

Thus from (E.24), the chemical potential of the solid component is

$$\frac{\mu_s}{RT} = \frac{(F^o - H_o^o)_s}{RT} + \frac{(H_o^o)_s}{RT} + \frac{F_s'}{RT}\ , \qquad (E.27)$$

$$\frac{F_s'}{RT} = \frac{P V_s - P^o V_s^o}{RT} - \frac{1}{RT}\int_{V_s^o}^{V_s} P\, dV\ ,$$

$$\frac{F_s'}{RT} = \frac{P V_s - P^o V_s^o}{RT} - \frac{1}{RT}\int_{V_s^o}^{V_s} [p_1(V) + a(V)T + b(V)T^2] dV\ . \qquad (E.28)$$

E.5 Solution of the BKW Equations

The first solutions of the BKW equations with chemical equilibrium were for CHNO explosives only and were performed by Cowan and Fickett[5] in 1954. The first attempt to write a computer code with any combination of elements was the BKW code written for the IBM 704 in 1956 and later used on the IBM 7090. The detonation properties of explosives composed of various combinations of the elements carbon, hydrogen, nitrogen, boron, aluminum, oxygen, and fluorine were calculated and found to compare favorably with the experimental observations.[21] The code STRETCH BKW[22] was written in 1961 for the IBM-7030 computer. The speed and versatile nature of the code and computer permitted a parameter study to determine whether it was possible to describe the detonation properties of most explosives to within experimental error with a single set of equation of state parameters. The study[10] indicated that it was not possible to find a set of BKW equation of state parameters that would reproduce the experimentally observed detonation properties of RDX and TNT with covolumes that reproduced the experimental shock Hugoniots of the individual detonation products.

In 1967, the BKW technique was written in FORTRAN[23] and became generally available for use on any computer that could use codes in the FORTRAN IV language. In the next decade, the code was used by the Center of Scientific and Technical Explosive Research, Sterrebeeck, Belgium; the Swedish Detonic Research Foundation; the Swedish Research Institute of National Defense; the Australian Materials Research Laboratories and Weapon Research Establishment; the Department of Reaction Chemistry of the University of Tokyo; and the National Chemical Laboratory for Industry in Kanagawa, Japan. The

Atomic Weapons Research Establishment of Aldermaston, England, used BKW for many explosive mixtures.

Other versions of the BKW code were written in the 1960s and 1970s. Cheret of the French Commission of Atomic Energy wrote a code called Arpege,[24] in the 1960s. The Lawrence Livermore code RUBY[25] was replaced by a code called TIGER[26] written by scientists at the Stanford Research Institute. It has been used extensively by various U.S.A. military laboratory scientists. Except for coding details, the various computer codes were essentially equivalent. The parameters of the BKW equation of state have been recalibrated by various groups of scientists for use with their particular types of explosives. Many of the parameter sets reproduce the performance of particular explosives, but do not reproduce the shock Hugoniots of the detonation products.

The BKW code has been operating on small, inexpensive personal computers for over a decade. Many explosive and propellant scientists and engineers routinely use BKW or similar codes to evaluate the performance of new chemical compounds or various mixtures of explosives, binders, and other components.

The calculations are performed in steps which we call systems and are generally coded as a "subroutine."

The Nomenclature

α	BKW Equation of State Constant = 0.5
β	BKW Equation of State Constant = 0.16
Θ	BKW Equation of State Constant = 400
κ	BKW Equation of State Constant = 10.9097784436
M	Number of Elements in the Explosive
N	Number of Gaseous Species in Detonation Products
NT	Total Number of Species in Detonation Products
T	Temperature in ^oK
P	Pressure in Mbar
SO or S^o	Entropy in cal/deg-mole
$(H - HO)$	Enthalpy in cal/mole
$(F - HO)/T$	Free energy in cal/deg-mole
IC	Integration Constant
$(\Delta H_f^o)_i$	Heat of Formation at $0^o K$ of Component i in cal/mole [Elements ($0^o K$) $\rightarrow$ Component ($0^o K$)]
V_o	$1/\rho_o$ where ρ_o is Density in g/cc
V_o'	$1/\rho_o$ where ρ_o is Density of Explosive in g/cc
X_i	Number of Moles of ith Species per Mole of Explosive
Y_i	X_i One Step Earlier in the Same Subroutine
α_{ik}	Input Detonation Product Elemental Composition Matrix
b_k	Input Explosive Elemental Composition Vector
T_v	Temperature in Volts
D	Detonation Velocity in cm/μsec
V_g	Volume of the Gas in cc/mole
V_s	Volume of the Solid in cc/g
ρ_s	Density of the Solid in g/cc
k_i	Covolume

R_1	1.98718
R_2	8.341439 x 10^{-5}
R_3	2.39004905 x 10^4
R_4	0.98692 x 10^6
R_5	11, 605.6
R_6	0.4342944819
f_i	Total Free Energy of Gas
G_i	Total Free Energy of Solid
F'_s	Imperfection Solid Free Energy
E'_s	Imperfection Solid Enthalpy
S'_s	Imperfection Solid Entropy
E_g	Total Enthalpy of Gas
E'_g	Imperfection Gas Enthalpy
S'_g	Imperfection Gas Entropy
$(E_s)_i$	Total Enthalpy of Solid i
$MOLWT$	Molecular Weight of a Solid
$AMOLWT$	Explosive Formula Weight
E_{Total}	Energy in cal/mole
V_{Total}	Volume in cc/mole of Explosive
VPG	Volume in cc/g of Explosive
E_o	Heat of Formation of Explosive [Elements ($0^o K$) $\rightarrow$ Explosive ($300^o K$)]
U_s	Shock Velocity
U_p	Particle Velocity
α'	Linear Coefficient of Thermal Expansion
K	Isothermal Compressiblity ($Mbar^{-1}$)
C_v	Heat Capacity (cal/g-^{o}C)

System I. Given P and T, compute V_g.

1. $$\bar{X} = \sum_{i=1}^{N} X_i \ .$$

2. $$Z = \kappa \sum_{i=1}^{N} \frac{X_i}{\bar{X}} k_i \ .$$

3. Linear feedback on V_g. Initial guess = 15, ratio = 1.1, error = 1 x 10^{-6}.

a. $$x = \frac{Z}{V_g(T + \Theta)^{\alpha}} \ .$$

b. $$F_{(x)} = 1 + xe^{\beta x} \ .$$

c. $F_{(x)} - \dfrac{PV_g}{R_2T} = 0$.

4. Find F_i^* for $i = 1$ to N.

a. Calculate the ideal gas thermodynamic function $(F - HO/T)_i$, for $(i = 1$ to $NT)$. Using $A, B, C, D, E,$ and IC from the thermodynamics code TDF described in Appendix F, the thermodynamic functions are calculated from

$$SO = A + BT + CT^2 + DT^3 + ET^4 \ .$$

$$H - HO = \frac{BT^2}{2} + \frac{2CT^3}{3} + \frac{3DT^4}{4} + \frac{4ET^5}{5} + IC \ ,$$

since $\int(dH/dT) = \int T(dS/dT)$.

$$\frac{(F - HO)}{T} = -\left(A + \frac{BT}{2} + \frac{CT^2}{3} + \frac{DT^3}{4} + \frac{ET^4}{5}\right) + \frac{IC}{T} \ ,$$

since $(F - HO)/T = (H - HO)/T - SO$.

b. Form F_i^* by solving

$$F_i^* = \left(\frac{F - HO}{R_1T}\right)_i + \frac{(\Delta H_f^o)_i}{R_1T} + ln(R_4P)$$

$$-\left[lnF_{(x)} - \left(\frac{e^{\beta x} - 1}{\beta}\right) - \kappa\, k_i \frac{F_{(x)} - 1}{Z}\right] .$$

5. Find $(V_s)_i$ and $(G^*)_i$ for $(i = N+1$ to $NT)$.
Use the Cowan solid equation of state to calculate V_s and then F_s'.

$$(G^*)_i = \left(\frac{F - HO}{R_1T}\right)_i + \frac{(H_f^o)_i}{R_1T} + \frac{F_s'}{R_2T} \ .$$

The Cowan equation of state is of the form

$$P = A + B\rho + C\rho^2 + D\rho^3 + E\rho^4 + (A1 + A2\rho)\, T_v$$

$$+\left[C_1 + \frac{(C_2\rho_o)}{\rho} + \frac{(C_3\rho_o)^2}{\rho^2}\right] T_v^2 \ .$$

Normally it is assumed that $(3\alpha'/K = A1 + A2\rho_o,)$ and that $A2 = 1.447404/$(Atomic Weight) .

Shock Hugoniot Temperature Calculation

The coefficients are generated using the STM code. The STM code calculates the Hugoniot pressures and volumes from linear fits of experimentally measured shock and particle velocity.

The temperatures are calculated in the STM code using the Walsh and Christian[27] technique.

$$T = T_o e^{b(V_o - V)} + \frac{(V_o - V)P}{2C_v} + \frac{e^{-bV}}{2C_v} \int_{V_o}^{V} P\, e^{bV} [2 - b(V_o - V)]\, dV \,,$$

with $b = 3\alpha'/K(C_v)$. Simpson's rule is used to evaluate the integral.

The STM code is also used to calculate the solid and liquid shock Hugoniot temperatures that are used to generate the coefficients needed by the HOM equation of state. The executable STM code is included on the CD-ROM. Also included are the data files for many materials.

a. To find the volume:

(1) If incompressible, $V_s = V_o$.

(2) Otherwise, linear feedback is used to find ρ_s, with $\rho_{guessed} = 1.1\,\rho_o$, until the error is less than 1.0 x 10^{-6}, by solving Equation (E.4).

$$T_v = (T)/R_5 \,.$$

$$0 = A_s + B_s \rho_s + C_s \rho_s^2 + D_s \rho_s^3 + E_s \rho_s^4 + (A1 + A2\,\rho_s)\, T_v$$

$$+ \left(C_1 + \frac{C_2}{\rho_s} + \frac{C_3}{\rho_s^2} \right) T_v^2 - P \,.$$

b. To find the imperfection free energy (F_s'):

(1) If incompressible, $F_s' = (P)(Vo)(MOLWT)$.

(2) Otherwise, solve Equation (E.28).

$$F_s' = (MOLWT)PV_s - (MOLWT)\left[A_s V_s + B_s \ln V_s - \frac{C_s}{V_s} - \frac{D_s}{2{V_s}^2} - \frac{E_s}{3V_s^3} \right]_{V_o}^{V}$$

$$- (MOLWT)\left[(A1\, V_s + A2\, \ln V_s) T_v + \left(C_1 V_s + \frac{C_2 V_s^2}{2} + \frac{C_3 V_s^3}{3} \right) T_v^2 \right]_{V_o}^{V} .$$

Units are Mbar-cc/mole.

c. To find the imperfection enthalpy (E_s'):

(1) If incompressible, $E_s' = 0$.

(2) Otherwise, solve Equation (E.15).

$$E'_s = -(MOLWT)\left[A_s V_s + B_s ln V_s - \frac{C_s}{V_s} - \frac{D_s}{2V_s{}^2} - \frac{E_s}{3V_s^3}\right]_{V_o}^{V}$$

$$+(MOLWT)\left[\left(C_1 V_s + \frac{C_2 V_s^2}{2} + \frac{C_3 V_s^3}{3}\right) T_v^2\right]_{V_o}^{V} .$$

Units are Mbar-cc/mole.

d. To find the imperfection entropy (S'_s):

(1) If incompressible, $S'_s = 0$.

(2) Otherwise, solve Equation (E.22).

$$S'_s = +(MOLWT)\left[(A1\ V_s + A2\ ln V_s) + 2\left(C_1 V_s + \frac{C_2 V_s^2}{2} + \frac{C_3 V_s^3}{3}\right) T_v\right]_{V_o}^{V} .$$

6. Solve the chemical equilibrium for the new free energies to obtain the new X_i.

The solution is based on a modified version of White, Johnson, and Dantzig's[28] minimization of free energy technique. The technique minimizes the Gibbs free energy at constant temperature and pressure subject to the conservation of mass constraint since at chemical equilibrium the free energy is at a minimum.

An example, $C_4H_8N_8O_8$ $b_k = 4, 8, 8, 8$

No	α_{ik}	C	H	N	O	FREE ENERGY (F_i)	NUMBER MOLES (Y_i)
1	H_2O	0.	2.	0.	1.	-2.28	4.0
2	H_2	0.	2.	0.	0.	9.56	0.01
3	O_2	0.	0.	0.	2.	7.14	0.01
4	CO_2	1.	0.	0.	2.	-1.25	2.0
5	CO	1.	0.	0.	1.	4.62	0.01
6	NH_3	0.	3.	1.	0.	9.79	0.01
7	H	0.	1.	0.	0.	18.42	0.01
8	NO	0.	0.	1.	1.	12.10	0.01
9	N_2	0.	0.	2.	0.	10.25	4.0
10	OH	0.	1.	0.	1.	13.35	0.01
11	C_{Solid}	1.	0.	0.	0.	2.29	2.0

$$\bar{X} = \sum_{i=1}^{N} X_i .$$

$$\bar{Y} = \sum_{i=1}^{N} Y_i \ .$$

$$f_i(y) = F_i^* + ln\frac{y_i}{\bar{y}} \quad \text{for } (i = 1 \text{ to } N) \ .$$

$$G_i(y) = G_i^* \quad \text{for } (i = N+1 \text{ to } NT) \ .$$

To form $[A]X = [B]$, the equations are

$$\frac{X_i}{Y_i} - \frac{\bar{X}}{\bar{Y}} + \sum_{k=1}^{M} \pi_k \, \alpha_{ik} = -f_i(y) \quad \text{for } (i = 1 \text{ to } N) \ ,$$

$$\sum_{k=1}^{M} \pi_k \, \alpha_{ik} = -G_i(y) \quad \text{for } (i = N+1 \text{ to } NT) \ ,$$

$$\sum_{k=1}^{M} X_i \, \alpha_{ik} = b_k \quad \text{for } (i = 1 \text{ to } NT) \ ,$$

$$\sum_{i=1}^{N} X_i - \bar{X} = 0 \ .$$

The Matrix $[A]X = [B]$

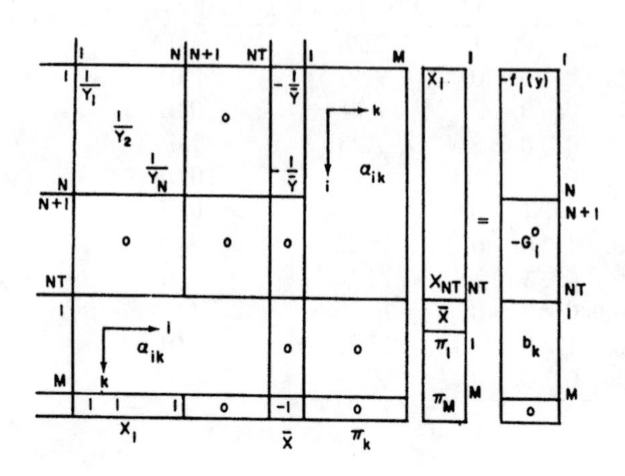

The Matrix is solved using the LASL Linear System Solver.

The system has converged when

$$\sum_{i=1}^{NT} |Y_i - X_i| < 1 \times 10^{-8} .$$

Otherwise, set X_i for ($i = 1$ to NT) to Y and resolve the matrix.

7. Test for convergence of System I.

$$\sum_{i=1}^{NT} |Y_i - X_i| < 2 \times 10^{-5} .$$

If not converged, return to step 1.

System II. Given P, T, and V, calculate E.

System I must have been performed.

1. $E'_g = R_1 T \left(\alpha T \dfrac{F_{(x)} - 1}{T + \Theta} \right) .$

2. Solve Equation (E.12)

$$E_g = \left[\sum_{i=1}^{N} \frac{X_i}{\bar{X}} [(H - HO)_i - R_1 T + (\Delta H_f^o)_i] \right] + E'_g ,$$

where $(H - HO)_i$, is obtained using the thermodynamic function subroutine for ($i =$ 1 to NT).

3. $(E_g)_i = (H - HO)_i + (\Delta H_f^o)_i + R_3 E'_s$

for ($i = N + 1$ to NT), where E'_s is obtained from the Cowan solid equation of state subroutine.

4. $E_{TOTAL} = \bar{X}_g E_g + \sum_{i=N+1}^{NT} X_i (E_s)_i .$

5. $V_{TOTAL} = \bar{X}_g V_g + \sum_{i=N+1}^{NT} X_i (V_s)_i (MOLWT)_i .$

6. $VPG = (V_{TOTAL})/(AMOLWT) .$

System II A. Given a P, compute Hugoniot Temperature.

Linear feedback on T. Initial guess of $3000^o K$, ratio of 1.1, error of 1×10^{-6}.

$$(1x10^{-5})[E_{TOTAL} - E_o - \frac{1}{2}(P + P_o)(V_o' - VPG)(R_3)(AMOLWT)] = 0 \ .$$

$$P_o = (PO) = 1 \times 10^{-6} \ .$$

Systems I and II are used to calculate the necessary values. 1×10^{-5} is a convenient scaling constant.

System III. Find the CJ values.

1. $P_{guessed} = 0.15 + 0.25(\rho_o - 1)$. Ratio = 0.8, error = 1×10^{-6}.

2. Use minimum of a parabola technique for finding minimum detonation velocity

$$D = V_o'\left(\frac{P - P_o}{V_o' - VPG}\right)^{\frac{1}{2}}$$

by approximating $D = f(P)$ with a parabola.

The formula for calculating P_{min} from three sets of D and P is

$$P_{min} = \frac{1}{2}\frac{P_1^2(D_3 - D_2) + P_2^2(D_1 - D_3) + P_3^2(D_2 - D_1)}{P_1(D_3 - D_2) + P_2(D_1 - D_3) + P_3(D_2 - D_1)} \ .$$

3. $$\gamma_{CJ} = \frac{\rho_o D^2}{P_{CJ}} - 1 \ .$$

4. $$U_{CJ} = [P_{CJ}(V_o' - VPG)]^{\frac{1}{2}} \ .$$

System IV. Given P, T, and V, calculate S.

System I must have been performed. Solve equation (E.20).

1. $$S_g' = -R_1\left[\sum_{i=1}^{N}\left(\frac{X_i}{\bar{X}} ln \frac{X_i}{\bar{X}}\right) + ln(R_4 P)\right]$$

$$+ R_1\left[ln\ F_{(x)} - \left(\frac{e^{\beta x} - 1}{\beta}\right) + \frac{\alpha T(F_{(x)} - 1)}{(T + \Theta)}\right] \ .$$

2. $$S_g = \sum_{i=1}^{N} \left(\frac{X_i}{\bar{X}} S_i^o \right) + S_g' ,$$

where S_i^o is obtained from the thermodynamic function subroutine.

3. $$(S_s)_i = S_i^o + \frac{R_3(S_s')_i}{R_5} \quad \text{for} \quad (i = N+1 \text{ to } NT) ,$$

where S_s' is obtained from the Cowan solid equation of state subroutine.

4. $$S_{TOTAL} = \bar{X}_g S_g + \sum_{i=N+1}^{NT} X_i (S_s)_i .$$

System IV A. Compute the C-J isentrope for the C-J pressure.

1. From the C-J pressure and temperature, the energy is found using System II, and the C-J entropy is found using System IV.

2. From the C-J point, P is first decreased by multiples of 0.175 until P reaches some minimum value (default is 1 x 10^{-4}). Then increase P^{CJ} by multiples of 1.15 until it reaches 1.0 Mbar.

3. Find the isentrope value by using linear feedback on T and the equation

$$S_{TOTAL} - S_{CJ} = 0 .$$

Initial guess = T_{CJ}, ratio = 0.9, error = 0.5 for $P < P_{CJ}$.

Initial guess = T_{CJ}, ratio = 1.1, error = 0.5 for $P > P_{CJ}$.

System V. Compute Hugoniot curve.

Use System II A to compute the necessary values.

1. P = AMHUGP - (n)(DELP) until it is less than DELP where n is 0 for first calculation, 1 for 2nd, etc. Default is P = 0.5 - n(0.05).

2. $$U_s = V_o' \left(\frac{P}{V_o' - VPG} \right)^{\frac{1}{2}} .$$

3. $$U_p = \sqrt{(P)(V_o - VPG)} .$$

E.6 ISPBKW - Propellant Performance

The version of BKW that calculates the propellant performance is called ISPBKW and the executable code is included on the CD-ROM. The calculation of the performance of propellants is discussed in Chapter 2.8.

Linear feedback on $T_{Chamber}$ using System 1, System 2, and System 4 to determine the rocket Chamber state values at 68.94733 bars (1000 psi). Initial guess of $3000^o K$, ratio of 1.1, error of 1 x 10^{-2}.

$$[E_{TOTAL} - E_o - (P)(V_o' - VPG)(R_3)(AMOLWT)] = 0 .$$

$$P = 6.89473 \times 10^{-5} .$$

Linear feedback on $T_{Exhaust}$ using System 1, System 2, and System 4 to determine the rocket Exhaust state values at 1.01325 bar (1 atmosphere). Initial guess of $3000^o K$, ratio of 0.9, error of 1 x 10^{-2}.

$$S_{Exhaust} - S_{Chamber} = 0 .$$

The Specific Impulse ISP is calculated using

$$ISP = 9.330\sqrt{E_{Chamber} - E_{Exhaust}} .$$

For propellants containing aluminum, ISPBKW changes the Al_2O_3 thermodynamic constants from solid constants to liquid constants at temperatures above 2250° K.

E.7 Linear Feedback Technique

The linear feedback solves F(X) = 0 for X by iteration.

The method assumes a given Xguessed, ratio, and max zero or error permitted.

1. Initial entry
   a. If Xguessed = 0, set it to 1.
   b. Set count = 1 and XP = Xguessed.
   c. Exit to get F(XP).
2. Second entry
   a. Set XN2 = XP, FN2 = F(XP), FN = F(XP), and count = 2.
   b. If |FN2| < max zero, set count = 0, Xguessed = XP, and exit with XP = the solution.
   c. Otherwise, set XP = (Xguessed)(ratio) and exit to get F(XP).
3. Third entry
   a. Set XN1 = XP, FN1 = F(XP), FN = F(XP), and count = 3.
   b. If |FN1| < max zero, set count = 0, Xguessed = XP, and exit with XP = the solution.
   c. Otherwise, set XP = XN1 - FN1(XN1 -XN2)/(FN1 -FN2) and exit to get F(XP).
4. Fourth and succeeding entries.
   a. If the count > 1000, exit with count = - count.
   b. Otherwise, set XN = XP, FN = F(XP), and count = count + 1.

c. If XN = XN1 or FN = FN1, exit with count = - count.

d. If FN < max zero, set count = 0, Xguessed = XP, and exit with XP = the solution.

e. Otherwise, set XP = XN -FN(XN - XN1)/(FN - FN1).

f. If FN and FN1 are of opposite signs, set XN2 = XN1, FN2 = FN1, XN1 = XN, FN1 = FN, and exit to get F(XP).

g. If FN and FN2 are of the same sign, set XN2 = XN1, FN2 = FN1, XN1 = XN, FN1 = FN, and exit to get F(XP).

h. If XP lies between XN and XN2, set XN1 = XN, FN1 = FN, and exit to get F(XP).

i. Otherwise, set XP = XN - FN(XN - XN2)/(FN - FN2), XN1 = XN, FN1 = FN, and exit to get F(XP).

E.8 Explosive Mixtures

The description of various components of explosive mixtures is given in Table E.5. Knowing the weight percent of the components of the mixture, the mixture formula, formula weight, and heat of formation may be calculated. This calculation is performed by the USERBKW code included on the CD-ROM. The information given in Table E.5 is included in the ZZZCOMPS file.

TABLE E.5 Components of Explosive Mixtures

Material	ρ_T (g/cc)	Formula	Formula Weight	ΔH^o_{fo}
HMX	1.90	$C_4H_8N_8O_8$	296.16	43.46
RDX	1.80	$C_3H_6N_6O_6$	222.12	33.97
TNT	1.64	$C_7H_5N_3O_6$	227.13	−1.44
Wax	-	CH_2	14.02	−2.67
Halo Wax	1.67	$C_{3.642}H_{1.36734}Cl_{1.54724}$	100.00	3.0
Kel-F	2.11	C_2F_3Cl	116.48	−145.0
Exon	-	$C_4Cl_2F_3H_3$	178.97	−155.0
DNPA	1.47	$C_6H_8N_2O_6$	204.14	−103.2
Teflon	2.20	C_2F_4	100.02	−145.0
Bis-2, 2-dinitropropyl formal	-	$C_7H_{12}N_4O_{10}$	312.19	−125.0
Cellulose Nitrate	-	$C_6H_7N_2O_9$	251.13	−15.0
CEF-Tris−β chloroethyl phosphate	-	$C_6H_{12}O_4Cl_3P$	285.50	−125.0
Viton A	1.85	$C_{17}F_{20}H_{14}$	598.3	−900.0
Nitroso rubber	1.93	C_3NOF_7	199.03	−300.0
Estane	-	$C_{5.1307}H_{7.291}N_{0.1557}O_{1.8031}$	100.0	−80.0
KF-820	-	$C_4H_2F_5Cl$	180.51	−230.0
Polystyrene	-	C_8H_8	104.14	10.0
DOP-Di-2-ethyl hexyl phthalate	-	$C_{24}H_{38}O_4$	390.54	0.0
FEFO	-	$C_5H_6N_4O_{10}F_2$	320.1	−154.0
Nylon	-	$C_{13}H_{23}N_3O_3$	269.35	−320.0

References

1. G. B. Kistiakowsky and E. B. Wilson, "Report on the Prediction of Detonation Velocities of Solid Explosives," Office of Scientific Research and Development report OSRD-69 (1941).

2. R. Becker, "Eine Zustandsgleichung für Stickstoff bei Grossen Dichten," Zeitschrift für Physik 4, 393 (1921).

3. R. Becker, "Physikalisches über Feste und Gasformige Sprengstoffe," Zeitschrift für Technische Physik 3, 249 (1922).

4. G. B. Kistiakowsky and E. B. Wilson, "The Hydrodynamic Theory of Detonation and Shock Waves," Office of Scientific Research and Development report OSRD-114 (1941).

5. R. D. Cowan and W. Fickett, "Calculation of the Detonation Products of Solid Explosives with the Kistiakowsky-Wilson Equation of State," Journal of Chemical Physics 24, 932 (1956).

6. J. Jeans, "The Dynamical Theory of Gases," Dover Publications, New York, page 134 (1925). Also Hirshfelder, Curtiss, and Bird, "Molecular Theory of Gases and Liquids," John Wiley & Sons, pages 31 and 157 (1954).

7. Melvin H. Rice and John M. Walsh, "Equation of State of Water to 250 Kilobars," Journal of Chemical Physics 26, 816 (1957). and L. V. Altshuler, A. A. Bakanova, and R. F. Trunen, "Phase Transformations of Water Compressed by Strong Shock Waves," Doklady Soviet Physics 3, 761 (1958).

8. Richard Dean Dick, "Shock Wave Compression of Benzene, Carbon Disulfide, Carbon Tetrachloride, and Liquid Nitrogen," Los Alamos Scientific Laboratory report LA-3915 (1968).

9. V. N. Zubarev and G. S. Telegin, "The Impact Compressibility of Liquid Nitrogen and Solid Carbon Dioxide," Doklady Akademii Nauk SSSR 142, 309 (1962), Soviet Physics Doklady, English Translation 7, 34 (1962).

10. Charles L. Mader, "Detonation Properties of Condensed Explosives Computed Using the Becker-Kistiakowsky-Wilson Equation of State," Los Alamos Scientific Laboratory report LA-2900 (1963).

11. Linus Pauling, "Nature of the Chemical Bond," Cornell University Press (1960).

12. Gerhard Herzberg, "Molecular Spectra and Molecular Structure, I. Spectra of Diatomic Molecules," D. Van Nostrand Co. (1950).

13. Wildon Fickett, "Detonation Properties of Condensed Explosives Calculated with an Equation of State Based on Intermolecular Potentials," Los Alamos Scientific Laboratory report LA-2712 (1962).

14. Jerry Wackerle, Los Alamos Scientific Laboratory, private communication.

15. Jerry Wackerle, W. L. Seitz, and John C. Jamieson, "Shock Wave Equation of State for High-Density Oxygen," Symposium on the Behavior of Dense Media Under High Dynamic Pressures (1967).

16. M. Van Thiel and M. Wasley, "Compressibility of Liquid Hydrogen to 40,000 Atmospheres at 1100°K," Lawrence Livermore Laboratory report UCRL-7833 (1964).

17. Richard Dick, Los Alamos Scientific Laboratory, private communication.

18. John B. Ramsay and W. C. Chiles, "Detonation Characteristics of Liquid Nitric Oxide," Sixth International Symposium on Detonation ACR-221, 723 (1976).

19. J. R. Reitz, "The Nuclear Contribution to the Equation of State of Metals," Los Alamos Scientific Laboratory Report LA-1454 (1952).

20. F. D. Rossini, "Chemical Thermodynamics," John Wiley & Sons, Inc., New York (1950).

21. Charles L. Mader, "Detonation Performance Calculations Using the Kistiakowsky-Wilson Equation of State," Los Alamos Scientific Laboratory report LA-2613 (1961).

22. Charles L. Mader, "STRETCH BKW: A Code for Computing the Detonation Properties of Explosives," Los Alamos Scientific Laboratory report LADC-5691 (1962).

23. Charles L. Mader, "FORTRAN BKW: A Code For Computing The Detonation Properties of Explosives," Los Alamos Scientific Laboratory report LA-3704 (1967).

24. Roger Cheret, "The Numerical Study of the Detonation Products of an Explosive Substance," French Commission of Atomic Energy report CEA-R-4122 (1971).

25. H. B. Levine and R. E. Sharples,"Operator's Manual for RUBY," Lawrence Livermore Laboratory report UCRL-6815 (1962).

26. M. Cowperthwaite and W. H. Zwisler, "Tiger Computer Program Documentation," Stanford Research Institute publication No. Z106 (1973).

27. John M. Walsh and Russell H. Christian, "Equation of State of Metals from Shock Wave Measurements," Physical Review 97, 1554 (1955).

28. W. B. White, S. M. Johnson, and G. B. Dantzig, "Chemical Equilibrium in Complex Mixtures," Journal of Chemical Physics 28, 751 (1958).

appendix F

Equations for Computing of Thermodynamic Functions of Gases and Solids

The calculation of ideal gas thermodynamic properties of detonation products is performed using a code called TDF.[1] It was first written in IBM 7090 machine language in the late 1950s. It was later written for the IBM 7030. A FORTRAN version was written in the late 1960s and made available through the COSMIC library for computer programs. The ideal gas thermodynamic properties of detonation products were compiled and made available as References 2 and 3.

For monatomic, diatomic, and polyatomic gases, the TDF code may be used to calculate the free energy, enthalpy, and entropy as functions of temperature. The moments of inertia can be calculated for the polyatomic case. For one-Debye Theta and two-Debye Theta solids, the heat capacity is computed as a function of temperature in addition to the free energy, enthalpy, and entropy. These functions are fit to a fourth degree polynominal by the method of least squares. The integration constant, IC, in the equation $H^o - H^o_o = \int T(\partial S/\partial T) + IC$ is computed from the fit of entropy as a function of temperature.

The use of fits introduces errors into any detonation or propellant equation of state calculation. Future detonation and propellant performance computer codes should include the direct calculation of the ideal gas thermodynamic parameters.

The executable TDF code is included on the CD-ROM. Also included are data files with the parameters for monatomic gases, diatomic gases, polyatomic gases, and solids for both detonation and propellant decomposition gases and solids.

The Nomenclature

Mw	Molecular Weight (grams/mole)
T	Temperature (^{o}K)
σ	Symmetry Number
h	Planck's Constant (6.62377 x 10^{-27}) erg-sec
c	Speed of Light (2.99790 x 10^{10}) cm/sec
k	Boltzmann's Constant (1.380257 x 10^{-16}) erg/deg
R	Gas Constant (1.98719) cal/degree mole
T_i	Term Value of ith Electronic State (1/cm)

g_i	Statistical Weight of ith Electronic State
We	Vibrational Frequency (1/cm)
$WeXe$	Anharmonic Coefficient (1/cm)
Be	Rotational Coefficient (1/cm)
Ae	Rotational-Vibrational Coefficient (1/cm)
ν_i	Fundamental Frequency (1/cm)
d_i	Degeneracy of Fundamental Frequency
$I_A,\ I_B,\ I_C$	Moments of Inertia (g-cm^2 x 10^{39})
m_i	Atomic Weight of ith Atom (amu)
$x_i,\ y_i,\ z_i$	Cartesian Coordinates of ith Atom (A)
N	Avogadro's Number (6.0238 x 10^{23}) (g-mole)$^{-1}$
P	One Atmosphere of Pressure
M	Dimension (1, 2, or 3)
S	No of Atoms per Cell
Θ	Debye Temperature (^{o}K)
Θ_t	Debye Temperature Transverse (^{o}K)
Θ_l	Debye Temperature Longitudinal (^{o}K)
C_V	Heat Capacity (cal/mole/^{o}K)

F.1 Monatomic Gas

The theory of the calculation is described by Rossini.[4]

Translation contribution

$$F_t \,/\, T \;=\; -(3/2) R\; ln\; Mw - (5/2) R\; ln\; T + 7.28295 \;.$$

$$H_t \,/\, T \;=\; -(5/2) R \;.$$

Electronic contribution

$$U_i \;=\; \frac{h\, c\, T_i}{kT} \;.$$

$$Q_e \;=\; g_o + \sum_i g_i\, e^{-U_i} \;.$$

$$\frac{F_e}{T} \;=\; -R\; ln\; Q_e \;.$$

$$\frac{H_e}{T} \;=\; \frac{R}{Q_e} \sum_i U_i\, g_i\, e^{-U_i} \;.$$

$$\frac{F^o - H^o_o}{T} = \frac{F_t}{T} + \frac{F_e}{T} \qquad \text{(Free Energy in cal/mole/}^oK) \; .$$

$$H^o - H^o_o = H_t + H_e \qquad \text{(Enthalpy in cal/mole)} \; .$$

F.2 Diatomic Gas

The method of calculation is that of Pennington and Kobe,[5] using Herzberg's units and symbolism.[6]

$$W_o = We - 2We\,Xe \; .$$

$$D_o = 4(Be)^3/(We)^2 \; .$$

$$\chi = We\,Xe\;/\;We \; .$$

$$B_o = Be - (1/2)Ae \; .$$

$$s = 2D_o\,k/((B_o)^2\,c\,h\,) \; .$$

$$r = (Ae/Be)[1 + (Ae/Be)] \; .$$

$$F_{ro+to}/T = (3/2)R\,ln\,Mw - R\,ln\,(h\,c\,B_o/k) - R\,ln\,\sigma - 7.28295 \; .$$

$$H_{ro+to}/T = (7/2)R \; .$$

Vibration contribution

$$U_o = \frac{h\,c\,W_o}{kT} \; .$$

$$\frac{F_v}{T} = -R\,ln\,(1 - e^{-U_o}) \; .$$

$$\frac{H_v}{T} = \frac{R\,U_o}{e^{U_o} - 1} \; .$$

Electronic contribution

$$U_i = \frac{h\,c\,T_i}{kT} \; .$$

$$Q_e = g_o + \sum_i g_i\, e^{-U_i} \ .$$

$$\frac{F_e}{T} = R\, ln\, Q_e \ .$$

$$\frac{H_e}{T} = \frac{R}{Q_e} \sum_i g_i\, U_i\, e^{-U_i} \ .$$

Pennington and Kobe corrections

$$\phi_1 = \frac{1}{e^{U_o} - 1} \ .$$

$$\phi_2 = \frac{U_o\, e^{U_o}}{(e^{U_o} - 1)^2} \ .$$

$$\phi_4 = \frac{2U_o}{(e^{U_o} - 1)^2} \ .$$

$$\phi_5 = \frac{2U_o(2U_o\, e^{U_o} - e^{U_o} + 1)}{(e^{U_o} - 1)^3} \ .$$

$$\frac{F_c}{T} = R(s\, T + r\, \phi_1 + \chi\, \phi_4) \ .$$

$$\frac{H_c}{T} = R(s\, T + r\, \phi_2 + \chi\, \phi_5) \ .$$

$$\frac{F_o - H_o^o}{T} = -(7/2) R\, ln\, T - \frac{F_c}{T} - \frac{F_{ro+to}}{T} - \frac{F_e}{T} - \frac{F_v}{T} \ .$$

$$H_o - H_o^o = H_v + H_{ro+to} + H_c + H_e \ .$$

F.3 Polyatomic Gas

The harmonic-oscillator rigid rotator approximation is used to compute the thermodynamic functions.[7]

$$U_i = \frac{h\, c\, \nu_i}{k\, T} \ .$$

$$\frac{F_v}{T} = R\sum_i d_i \ln\,(1 - e^{-U_i}) \ .$$

$$H_v = RT\sum_i \frac{d_i\, U_i\, e^{-U_i}}{1 - e^{-Ui}} \ .$$

Translation contribution

$$\frac{F_t}{T} = -(5/2)R \ln T - (3/2)R \ln Mw - R \ln\left[\left(\frac{2\pi}{N}\right)^{(3/2)} \frac{k^{(5/2)}}{h^3}\right] + R \ln P \ .$$

$$H_t = (5/2)RT \ .$$

Rotation contribution

Nonlinear

$$\frac{F_r}{T} = -(3/2)R \ln T - (R/2) \ln\,(I_A I_B I_C) + R \ln \sigma + 3.01407 \ .$$

$$H_r = (3/2)RT \ .$$

Linear

$$\frac{F_r}{T} = -R \ln T - R \ln\,(I_A) + R \ln \sigma + 2.76764 \ .$$

$$H_r = RT \ .$$

$$\frac{F^o - H^o_o}{T} = \frac{F_r}{T} + \frac{F_v}{T} + \frac{F_t}{T} \ .$$

$$H^o - H^o_o = H_r + H_v + F_t \ .$$

F.4 Moment of Inertia

The moments of inertia are calculated using J. O. Hirshfelder's method.[8]

The moments of inertia are the eigenvalues of the symmetric matrix

$$\begin{pmatrix} A & -D & -E \\ -D & B & -F \\ -E & -F & C \end{pmatrix} .$$

where

$$Mw = \sum_i m_i .$$

$$A = \sum_i m_i \, (y_i^2 + z_i^2) - \frac{1}{Mw}\left(\sum_i m_i y_i\right)^2 - \frac{1}{Mw}\left(\sum_i m_i \, z_i\right)^2 .$$

$$B = \sum_i m_i \, (x_i^2 + z_i^2) - \frac{1}{Mw}\left(\sum_i m_i x_i\right)^2 - \frac{1}{Mw}\left(\sum_i m_i \, z_i\right)^2 .$$

$$C = \sum_i m_i \, (x_i^2 + y_i^2) - \frac{1}{Mw}\left(\sum_i m_i x_i\right)^2 - \frac{1}{Mw}\left(\sum_i m_i \, y_i\right)^2 .$$

$$D = \sum_i m_i \, x_i \, y_i - \frac{1}{Mw}\left(\sum_i m_i x_i\right)\left(\sum_i m_i \, y_i\right) .$$

$$E = \sum_i m_i \, x_i \, z_i - \frac{1}{Mw}\left(\sum_i m_i x_i\right)\left(\sum_i m_i \, z_i\right) .$$

$$F = \sum_i m_i \, y_i \, z_i - \frac{1}{Mw}\left(\sum_i m_i y_i\right)\left(\sum_i m_i \, z_i\right) .$$

The moments of inertia are then multiplied by 0.166035 to get the proper mass units for use in the polyatomic thermodynamic function code.

F.5 One-Debye Theta Solid

The formulas for the Debye approximation are given by Menzel[9] and Tarassov.[10] The Debye integral is done numerically using Simpson's Rule and 1000 intervals.

$$x = \Theta / T .$$

$$D(x) = \frac{M}{x^M} \int_o^x \frac{x^M\,dx}{e^x - 1} \ .$$

$$\frac{F^o - H_o^o}{T} = S\,R\,[M\,ln\,(1 - e^{-x}) + (2 - M)\,D(x)] \ .$$

$$H^o - H_o^o = 3S\,R\,T\,D(x) \ .$$

$$C_v = 3S\,R\left[(M + 1)\,D(x) - \frac{Mx}{e^x - 1}\right] \ .$$

F.6 Two-Debye Theta Solid

The theory of the calculation is described by Krumhansl and Brooks.[11]

$M = 2,$ (2 dimensional)

$$x_t = \frac{\Theta_t}{T} \ .$$

$$x_l = \frac{\Theta_l}{T} \ .$$

$$D(x) = \frac{M}{x^M} \int_o^x \frac{x^M\,dx}{e^x - 1} \ .$$

$$\frac{F^o - H_o^o}{T} = \frac{1}{2} S\,R\big[(1 - M)\,D(x_t) - M\,ln\,(1 - e^{-x_t})\big]$$

$$+ S\,R\big[(1 - M)\,D(x_l) - M\,ln\,(1 - e^{-x_l})\big] \ .$$

$$H^o - H_o^o = 2S\,R\,T\left(\frac{1}{2} D(x_t) + D(x_l)\right) \ .$$

$$C_v = 2S\,R\left[\frac{1}{2}\left((M + 1)\,D(x_t) - \frac{Mx_t}{e^{x_t} - 1}\right) + (M + 1)\,D(x_l) - \frac{Mx_l}{e^{x_l} - 1}\right] \ .$$

F.7 General

In all cases, the entropy (cal/mole/^{o}K) is calculated from the equation

$$S^o = \frac{H^o - H^o_o}{T} - \frac{F^o - H^o_o}{T} \; .$$

The integration constant is calculated from the fit of S^o

$$IC = H^o - H^o_o - \int T \frac{\partial S}{\partial T} dT \; .$$

S^o is fit to $S^o = A + B\,T + C\,T^2 + D\,T^3 + E\,T^4$, then

$$IC = H^o - H^o_o - \left(\frac{B\,T^2}{2} + \frac{2C\,T^3}{3} + \frac{3}{4} D\,T^4 + \frac{4}{5} E\,T^5 \right) \; .$$

References

1. William R. Gage and Charles L. Mader, "STRETCH TDF: A Code for Computing Thermodynamic Functions," Los Alamos Scientific Laboratory report LADC-6575 (1964).

2. Charles L. Mader, "Thermodynamic Properties of Detonation Products" (Supplement to GMX-2-R-59-3), Los Alamos Scientific Laboratory report GMX-2-R-60-1 (1959).

3. Charles L. Mader, "Ideal Gas Thermodynamic Properties of Detonation Products," Los Alamos Scientific Laboratory unpublished report AECU-4508 (1959).

4. F. D. Rossini, "Chemical Thermodynamics," J. Wiley & Sons, New York (1950).

5. R. E. Pennington and K. A. Kobe, "Contributions of Vibrational Anharmonicity and Rotation-Vibration Interaction to Thermodynamic Functions," Journal of Chemical Physics 22, 1442 (1954).

6. Gerhard Herzberg, "Molecular Spectra and Molecular Structure, I. Spectra of Diatomic Molecules," D. Van Nostrand Co. (1950).

7. Gerhard Herzberg, "Molecular Spectra and Molecular Structure, II. Infrared and Raman Spectra of Polyatomic Molecules," D. Van Nostrand Co. (1950).

8. J. O. Hirshfelder, "Simple Method for Calculating Moments of Inertia," Journal of Chemical Physics 8, 430 (1940).

9. Donald H. Menzel, "Fundamental Formulas of Physics," Dover (1960).

10. V. V. Tarassov, "On a Theory of Low Temperature Heat Capacity of Linear Macromolecules," Comptes Rendus (Doklady) of the Academy of Sciences USSR 45, 20 (1945).

11. J. Krumhansl and H. Brooks, "The Lattice Vibration Specific Heat of Graphite," Journal of Chemical Physics 21, 1663 (1953).

appendix G

General Derivation of Flow Equations Used

The three-dimensional partial differential equations for viscous, conducting, elastic-plastic, compressible flow are:

G.1 Eulerian Conservation of Mass

$$\left(\frac{\partial}{\partial t} + U \cdot \nabla\right)\rho = -\rho \nabla \cdot U$$

or

$$\frac{\partial \rho}{\partial t} + U_x\left(\frac{\partial \rho}{\partial x}\right) + U_y\left(\frac{\partial \rho}{\partial y}\right) + U_z\left(\frac{\partial \rho}{\partial z}\right) = -\rho\left(\frac{\partial U_x}{\partial x} + \frac{\partial U_y}{\partial y} + \frac{\partial U_z}{\partial z}\right) , \qquad (G.1)$$

since

$$\nabla = \frac{\partial}{\partial x}\hat{i} + \frac{\partial}{\partial y}\hat{j} + \frac{\partial}{\partial z}\hat{k} ,$$

$$U \cdot = U_x\hat{i} + U_y\hat{j} + U_z\hat{k} ,$$

and

$$\hat{i} \cdot \hat{i} = 1, \quad \hat{i} \cdot \hat{j} = 0, \quad \hat{i} \cdot \hat{k} = 0, \ etc.$$

G.2 Eulerian Conservation of Momentum

$$\rho\left(\frac{\partial}{\partial t} + U \cdot \nabla\right)U = - \nabla \cdot \sigma + \rho g \ .$$

$$\sigma = \partial_{ij} P - S_{i,j} \quad .$$

$$\rho\left[\frac{\partial U_x}{\partial t} + U_x\left(\frac{\partial U_x}{\partial x}\right) + U_y\left(\frac{\partial U_x}{\partial y}\right) + U_z\left(\frac{\partial U_x}{\partial z}\right)\right] = -\frac{\partial(P - S_x)}{\partial x}$$

$$+\frac{\partial S_{xy}}{\partial y} + \frac{\partial S_{xz}}{\partial z} + \rho g_x \quad . \tag{G.2}$$

$$\rho\left[\frac{\partial U_y}{\partial t} + U_x\left(\frac{\partial U_y}{\partial x}\right) + U_y\left(\frac{\partial U_y}{\partial y}\right) + U_z\left(\frac{\partial U_y}{\partial z}\right)\right] = -\frac{\partial(P - S_y)}{\partial y}$$

$$+\frac{\partial S_{xy}}{\partial x} + \frac{\partial S_{yz}}{\partial z} + \rho g_y \quad .$$

$$\rho\left[\frac{\partial U_z}{\partial t} + U_x\left(\frac{\partial U_z}{\partial x}\right) + U_y\left(\frac{\partial U_z}{\partial y}\right) + U_z\left(\frac{\partial U_z}{\partial z}\right)\right] = -\frac{\partial(P - S_z)}{\partial z}$$

$$+\frac{\partial S_{xz}}{\partial x} + \frac{\partial S_{yz}}{\partial y} + \rho g_z \quad .$$

G.3 Eulerian Conservation of Energy

$$\rho\left(\frac{\partial}{\partial t} + U \cdot \nabla\right)I = -\sigma : \nabla U + \lambda \nabla^2 T \quad ,$$

where

$$\sigma : \nabla U = \sigma_{i,j}\frac{\partial U_i}{\partial X_j} \quad .$$

$$\rho\left[\frac{\partial I}{\partial t} + U_x\left(\frac{\partial I}{\partial x}\right) + U_y\left(\frac{\partial I}{\partial y}\right) + U_z\left(\frac{\partial I}{\partial z}\right)\right]$$

$$= -(P - S_x)\frac{\partial U_x}{\partial x} - (P - S_y)\frac{\partial U_y}{\partial y} - (P - S_z)\frac{\partial U_z}{\partial z}$$

$$+ S_{xy}\left(\frac{\partial U_y}{\partial x} + \frac{\partial U_x}{\partial y}\right) + S_{xz}\left(\frac{\partial U_z}{\partial x} + \frac{\partial U_x}{\partial z}\right) + S_{yz}\left(\frac{\partial U_z}{\partial y} + \frac{\partial U_y}{\partial z}\right)$$

$$+ \lambda\left(\frac{\partial^2 T}{\partial x^2} + \frac{\partial^2 T}{\partial y^2} + \frac{\partial^2 T}{\partial z^2}\right) . \quad (G.3)$$

G.4 Stress and Viscosity Deviators

$$\frac{\partial S_x}{\partial t} = 2\mu\left(\frac{\partial U_x}{\partial x} - \frac{1}{3V}\frac{\partial V}{\partial t}\right) .$$

$$\frac{\partial S_y}{\partial t} = 2\mu\left(\frac{\partial U_y}{\partial y} - \frac{1}{3V}\frac{\partial V}{\partial t}\right) .$$

$$\frac{\partial S_z}{\partial t} = 2\mu\left(\frac{\partial U_z}{\partial z} - \frac{1}{3V}\frac{\partial V}{\partial t}\right) .$$

$$f = 2\left[(S_x)^2 + (S_y)^2 + (S_z)^2\right] .$$

If $f \leq \frac{2}{3}(Y_o)^2$, the S_x, S_y, and S_z are scaled by F where

$$F = (\sqrt{((2/3)Y_o^2) / f}) .$$

For real viscosity

$$q_x = 2\mu_V \frac{\partial U_x}{\partial x} - \frac{2}{3}\mu_V\left(\frac{1}{V}\frac{\partial V}{\partial t}\right) .$$

$$q_y = 2\mu_V \frac{\partial U_y}{\partial y} - \frac{2}{3}\mu_V\left(\frac{1}{V}\frac{\partial V}{\partial t}\right) .$$

$$q_z = 2\mu_V \frac{\partial U_z}{\partial z} - \frac{2}{3}\mu_V\left(\frac{1}{V}\frac{\partial V}{\partial t}\right) .$$

$$q = \left(\lambda + \frac{2}{3}\mu_V\right)\left(\frac{1}{V}\frac{\partial V}{\partial t}\right) .$$

$$S_x^n = S_x + q_x - q .$$

$$S_y^n = S_y + q_y - q .$$

$$S_z^n = S_z + q_z - q \quad .$$

G.5 Lagrangian to Eulerian Transformation

$$\frac{\partial U}{\partial t} = \left(\frac{\partial}{\partial t} + U \cdot \nabla\right) U \quad .$$

$$\frac{\partial \rho}{\partial t} = \left(\frac{\partial}{\partial t} + U \cdot \nabla\right) \rho \quad .$$

$$\frac{\partial I}{\partial t} = \left(\frac{\partial}{\partial t} + U \cdot \nabla\right) I \quad .$$

G.6 Conversion to Radial Geometry

$$U_x = U_r \cos \alpha \quad .$$

$$U_y = U_r \cos \beta \quad .$$

$$U_z = U_r \cos \gamma \quad .$$

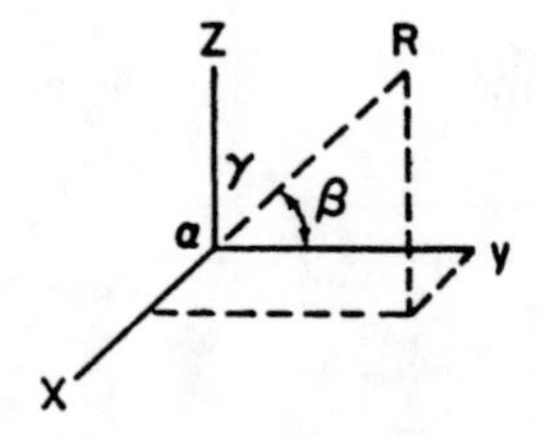

$$R = \sqrt{x^2 + y^2 + z^2} \quad .$$

$$\frac{\partial R}{\partial x} = \frac{x}{R} = \cos \alpha \quad .$$

$$\frac{\partial R}{\partial y} = \frac{y}{R} = \cos \beta \quad .$$

$$\frac{\partial R}{\partial z} = \frac{z}{R} = \cos \gamma \quad .$$

$$\frac{\partial U_x}{\partial x} = \frac{\partial U_r \cos \alpha}{\partial x} = U_r \frac{\partial \cos \alpha}{\partial x} + \cos \alpha \left(\frac{\partial U_r}{\partial x}\right)$$

$$= \frac{R^2 - x^2}{R^3} U_r + \cos\alpha \frac{\partial U_r}{\partial R} \frac{x}{R} \quad .$$

Substituting into Equation (G.1) gives

$$\frac{\partial \rho}{\partial t} + U_r \cos\alpha \frac{\partial \rho}{\partial R} \frac{x}{R} + U_r \cos\beta \frac{\partial \rho}{\partial R} \frac{y}{R} + U_r \cos\gamma \frac{\partial \rho}{\partial R} \frac{z}{R}$$

$$= -\rho \left(\cos\alpha \frac{x}{R} \frac{\partial U_r}{\partial R} + \cos\beta \frac{y}{R} \frac{\partial U_r}{\partial R} + \cos\gamma \frac{z}{R} \frac{\partial U_r}{\partial R} \right)$$

$$+ \frac{U_r}{R} \left(\frac{R^2 - x^2}{R^2} + \frac{R^2 - y^2}{R^2} + \frac{R^2 - z^2}{R^2} \right) \quad .$$

$$\frac{\partial \rho}{\partial t} + U_r \frac{\partial \rho}{\partial R} = -\rho \left(\frac{\partial U_r}{\partial R} + \frac{2U_r}{R} \right) = -\frac{\rho}{R^2} \frac{\partial R^2 U_r}{\partial R} = \frac{\rho}{R^{\alpha-1}} \frac{\partial R^{\alpha-1} U_r}{\partial R} \quad ,$$

which in Lagrangian space is

$$\frac{\partial \rho}{\partial t} = -\frac{\rho}{R^{\alpha-1}} \frac{\partial R^{\alpha-1} U_r}{\partial R} \quad .$$

Since $dm = \rho_o R_o^{\alpha-1} dR_o = \rho R^{\alpha-1} dR$, and

$$V = R^{\alpha-1} \frac{dR}{dm} \quad ,$$

$$\frac{\partial \rho}{\partial t} = -\frac{1}{V^2} \frac{\partial R^{\alpha-1} U_r}{\partial m} \quad ,$$

$$\frac{\partial V}{\partial t} = \frac{\partial R^{\alpha-1} U_r}{\partial m} \quad .$$

If elastic-plastic and gravity terms are ignored, similar transformations of Equation (G.3) to radial geometry give for conservation of energy

$$\rho \left(\frac{\partial I}{\partial t} + U_r \frac{\partial I}{\partial R} \right) = -\frac{P}{R^{\alpha-1}} \frac{\partial R^{\alpha-1} U_r}{\partial R} \quad ,$$

which in Lagrangian space is

$$\rho \frac{\partial I}{\partial t} = -\frac{P}{R^{\alpha-1}} \frac{\partial R^{\alpha-1} U_r}{\partial R} \quad .$$

Taking $E = I + 1/2U_r^2, \;\; dm = R^{\alpha-1} dR/V,$ and substituting in the momentum equation

$$\frac{\partial E}{\partial t} = -\frac{\partial P U_r R^{\alpha-1}}{\partial m} \quad .$$

If elastic-plastic, heat conduction, and gravity terms are ignored, transformation of Equation (G.2) to radial geometry gives for conservation of momentum.

$$\rho \left(\frac{\partial U_r}{\partial t} + U_r \frac{\partial U_r}{\partial R} \right) = -\frac{\partial P}{\partial R} \quad ,$$

which in Lagrangian space is

$$\rho \frac{\partial U_r}{\partial t} = -\frac{\partial P}{\partial R} \quad ,$$

and with $dm = R^{\alpha-1} dR/V$

$$\frac{\partial U_r}{\partial t} = -R^{\alpha-1} \frac{\partial P}{\partial m} \quad .$$

Converting Equation (G.1) to R and z (cylindrical)

$$\frac{\partial \rho}{\partial t} + U_r \cos \alpha \frac{\partial \rho}{\partial R} \frac{x}{R} + U_r \cos \beta \frac{\partial \rho}{\partial R} \frac{y}{R} + U_z \frac{\partial \rho}{\partial z}$$

$$= -\rho \left(\cos \alpha \frac{x}{R} \frac{\partial U_r}{\partial R} + \cos \beta \frac{y}{R} \frac{\partial U_r}{\partial R} + \frac{\partial U_z}{\partial z} \right)$$

$$- \rho \frac{U_r}{R} \left(\frac{R^2 - x^2}{R^2} + \frac{R^2 - y^2}{R^2} \right) \quad .$$

$$\frac{\partial \rho}{\partial t} + U_r \frac{\partial \rho}{\partial R} + U_z \frac{\partial \rho}{\partial z} = -\rho \left(\frac{\partial U_r}{\partial R} + \frac{\partial U_z}{\partial z} + \frac{U_r}{R} \right) \quad .$$

Equation (G.2) becomes for conservation of momentum

$$\rho \left(\frac{\partial U_x}{\partial t} + U_r \frac{\partial U_r}{\partial R} + U_z \frac{\partial U_x}{\partial z} \right) = \frac{\partial P}{\partial R} \quad ,$$

$$\rho\left(\frac{\partial U_z}{\partial t} + U_r \frac{\partial U_z}{\partial R} + U_z \frac{\partial U_z}{\partial z}\right) = \frac{\partial P}{\partial z} \quad ,$$

and Equation (G.3) becomes for conservation of energy

$$\rho\left(\frac{\partial I}{\partial t} + U_r \frac{\partial I}{\partial R} + U_z \frac{\partial I}{\partial z}\right) = -P\left(\frac{\partial U_r}{\partial R} - \frac{\partial U_z}{\partial z} - \frac{U_r}{R}\right) \quad .$$

Author Index

Numbers in *italic* type indicate pages where References are listed in full.

Subject Index

NUMERICAL MODELING OF EXPLOSIVES AND PROPELLANTS CD-ROM CONTENTS

COMPUTER CODES for MODELING EXPLOSIVES and PROPELLANTS on PERSONAL COMPUTERS

BASIC PROGRAMS

- **BKW** Code for computing detonation performance
- **USERBKW** Code for helping the user develop input BKW files
- **SIN** Code for one dimensional modeling of explosive hydrodynamics
- **USERSIN** Code for helping the user develop input SIN files
- **GLAW** Code for computing gamma law equation of state for HE

SUPPORT PROGRAMS

- **TDF** Code for computing thermodynamic functions for BKW
- **SEQS** Code for computing single shock Hugoniot temperatures
  Forms fits required by HOM equation of state in FIRE, SIN and TDL
  Forms fits required for solid products in BKW code
- **FIRE** Code for computing Forest Fire rate of decomposition of
  shocked heterogeneous explosives and propellants.
  Forms decomposition rate fits for SIN and TDL codes

PROPELLANT PERFORMANCE CODE

- **ISPBKW** Code for computing propellant performance

This code calculates the propellant temperature, specific volume, enthalphy and entropy for the combustion chamber and the exhaust. The Specific Impulse (ISP) of the propellant is calculated.

TWO DIMENSIONAL LAGRANGIAN HYDRODYNAMIC CODE

- **TDL** Code for computing reactive hydrodynamic flow

This code calculates reactive hydrodynamic flow with chemical reactions described by an Arrhenius law, C-J volume burn, sharp-shock burn, or the Forest Fire heterogeneous shock initiation burn. Elastic- plastic flow and spallation are included in the description of solids. Files are furnished with equation of state and Forest Fire rate constants. Graphic capabilities include cross sections, contour plots, wire frame and picture plots of the state variables using the MCGRAPH libraries.

- **USERTDL** Code for helping the user develop input TDL files

GENERAL INFORMATION

The Personal Computer codes require an IBM compatible Personal Computer with or without a math co-processor for the executable codes. The codes include MCGRAPH graphics for HP Laserjet or PaintJet printers. The codes operate on the DOS or OS/2 operating systems. They operate under the Windows DOS shell.

ANIMATIONS

Initiation of PBXN-15 by Aluminum Flying Plates

KEN\SIN15 - Initiation by a 0.15 cm/μsec Aluminum Flying Plate

KEN\TDL15AL - PBXN-5 Cylinder Confined by Aluminum

KEN\TDL15FE - PBXN-5 Cylinder Confined by Steel

KEN\SIN13 - Initiation by a 0.13 cm/μsec Aluminum Flying Plate

KEN\TDL13AL - PBXN-5 Cylinder Confined by Aluminum

KEN\TDL13FE - PBXN-5 Cylinder Confined by Steel

Shocks From Underwater Detonations Interacting with Steel Plates

LEESIN.MVE - 35 kb TNT Sphere/ 60 cm Water/ 1.9 cm Steel

LEETDL.MVE - 35 kb TNT Sphere/ 30 cm Water/ 1.9 cm Steel

Hydrazine Shock Hugoniot Experiment - TDL Calculation

GARCIA.MVE - Baratol/2.54 Al/ Hydrazine in Al Cup

Plane Wave Lens - TDL Calculation

P100.MVE - TNT/PBX-9501 Plane Wave Lens

Hemispherical Detonator - TDL Calculation

DET.MVE - 9404 Detonator Initiation of PBX-9502

Explosively Formed Penetrator - ZeuS Calculation

ZEUS.MVE - An Explosively Formed Penetrator

Aquarium Test - TDL Calculation

AQUAR.MVE - A Cylinder of PBX-9502 in Water

Failure Diameter - TDL Calculation with Forest Fire

FDIA.MVE - 0.6 and 0.3 cm Dia PBX-9502 Cylinders

Homogeneous Shock Initiation of Nitromethane - SIN Calculation
NM.MVE - 90 kbar Shock Initiation of Nitromethane

Heterogeneous Shock Initiation of PBX-9502 - SIN Calculation
FOREST.MVE - Forest Fire Burn of PBX-9502

Flying Aluminum Foil Initiation of PBX-9404 - SIN Calculation
FFOIL.MVE - Short Shock Pulse Initiation of PBX-9404

Energy From Shocked But Not Detonated PBX-9404 - SIN Calculation
FFENG.MVE - 30 Kbar, 0.64 cm 9404, 0.2 cm Plexiglas

PBX-9404 Shocking an Aluminum Plate - SIN Calculation
9404Al.MVE - 5 cm PBX-9404/1 cm Aluminum

PBX-9404 Shocking a Steel Plate - SIN Calculation
9404FE.MVE - 5 cm PBX-9404/1 cm Steel

PBX-9404 Shocking a Plexiglas Plate - SIN Calculation
9404PXG.MVE - 5 cm PBX-9404/1 cm Plexiglas

A Sphere of PBX-9404 Shocking an Aluminum Plate - A SIN Calculation
CONVERG.MVE - Converging 5 cm PBX-9404/1 cm Aluminum/4 cm Air

A Sphere of PBX-9404 Shocking an Aluminum Plate - A SIN Calculation
DIVERG.MVE - Diverging 5 cm PBX-9404/1 cm Aluminum/4 cm Air

An Aluminum Driver Shocking an Aluminum Target - SIN Calculation
ALAL.MVE - 0.5 cm Al Target at 0.1 cmμsec/1 cm Al Target

An Aluminum Driver Shocking a Plexiglas Target - SIN Calculation
ALPXG.MVE - 0.5 cm Al Target at 0.1 cmμsec/1 cm Plexiglas Target

An Aluminum Driver Shocking a Gold Target - SIN Calculation
ALAU.MVE - 0.5 cm Al Target at 0.1 cmμsec/1 cm Gold Target

An Aluminum Driver Shocking an Aluminum Target - SIN Calculation
ALALEP.MVE - 0.3 cm Al Target at 0.04 cmμsec/1.2 cm Al Target

High Pressure Steam Shocking Air - SIN Calculation
STEAM.MVE - 1000 psi Steam Shocking Air -Slab, Cylinder, Sphere

DATA FILES

- **USERBKW\ZZZTHERC** - Thermodynamic Fit Coefficients for Detonation Products
- **USERBKW\ZZZSOLEQS** - Cowan Equation of State Constants for Solid Detonation Products
- **USERBKW\ZZZCOMPS** - Explosive and Binder Composition, Density, ΔH_f^o
- **SIN\EOSDATA** - HOM Equation of State Constants
- **SIN\FFDATA** - Forest Fire Rate Constants
- **TDF\TDFFILES** - Thermodynamic Gas and Solid Parameters
- **SEQ\SEQSFILES** - Solid Equation of State Parameters
- **HUGONIOT** - Shock Hugoniot Data from LASL Shock Hugoniot Data Volume
  HUG Data Processing and Graphing Code
  Output Files and Graphs for HUGDATA File
- **AQUARIUM** - Aquarium Data from
  Los Alamos Explosive Property Data Volume
  AQUAR Data Processing and Graphing Code
  Output Files and Graphs for AQUARDAT File

MISCELLANEOUS

- **MISC\JWL** - Code for Generating JWL Coefficients from HOM Fits
- **MISC\EDITOR** - A Line Editor for DOS and OS/2
- **MISC\DETSYP** - Index to Detonation Symposiums

Includes Explosive Composition Index

- **MISC\MATCH** - Shock Matching Code
- **MISC\HOM** - HOM Material Constants

LISTINGS AND GRAPHS

- **RUN\BKW.OUT** - BKW Output Files and Graphs for Various Explosives
- **RUN\GLAW.OUT** - GLAW Output Files and Graphs for Various Explosives
- **RUN\SEQS.OUT** - SEQS Output Files and Graphs for Various Materials
- **RUN\TDF.OUT** - TDF Output Files and Graphs for Various Species
- **RUN\FIRE.OUT** - FIRE Output Files and Graphs for Various Explosives
- **RUN\ISP.OUT** - ISPBKW Output Files and Graphs for Various Propellants